Developments in Clay Science

Surface and Interface Chemistry of Clay Minerals

Volume 9

Developments in Clay Science

Surface and Interface Chemistry of Clay Minerals

Volume 9

Series Editor

F. Bergaya

Edited by

R. Schoonheydt
Center for Surface Chemistry and Catalysis, Catholic University of Leuven, Leuven, Belgium

C. T. Johnston
Departments of Earth, Atmospheric, and Planetary Sciences and Agronomy, Purdue University, West Lafayette, IN, United States

F. Bergaya
Interfaces, Confinement, Matériaux et Nanostructures (ICMN), Centre National de la Recherche Scientifique (CNRS), Orléans, France

ELSEVIER

Elsevier
Radarweg 29, PO Box 211, 1000 AE Amsterdam, Netherlands
The Boulevard, Langford Lane, Kidlington, Oxford, OX5 1GB, United Kingdom
50 Hampshire Street, 5th Floor, Cambridge, MA 02139, United States

ISBN: 978-0-08-102432-4
ISSN: 1572-4352

Publisher: Candice Janco
Acquisition Editor: Amy Shapiro
Editorial Project Manager: Tasha Frank
Production Project Manager: Omer Mukthar
Cover Designer: Greg Harris

Typeset by SPi Global, India

Dedication

Robert Schoonheydt has only nice memories of clay mineral science with excellent PhD students and postdocs in the lab, with truly international scientific collaboration with universities and laboratories in China, India, Japan, the United States, Hungary, Italy, Romania, Spain, and France, to name the most important ones only. He hopes that clay mineral science continues to develop with the positive spirit that he experienced over the 45 years of his career.

Contents

Contributors

B. Baeyens Laboratory for Waste Management, Paul Scherrer Institute, Villigen, Switzerland

Sébastien Balme Institut Europeen des Membranes, UMR 5635, Universite de Montpellier, ENSCM, CNRS, Montpellier Cedex 5, France

Faïza Bergaya Interfaces, Confinement, Matériaux et Nanostructures (ICMN), Centre National de la Recherche Scientifique (CNRS), Orléans, France

Sergey V. Churakov Paul Scherrer Institute, Laboratory for Waste Management, Villigen; University of Bern, Institute of Geological Sciences, Bern, Switzerland

Dipak Kumar Dutta Advanced Materials Group, Materials Science and Technology Division, CSIR-North East Institute of Science and Technology, Jorhat, Assam, India

Maguy Jaber Sorbonne Université, Laboratoire d'Archéologie Moléculaire et Structurale, UMR 8220, BP225, Paris Cedex 05, France

Cliff T. Johnston Departments of Earth, Atmospheric, and Planetary Sciences and Agronomy, Purdue University, West Lafayette, IN, United States

Jun Kawamata School of Chemistry, Yamaguchi University, Yamaguchi, Japan

Jean-François Lambert Laboratoire de Réactivité de Surface, Sorbonne Université- CNRS (UMR 7197), Paris, France

Xiandong Liu State Key Laboratory for Mineral Deposits Research, School of Earth Sciences and Engineering, Nanjing University, Nanjing, PR China

M. Marques Fernandes Laboratory for Waste Management, Paul Scherrer Institute, Villigen, Switzerland

Laurent J. Michot Physicochimie des Electrolytes et Nanosystèmes Interfaciaux, Phenix, UMR 8234, CNRS/Sorbonne Université, Paris Cedex, France

Hisako Sato Graduate School of Science and Engineering, Ehime University, Matsuyama, Japan

Robert A. Schoonheydt Centre for Surface Chemistry and Catalysis, Catholic University of Leuven, Leuven, Belgium

Yasutaka Suzuki School of Chemistry, Yamaguchi University, Yamaguchi, Japan

Makoto Tominaga Lehrstuhl für Anorganische Chemie I, University of Bayreuth, Bayreuth, Germany

Yasushi Umemura Department of Applied Chemistry, National Defense Academy, Yokosuka, Kanagawa, Japan

Fernando Wypych Department of Chemistry, Universidade Federal do Paraná, Curitiba, Brazil

Akihiko Yamagishi Faculty of Medicine, Toho University, Tokyo, Japan

Preface

Clay minerals are the most important components of clay, probably the first material that was exploited and used by mankind several thousand years ago. Since these early times, the areas of application of clays and clay minerals have continuously expanded and continue to do so. These applications are summarized in the table below and are somewhat arbitrarily divided into traditional and more recent application areas.

Table: Areas of application of clays and clay minerals

Traditional applications	**Some more recent applications**
Paper filling and coating	Catalysts and catalyst carriers
Ceramics	Pesticide carriers
Civil engineering	Environmental technology
Foundry sands binders	Food industry
Well drilling	Fracking
Bleaching and refining of oil	Water and sewage treatment
Clarification of wines and beers	(Radio)active waste containment
Soaps and detergents	Clay polymer nanocomposites fillers
Soil stabilization	Fuel cells, wires, cables, etc.
Pet litter adsorbents	Cosmetics and pharmaceuticals
Animal feed	Health/medicine

All together, these areas of application represent a multibillion kilogram business every year. This success has been made possible by (1) clay's easy availability, with clay deposits in almost every country of the world and (2) the unique physical and chemical properties of the clay minerals.

Clay minerals, a group of phyllosilicates, are two-dimensional silicates with sizes <2 μm. This small size means large surface areas. As a consequence, surface chemistry is at the core of the development of clay mineral applications. Surface chemistry is defined as the interaction between adsorbed molecules or ions and the surface atoms of clay minerals. Any molecule or ion can be adsorbed, from small, inert molecules such as N_2 all the way to (bio)polymers. The interaction with the surface might be physical, that is, the adsorbed molecule retains its identity, or chemical, that is, a chemical bond is established between a surface atom and the adsorbed molecule.

To achieve the optimum combination of adsorbed molecule and the clay mineral surface, the latter can be manipulated. Clay mineral surfaces can be realized which fully cover the hydrophilic-to-hydrophobic range. Layer charge, and thus, chemical composition, exchangeable cations, and ion exchange are the most important for realizing this hydrophilic–hydrophobic range. But, as clay minerals occur in nature, water is the most important adsorbed molecule.

This book starts with an introductory chapter on clay minerals, their internal and external surfaces, surface atoms, and the forces between adsorbed molecules and surface atoms (Chapter 1). This chapter is followed by a fundamental discussion on surface area determination (Chapter 2); the surface area is the basis for surface chemistry and development of materials. Quantum chemical modelling accompanies the interpretation of experimental data and allows extrapolation beyond the experiment (Chapter 3). As clay minerals are found in nature, they contain water, and the study of adsorbed water is as old as the study of clay minerals themselves (Chapter 4).

The two following chapters treat inorganic cations. In Chapter 5, the emphasis is on radioactive cations; it forms the basic knowledge for radioactive waste containment. With ion exchange of chiral metal ion complexes, chiral clay minerals can be produced with interesting applications in chromatographic separation and chiral catalysis (Chapter 6).

The following four chapters deal with the interaction of clay mineral surfaces with organic molecules, which is an extremely important research area. This knowledge is necessary for pollution control and pollution abatement (Chapter 7 and Chapter 8). Organic molecules can be synthesized with clay minerals as catalysts or carriers of catalytic particles (Chapter 9). Organic polymers, including biopolymers, interact strongly with clay mineral surfaces; this leads to clay polymer nanocomposites with improved properties, at least when compared to those of the neat polymers (Chapter 10).

The final two chapters give a hint about possible new developments: the transformation of clay minerals into optical materials (Chapter 11) and clay mineral films (Chapter 12). In both chapters, the organization of adsorbed molecules and of the individual clay mineral layers in a nanometre thick film has to be precisely controlled so as to generate clay mineral nanocomposites with a range of possible new application areas.

The editors are convinced that the readers will find knowledge and data on the surface and interface chemistry of clay minerals in this book and inspiration for new research.

R. Schoonheydt, C. T. Johnston, F. Bergaya
July 2018

Acknowledgements

The three volume editors (R. Schoonheydt, C. T. Johnston, and F. Bergaya) would like to thank all the authors and their coworkers for their contributions to this book and for delivering chapters with a high scientific level on time. Extensive written and oral discussions between the authors and the editors allowed improved chapters as a result.

Cliff Johnston would like to express his deep appreciation to both Robert Schoonheydt and Faiza Bergaya for their tireless work on this project. As a junior faculty member, he first met Robert Schoonheydt about 30 years ago. Through his collaboration and sabbatical stays at K.U. Leuven with Robert, he was introduced to many excellent European clay scientists, including Faiza Bergaya in 1995, and to a number of outstanding Japanese clay scientists that included Akihiko Yamagishi and Hisako Sato, all of whom challenged and shaped his thinking about clay science.

Faiza Bergaya, who is also acting as series editor, would particularly like to thank Amy Shapiro (the acquisitions editor, Earth and Planetary Sciences, Elsevier) and Tasha Frank (the senior editorial project manager, Elsevier) for their patience and their continuous help during the achievement of this book. She thanks Robert Schoonheydt and Cliff Johnston for having accepted the huge editorial task of this Volume 9 and for the friendly relationship established during their close collaboration in these past months. Moreover, she would like to thank Susan Dennis (the acquisitions editor, Physical Sciences Series, Elsevier) for her confidence for several years.

The author of Chapter 9 is grateful to Dr D. Ramaiah, Director, CSIR-North East Institute of Science and Technology, Jorhat-785006, Assam, India, for his kind support. The author is also grateful to CSIR, New Delhi, for sanctioning the Emeritus Scientist Scheme (No. 21(1029)/16/EMR-II).

Finally, the support of the Elsevier team is highly appreciated by the three editors.

Clay minerals and their surfaces

1

Robert A. Schoonheydt, Cliff T. Johnston†, Faïza Bergaya‡*
*Centre for Surface Chemistry and Catalysis, Catholic University of Leuven, Leuven, Belgium, †Departments of Earth, Atmospheric, and Planetary Sciences and Agronomy, Purdue University, West Lafayette, IN, United States, ‡Interfaces, Confinement, Matériaux et Nanostructures (ICMN), Centre National de la Recherche Scientifique (CNRS), Orléans, France

1.1 TO or 1:1 and TOT or 2:1 clay minerals

Clay minerals belong to the phyllosilicate family (Bailey, 1988b; Brigatti et al., 2013). Each layer of these layered silicates consists of T sheets of corner-sharing Si tetrahedra, $[SiO_4]$, and O sheets of edge-sharing Al octahedra, $[Al(OH)_2O_4]$. The T and O sheets are connected with a common oxygen atom (Fig. 1.1).

The possible combinations of T and O sheets among both TO and TOT layers lead to a classification of nine distinguished main groups of clay minerals reported in Table 1.1. The 2D connectivity of the planar layer at the molecular level should be distinguished from the morphology at an upper level of organization (Ras et al., 2007).

The structure of TO (or 1:1) clay minerals layers, as shown in Fig. 1.1, is representative of the kaolinite structure. The chemical structural formula of the unit cell is $[Si_2Al_2O_5(OH)_4]$ with two types of structural OH groups. The inner surface OH groups are located on the terminating surface of the Al-octahedral sheet. The H atoms are bonded to O atoms of the O sheet and point towards the basal oxygen atoms of the opposing T sheet (Fig. 1.1) of the next layer with which they engage in H-bonding (Johnston, 2010). The second type of structural OH group is the inner OH. It is the Al–OH group located on the common plane of oxygen atoms joining the O and T sheets (Fig. 1.1). The sum of the formal charges on the atoms in the TO structure equals zero (Si = +4; Al = +3; H = +1, O = −2).

The structure of TOT of 2:1 clay minerals layers (Fig. 1.2) is composed of an O sheet (composed of Al^{3+}, Fe^{3+} or Mg^{2+} cations), sandwiched between two Si^{4+} T sheets. The chemical formula of the unit cell is $[Si_8Al_4O_{20}(OH)_4]$ or $[Si_4Al_2O_{10}(OH)_2]$ for a half unit cell, and is represented by the clay mineral pyrophyllite. In contrast to the TO structure (e.g., kaolinite), the OH groups of TOT structures (e.g., pyrophyllite) are all inner OH groups. When Al^{3+} cations occupy the centre of the octahedral cavities in the O sheet, the TOT or TO layer is called dioctahedral. If the O sheet is filled with Mg^{2+} cations, the TOT or TO layer is called trioctahedral, and the chemical formula of the TOT unit cell is $[Si_8Mg_6O_{20}(OH)_4]$. A representative example of the TOT clay mineral is talc. An example of a trioctahedral TO clay mineral is serpentine, with $[Si_2Mg_3O_5(OH)_4]$ as the chemical formula of the unit cell. All these reported unit cell compositions are ideal as they do not carry any permanent structural charge on the basal surfaces (Brigatti et al., 2013).

Developments in Clay Science, Vol. 9. https://doi.org/10.1016/B978-0-08-102432-4.00001-9

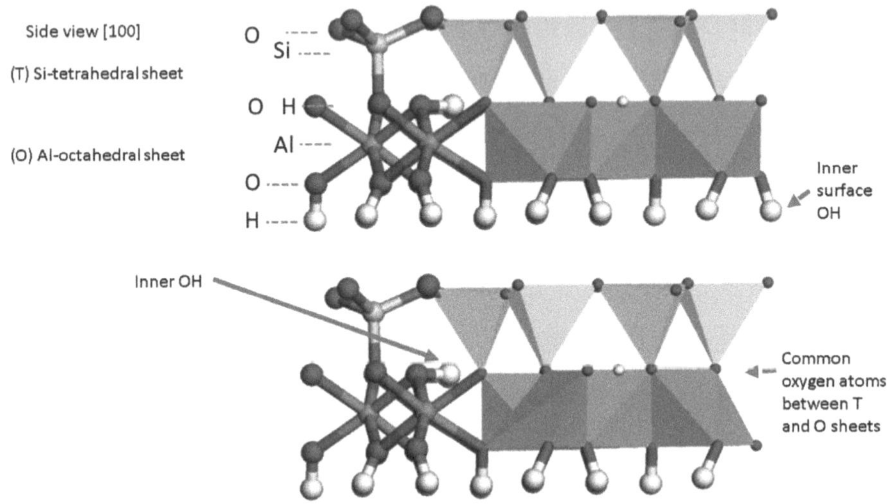

Fig. 1.1 Association of T and O sheets (example: schematic structure of kaolinite). *Red balls* (*dark grey* in the print version) are oxygen; *purple balls* (*grey* in the print version) are aluminium, and *grey balls* are hydrogen atoms.

Table 1.1 The nine main groups of clay minerals

Group of clay minerals	T and O associations with more typical examples of clay minerals
Kaolinite and serpentine	TO Kaolinite (*laths*) and halloysite (*nanotubes*)
Pyrophyllite and Talc	TOT
Smectites	TOT
Vermiculites	TOT
Micas	TOT
Chlorites	TOT
Interstratified clay minerals	Regular or irregular combinations of TO and TOT and/or of different TOT
Sepiolites and palygorskites	TOT with a discontinuous O sheet *Fibrous morphology*
Allophanes and imogolites	TO of short range order Poorly crystalline clay minerals (X-ray amorphous)

1.2 Structural considerations

The Si tetrahedra in the tetrahedral sheet each have three corner-shared oxygen atoms. The fourth or apical oxygen atom points towards the octahedral sheet. The chemical representation of the tetrahedral sheet is $[(Si_2O_5)^{2-}]_n$. The arrangement

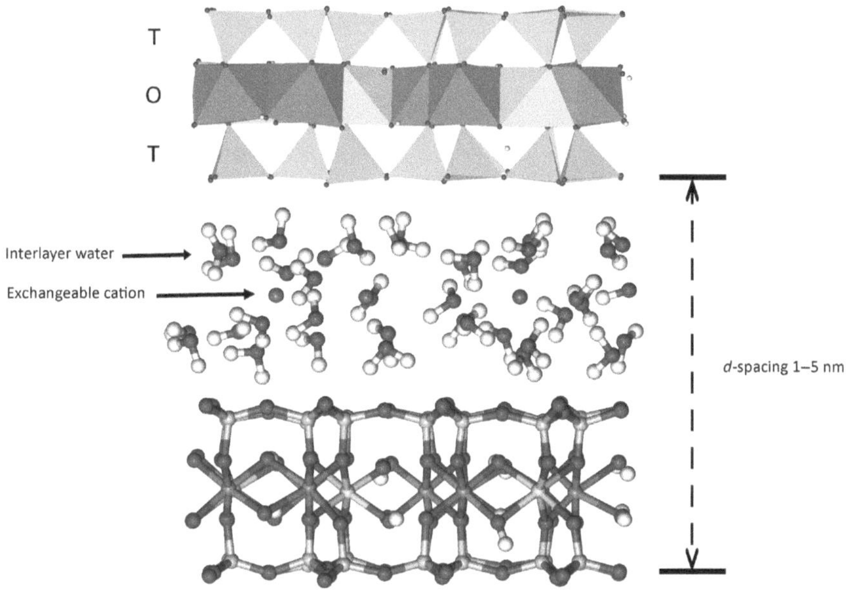

Fig. 1.2 Representation of a TOT layer with an O sheet sandwiched between two T sheets: (example: structure of dioctahedral montmorillonite showing exchangeable cations and interlayer water). *Red balls* (*grey* in the print version) are oxygen; *gold balls* (*light grey* in the print version) are silicon, *green balls* (*light grey* in the print version) are magnesium, *purple balls* (*grey* in the print version) are aluminium, and *grey balls* are hydrogen atoms.

of these dimers in one sheet leads to the hexagonal arrangement of the Si tetrahedra, as shown in Fig. 1.3.

The Si—O bond length is on average 0.161 nm, leading to the 'ideal' unit cell dimensions given in Table 1.2.

Similarly, the Al octahedra form a sheet by sharing their edges (Fig. 1.3), and the cation positions form regular hexagons. The unit cell dimensions are given in Table 1.2. Such a sheet is essentially the same as the $Al(OH)_3$ sheet of the gibbsite lattice. In the case of Al^{3+} octahedra, two-thirds of the cationic positions are occupied. The TOT and TO clay minerals with Al in the octahedral sheet are called dioctahedral. For divalent cations such as Mg^{2+}, the octahedral sheet is identical to the $Mg(OH)_2$ sheet of brucite, where all of the cationic positions are filled. The corresponding TOT and TO clay minerals are called trioctahedral. The unit cell dimensions of the tetrahedral and octahedral sheets are close (Table 1.2). They can be linked by sharing the apical oxygens of the Si tetrahedra. There is, however, a small mismatch which is compensated by rotation of the Si tetrahedra. The hexagonal array is distorted from an ideal hexagonal structure to a ditrigonal structure where three O atoms move inwards, three O atoms move outwards, and the Si–O–Si angle takes a value between 180 and 120 degrees, 180 degree being for the ideal hexagonal array (Fig. 1.4) (Bailey, 1988a).

Other factors influencing this angle and deformations in octahedral and tetrahedral sheets result from isomorphous substitution and defects.

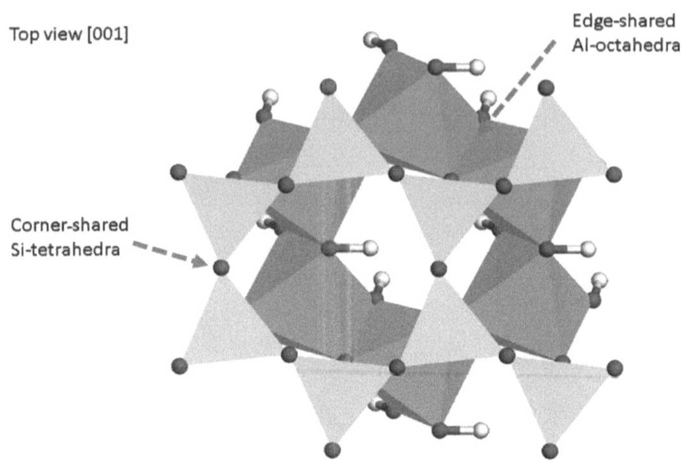

Fig. 1.3 [001] projection of T and O sheets (example: schematic structure of kaolinite). Silicon tetrahedrons shown in *gold* (*light grey* in the print version) (corner shared), and aluminium octahedrons shown in *purple* (*dark grey* in the print version) (edge shared).

Table 1.2 Unit cell dimension of the tetrahedral and octahedral sheets

	a (nm)	*b* (nm)
Si tetrahedral sheet	0.53	0.916
Al octahedral sheet	0.46	0.808–0.868

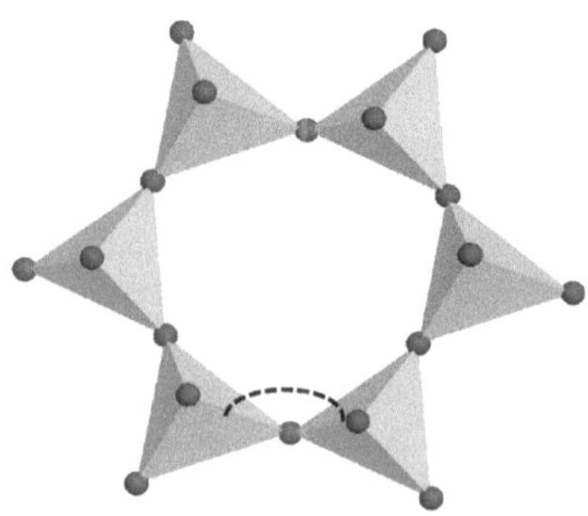

Fig. 1.4 Hexagonal array of Si-tetrahedrons showing ditrigonal symmetry.

1.3 Isomorphous substitution

Heterovalent isomorphous substitution (referred to as isomorphous substitution) is the substitution of a cation in the lattice by a cation of lower charge and about equal size. Examples are Si^{4+} by Al^{3+} in tetrahedral sheets, Al^{3+} by Mg^{2+} in octahedral sheets of

Table 1.3 Ionic radii

Ion	Coordination number 4 (T_d) and 6 (O_h)	Radius (pm)	r_c/r_o
Al^{3+}	4	39	0.287
Al^{3+}	6	53	0.384
Si^{4+}	4	26	0.193
Mg^{2+}	6	72	0.521
Fe^{2+}	4	63	0.457
Fe^{2+}	6	78	0.565
Fe^{3+}	4	49	0.355
Fe^{3+}	6	65	0.471
Li^{+}	4	59	0.536
Li^{+}	6	74	
O^{2-}	2	135	
O^{2-}	4	138	
O^{2-}	6	140	

From Huheey, J.E., Keiter, E.A., Keiter, R.L., 1997. Inorganic Chemistry. Principles of Structure and Reactivity. Prentice Hall, New York, pp. 737.

dioctahedral layers, and Mg^{2+} by Li^{+} in octahedral sheets of trioctahedral layers (Schoonheydt and Johnston, 2013). Fe is invariably present in nature and in the structure of clay minerals. Fe^{3+} can substitute for Si^{4+} in tetrahedral sheets; Fe^{2+} and Fe^{3+} can substitute for Al^{3+} in octahedral sheets. Ionic radii of the cations and O atoms, the ratios of the radii of the cations, r_c to that of the oxygens, r_o (r_c/r_o) and the preferred coordination environment are given in Table 1.3 (Huheey et al., 1997).

The sheets comprising the structures of clay minerals can be considered as sheets of closely packed O atoms. The cations merely fill the holes between the O atoms. This is illustrated in Fig. 1.5 for the Al octahedron and Si tetrahedra.

When the atoms are considered as hard spheres one obtains $2r_o/(2r_c+2r_o)=\cos$ 45 degree $=0.707$ or $r_c/r_o=0.414$ in the case of the octahedron. For tetrahedral coordination, one obtains the radius ratio $r_c/r_o=0.225$. Cationic radii with r_c/r_o between 0.414 and 0.225 prefer tetrahedral coordination and substitute preferentially in the tetrahedral sheet according to Pauling's 1st rule. If r_c/r_o is larger than 0.414, octahedral coordination is preferred or substitution in the octahedral sheets. This is given in the second column of Table 1.3. In reality, the ions are not hard spheres, and the r_c/r_o ratios do not strictly obey the rule but must be close. The most common TOT clay minerals with isomorphous substitution are reported in Table 1.4 with their unit cell formula and their names.

Often isomorphous substitution occurs simultaneously in the tetrahedral and octahedral sheets. The extent of isomorphous substitution in the tetrahedral sheet is represented by x, and that in the octahedral sheet by y. The total amount of isomorphous substitution is then $x+y$. The consequence of isomorphous substitution is the presence

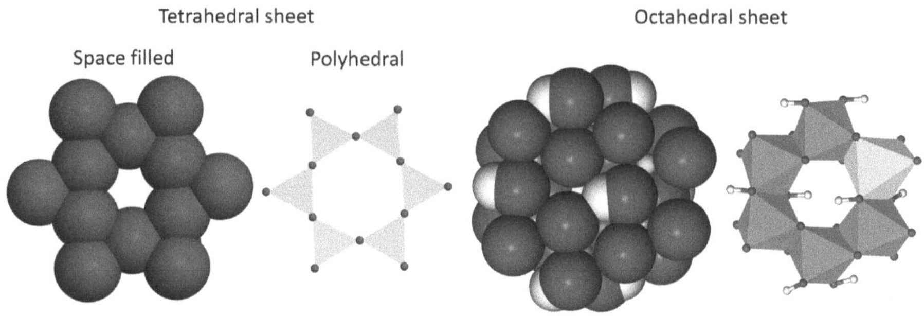

Fig. 1.5 Packed oxygen atoms of (1) a tetrahedral sheet (*left*) shown in space filled (*left side*) and polyhedral representation. Similar depiction of the (2) octahedral sheet shown on the *right* in both space filled and polyhedral representation. The polyhedron shown in *green* represents isomorphic substitution of Mg for Al.

Table 1.4 Clay mineral names with their corresponding unit cell formulae

Unit cell formula	Clay mineral name
$Si_8^{IV}\ Al_4^{VI}\ O_{20}(OH)_4$	Pyrophyllite
$Si_8^{IV}Mg_6^{VI}O_{20}(OH)_4$	Talc
$(Si_{8-x}Al_x)^{IV}Al_4^{VI}O_{20}(OH)_4M^+_x$	Beidellite
$Si_8^{IV}(Al_{4-y}Mg_y)^{VI}O_{20}(OH)_4M^+_y$	Montmorillonite
$Si_8^{IV}(Mg_{6-y}Li_y)^{VI}O_{20}(OH)_4M^+_y$	Hectorite
$(Si_{8-x}Al_x)^{IV}Mg_6^{VI}O_{20}(OH)_4M^+_x$	Saponite

IV and VI refer, respectively, to the tetrahedral and octahedral sheets.

of excess negative charge: the sum of the negative charges in the unit cell is larger than the sum of the positive charges. The excess negative charge in the structure is compensated by cations, indicated by M^+ in Table 1.4. These cations do not belong to the structure and are exchangeable by other cations. They are called exchangeable cations. For the smectites (or swelling clay minerals) of Table 1.4, the amount of negative charges per unit cell is $x+y=0.4$–1.2. Higher charges occur for vermiculites ($x+y=1.2$–1.8) and micas ($x+y=2$–4).

Layer charge density is the average charge per unit surface. It is $0.4/nm^2$–$1.3/nm^2$ for smectites. This is an important quantity, as it determines the amount and molecular arrangement of organic exchangeable cations, especially in the case of bulky cations such as long chain alkylammoniums and cationic dyes. Thus, if the size of the monovalent exchangeable organic cation exceeds the average surface area per charge in the structure, the exchangeable organic cations must adopt special configurations in order to compensate the negative lattice charge sites (Laird et al., 1989). These configurations might include bilayers, dimers, trimers, and higher aggregates, and even crystalline compaction. Layer charge has two important consequences that has 3 important consequences that are discussed in Section 1.4.

1.4 Consequences of isomorphous substitution

1.4.1 Cation exchange

Cation exchange is the substitution of exchangeable cations originally present, B, by other types of cations in solution (A) (Bergaya et al., 2013). The reaction can formally be written as

$$\left(B_{CM}^{n+}\right)_m + n(A^{m+}) \leftrightarrow \left(A_{CM}^{m+}\right)_n + m(B^{n+}) \tag{1.1}$$

where CM means clay mineral.

This equilibrium reaction is characterized by a thermodynamic equilibrium constant, K_c given by

$$K_c = \frac{\left(A_{CM}^{m+}\right)^n (B^{n+})^m}{\left(B_{CM}^{n+}\right)^m (A^{m+})^n} \tag{1.2}$$

where () are activities.

Activities are expressed as the product of the concentration and activity coefficients. In solution, the activity of the aqueous species, (A^{m+}) can be represented by $(A^{m+})^n = [A^{m+}]^n \gamma_M^n$ where $[A^{m+}]$ is the concentration of A in moles dm^{-3} and γ_M^n is the single ion activity coefficient. There are no simple expressions for activities of ions on solid surfaces. Gaines and Thomas (1953) assumed that the activity of the cation on the solid phase was represented by its equivalent fractions, while Vanselow used molar fractions of the ions on the solid (Sposito, 1981a,b):

$$X_A^{eq} = \frac{m[A-CM]}{m[A-CM]+n[M-CM]} \text{ and } X_M^{eq} = \frac{n[M-CM]}{m[A-CM]+n[M-CM]} \tag{1.3}$$

$$X_A = \frac{[A-CM]}{[A-CM]+[M-CM]} \text{ and } X_B = \frac{[M-CM]}{[A-CM]+[M-CM]} \tag{1.4}$$

$$K_c^{GT} = \frac{\left(X_A^{eq}\right)^n (B^{n+})^m}{\left(X_B^{eq}\right)^m (A^{m+})^n} \quad K_c^V = \frac{(X_A)^n (B^{n+})^m}{(X_B)^m (A^{m+})^n} \tag{1.5}$$

If $m=n$, the Gaines–Thomas and Vanselow selectivity coefficients are identical.

1.4.2 Cation exchange capacity

Cation exchange capacity (CEC) is the amount of a cation that can be exchanged by another cation on the surface of a clay mineral. It is expressed in cmol(+)/kg, which is numerically equivalent to meq/100 g, where mol(+) represents moles of electrical charge. Theoretically, it can be obtained from the amount of isomorphous substitution, as shown by the structural formulae of the unit cells in Table 1.4. For instance, for beidellite, with Na^+ as the exchangeable cation, the CEC is 0.27 mmol/g if $x=0.2$ and 0.81 mmol/g if $x=0.6$. This is called the pH-independent CEC.

At the edges of the clay mineral layers (Figs 1.1 and 1.2), the O atoms are coordinatively unsaturated and pick up protons from the aqueous environment, thus forming surface OH groups. They can pick up a proton or lose their proton, depending on the pH conditions. In acidic conditions, the edges are positively charged; at alkaline pH, they are negatively charged. In between, there is a pH at which the sum of the positive charges equals the sum of the positive charges. This is called the point of zero charge. The OH groups at the edges are bonded to Si, Al, or Mg and the point of zero charge depends on the chemical composition of the clay mineral layers, particle size, presence of impurities, and method of determination of the point of zero charge. Values reported in the literature can vary over the range pH 3–7 (Lagaly, 2006).

The protons at the edges are exchangeable, and charge resulting from these edge sites is pH-dependent and increases with increasing pH. This contribution to the CEC is called the pH-dependent CEC. Factors influencing this pH-dependent CEC are particle size, degree of swelling (see Section 1.4.3), and type of cation. The sum of the pH-independent and pH-dependent exchange capacities is the total CEC.

The CEC is dependent on the type of exchangeable cation for two reasons:

(1) Not all the exchange sites are available or accessible. This is because the dilute clay mineral dispersion is composed of single layers, particles, and aggregates of particles. In these particles, not all exchange sites might be accessible.
(2) The charge density of the particles might be different from particle to particle. Exchange sites are accessible dependent on this charge density. If it is too high, swelling is limited, and cations might not be able to exchange on these high charge density particles.

The exchange of bulky cations, such as metal ion complexes, organic cations, and cationic polymers, including cationic proteins, renders the surface hydrophobic, thus creating an environment for adsorption of these cations in excess of the CEC. This means that the bulky cations with their associated anions prefer to be located in the hydrophobic environment of the clay mineral particles and not in the hydrophilic water environment of the solution. The amount of adsorbed cations exceeds the CEC.

1.4.3 Intercalation and swelling

Swelling is the spontaneous increase of the interlayer distance of clay mineral particles, due to the adsorption of molecules in the interlayer space. As a consequence, the *c* dimension of the unit cell increases, while the *a* and *b* dimensions remain constant. It is observed in the XRD spectrum by a shift of the d_{001} line to lower 2θ values. The process is governed by (i) the charge density of the clay mineral layers; (ii) the type of exchangeable cation in the interlayer space; and (iii) the type of adsorbing molecule or ion.

If the charge density of the clay mineral layers is zero or close to zero, there are no exchangeable cations in the interlayer space, and the clay mineral does not swell. Such is the case of kaolinite, pyrophyllite, and talc. The charge density of smectites, $\sigma=0.4$–1.2e per unit cell, is ideal for swelling in water with Li^+ and Na^+ as the exchangeable cations in the interlayer space. The interlayer distance can increase to infinity, or the smectite particles are disassembled into single layers, diffusing randomly in the liquid water. If the charge density is above 1.2e per unit cell, swelling in water is limited to one or two layers of water molecules in the interlayer space. This is the case of vermiculites. When the charge density is 2 or higher, as for micas, swelling does not occur.

In general terms, swelling occurs because the attractive forces between the exchangeable cations and the adsorbing molecules overcome the attractive forces between the negative charge density of the clay mineral layers and the exchangeable cations. These attractive forces are influenced by the charge density of the clay mineral particles, by the size and shape of these particles, and by the charge of the exchangeable cations.

There are two limiting cases of intercalation of molecules in the interlayer space.

(1) Random intercalation: there is no preference of the incoming molecules for any of the clay mineral particles in the sample, and within one particle, the molecules are randomly distributed in the interlayer space.

(2) Interstratification: the interlayer spaces are filled up successively. Interlayer spaces might be filled up randomly or in organised fashion (Fig. 1.6).

Random intercalation is only possible if all clay mineral particles in the sample are identical in terms of charge density, number of layers per particle, size, and shapes of the particles. Such an ideal situation does not occur. In the case of interstratification, the incoming molecules favour the particles that offer the largest attractive forces towards the molecules. Once one molecule has entered the interlayer space, it has opened an easy pathway for subsequent molecules, until at some loading, the molecule–molecule repulsion prohibits the entry of additional molecules. In reality, intercalation is a mix of random intercalation and interstratification.

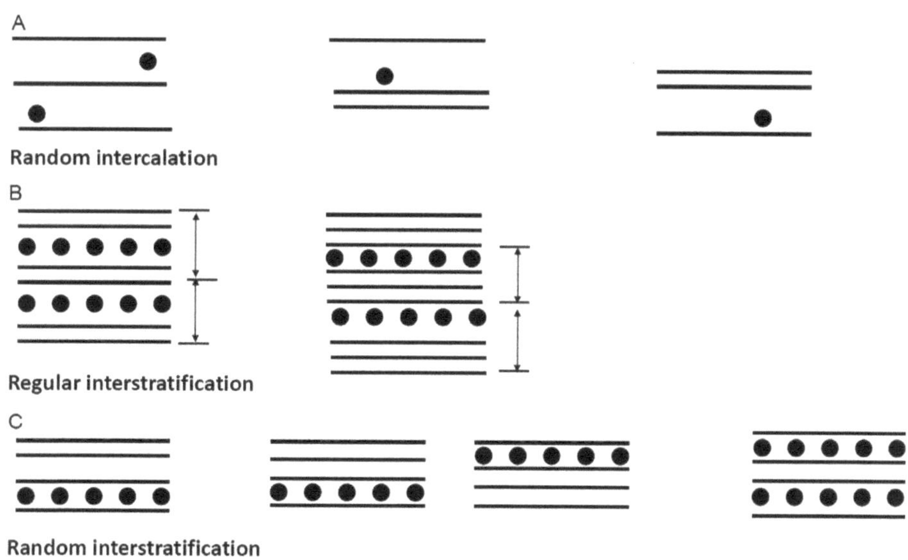

Fig. 1.6 Possible organization of molecules (*black balls*) in the interlayer space.

XRD experiments give some hints as to what is happening. The d_{001} line reflects the distance between two successive clay mineral layers in the clay mineral particles of the sample. This distance is the sum of the height of the clay mineral layer or 0.96 nm in the case of smectites, and the thickness of the interlayer space. Thus, as the thickness of the interlayer space increases with adsorption of molecules in the interlayer space, so will the d_{001} line shift to smaller values. In the case of random intercalation, the d_{001} line shifts gradually as more molecules are intercalated. In the case of interstratification, different d_{001} lines may be observed dependent of the degree and the order of interstratification.

1.5 Surfaces, surface areas, and surface sites

In a clay mineral sample, there are single layers, particles, and aggregates of particles. A single layer has a planar surface and an edge surface. A particle has an external surface and the interlayer space. An aggregate has an external surface, interlayer surface, and pores (Fig. 1.7).

There are then two types of porosity in a clay mineral sample: the interlayer space and the pores between the particles in the aggregates. The interlayer space is available for adsorption if the molecules or pillars keeping the clay mineral layers apart have free space between them for adsorption of incoming molecules or can be replaced/substituted by the incoming molecules. The interlayer space is usually microporous; that is, the free space between the layers is less than 2 nm. The pores in aggregates are interparticle pores. They can be mesoporous (2 nm $< d <$ 50 nm) or macroporous ($d >$ 50 nm).

The surface area of a clay mineral is the surface which is occupied by the probe molecule used to determine the surface area. This is usually N_2. One N_2 molecule occupies 0.16 nm^2 (i.e., cross-sectional area). If one can determine the number of adsorbed N_2 molecules per unit mass the surface area can be obtained. Several situations can be envisaged. (1) The interlayer space is filled up with molecules, usually water. In this case, N_2 cannot substitute for water in the interlayer space, and one measures only the external surface area of the clay mineral sample. (2) The interlayer

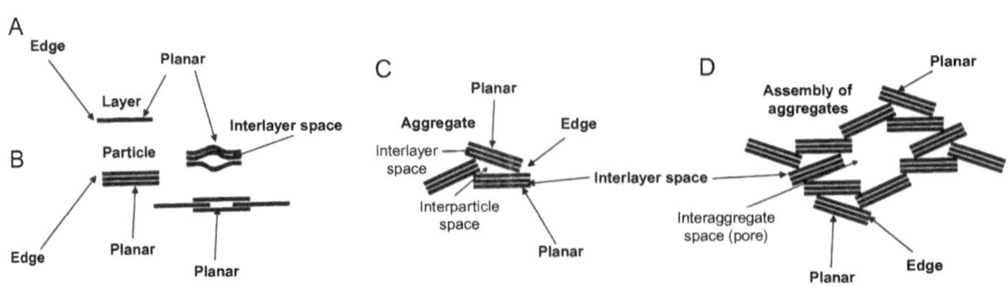

Fig. 1.7 Definition of planar surfaces, edge surfaces and of interlayer space
From Bergaya, F., Theng, B.K.G., Lagaly, G., 2006. Handbook of Clay Science. Elsevier, Amsterdam.

space is filled with pillars or bulky molecules, and the distance between the molecules or pillars is larger than the size of the N_2 molecule. Upon adsorption, the N_2 molecules occupy the external surface area and the free interlayer surface area not occupied by the pillars/bulky molecules.

If the goal is to measure the interlayer surface area, a probe molecule is necessary which is strongly/preferentially adsorbed in the interlayer space and forms a closely packed monolayer. From the amount adsorbed and the size of the molecule, one calculates the interlayer surface area (see Chapter 2). A popular molecule with which to do so is ethylene glycol.

1.6 Surface atoms

The overwhelming majority of the surface atoms of clay minerals are oxygen atoms (Figs 1.1 and 1.2). The O atoms of the Si tetrahedral sheets of smectites and kaolinite are coordinatively saturated. They are attractive towards adsorbing molecules by dispersion or nonpolar interactions and through electrostatic interaction for clay minerals with isomorphous substitution, because they carry a negative charge. The Al octahedral sheet of a kaolinite layer is terminated with OH groups (see Fig. 1.1). These OH groups are polar and capable of H-bonding between layers. These bonds keep the kaolinite layers together by H-bonding with the O atoms of the Si tetrahedral sheet of the opposing layer in the particle. Consequently, these OH groups are not accessible for adsorption of molecules in the interlayer region of kaolinite. Adsorption is only possible with strongly polar molecules that are capable of breaking up the H-bonding between successive layers of kaolinite in a particle. Examples of these types of polar molecules include hydrazine, dimethyl sulfoxide, potassium acetate, and urea.

The O atoms at the edges of clay mineral particles are coordinatively unsaturated and carry a negative charge. The interaction of these O atoms with an aqueous environment is pH dependent:

$$-OH_2^+ \longleftrightarrow -OH \longleftrightarrow -O^-$$

$$\xrightarrow{\hspace{4em}}$$

Increasing pH

The OH groups are polar and interact with adsorbing molecules by polar–polar, polar–apolar, and dispersion interactions. They may also adsorb cations by a surface complexation reaction:

$$-OH + M^{n+} > [-OM]^{(n-1)} + + H^+ \qquad (1.6)$$

In the interlayer space of smectites, the O atoms of the Si tetrahedral sheets and the exchangeable cations are the main adsorption sites (Fig. 1.2). Polar molecules such as water complete the coordination sphere of the exchangeable cations and have H-bonding interactions with the O atoms of the Si tetrahedral sheets and with each other, thus forming one or two layers hydrates. Traditionally, smectites have been considered to be hydrophilic materials. However, if the exchangeable cations are exchanged with organic cations such as alkylammonium cations ($[H_{4-x}N\text{-}(R)_x]^+$, $x=1$–4; $R=C_nH_{2n+1}$), the surface is rendered hydrophobic. In this case, the positively charged headgroup is adsorbed on the siloxane surface to effectively neutralize the negative charges of the smectite layers. The alkyl chains can adopt different configurations, depending on the length of the chain and the charge density of the clay mineral layers. Hydrophilic and hydrophobic clay minerals are capable of adsorbing different types of molecules: polar molecules in the first case, apolar molecules in the second case. Intermediate situations can be created by exchanging bulky inorganic cationic complexes, cationic dyes, and cationic polymers in the interlayer space of smectites. These cations do not leave much space for adsorption in the interlayer space and create an amphiphilic environment.

The adsorption phenomena discussed so far are examples of physical adsorption (or physisorption) phenomena. Upon adsorption, the adsorbing molecules and the surface atoms retain their identity, and no chemical bond is established. Chemisorption occurs when a chemical bond is established between the surface atoms and the adsorbing molecules. Both the surface and the molecule must carry reactive groups. Examples are:

$$\text{-surface-OH} + \text{HO-PO}_3 > \text{surface-O-PO}_3 + H_2O \quad (1.7)$$

$$\text{-surface-OH} + \text{Cl}- > \text{-surface-OCl} + H_2O \quad (1.8)$$

A comparison between physisorption and chemisorption is given in Table 1.5. Physisorption is a weak interaction between the adsorbed molecules and the surface and leads to the formation of adsorbed mono- and multilayers. Such an ordering phenomenon is counteracted by temperature. Thus, the amount of physisorbed molecules decreases with increasing temperature. Physisorption and chemisorption are exothermic processes. A continuous scale of enthalpy of adsorption can be established from physisorption to chemisorption. However, it is difficult and dangerous to distinguish physisorption from chemisorption on the basis of enthalpy changes alone accompanying these phenomena. Other criteria must be taken into account, such as reversibility and temperature dependence. For example, a molecule might physisorb at low temperatures and chemisorb at high temperatures. Above all, the chemical bond between the surface and the molecule must be clearly established to conclude that the adsorption process is chemisorption.

Table 1.5 Comparison of physisorption and chemisorption

	Physisorption	**Chemisorption**
Enthalpy of adsorption	Small (<20 kJ/mol)	Large (>80 kJ/mol)
Specificity	Nonspecific	Specific
Reversibility	Reversible	Nonreversible
Thickness of adsorbed layer	Multilayers	Monolayers
Conditions	Room temperature or less	High temperature

1.7 Molecule–molecule and molecule–surface interactions

Physisorption or physicochemical interactions between molecules and surfaces involve molecule–molecule interactions and molecule–surface interactions. In the case of chemisorption, the molecule–surface interactions result in the formation of a chemical bond between the surface atoms and the adsorbed molecules. Additional physical interactions between the chemisorbed molecule and the surface atoms and intermolecular interactions between chemisorbed molecules may occur too, but they are weak when compared with the molecule–surface chemical bond formation and are usually neglected. In this chapter, three types of interactions are discussed: (1) molecule–molecule, (2) molecule–surface, and (3) surface-surface.

1.7.1 Molecule–molecule interactions

1.7.1.1 Ion-ion interactions

If the molecules are ions, they will undergo electrostatic attraction and repulsion according to Coulomb's law:

$$E_{ij} = \frac{Q_i Q_j}{4\pi\varepsilon_0 \varepsilon r_{ij}} \tag{1.9}$$

With Q_i and Q_j representing the point charges of the ions and r_{ij} is the distance between the ions i and j. ε is the dielectric constant of the medium in which the ions reside, and ε_0 is the vacuum permittivity. The use of ε implies that this medium is considered a continuum. The dielectric constant of the medium is required, because Coulomb's energy is due to a long-range electrostatic interaction, which decays with distance as r^{-1}. For all charges Q_i and Q_j in the medium, the total Coulombic interaction energy is

$$E_{ij} = \Sigma_i \Sigma_{j \neq I} \frac{Q_i Q_j}{4\pi\varepsilon_0 \varepsilon r_{ij}} \tag{1.10}$$

Charge–dipole interaction is the electrostatic interaction between a point charge Q_i and a permanent point dipole as shown in Fig. 1.8. The electrostatic interaction energy is the sum of the Coulombic interactions between the point charge Q_i and the charged q_j and $-q_j$ on the permanent dipole:

$$E_{cd} = \frac{q_j Q_i}{4\pi\varepsilon_0 \varepsilon r'_{ij}} + \frac{-q_j Q_i}{4\pi\varepsilon_0 \varepsilon r_{ij}} \tag{1.11}$$

This expression can be rearranged into

$$E_{cd} = \frac{-Q\mu cos\vartheta}{4\pi\varepsilon_0 r^2} \tag{1.12}$$

In this case, the negative pole of the dipole is attracted towards the point charge; the positive pole is repelled. The result is that the dipole is aligned with the point charge as shown in Fig. 1.8. Such an orientation is counteracted by increasing temperature as the thermal energy induces a random orientation of the dipoles in the system. Eq. (1.12) then becomes

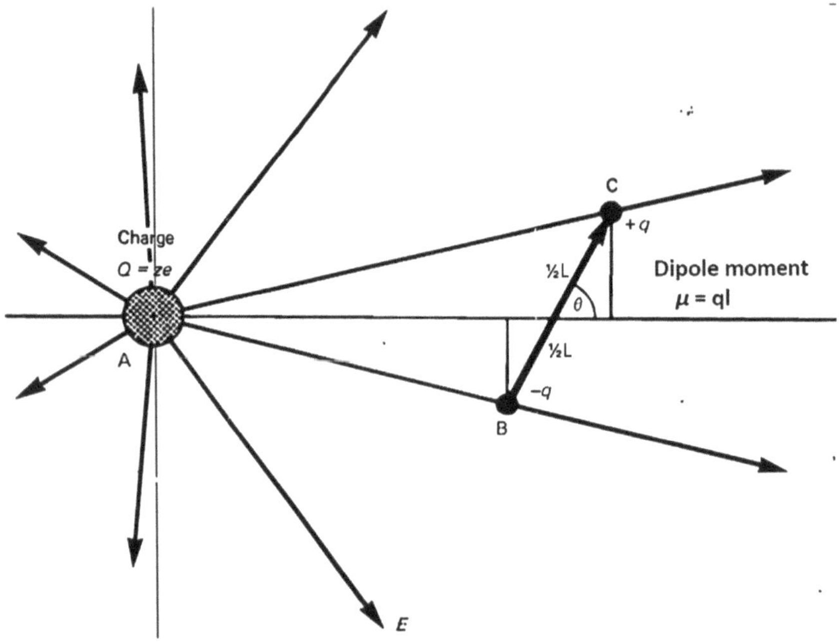

Fig. 1.8 Ion–dipole interactions
From Israelachvili, J.N., 1989. Intermolecular and Surface Forces. Academic Press, London.

$$E_{cd} = \frac{-Q^2\mu^2}{6kT(4\pi\varepsilon_0)^2 r^4} \tag{1.13}$$

The charge–dipole interaction energy decreases with distance as r^4. It is a short range interaction between neighbouring charges and dipoles. With short range interactions, the dielectric constant of the medium is usually set equal to 1. If one wants to express this interaction per mole of dipoles or per mole of charges, the interaction energy of Eq. (1.13) has to be multiplied by Avogadro's number N_a.

1.7.1.2 Dipole–dipole interactions

The interaction between two fixed dipoles is the sum of the electrostatic attractions between the poles with opposite charges and the electrostatic repulsion between poles of like charges. The result is an alignment or organization of the dipoles as shown in Fig. 1.9. The maximum attractive interaction is obtained for $\theta_1 = \theta_2 = 0$: $E_{dd} = -\mu_1\mu_2/4\pi\varepsilon_0 r^3$.

For $\vartheta_1 = 90$ degree and $\vartheta_2 = 0$ as well as for $\vartheta_1 = 0$ and $\vartheta_2 = 90$ degree $E_{dd} = 0$. For $\vartheta_1 = \vartheta_2 = 90$ degree, E_{dd} is repulsive: $\mu_1\mu_2/4\pi\varepsilon_0 r^3$. If freely rotating dipoles are taken into account together with the thermal energy, the dipole–dipole interaction energy becomes

$$E_{dd} = \frac{-\mu_1^2\mu_2^2}{3(4\pi\varepsilon_0)^2 kT\, r^6} \tag{1.14}$$

$1/r^6$ indicates that E_{dd} is a short range interaction: only neighbouring dipoles interact significantly. In this case, the dielectric constant is usually set equal to 1.

$\theta_1 = \theta_2 = 0°$: $w = -\mu_1\mu_2/(2\pi\varepsilon_0\varepsilon r^3)$

$\theta_1 = 0°$; $\theta_2 = 90°$; $w = 0$

$\theta_1 = 0°$; $\theta_2 = 180°$; $w = \mu_1\mu_2/(2\pi\varepsilon_0\varepsilon r^3)$

$\theta_1 = 0°$; $\theta_2 = 270°$; $w = 0$

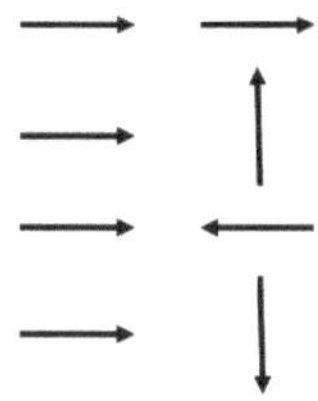

Fig. 1.9 Possible arrangement of dipoles.

1.7.1.3 Charge–nonpolar interaction

A charge $+Q$ interacts with a neighbouring nonpolar molecule as it attracts the electrons and repels the positively charged nuclei. Mainly, the electrons in the outer molecular orbitals or bonding molecular orbitals make a significant contribution to this interaction. The result is that a small dipole moment is induced in the nonpolar molecule, μ_i. The induced dipole moment is proportional to the electric field E of the charge $+Q$: $\mu_i = \alpha E$. The proportionality constant α is the polarizability of the molecule. With the induced dipole moment in *Cm* and E in volt, the dimensions of the polarizability are m^3. The charge–nonpolar interaction energy is then

$$E_{ci} = \frac{-Q^2\alpha^2}{2(4\pi\varepsilon_0)^2 r^4} \tag{1.15}$$

Also, dipoles will undergo a shift of their bonding electrons under the influence of an electric field of a charge Q. This means that a dipole moment will be induced, and that the total dipole moment is $\mu_T = \mu + \mu_i$. In general, $\mu_i \ll \mu$, and for dipolar molecules, the induced dipole moment is usually neglected.

1.7.1.4 Dipolar–nonpolar interaction

A nonpolar molecule in the immediate neighbourhood of a dipolar molecule is polarized in the same way as in the vicinity of a charge Q. The positive pole of the dipole attracts the bonding electrons and repels the nuclei. The negative pole does the reverse. For a fixed dipole, the interaction energy is given by

$$E_{dn} = \frac{-\mu^2\alpha(1 + 3cos\vartheta^2)}{2(4\pi\varepsilon_0)^2 r^6} \tag{1.16}$$

Averaged over all angles, that is, taking the thermal energy into account

$$E_{dn} = \frac{-\mu^2\alpha}{(4\pi\varepsilon_0)^2 r^6} \tag{1.17}$$

1.7.1.5 Nonpolar–nonpolar interactions

Two neighbouring nonpolar molecules attract each other. This is explained as follows. In a nonpolar molecule, the time-averaged distributions of negative charges (electrons) and positive charges (nuclei) coincide, and there is no dipole moment. However, at any time t, the distributions of the positive and negative charges do not coincide, and an instantaneous dipole moment is generated. The instantaneous dipoles of neighbouring molecules interact and are at the basis of the nonpolar–nonpolar interaction energy, also called dispersion energy of van der Waals energy. Empirical formulae for these interactions are:

For two identical spherical molecules:

$$E_{nn} = \frac{-3\alpha^2 I}{4(4\pi\varepsilon_0)^2 r^6} \tag{1.18}$$

For two different spherical molecules:

$$E_{nn} = \frac{-3\alpha_1\alpha_2}{2(4\pi\varepsilon_0)^2 r^6} \frac{I_1 I_2}{I_1 + I_2} \tag{1.19}$$

where I represents the ionization potential of the molecules. One remarks that the dispersive energy is attractive, short range, and independent of temperature.

Table 1.5 summarizes the expressions of the different types of intermolecular interaction energies. They have been calculated for a charge $Q = 1.6 \times 10^{-19} C$, a dipole moment of $6.17 \times 10^{-30} Cm$ or $1.85D$ (D = Debye) $T = 298$ K, $r = 0.4 \times 10^{-9}$ nm (4 Å), and $\alpha = 2.886 \times 10^{-40}$ m^3. The ionization energy is that of methane 12.6 eV or 1215 kJ/mol, and the dispersion energy is that of the methane–methane interactions. These energies have to be compared with those of the thermal energy, kT, which is 2.5 kJ/mol for $T = 298$ K. The following remarks have to be made.

1. All the intermolecular interactions are attractive and tend to induce organization in the molecular systems in which they are active. Thermal energy, kT, induces disorder and counteracts the intermolecular interactions. At room temperature, kT equals 4.21×10^{-21} J or 2.5 kJ/mol. This is larger than E_{dd}, E_{dn}, and E_{nn}, and disorder is prevalent. Intermolecular interactions with ions are larger than the thermal energy, and order is to some extent present in these systems at short distance; that is, the ions organize the molecules in their immediate environment. In the case of water, the ions acquire a hydration shell. At 10 K, kT equals 1.38×10^{-22} J or 73.8 J/mol. Organization will be prevalent in all molecular systems.
2. Intermolecular interactions are additive. A molecule with dipole moment μ_1 in the immediate environment of an ion with charge Q or another molecule with dipole moment μ_2 undergoes small electron displacements, which give rise to an induced dipole moment μ_i. All molecules, whether they have a dipole moment or not, have temporary displacements of nuclei and electrons, giving rise to temporary dipole moments and leading to dispersive interactions. Thus, interacting molecules with a dipole moment have three types of intermolecular interactions: E_{dd}, E_{dn}, and E_{nn}, the former being the largest (Table 1.6).
3. As the intermolecular interactions are attractive, they bring the molecules towards each other. This cannot continue indefinitely, because at short distance, electron–electron repulsion increases sharply, and at some characteristic distance, a minimum in energy is obtained at the equilibrium distance between the molecules. The attractive intermolecular interactions and the electron–electron repulsion are combined in the Lennard-Jones potential:

Table 1.6 Molecular interactions

Type	Symbol	Expression	Example (kJ/mol)
I_{ion}–I_{ion}	E_{ii}	$\frac{Q_iQ_j}{4\pi\epsilon_0\epsilon r_{ij}}$	4.43(ε = 78) 345(ε = 1)
I_{ion}–dipole	E_{id}	$\frac{-Q^2\mu^2}{6kT(4\pi\epsilon_0)^2r^4}$	75.3(ε = 1)
Dipole–dipole	E_{dd}	$\frac{-\mu_1^2\mu_2^2}{3(4\pi\epsilon_0)^2kTr^6}$	1.4
I_{ion}–nonpolar	E_{in}	$\frac{-Q^2\alpha^2}{2(4\pi\epsilon_0)^2r^4}$	5.6
Dipole–nonpolar	E_{dn}	$\frac{-\mu^2\alpha}{(4\pi\epsilon_0)^2r^6}$	0.13
Nonpolar–nonpolar	E_{nn}	$\frac{-3\alpha_1\alpha_2}{2(4\pi\epsilon_0)^2r^6}\frac{I_1I_2}{I_1+I_2}$	0.15

Examples are calculated with the following parameters:
$Q = 1.6\times10^{-19}$ C (electron charge); $r = 0.4$ nm; $\mu = 1.85$D or 6.17×10^{-30} Cm;
$\alpha = 2.88\times10^{-40}\,\text{m}^3$; $4\pi\epsilon_0 = 1.11\times10^{-10}$; $T = 298$°K; $I_1 = I_2 = 12.5\,\text{eV} = 2\times10^{-18}$ J;
$k = 1.38\times10^{-23}\,\text{JK}^{-1}$; $\varepsilon = 78$.

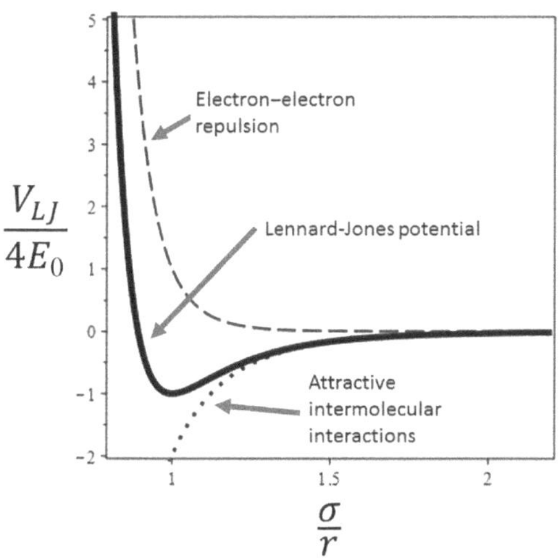

Fig. 1.10 The Lennard-Jones potential showing energy as a function of distance.

$$E_{LJ} = 4E_0\left[\left(\frac{\sigma}{r}\right)^{12} - \left(\frac{\sigma}{r}\right)^6\right] \tag{1.20}$$

Its graphical representation is shown in Fig. 1.10. $E_{LJ} = 0$ if $r = \sigma$, and the minimum $E_{LJ} = E_0$ is obtained for $r = 2^{1/6}\sigma = 1.12\sigma$.

1.7.1.6 H-bonding: X–H-----Y

H-bonding is the—mainly—electrostatic interaction between a H atom bonded to an electronegative atom X such as F, O, and N, and the electronegative atom Y of the same molecule, a neighbouring molecule or a surface. Typical H-bonding interaction energies and electronegativities are given in Table 1.7.

Typical bond lengths are 110 pm for X–H and 160–200 pm for H----Y. C—H bonds can also be involved in H-bonding, but then, electronegative elements must be bonded to C, such as in chloroform, Cl_3CH. One of the major consequences of H-bonding is the lengthening of the X–H bond length. This means that the X–H bond is weakened, and the dipole moment of the X–H bond is increased. This has three consequences:

(1) In vibrational spectroscopy, the force constant of the X–H vibration is decreased, resulting in a shift of the X–H stretching frequency to lower wavenumbers. This is evident from the expression of the harmonic stretching frequency of the diatomic molecule X–H:

$$\nu = \frac{1}{2\pi}\sqrt{\frac{k}{\mu_{X-H}}} \text{ or } 2\pi\nu = \omega = \sqrt{\frac{k}{\mu}}$$

In terms of wavenumbers $\tilde{\nu}$ one obtains $\tilde{\nu} = \frac{1}{c}\sqrt{\frac{k}{\mu}}$ with c the speed of light, k is the force constant of the X−H bond and $\mu = \frac{m_H m_x}{m_H + m_x}$ the reduced mass.

(2) The transition dipole moment $\Delta\mu$ increases, and so does the intensity of the X–H stretching vibration, as the latter is proportional to $|\Delta\mu|^2$.

(3) The proton involved in H-bonding becomes less shielded. Its chemical shift δ with respect to that of $Si(CH_3)_4$ ($\delta = 0$) shifts to positive values, or the resonance of the H-bonded H atom is downfield with respect to that of the reference, tetramethylsilane. This is illustrated in Table 1.8.

1.7.2 Molecule-surface and surface-surface interactions

The dispersion or van der Waals energy is of the form

$$E_{nn} = \frac{-C}{r^6} \tag{1.21}$$

Table 1.7 H-bonds and their strength, related to the electronegativity of the atoms

H-bond	Strength (kJ/mol)	Electronegativity (eV) (Sanderson)
F–H-----F	162	F: 3.92
O–H----N	29	N: 2.93
N–H----N	13	O: 3.46
O–H-----O	21	
N–H-----O	8	

From Huheey, J.E., Keiter, E.A., Keiter, R.L., 1997. Inorganic Chemistry. Principles of Structure and Reactivity. Prentice Hall, New York, pp. 737.

Table 1.8 Typical proton chemical shifts

Molecule	Chemical shift (ppm)
$Si(CH_3)_4$	0
H_2	4.34
H atom	13.16
Bare proton	30.94

For an atom interacting with a surface, the dispersion energy is the sum of the interactions of the atoms with all the surface atoms. For an adsorbed molecule, all the atoms of the adsorbed molecule interact in principle with all the surface atoms. For two interacting surfaces, all atoms of one surface interact with all atoms of the other surface. One simply has to add all these interactions together to obtain the total dispersion energy. These summations lead to dispersion, or van der Waals, energies, which are expressed in terms of the Hamaker constant A

$$A = \pi^2 C \rho_1 \rho_2 \tag{1.22}$$

where ρ_1 and ρ_2 are the number densities (m^{-3}) of atoms in, for example, surface 1 and surface 2. C is a constant (Jm^6) and A is expressed in Joule (J). Fig. 1.11 compares the dispersion interaction between two atoms with the dispersion interaction between two flat surfaces. Other geometries can be found in Israelachvili's book (1989). Thus, for two flat surfaces in close contact, the dispersion energy per unit surface area is given by

$$E_{nn} = \frac{-A}{12\pi D^2} \tag{1.23}$$

where D is the distance between the two surfaces. The units of E_{nn} are now $J\,m^{-2}$. A is almost constant and falls in the range $(0.4–4) \times 10^{-19}$ J whatever the material or the type of molecule. When the two surfaces are in contact, D is the diameter

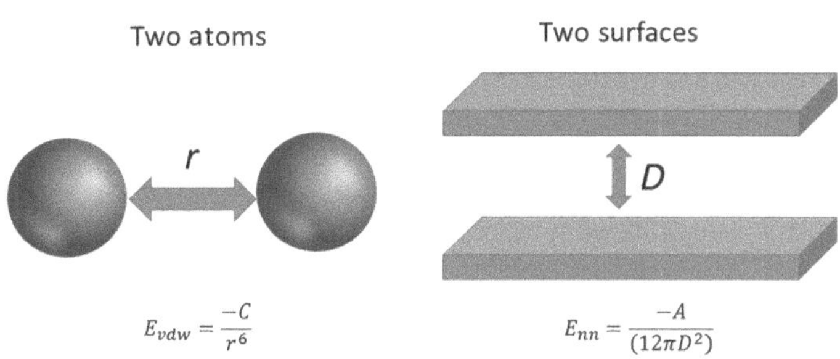

Fig. 1.11 Survey of different interaction geometries and expressions of dispersion energies. From Israelachvili, J.N., 1989. Intermolecular and Surface Forces. Academic Press, London.

of an O atom, 0.28 nm. One obtains $E_{nn} = -33.85\,\mathrm{mJ\,m^{-2}}$. The corresponding adhesive pressure is given by

$$P = \frac{A}{6\pi D^3} \tag{1.24}$$

Or $2.4 \times 10^8\,\mathrm{Nm^{-2}}$ or 2400 bar. It is not easy, maybe impossible, to separate these surfaces such as in particles of talc and pyrophyllite.

In this reasoning, the following assumptions have been made: (1) the interacting bodies, be they atoms, molecules, or particles, are rigid and (2) pairwise additivity or neighbouring atoms in a body do not influence the interacting atoms. In reality, atoms in surfaces and in molecules are vibrating and do not fulfil the two assumptions. More elaborated theories can be found in the book of Israelachvili (1989).

References

Bailey, S.W., 1988a. Hydrous Phyllosilicates (Exclusive of Micas). Mineralogical Society of America, Washington D.C.

Bailey, S.W., 1988b. Reviews in mineralogy. Hydrous Phyllosilicates (Exclusive of Micas), vol. 19. Mineralogical Society of America, Washington DC.

Bergaya, F., Lagaly, G., Vayer, M., 2013. Cation and anion exchange chapter 2.11 handbook of clay science. second ed. Developments in Clay Science Series, vol. 5. Elsevier, Amsterdam. Part 5B: techniques and applications, 333–359.

Brigatti, M.F., Galan, E., Theng, B.K.G., 2013. Chapter 2—Structure and mineralogy of clay minerals. In: Bergaya, F., Lagaly, G. (Eds.), Developments in Clay Science. Elsevier.

Gaines, G.L., Thomas, H.C., 1953. Adsorption studies on clay minerals. II. Formulation of the thermodynamics of exchange adsorption. J. Chem. Phys. 21, 714.

Huheey, J.E., Keiter, E.A., Keiter, R.L., 1997. Inorganic Chemistry. Principles of Structure and Reactivity. Prentice Hall, New York, p. 737.

Israelachvili, J.N., 1989. Intermolecular and Surface Forces. Academic Press, London, p. 296.

Johnston, C.T., 2010. Probing the nanoscale architecture of clay minerals. Clay Miner. 45, 245–279.

Lagaly, G., 2006. Colloid clay science. In: Bergaya, F., Theng, B.K.G., Lagaly, G. (Eds.), Handbook of clay science. Development in Clay Science, vol. 1. Elsevier, Amsterdam, pp. 141–245.

Laird, D.A., Scott, A.D., Fenton, T.E., 1989. Evaluation of the alkylammonium method of determining layer charge. Clays Clay Miner. 37, 41–46.

Ras, R.H.A., Umemura, Y., Johnston, C.T., Yamagishi, A., Schoonheydt, R.A., 2007. Ultrathin hybrid films of clay minerals. Phys. Chem. Chem. Phys. 9, 918–932.

Schoonheydt, R.A., Johnston, C.T., 2013. Surface and interface chemistry of clay minerals. In: Bergaya, F., Theng, B.K.G., Lagaly, G. (Eds.), Handbook of Clay Science. second ed. Part A. Fundamentals, Elsevier, Amsterdam.

Sposito, G., 1981a. Cation exchange in soils: an historical and theoretical perspective. In: Chemistry in the Soil Environment. American Society of Agronomy-Soil Science Society of America, Madison, WI.

Sposito, G., 1981b. The Thermodynamics of Soil Solutions. Oxford University Press, Oxford.

Further reading

Bergaya, F., Lagaly, G., 2013. Handbook of clay science. In: second ed. Developments in Clay Science Series, vol. 5. Elsevier, Amsterdam. Part 5A: fundamentals pp. 874, part 5B: techniques and applications, pp. 813.

Determination of surface areas and textural properties of clay minerals

2

Laurent J. Michot
Physicochimie des Electrolytes et Nanosystèmes Interfaciaux, Phenix, UMR 8234, CNRS/Sorbonne Université, Paris Cedex, France

2.1 Introduction

For many applications taking advantage of the various surface properties of clay minerals, it is of prime importance to be able to precisely assess the total amount of available surface area and to distinguish between the various types of surfaces and pores present in the sample (i.e. micropores, mesopores, and external surface area). In this chapter, the use of the terms micropores, mesopores, and macropores is based on the IUPAC definition for gas adsorption (Sing et al., 1985), that is, micropores <2 nm, mesopores between 2 and 50 nm, and macropores >50 nm.

The aim of this chapter is to present the various methods currently available for obtaining such information and to discuss their applicability and limits. Furthermore, in view of the specific features of clay minerals and related materials, dedicated procedures need to be defined that depend on both the nature of the investigated mineral and its use. For example, in the case of swelling clay minerals, the surface area to be analysed depends on the conditions in which the material is employed (in aqueous or dried conditions). This chapter will then be organised in three subsections dealing with (i) the measurement of specific surface area (SSA) for nonswelling and nonmicroporous clay minerals; (ii) the case of microporous clay minerals, and (iii) the specific case of swelling clay minerals. Some concluding remarks on new materials and new techniques will finally be given.

2.2 Nonswelling and nonmicroporous clay minerals

In the case of nonswelling clay minerals, the adsorption of ethylene glycol monoethylether (EGME) or methylene blue (MB) is sometimes employed for determining surface areas. Still, the most widely used method is based on the application of the Brunauer–Emmett–Teller (BET) equation for nitrogen gas adsorption (Gregg and Sing, 1982). This is the basis of the standard method ISO 9277 confirmed in 2010. In this method, the evolution of the adsorbed amount of nitrogen gas is measured as a function of the relative pressure (P/P_0) of nitrogen, thus yielding an adsorption isotherm. Assuming that multilayer adsorption occurs, that all surface sites are equivalent, that the adsorption energy for all layers after the first one are equal

Developments in Clay Science, Vol. 9. https://doi.org/10.1016/B978-0-08-102432-4.00002-0

to the liquefaction energy of the adsorbate, and neglecting lateral interactions between adsorbed molecules, the adsorption isotherm can be expressed as (Brunauer et al., 1938):

$$\frac{V_{ads}}{V_m} = \frac{C\frac{P}{P_0}}{\left(1-\frac{P}{P_0}\right)\left(1-\frac{P}{P_0}+C\frac{P}{P_0}\right)} \tag{2.1}$$

where V_{ads} is the adsorbed volume at a relative pressure P/P_0, V_m is the adsorbed volume for one monolayer, and C and the energetic constant defined as $C = \exp^{\frac{E_1-E_L}{RT}}$, where E_1 is the adsorption energy for the first layer in $\mathrm{J\,mol^{-1}}$, E_L the liquefaction energy in $\mathrm{J\,mol^{-1}}$, and R the ideal gas constant $\mathrm{J\,mol^{-1}\,K^{-1}}$.

Eq. (2.1) can be also written as:

$$\frac{\frac{P}{P_0}}{\left(1-\frac{P}{P_0}V_{ads}\right)} = \left(\frac{C-1}{V_mC}\right)\frac{P}{P_0} + \frac{1}{V_mC} \tag{2.2}$$

This is the form that is used for determining SSA. Indeed, plotting the left-hand term as a function of relative pressure yields a linear plot in which the slope and ordinate at origin allow calculating both C and V_m. When V_m is expressed in STP $\mathrm{cm^3\,g^{-1}}$, assuming a cross-sectional area of 0.162 $\mathrm{nm^2}$ for adsorbed nitrogen (Sing et al., 1985), the SSA of the sample expressed in $\mathrm{m^2\,g^{-1}}$ is obtained by the formula $SSA = 4.37*V_m$.

As an illustration, the adsorption–desorption isotherm obtained by nitrogen adsorption on a commercial kaolin obtained from Imerys is reported in Fig. 2.1A; and the corresponding application of the BET equation that yields an energetic constant C of 92 and a SSA of 24 $\mathrm{m^2\,g^{-1}}$ is reported in Fig. 2.1B.

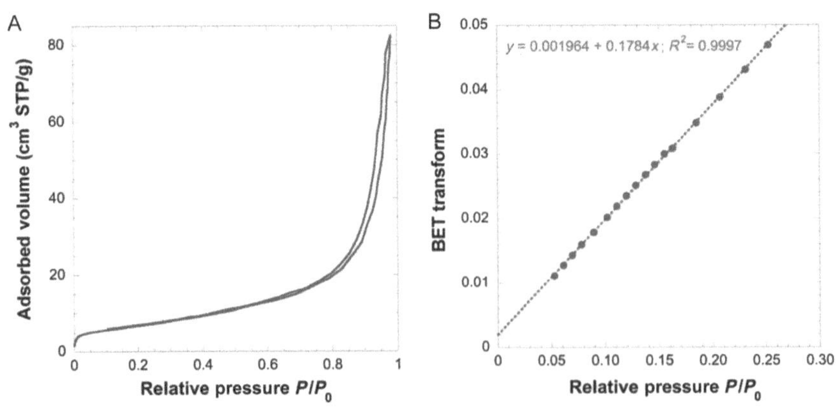

Fig. 2.1 Nitrogen adsorption isotherm and BET transform on a kaolin obtained from Imerys.

In principle, the BET equation should be applicable for relative pressures ranging between 0.05 and 0.30. However, it is necessary to examine in detail the linearity range of the equation, as it can be shifted towards lower relative pressures, especially when the investigated solid contains micropores. It must also be pointed out that the BET analysis cannot be applied to materials that display a low affinity towards nitrogen. This could be the case, for example, when clay minerals are covered with apolar organic molecules. One of the main interests of gas adsorption techniques is that the adsorption isotherm bears much more textural information than the SSA.

First, the presence of micropores can be assessed from the shape of the isotherm. For example, strong uptake at low relative pressure, less than 0.05, is often indicative of the presence of micropores. Second, various data treatment procedures have been proposed in the literature for evaluating the amount of micropores in a sample. The most common method is the '*t*-plot' (de Boer et al., 1966). The principle of the method is based on the comparison of the experimental isotherm, with a reference isotherm obtained for a nonporous solid. For the nonporous material, the isotherm can be replotted as *t* vs relative pressure P/P_0 where *t* is the statistical thickness of the adsorbed nitrogen layer, the value of 0.354 nm being that of a monolayer. Plotting for the same P/P_0, the adsorbed amount vs *t* yields a curve that can be used for microporosity assessment. Fig. 2.2 displays the adsorption–desorption isotherm obtained for a Ni–Al LDH (LDH, layered double hydroxide) sample exchanged with fluorine anions (Fig. 2.2A) and the corresponding *t*-plot. The calculated BET surface area is 63.2 $m^2\,g^{-1}$, with a corresponding *C* constant of 134. Classically, *C* values greater than 100 are considered indicative of the presence of micropores, with non-microporous materials typically displaying *C* values between 30 and 100. The *t*-plot treatment (Fig. 2.2B) yields a total surface area of 63.4 $m^2\,g^{-1}$, a micropore volume of 2 $cm^3\,g^{-1}$ corresponding to an equivalent surface area of 8.7 $m^2\,g^{-1}$, and a non-microporous surface area of 54.1 $m^2\,g^{-1}$. It must be pointed out that the upward deviation in the *t*-plot observed for *t*-values ≥5 Å indicates the presence of

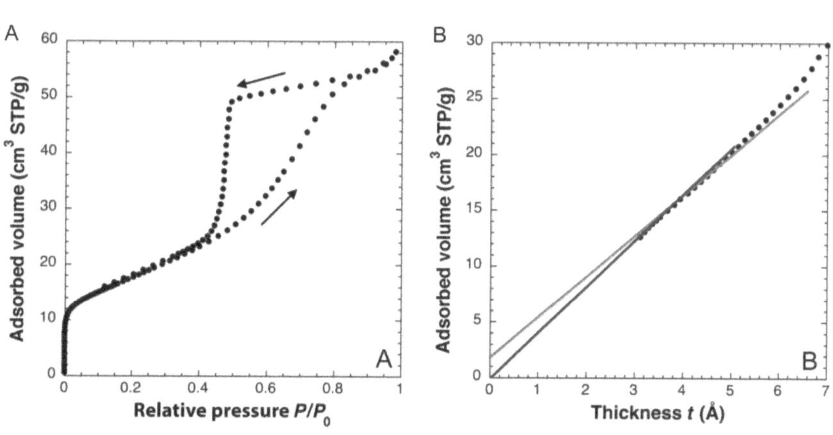

Fig. 2.2 Nitrogen adsorption–desorption isotherm and *t*-plot obtained on a Ni–Al LDH.

mesopores, which concurs with the shape of the adsorption–desorption isotherm that exhibits a strong hysteresis.

In principle, the *t*-plot method is valid only when, for the experimental system, the affinity of the adsorbate towards the adsorbent is similar to that for the reference solid. For this reason, *t*-curves for various energetic constants of the adsorbate–adsorbent couple have been published in the literature (Lecloux and Pirard, 1979) and are sometimes helpful. For a series of samples, or in cases where gaseous probes other than nitrogen are used, it can be relevant to implement a closely related method, the α_s method (Gregg and Sing, 1982), in which the user designs his or her own reference isotherm. This can be particularly useful when a particular treatment (chemical, grinding, thermal, etc.) is applied to a starting solid. The adsorption isotherm on the initial material is acquired with high accuracy, and α is then defined as V_{ads}/V_x where x is a chosen value of P/P_0. A reference isotherm α vs P/P_0 can then be generated. The isotherms of the treated solids can then, as in the *t*-plot method, be plotted as V_{ads} vs α. Deviations from linearity in these plots can be interpreted as changes in the adsorbent/adsorbate interactions that can be linked to chemical or textural changes.

Using the medium and high relative pressure range of the adsorption–desorption isotherm, that is, $0.4 \leq P/P_0 \leq 0.98$, information on the presence of mesopores and on their size distribution can be derived on the basis of the Kelvin equation that relates pore diameter with the relative pressure at which capillary condensation occurs. In general, the analysis uses the desorption branch of the isotherm. Using assumptions about pore geometry, it is then possible to derive a mesopore size distribution. When cylindrical pore geometry is considered, the so-called BJH method proposed by Barrett et al. (1951) is used. Modifications of this method for taking into account the layered structure of clay minerals have been proposed (e.g. Delon and Dellyes, 1967). Still, it must be pointed out that all these methods provide estimates of pore size distribution and do not truly capture the full complexity of pore size distributions in clay materials.

Advanced gas adsorption techniques can also provide additional information that is of prime importance for understanding the surface properties of clay minerals. In particular, it provides ways of determining the aspect ratio of these minerals, which is particularly relevant. Indeed, this parameter controls, among others, the mechanical properties of clay-filled polymers (e.g. Michot et al., 1993), the barrier properties of clay polymer nanocomposites (e.g. Tan and Thomas, 2017), the adsorption of metals and metalloids in the environment (e.g. Rajec et al., 1999; Auang et al., 2015), the properties of filled papers (e.g. Adams, 1993), the porosity of dense packings of anisotropic particles (e.g. Ferrage et al., 2015) or the rheological properties of clay mineral dispersions (e.g. Blachier et al., 2014). Various techniques can be applied to measure the aspect ratio of nonswelling clay minerals, and in recent years, atomic force microscopy (AFM) has proven very useful for such purposes (e.g. Bickmore et al., 1999, 2002; Balnois et al., 2003; Tournassat et al., 2003; Cadene et al., 2005; Sayed-Hassan et al., 2006).

Still, microscopic techniques require recording numerous images to obtain statistically reliable data, whereas gas adsorption analyses a large number of particles, thus yielding average values, which represents a definite advantage of this technique.

The principle supporting aspect ratio determination by gas adsorption lies in the fact that in an adsorption experiment, the adsorbate probes the surface heterogeneity of the sample, more energetic sites being filled at lower under-saturation than the less energetic ones. For clay minerals, edge and basal faces have different surface chemistry, and should then behave differently towards the adsorbate. In the case of gas adsorption, as the filling of one monolayer is complete for low relative pressures (typically less than 0.1), analysing surface heterogeneity requires being able to obtain numerous experimental data points at low relative pressure. This can be achieved using quasi-equilibrium setups equipped with accurate pressure sensors (Michot et al., 1990; Villiéras et al., 1992, 1997b, 2002). Using such setups, 2000–3000 experimental data points can be collected at low relative pressures. Consequently, the derivative of the adsorbed amount as a function of the logarithm of relative pressure, $\ln(P/P_0)$, can be accurately calculated. This latter parameter corresponds to the free energy of adsorption in kT scale. Fig. 2.3 presents a typical derivative curve obtained by treating the adsorption isotherm of argon at 77 K on an illite sample exchanged with Na^+ cations (Bardot et al., 1998; Villiéras et al., 2002). The derivative isotherm thus obtained clearly displays various peaks and shoulders, the meaning of which can be interpreted on the basis of the derivative isotherm summation (DIS)

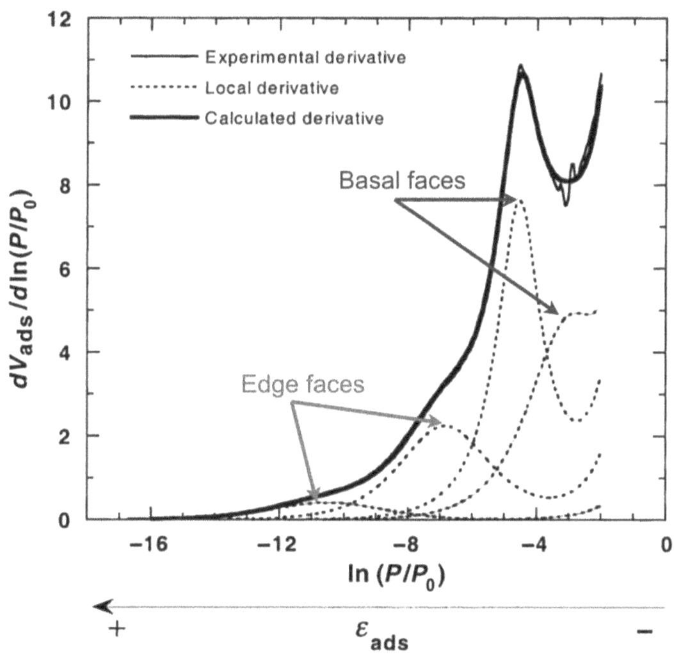

Fig. 2.3 Derivative argon adsorption isotherm and DIS treatment of an illite exchanged with Na^+ cations.
Adapted from Villiéras, F., Michot, L.J., Bardot, F., Chamerois, M., Eypert-Blaison, C., Gérard, G., Cases, J.M., 2002. Surface heterogeneity of minerals. C. R. Geosci. 334, 597–609.

technique (Villiéras et al., 1992, 1997a,b, 2002). In short, the adsorption isotherm of an adsorbent with n distinct domains (faces, sites, etc.) is written as:

$$\theta = \sum_{i=1}^{n} X_i \theta_{it} \quad \text{with} \quad \theta_{it} = \int_{\Omega} \theta_i(\varepsilon)\chi_i(\varepsilon)d\varepsilon \tag{2.3}$$

where θ_{it} is the adsorption isotherm on the ith domain, X_i its contribution to θ_t, $\theta_i(\varepsilon)$ is a 'local' theoretical adsorption isotherm, and $\chi_i(\varepsilon)$ the dispersion of the adsorption energy ε on the ith domain. In a first approximation, each domain can be considered as energetically homogeneous (Villiéras et al., 1992).[a] In that case, $\chi_i(\varepsilon) = 1$ and θ_{it} can then be described as a 'local' theoretical one-layer (Langmuir–Temkin) or multilayer adsorption isotherm (BET–Hill–Halsey) on a homogeneous surface. The experimental derivative isotherm is then fitted as the sum of theoretical DIS domains. Each local isotherm θ_{it} is fully characterised by three parameters: the interaction between the surface and an adsorbed molecule, the lateral interactions between two neighbouring adsorbed molecules, and the monolayer capacity (V_m) of the domain.

According to such a treatment, the local isotherms such as those displayed in Fig. 2.3 can yield the relative proportions of the various sites/faces to the total adsorption isotherm. In particular, assuming that the adsorption energy on basal faces is lower than that on edge faces, the different local isotherms can be assigned to either edge or basal faces, which directly yields the aspect ratio of the investigated mineral.

Analysis of low-pressure gas adsorption features on numerous clay minerals reveals the following trends.

- Whatever the clay mineral used, nitrogen is not a suitable probe for assessing the aspect ratio of particles. Indeed, due to their polarizability (quadrupolar momentum) nitrogen molecules have specific interactions with surface sites (hydroxyl groups and/or exchangeable cations).
- For uncharged nonswelling clay minerals, low-pressure argon adsorption yields accurate aspect ratios as verified by cross-validation between AFM measurements and argon adsorption data (Sayed Hassan et al., 2005, 2006). Talc is a specific case, as it possesses high-energy adsorption sites that can be identified as the ditrigonal holes in the tetrahedral sheet (Michot et al., 1994). Once these sites are specifically taken into account, valid aspect ratios can be derived from low-pressure argon adsorption isotherms. A similar situation is encountered with layered silicic acids (Eypert-Blaison et al., 2002).
- For charged nonswelling clay minerals, low-pressure argon adsorption yields correct basal and lateral surface areas only when the surface exchangeable cations are small and monovalent, such as lithium and sodium. Indeed, larger cations or divalent cations interfere with the formation of an adsorbed film, which affects the quantitative measurements of the various surface types (Bardot et al., 1998).

[a] Further refinement of the model can be implemented by considering an energy distribution on each domain (Villiéras et al., 1997a,b). This will not be discussed here, as it is beyond the scope of this chapter.

2.3 Microporous clay minerals

Various clay minerals are structurally microporous. This is, for example, the case of the fibrous clay minerals sepiolite and palygorskite,[b] and of the family of allophane/imogolites. Modified clay mineral samples, such as pillared clays, also bear this characteristic. This microporous nature leads to specific problems regarding their surface area determination by gas adsorption.

First, and this is mainly the case for structurally microporous samples, it can be tricky to properly define the outgassing conditions that must be applied prior to the analysis by gas adsorption. Indeed, in the natural state, structural micropores are generally filled with water molecules that need to be removed before adsorption experiments while maintaining the structural integrity of the material. In the case of sepiolite (Grillet et al., 1988) and palygorskite (Cases et al., 1991), detailed adsorption studies were carried out for various outgassing temperatures. For the first mineral, when outgassing temperatures $\leq$100°C were used, the porous structure was maintained, and high SSA and micropore volumes were evidenced. For the second mineral, similar results were obtained with a shift of the maximal outgassing temperature towards lower values, that is, around 60°C. In the case of imogolite, various studies showed that heat treatment under vacuum at 250–300°C appears as the optimal pretreatment condition for maximising the accessible pore volume, while maintaining the integrity of the nanotubular structure (Fernandez-Martinez and Michot, 2016 and references within). For structurally microporous clay minerals, defining the proper outgassing conditions is then of prime importance for gas adsorption measurements. This generally requires carrying out thermal analysis investigations. In that regard, sample-controlled thermal analysis techniques such as CRTA (Rouquerol, 1970, 1989; Rouquerol et al., 1992) present definite advantages. In this technique, the temperature increase is controlled by the pressure above the sample, that is, when the pressure reaches a certain threshold defined by the user, heating is stopped until the pressure drops lower than the threshold. This overcomes the problems associated with desorption kinetics. Furthermore, as the technique is carried out under vacuum, the results obtained can be directly transferred in terms of outgassing conditions. The aforementioned studies on sepiolite (Grillet et al., 1988) and palygorskite (Cases et al., 1991) clearly show the benefits associated with the use of CRTA.

Second, it must be emphasised that in the case of microporous samples, the BET surface area must, in fact, be considered as an equivalent surface area, as the concept of surface area does not apply to micropores. Techniques such as the *t*-plot or α_s method must be employed in that case. The microporous volumes thus determined should be expressed in liquid volumes, taking into account the density of the liquid adsorbate, as this representation is much more meaningful than the gaseous volume. An alternate technique for purely microporous samples is to use the equation developed by Dubinin (1966)

[b] This microporous character is clearly one of the key factors of the stability of the so-called Maya blue pigment (e.g., JoseYacaman et al., 1996; Hubbard et al., 2003).

$$\log V = \log V_0 \left[\log \left(\frac{P_0}{P} \right) \right]^2 \tag{2.4}$$

where V is the adsorbed volume and V_0 the total micropore volume. The ordinate at the origin of the plot of log V vs $[\log(P/P_0)]^2$ then yields the total micropore volume.

Third, the measured micropore volume can be highly dependent on the nature/size of the adsorbate used. For example, in the case of imogolite, the micropore volume determined varies with the nature of the gaseous probe, diminishing from 0.061 $cm^3\,g^{-1}$ for *n*-butane to 0.040 $cm^3\,g^{-1}$ for cyclohexane and 0.0065 $cm^3\,g^{-1}$ for perfluorobutylamine, a bulky molecule with a diameter of 1.02 nm (Adams, 1980; Ackerman et al., 1993). Similarly in sepiolite, the micropore volumes determined with nitrogen and CO_2 are 0.046 and 0.129 $cm^3\,g^{-1}$, respectively (Grillet et al., 1988). Furthermore, classical point-by-point setups sometimes fail in assessing the total micropore volume. This was, for example, shown in the case of sepiolite, where CO_2 adsorption in a quasi-equilibrium setup yielded a micropore volume of 0.253 $cm^3\,g^{-1}$ (Michot et al., 1990), instead of 0.129 $cm^3\,g^{-1}$ using a classical setup. The quasi-equilibrium value concurs with the structural one that yields a pore volume of 0.248 $cm^3\,g^{-1}$(Rautureau and Tchoubar, 1976).

Finally, no treatment is currently available for obtaining truly reliable micropore size distributions. Therefore, properly assessing microporosity relies on either using various adsorbates or series of adsorbents. A typical example of this is provided by the study of Al_{13}-pillared synthetic saponite samples with increasing layer charge (Bergaoui et al., 1995; Michot et al., 1998). Combining low-pressure argon, nitrogen, and CO_2 adsorption allowed obtaining a clear picture of the phenomena occurring in such samples: argon and nitrogen adsorb around the pillars, whereas CO_2 really fills the micropore volumes.

It is worth pointing out that numerous systems can be microporous and mesoporous at the same time, and that for synthetic systems, an evolution between those two types of textural features can sometimes be observed. An illustration of such a case is provided by synthetic allophanes (Fig. 2.4). For increasing Al/Si ratios, the samples evolve from meso–macroporous, to micro–mesoporous to purely microporous (Montargès-Pelletier et al., 2005). This textural evolution can be linked to structural changes in the material, which reveals once again that textural analysis is a very efficient tool for understanding clay minerals.

2.4 Swelling clay minerals

In the case of nonswelling clay minerals, the SSA determined in the dry state remains generally valid when the materials are dispersed in a solvent, provided that a proper assessment of microporosity and mesoporosity has been carried out. Consequently, when dealing with adsorption phenomena at the solid liquid interface, it is possible to reason on the basis of surface area determinations performed on dry systems. In the case of swelling clay minerals, the situation is quite different, as significant differences in surface properties can be present between 'dry' and 'wet' conditions.

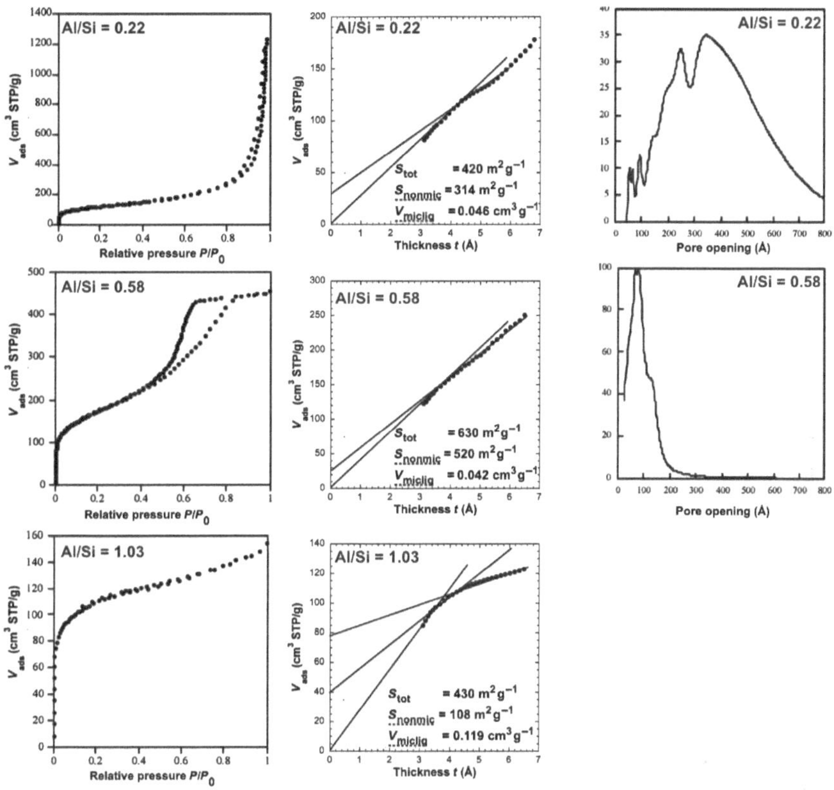

Fig. 2.4 Nitrogen adsorption isotherms, *t*-plot treatment and mesopore size distribution of synthetic allophane samples with varying Al/Si ratios.
Adapted from Montargès-Pelletier, E., Bogenez, S., Pelletier, M., Razafitianamaharavo, A., Ghanbajaa, J., Lartiges, B., Michot, L.J., 2005. Synthetic allophane-like particles: textural properties. Colloids Surf. A. 255, 1–10.

Then surface area determinations for 'dry' samples will be considered first before developing how the data thus obtained must be treated when the samples are used as dispersions.

2.4.1 The dry state

2.4.1.1 Gas adsorption techniques using 'classical' adsorbates

In the dry state, nitrogen or argon adsorption can be used for characterising swelling clay minerals. All the techniques described in the previous paragraphs can then be applied to measure SSA, micropores, and mesopores in such samples. Fig. 2.5 presents the adsorption–desorption isotherm of nitrogen at 77 K on a montmorillonite (Mt) from Sardinia together with the BET transform and *t*-plot treatment. The curves thus obtained appear to be fully exploitable. However, it was long recognised that various

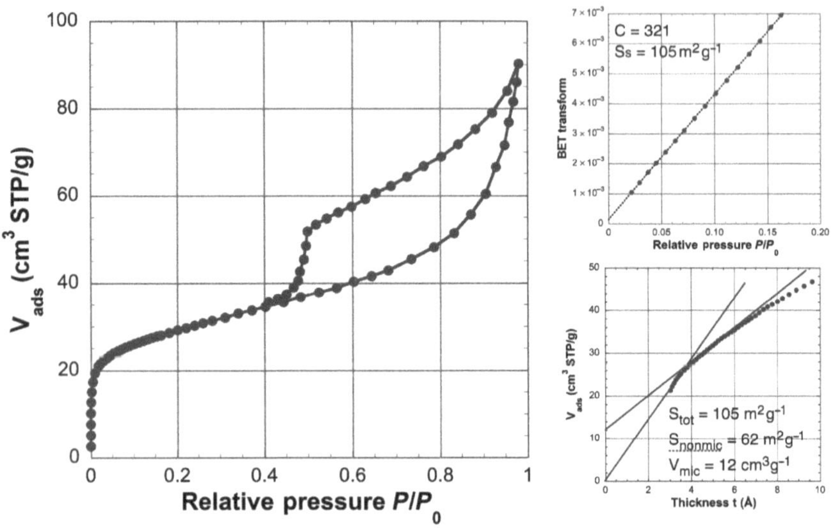

Fig. 2.5 Nitrogen adsorption–desorption isotherm, BET treatment and *t*-plot treatment on a Mt from Sardinia.

parameters influence the results obtained by gas adsorption and that the results obtained must be interpreted with caution.

1. *The preparation method of the sample affects the results.* Greene-Kelly (1964) carried out a rather extensive study on the surface area of Mt. This author showed, for example, that the SSA of films of Na^+-Mt from Wyoming was around 5 $m^2\ g^{-1}$, but that powdering the film could yield surface areas as high as 67 $m^2\ g^{-1}$. A freeze-dried powder of a similar Wyoming sample leads to an SSA of around 40 $m^2\ g^{-1}$ (Berend et al., 1995), but values of 19 $m^2\ g^{-1}$ (Rutherford et al., 1997), 23 $m^2\ g^{-1}$ (Dogan et al., 2006), or 40 $m^2\ g^{-1}$ (Aylmore et al., 1970) are also reported for powders. Obviously, the differences thus observed in BET surface areas are also transferred in terms of micropore volume or mesopore size distributions, and variable results have been obtained, depending on the study. In terms of preparation, it was also recognised long ago that the outgassing temperature also has some influence on the measured surface area (Thomas and Bohor, 1968; Aylmore et al., 1970).
2. *The nature of the exchangeable cation has a strong influence on the measured surface areas.* This was mentioned as early as 1952 (Mooney et al., 1952a,b), and has been repeatedly reported since (e.g. Thomas and Bohor, 1968; Aylmore et al., 1970; Berend et al., 1995; Cases et al., 1997; Rutherford et al., 1997). In general, for monovalent cations, the nitrogen BET surface area increases with the radius of the cation. This trend is less clear for divalent cations that have also been much less studied. A rather specific case is that of Cs^+-exchanged samples for which the SSA significantly increases. This is generally attributable to the fact that, for this sample, the large ionic radius of the Cs^+ ion (also indicated by the basal spacing of 1.12 nm for dry samples), could allow access of some nitrogen molecules to the interlayer space (Thomas and Bohor, 1968; Aylmore et al., 1970; Berend et al., 1995; Rutherford et al., 1997). Such an interpretation is supported by adsorption using other gases, CO_2 particularly

(Thomas and Bohor, 1968), which is known to be more efficient for filling small micropores (see Section 2.3). It is also confirmed by using classical micropore assessment techniques such as the t-plot or α_s plot (Rutherford et al., 1997). For the other cations, the interpretation of the observed differences is much less clear. It may be due to changes in the textural properties of the powders at low spatial scales, which could modify the amounts of micropores in the sample.

However, understanding the microporosity of dry swelling clay minerals remains a rather complex issue. Indeed, as mentioned in the section dealing with nonswelling clay minerals, low-pressure adsorption measurements have revealed that the surfaces of swelling clay minerals bear some high-energy sites with which nitrogen can enter in specific interactions. These are, on one hand, the empty ditrigonal holes for trioctahedral minerals, such as saponite (Michot and Villiéras, 2002), and on the other hand, surface cations (Villiéras et al., 2002). Such high-energy sites will appear as micropores when classical procedures, such as t-plots or α_s plots, are used, and will then be mixed with 'true' micropores resulting from the turbostratic arrangement of clay mineral layers, giving rise to slit-shaped pores at particle edges. Therefore, assessing the textural features of dry swelling clay minerals in the micropore region is not an easy task, especially considering the sample preparation issues previously described.

The mesoporous structure of dry swelling clay minerals is also rather complex, and highly dependent on both the nature of the exchangeable cation and sample preparation procedures. As shown by Fig. 2.5, the gas adsorption isotherms of swelling clay minerals often display a significant hysteresis loop, revealing the presence of an important mesoporosity. However, this hysteresis loop is highly dependent on preparation procedures. Furthermore, the mesopore size distribution is generally extremely polydispersed, and gas adsorption procedures are not always the best-suited techniques for measuring it, as accessibility issues are far from being solved. Some attempts based on small-angle scattering techniques using either X-rays or neutrons have been proposed in the literature (Bihannic et al., 2001, 2008). The idea is either to model the small-angle scattering curves, or to invert them, considering a biphasic structure (pores and solids) to derive a pore size distribution. However, the results remain difficult to interpret, mainly because assumptions on pore shapes have to be implemented in the procedure.

Finally, it must be mentioned that despite all the difficulties associated with the measurement of SSA for swelling clay minerals, low-pressure adsorption techniques can provide valuable information on the texture of these clay minerals. Indeed, as shown by various studies (Tournassat et al., 2003; Perronnet et al., 2007; Decarreau et al., 2014), low-pressure argon adsorption on clay minerals provides reliable estimates of the edge surface area of swelling clay minerals, a parameter that can be of prime importance for understanding the reactivity of these minerals in various environments.

The various problems described herein were already recognised a long time ago by Greene-Kelly (1964), stating that gas adsorption was not suitable for measuring the SSA of Mt. Such a bold statement needs some softening, as gas adsorption techniques provide valuable information on the behaviour of swelling clay minerals. Still, gas adsorption does not allow assessing the full surface area of swelling clay minerals, as gaseous probes such as N_2 and Ar do not access the interlayer space. To try to overcome this difficulty, various techniques based on the use of polar adsorbates that can access the interlayer space were proposed in the literature, especially by soil scientists.

2.4.1.2 Adsorption techniques using polar adsorbates

In view of the above-described difficulties, various techniques have been proposed in the literature for assessing the full surface area of swelling clay minerals. It was, for example, proposed to use *para*-nitrophenol for such a purpose (Theng, 1995; Hedley et al., 2000). At present, three main adsorbates are mainly used for this purpose, that is, methylene blue (MB), ethylene glycol monoethylether (EGME) and water. The case of these three molecules will be examined in the next paragraphs.

The absorption of MB by clay minerals has been used since the 1950s for determining either cation exchange capacities or surface areas (Hang and Brindley, 1970 and references within), and is still extensively used in industry and in the civil engineering community (e.g. Yukselen and Kaya, 2008). The amount of MB absorbed by or exchanged onto clay from an aqueous solution is usually measured using the rest method, that is, determining the amount left in the solution after solid/liquid separation by visible spectroscopy. Then, using an assumption about the cross-sectional area of adsorbed MB molecules, SSA can be obtained. For swelling clay minerals, the method leads to values that are significantly higher than the values measured by gas adsorption, which shows that MB has some access to internal clay mineral surfaces. Still, the values obtained can be highly variable, depending on the system and experimental conditions. Layer charge and the nature of the exchangeable cations significantly affect the results. For example, it was shown that the results obtained with Ca^{2+}-exchanged Mt were significantly different from those obtained with the Na^{+}-exchanged form (Hang and Brindley, 1970). Furthermore, clay mineral layer charge and pH significantly affect the results. For example, Kahr and Madsen (1995) have shown that reliable surface areas can only be obtained by adsorption of MB when the average distance between charges is compatible with the size of adsorbed MB. This corresponds to a layer charge between 0.28 and 0.33 per half unit cell. Similar conclusions were reached by Neumann et al. (2002), or more recently by Hegyesi et al. (2017), who also showed that the pH of the dispersion had to be considered for significant results to be obtained.

Methods based on the retention of ethylene glycol (EG) or of EGME have been used in the soil science community for decades to assess the total SSA of clays (e.g. Dyal and Hendricks, 1950) or clayey soils (e.g. Bower and Gschwend, 1952). These techniques are based on the retention of a monolayer of either EG or EGME at the sample surface. Knowing the cross-sectional area of the molecule, SSA can be derived. In practice, excess liquid is placed in contact with the dry solid to be analysed. The solid is then placed under vacuum to eliminate the excess liquid and when the weight of the sample does not evolve any longer, it is assumed that the solid is covered with one monolayer. Historically, EG was first used, but due to its relatively low vapour pressure, long equilibration times were required, which prompted the replacement of EG by EGME, which is more volatile (Heister, 2014 and references within). The EGME method suffers from serious drawbacks in terms of surface area determination. The most crucial one is that various studies have revealed that the cross-sectional area of EGME molecules varies with both the nature of the exchangeable cation and the magnitude of layer charge (Kellomäki et al., 1987, 1989; Tiller and Smith, 1990; Churchman et al., 1991; Chiou and Rutherford, 1993, 1997; Heister, 2014). Consequently, for pure swelling clay minerals, reliable surface areas cannot be measured using this method. In addition, in soils, the situation is even more complex, as several factors converting EGME retention to surface area are needed (Tiller and Smith, 1990; Heister, 2014). The presence of organic molecules also affects the results, as strong interaction between soil organic matter and EGME often occurs. As stated by Tiller and Smith (1990), it appears that 'The widespread use of the EGME procedures for measurement of total surface area of soils does not appear to be justified' and that 'EGME procedures can still usefully contribute to the interpretation of soil properties and to clay mineral characterisation, notwithstanding the limitations of quantitative measurement of total surface area'.

Water adsorption by swelling clay minerals is a full subject by itself and is analysed in detail in Chapter 4 of this book. In the forthcoming paragraphs, the discussion will be restricted to knowing if water adsorption can provide reliable textural information (SSA and pores) for swelling clay minerals. A few recent reports in the literature suggest that this is the case, on the basis of mixtures of Freundlich, BET, and Frenkel–Halsey–Hill (FHH) isotherms (Hatch et al., 2012; Millan et al., 2013; Revil and Lu, 2013). It will be shown in the following that such claims are somehow unfounded. When water vapour is used as an adsorbate, some of the parameters mentioned earlier for gas adsorption have to be considered. In particular, a proper definition of the outgassing conditions has to be achieved, if possible using preliminary thermal analysis measurements. Still, the most striking feature of water adsorption on clay minerals is the strong dependency on the nature of the exchangeable cation. Fig. 2.6 presents the water adsorption/desorption isotherms of water vapour obtained for a series of homionic Wyoming Mt samples (Berend et al., 1995; Cases et al., 1997).

Without even making any calculation, it clearly appears that each cation-exchanged sample would lead to a different SSA if a BET type treatment were applied. It was shown long ago (Mooney et al., 1952b) and confirmed in numerous publications since (e.g. Berend et al., 1995; Cases et al., 1997; Ferrage et al., 2010, 2011) that these water hydration isotherms can only be interpreted by considering the evolution of interlayer distances with water relative pressure derived from diffraction measurements using either X-rays or neutrons. A proper analysis of these diffraction experiments allow, in addition, obtaining information on water adsorption in the interlayer space and on the external surfaces of swelling clay minerals. This is due to the fact that the main engine for water adsorption in swelling clay minerals is cation hydration, which has been confirmed in numerous spectroscopic studies, and in numerous computer simulations (see Chapter 3 in this book). Consequently, the amount of layer charge also influences the adsorbed amount of water in swelling clay minerals (e.g. Suquet and Pézerat, 1987; Laird, 1999; Michot et al., 2005). An interesting and somewhat peculiar

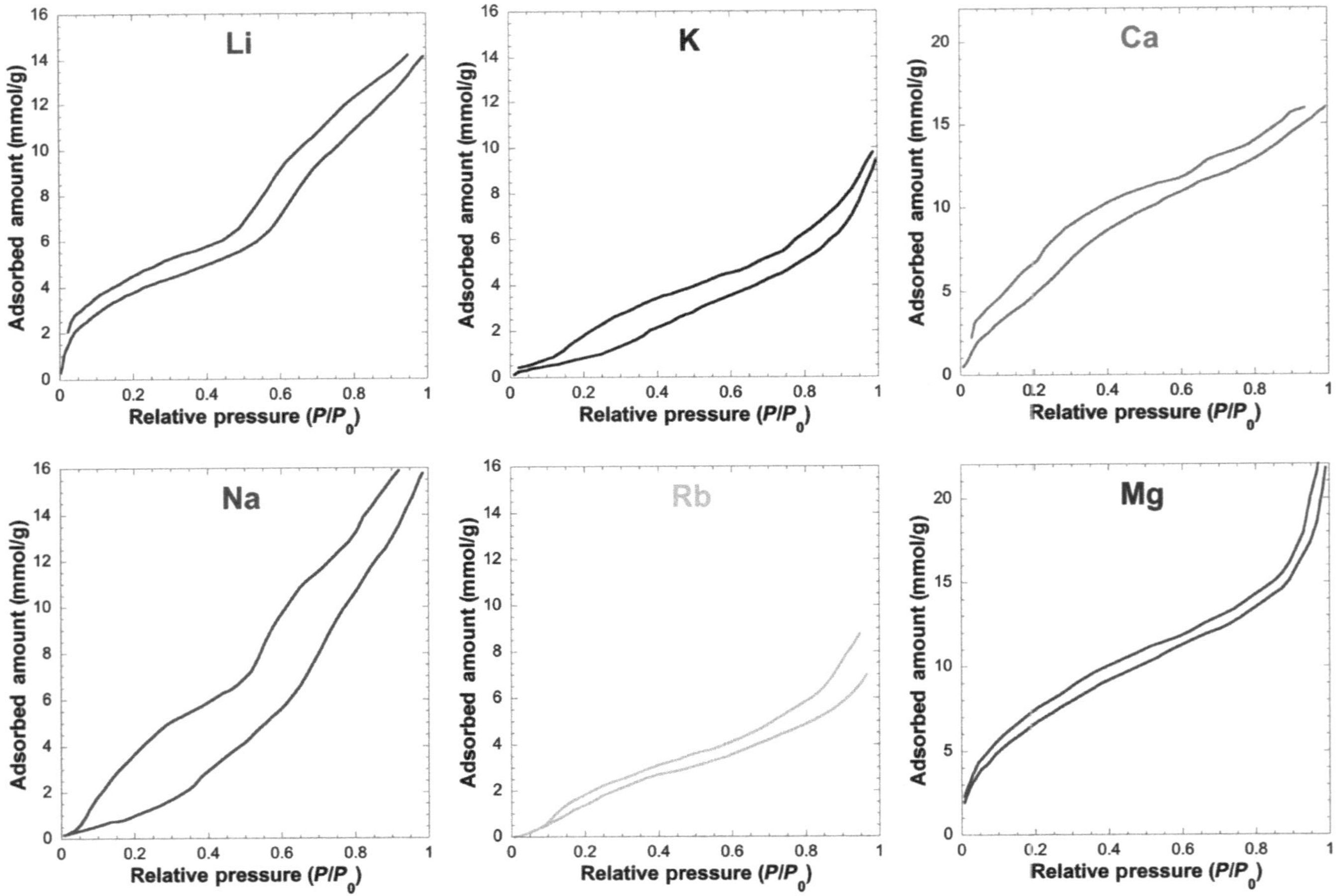

Fig. 2.6 Water vapour adsorption–desorption isotherms obtained on Wyoming Mt exchanged with various monovalent and divalent cations. Adapted from Berend, I., Cases, J.M., François, M., Uriot, J-P., Michot, L.J., Masion, A., Thomas, F., 1995. Mechanism of adsorption of water vapor by homoionic montmorillonite. 2. The Li+, Na+, K+, Rb+ and Cs+ exchanged forms. Clays Clay Miner. 43, 324–336 and Cases, J.M., Berend, I., François, M., Uriot, J-P., Michot, L.J., Thomas, F., 1997. Mechanism of adsorption of water vapor by homoionic montmorillonite. 3. The Mg^{2+}, Ca^{2+}, Sr^{2+} and Ba^{2+} exchanged forms. Clays Clay Miner. 45, 8–22.

case in that regard is provided by the comparison between water adsorption on two synthetic clay minerals with the same layer charge and exchangeable cation, but with different layer structure, that is, OH-hectorite and F-hectorite (Dazas et al., 2013). The uptake of water on the fluorinated sample is significantly lower than that on the hydroxylated one, which shows that cation hydration is the dominant factor explaining water adsorption in swelling clay minerals, but not the only one. In any case, the extended body of literature on swelling clay mineral hydration clearly shows that water adsorption is not an appropriate technique for determining the textural properties of the materials.

For other swelling-inducing adsorbates such as methanol and isopropanol (Annabi-Bergaya et al., 1979, 1980), the nature of the interlayer cation strongly influences adsorption features. For similar reasons as those described herein, the use of such adsorbates is then not appropriate for determining surface areas.

2.4.2 Swelling clay minerals dispersions

One key question in the determination of surface area for swelling clay minerals is the transferability of the data obtained in the dry state to understand the behaviour of these minerals in liquid dispersions. It must be recalled here that for fully isolated single layers, the surface area can be approximated from the unit cell of the material and its formula weight FW. Indeed, if a, b and c are the unit cell parameters, neglecting the edge surface area, the SSA in $m^2 g^{-1}$ can be approximated as $2^*[ab]^*N_A/FW$. Such a calculation yields values ranging between 700 and 900 $m^2 g^{-1}$.

A situation where the whole surface area is available is encountered in the case of dilute dispersions of swelling clay minerals exchanged with Na^+ or Li^+ cations. In that case, the layers are fully exfoliated and the geometrical dimensions of one single layer can then be used to derive the SSA available in the dispersion. Therefore, if the average dimensions of the layer have been determined by another technique, for example, AFM or TEM, and the total solid SSA per unit volume is obtained from the surface of a single layer $S = 2^*\pi r^2 + 2^*\pi rh$ (where h is the thickness and r the equivalent diameter) * the number of objects per unit volume. In such a case, scattering techniques such as small-angle X-ray scattering (SAXS) or small-angle neutron scattering (SANS) can be used to confirm full dispersion of the layers in the surrounding medium. In such experiments, one measures the evolution of the intensity of the scattered X-rays or neutrons as a function of the wave vector q ($q = 4\pi/\lambda \sin\theta$, where 2θ is the scattering angle and λ the wavelength). As an example, Fig. 2.7 (Balnois et al., 2003) displays the normalised small-angle scattering curve of a dilute Laponite dispersion (8 gL, grey circles) together with the calculation based on the form factor of a monodisperse disk 14 nm in radius and 1 nm in thickness (continuous grey line) and a direct computation from AFM data (continuous black line). An excellent agreement between the three methods is obtained, which confirms that the clay mineral is fully dispersed.

At low ionic strength for Na^+ or Li^+ cations, in aqueous dispersion, the whole surface area of swelling clay minerals appears to remain accessible in a wide range of solid contents. Indeed, SAXS measurements in such dispersions systematically display a −2

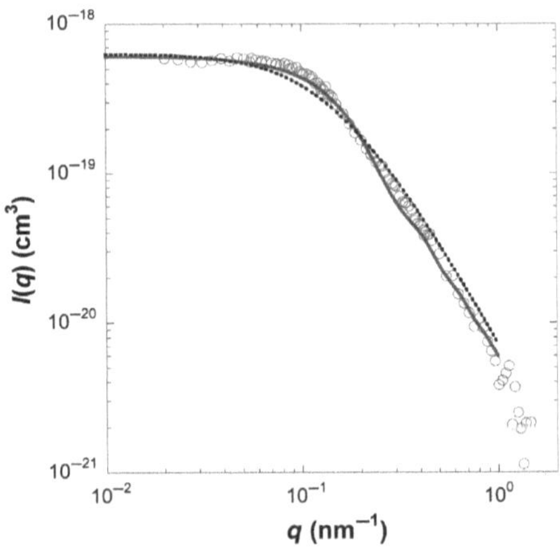

Fig. 2.7 Experimental and calculated SAXS curves for a Laponite dispersion at 8 g L. Adapted from Balnois, E., Durand-Vidal, S., Levitz, P., 2003. Probing the morphology of Laponite clay colloids by atomic force microscopy. Langmuir 19, 6633–6637.

slope at high and medium q values, and correlation peaks, the position of which depends on the solid concentration (e.g. Michot et al., 2006, 2009; Paineau et al., 2011).

The situation becomes more complex when the clay minerals are either exchanged with divalent cations, or flocculated at higher ionic strength. Still, such situations are relevant in numerous natural and industrial situations, such as the adsorption of nucleobases in reconstituted seawater (Pedreira-Segade et al., 2018). Indeed in these cases, ion–ion correlations lead to a very weak repulsive or even attractive double layer interaction, which results in the formation in the dispersion of so-called tactoids[c] (Blackmore and Miller, 1961). The nature and structure of these particles is highly dependent on the nature of the cations and on the ionic strength (Hight et al., 1960, 1962; Shramm and Kwak, 1982; Segad et al., 2012a,b, 2015; Michot et al., 2013). Furthermore, the formation of these larger particles in dispersion leads to significant changes of the slope of the small-angle scattering curves, the interpretation of which remains, up to now, elusive. As an illustration, Fig. 2.8 displays the WAXS–SAXS curves obtained on various coagulated or exchanged Na^+-Mt samples.

In all cases, a correlation peak corresponding to an interlayer distance of around 2 nm is observed, revealing particle formation in the dispersion. In such conditions, determining the accessible surface area becomes extremely difficult. One solution is to use Scherrer's formula to yield the average size of the particles formed by the coherent stacking of individual layers (Scherrer, 1912; Patterson, 1939). This equation writes as $S = \frac{0.89\lambda}{\mathrm{FWHM}\cos(\theta)}$, where S is the average particle size in nm, λ the X-ray wavelength in nm, FWHM the full width at half maximum of the correlation peak in radians, and θ the Bragg angle, the shape factor of 0.89 being classically used for clay

[c] This term has often been used in the literature but should be avoided and replaced by the term 'particles' (Lagaly and Dekany, 2013).

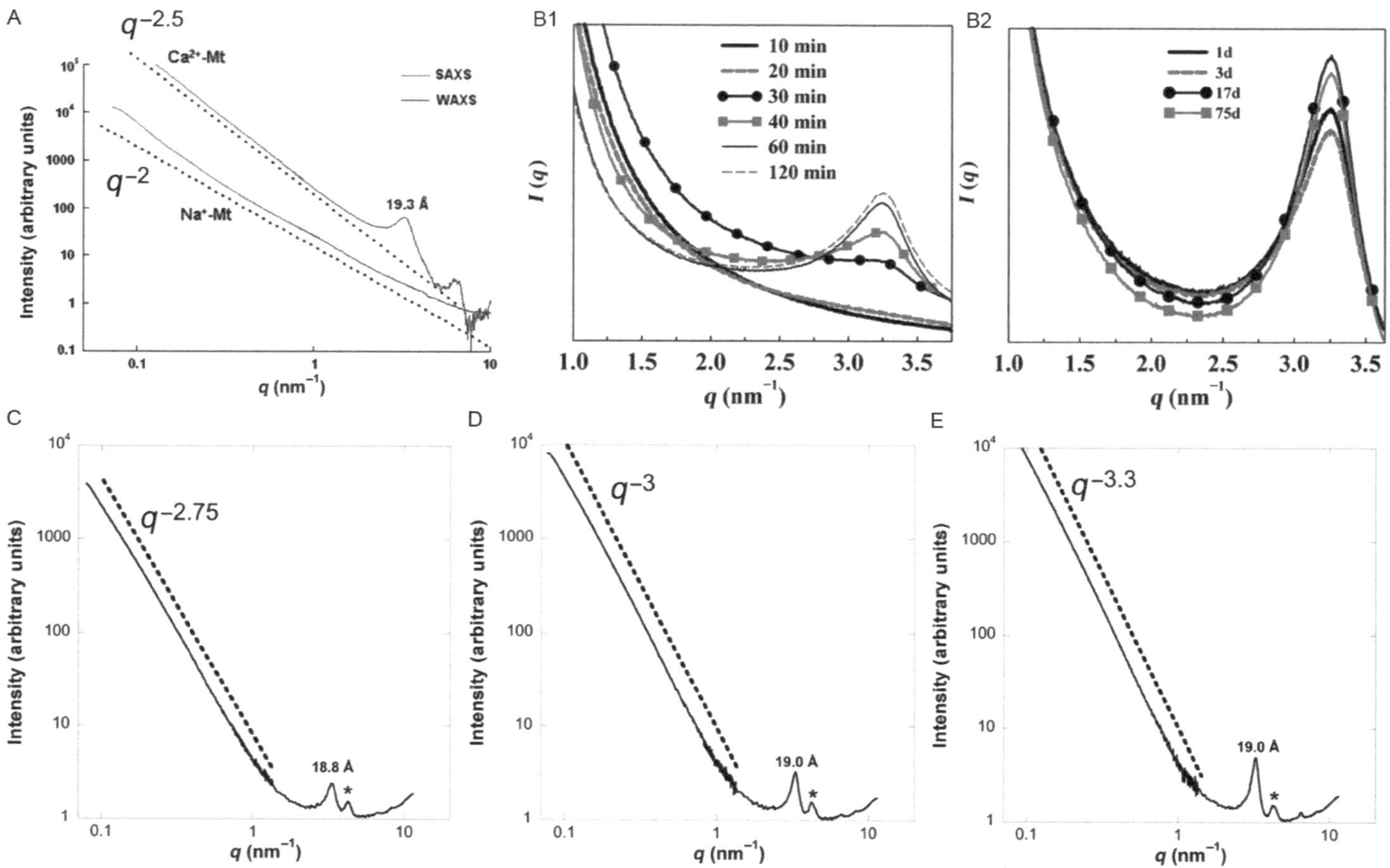

Fig. 2.8 See legend on next page.

mineral particles (Drits et al., 1997). Once the thickness is thus obtained, the SSA can be determined on the basis of geometrical considerations. However, such an approach suffers from numerous drawbacks: (i) Scherrer's equation provides an average value, whereas significant dispersion in thickness can be expected; (ii) the formula yields a coherent stacking domain (CSD) and it is highly debatable to assimilate the CSD to particles. For example, in the case of flocculated systems, flocs are much larger than the CSD (Michot et al., 2013); and (iii) the stacking may not be regular, and in that case, the degree of covering between adjacent layers significantly affects the determined surface area values. To sum up, obtaining reliable surface area values in swelling clay minerals dispersions for Na^+ or Li^+ clay minerals is relatively straightforward, but no technique is currently available for obtaining this information in the case of other dispersions.

2.5 Concluding remarks

As shown in this chapter, determining SSA and textural properties of clayey materials can be a rather complex task, especially in the case of swelling clay minerals. Special care has to be taken in sample preparation, and in the case of these complex hierarchically evolving and reactive materials, interpreting the results requires a deep understanding of what is actually measured. In that regard, the questions that have arisen in this chapter can be applied to other complex systems, such as construction materials. It is also worth mentioning that in recent years, significant research activity has focused on complex structures involving clay minerals, that is, gas shales and oil shales, where similar problems to those described for pure clay minerals have been encountered. For example, the true meaning of surface areas and pore size distributions in such materials have been repeatedly raised. Various techniques has then been implemented for a better understanding of these materials with very high industrial relevance. One may quote (i) dual liquid porosimetry, where both kerosene and water are used

Fig. 2.8 (A). WAXS–SAXS curves of a 50 g L dispersion of a Mt from Sardinia exchanged with Na^+ and Ca^{2+} cations. (B). Time evolution of the WAXS curves of a 10 g L Na^+-Wyoming Mt dispersion upon dialysis against 5 mmol Ca^{2+} Cl_2. (C). WAXS–SAXS curve of a 2 g L Na^+-Wyoming Mt dispersion coagulated with 1 M $NaNO_3$. (D). WAXS–SAXS curve of a 2 g L Na^+-Wyoming Mt dispersion coagulated with 0.1 M $Ca(NO_3)_2$. (E). WAXS–SAXS curve of a 2 g L Na^+-Wyoming Mt dispersion coagulated with 0.1 M La $(NO_3)_3$.
Adapted from Segad, M., Jönsson, B., Cabane, B., 2012. Tactoid formation in montmorillonite. J. Phys. Chem. C. 116, 25425–25433; Michot, L.J., Bihannic, I., Thomas, F., Lartiges, B.S., Waldwogel, Y., Caillet, C., Thieme, J., Funari, S.S., Levitz, P., 2013. Coagulation of Na-montmorillonite by inorganic cations. A combined transmission X-ray microscopy, small angle and wide angle X-ray scattering study. Langmuir 29, 3500–3510; Blachier, C., Jacquet, A., Mosquet, M., Michot, L., Baravian, C., 2014. Impact of clay mineral particle morphology on the rheological properties of dispersions: a combined X-ray scattering, transmission electronic microscopy and flow rheology study. Appl. Clay Sci. 86, 87–96.

(Topór et al., 2016) as a complement of other, more classical techniques, such as XRD, nitrogen, and methane adsorption (Kuila et al., 2014); (ii) ultra-small angle X-ray scattering (USAXS) (Lee et al., 2014); (iii) SAXS and WAXS microscopy (Leu et al., 2016); (iv) SANS coupled with gas adsorption (Clarkson et al., 2013), or focused ion beam scanning electron microscopy (FIB SEM) (Gu et al., 2015) or fluid-invasion methods (Zhao et al., 2017); (v) ^{1}H NMR spectroscopy (Ohkubo et al., 2016) and NMR relaxometry (Korb, 2018). It may be anticipated that all these techniques could be successfully adapted to study pure clay mineral systems, and more complex environments, such as soils, could significantly improve our knowledge of the textural properties of these materials.

References

Ackerman, W.C., Smith, D.M., Huling, J.C., Kim, Y.W., Bailey, J.K., Brinker, C.J., 1993. Gas vapor adsorption in imogolite—a microporous tubular aluminosilicate. Langmuir 9, 1051–1057.

Adams, M.J., 1980. Gas chromatographic adsorption studies on synthetic imogolite. J. Chromatogr. A 188, 97–106.

Adams, J.M., 1993. Particle-size and shape effects in materials science—examples from polymer and paper systems. Clay Miner. 28, 509–530.

Annabi-Bergaya, F., Cruz, M.I., Gatineau, L., Fripiat, J.J., 1979. Adsorption of alcohols by smectites: I. Distinction between internal and external surfaces. Clay Miner. 14, 249–258.

Annabi-Bergaya, F., Cruz, M.I., Gatineau, L., Fripiat, J.J., 1980. Adsorption of alcohols by smectites: II. Role of the exchangeable cations. Clay Miner. 15, 219–223.

Auang, L.L., Tertre, E., Petit, S., 2015. Effect of the morphology of kaolinites on their sorption properties. J. Colloid Interface Sci. 443, 177–186.

Aylmore, L.A.G., Sills, I.D., Quirk, J.P., 1970. Surface area of homoionic illite and montmorillonite clay minerals as measured by the sorption of nitrogen and carbon dioxide. Clays Clay Miner. 18, 91–96.

Balnois, E., Durand-Vidal, S., Levitz, P., 2003. Probing the morphology of Laponite clay colloids by atomic force microscopy. Langmuir 19, 6633–6637.

Bardot, F., Villiéras, F., Michot, L.J., François, M., Gérard, G., Cases, J.M., 1998. High resolution gas adsorption study on illites permuted with various cations: assessment of surface energetic properties. J. Dispers. Sci. Technol. 19, 739–759.

Barrett, E.P., Joyner, L.G., Halenda, P.H., 1951. Determination of pore volumes and area distributions in porous substances. I. Computation from nitrogen isotherms. J. Am. Chem. Soc. 73, 373–380.

Berend, I., Cases, J.M., François, M., Uriot, J.-P., Michot, L.J., Masion, A., Thomas, F., 1995. Mechanism of adsorption of water vapor by homoionic montmorillonite. 2. The Li^+, Na^+, K^+, Rb^+ and Cs^+ exchanged forms. Clays Clay Miner. 43, 324–336.

Bergaoui, L., Lambert, J.F., Vicente-Rodriguez, M.A., Michot, L.J., Villiéras, F., 1995. Porosity of synthetic saponites with variable layer charge pillared by Al_{13} polycations. Langmuir 11, 2849–2852.

Bickmore, B.R., Hochella, M.F., Bosbach, D., Charlet, L., 1999. Methods for performing atomic force microscopy imaging of clay minerals in aqueous solutions. Clays Clay Miner. 47, 573–581.

Bickmore, B.R., Nagy, K.L., Sandlin, P.E., Crater, T.S., 2002. Quantifying surface areas of clays by atomic force microscopy. Am. Mineral. 87, 780–783.

Bihannic, I., Tchoubar, D., Lyonnard, S., Besson, G., Thomas, F., 2001. X-ray scattering investigation of swelling clay fabric. I. The dry state. J. Colloid Interface Sci. 240, 211–218.
Bihannic, I., Delville, A., Demé, B., Plazanet, M., Villiéras, F., Michot, L.J., 2008. Clay swelling: new insights from neutron based techniques. In: Liang, L., Rinaldi, H., Schöber, H. (Eds.), Neutron Applications in Earth, Energy, and Environmental Sciences. Springer, pp. 521–546.
Blachier, C., Jacquet, A., Mosquet, M., Michot, L., Baravian, C., 2014. Impact of clay mineral particle morphology on the rheological properties of dispersions: a combined X-ray scattering, transmission electronic microscopy and flow rheology study. Appl. Clay Sci. 86, 87–96.
Blackmore, A.V., Miller, R.D., 1961. Tactoid size and osmotic swelling in calcium montmorillonite. Soil Sci. Soc. Am. Proc. 24, 169–173.
Bower, C.A., Gschwend, F.B., 1952. Ethylene glycol retention by soils as a measure of surface area and interlayer swelling. Soil Sci. Soc. Am. J. 16, 342–345.
Brunauer, S., Emmett, P.H., Teller, E., 1938. Adsorption of gases in multimolecular layers. J. Am. Chem. Soc. 60, 309–319.
Cadene, A., Durand-Vidal, S., Turq, P., Brendle, J., 2005. Study of individual Na-montmorillonite particles size, morphology, and apparent charge. J. Colloid Interface Sci. 285, 719–730.
Cases, J.M., Grillet, Y., François, M., Michot, L., Villiéras, F., Yvon, J., 1991. Evolution of porous structure and surface area of palygorskite under vacuum thermal treatment. Clays Clay Miner. 39, 191–201.
Cases, J.M., Berend, I., François, M., Uriot, J.-P., Michot, L.J., Thomas, F., 1997. Mechanism of adsorption of water vapor by homoionic montmorillonite. 3. The Mg^{2+}, Ca^{2+}, Sr^{2+} and Ba^{2+} exchanged forms. Clays Clay Miner. 45, 8–22.
Chiou, C.T., Rutherford, D.W., 1993. Sorption of N_2 and EGME on some soils, clays and mineral oxides and determination of sample surface area by use of sorption data. Environ. Sci. Tech. 27, 1587–1594.
Chiou, C.T., Rutherford, D.W., 1997. Effects of exchanged cation and layer charge on the sorption of water and EGME vapors on montmorillonite clays. Clays Clay Miner. 45, 867–880.
Churchman, G.J., Burke, C.M., Parfitt, R.L., 1991. Comparison of various methods for the determination of specific surfaces of subsoils. J. Soil Sci. 42, 449–461.
Clarkson, C.R., Solano, N., Bustin, R.M., Chalmers, G.R.L., He, L., Melnichenko, Y.B., Radlinski, A.P., Blach, T.P., 2013. Pore structure characterization of North American shale gas reservoirs using USANS/SANS, gas adsorption, and mercury intrusion. Fuel 103, 606–616.
Dazas, B., Lanson, B., Breu, J., Robert, J.-L., Pelletier, M., Ferrage, E., 2013. Smectite fluorination and its impact on interlayer water content and structure: a way to fine tune the hydrophilicity of clay surfaces? Microporous Mesoporous Mater. 181, 233–247.
de Boer, J.H., Lippens, B.C., Linsen, B.G., Broekhollf, J.C.P., van der Heuvel, A., Osinga, T.J., 1966. The *t*-curve of multiplayer N_2-adsorption. J. Colloid Interface Sci. 21, 405–414.
Decarreau, A., Petit, S., Andrieux, P., Villieras, F., Pelletier, M., Razafitianamaharavo, A., 2014. Study of low-pressure argon adsorption ion synthetic nontronite: implications for smectite crystal growth. Clays Clay Miner. 62, 102–111.
Delon, J.F., Dellyes, R., 1967. Calculation of spectrum of porosity of phyllitous minerals. C. R. Acad. Sci. D 265, 1661–1665.
Dogan, A.U., Dogan, M., Onal, M., Sarikaya, Y., Aburub, A., Wurster, D.A., 2006. Baseline studies of the clay minerals society source clays: specific surface area by the Brunauer-Emmett-Teller (BET) method. Clays Clay Miner. 54, 62–66.

Drits, V., Srodon, J., Eberl, D.D., 1997. XRD measurement of mean crystallite thickness of illite and illite/smectite: reappraisal of the Kubler index and the Scherrer equation. Clays Clay Miner. 45, 461–475.

Dubinin, M.M., 1966. Modern state of the theory of gas and vapor adsorption by microporous adsorbents. Pure Appl. Chem. 10, 309–321.

Dyal, R.S., Hendricks, S.B., 1950. Total surface of clays in polar liquids as a characteristic index. Soil Sci. 69, 421–432.

Eypert-Blaison, C., Villiéras, F., Michot, L.J., Pelletier, M., Humbert, B., Ghanbaja, J., Yvon, J., 2002. Surface heterogeneity of kanemite, magadiite and kenyaite: a high resolution gas adsorption study. Clay Miner. 37, 531–542.

Fernandez-Martinez, A., Michot, L.J., 2016. Physicochemical properties of imogolite. In: Yuan, P., Thill, A., Bergaya, F. (Eds.), Developments in Clay Science. Nanosized Tubular Clay Minerals Halloysite and Imogolite, vol. 7. Elsevier, pp. 202–222.

Ferrage, E., Lanson, B., Michot, L.J., Robert, J.-L., 2010. Hydration properties and interlayer organization of water and ions in synthetic Na-smectite with tetrahedral layer charge. Part 1. Results from X-ray diffraction profile modeling. J. Phys. Chem. C 114, 4515–4526.

Ferrage, E., Sakharov, B., Michot, L.J., Delville, A., Bauer, A., Lanson, B., Grangeo, N.S., Frapper, G., Jimenez-Ruiz, M., Cuello, G.J., 2011. Hydration properties and interlayer organization of water and ions in synthetic Na-smectite with tetrahedral layer charge. Part 2. Towards a precise coupling between molecular simulations and diffraction data. J. Phys. Chem. C 115 (5), 1867–1881.

Ferrage, E., Hubert, F., Tertre, E., Delville, A., Michot, L.J., Levitz, P., 2015. Modeling the arrangement of particles in natural swelling-clay porous media using three-dimensional packing of elliptic disks. Phys. Rev. E. 91. Article # 062210.

Greene-Kelly, R., 1964. The specific surface areas of montmorillonites. Clay Miner. Bull. 5, 392–400.

Gregg, S.J., Sing, K.S.W., 1982. Adsorption, Surface Area and Porosity, second ed. Academic Press, London.

Grillet, Y., Cases, J.M., François, M., Rouquerol, J., Poirier, J.E., 1988. Modification of the porous structure and surface area of sepiolite under vacuum thermal treatment. Clays Clay Miner. 36, 233–242.

Gu, X., Cole, D.R., Rother, G., Mildner, D.F.R., Brantley, S.L., 2015. Pores in Marcellus shales: a neutron scattering and FIB-SEM study. Energy Fuel 29, 1295–1308.

Hang, P.T., Brindley, G.W., 1970. Methylene blue absorption by clay minerals. Determination of surface areas and cation exchange capacities (clay-organic studies XVIII). Clays Clay Miner. 18, 203–221.

Hatch, C.D., Wiese, J.S., Crane, C.C., Harris, K.J., Kloss, H.G., Baltrusaitis, J., 2012. Water adsorption on clay minerals as a function of relative humidity: application of BET and Freundlich adsorption models. Langmuir 28, 1790–1803.

Hedley, C.B., Saggar, S., Theng, B.K.G., Whitton, J.S., 2000. Surface area of soils of contrasting mineralogies using para-nitrophenol adsorption and its relation to air-dry moisture content of soils. Aust. J. Soil. Res. 38, 155–167.

Hegyesi, N., Vad, R.T., Pukanszky, B., 2017. Determination of the specific surface area of layered silicates by methylene blue adsorption: the role of structure, pH and layer charge. Appl. Clay Sci. 146, 50–55.

Heister, K., 2014. The measurement of the specific surface area of soils by gas and polar liquid adsorption methods—limitations and potentials. Geoderma 216, 75–87.

Hight Jr., R., Higdon, W.L., Schmidt, P.W., 1960. Small angle scattering study of sodium montmorillonite clay suspensions. J. Chem. Phys. 33, 1656–1661.

Hight Jr., R., Higdon, W.L., Darley, H.C.H., Schmidt, P.W., 1962. Small angle X-ray scattering from montmorillonite clay suspensions. II. J. Chem. Phys. 37, 502–510.
Hubbard, B., Kuang, W.X., Moser, A., Facey, G.A., Detellier, C., 2003. Structural study of Maya Blue: textural, thermal and solid-state multinuclear magnetic resonance characterization of the palygorskite-indigo and sepiolite-indigo adducts. Clays Clay Miner. 51, 218–226.
JoseYacaman, M., Rendon, L., Arenas, J., Puche, M.C.S., 1996. Maya blue paint: an ancient nanostructured material. Science 273, 223–225.
Kahr, G., Madsen, F.T., 1995. Determination of the cation exchange capacity and the surface area of bentonite, illite and kaolinite by methylene blue adsorption. Appl. Clay Sci. 9, 327–336.
Kellomäki, A., Nieminen, P., Ritamäki, L., 1987. Sorption of ethylene glycol monoethyl ether (EGME) on homoionic montmorillonites. Clay Miner. 22, 297–303.
Kellomäki, A., Kuula-Väisänen, P., Nieminen, P., 1989. Sorption and retention of ethylene glycol monoethyl ether (EGME) on silicas. J. Colloid Interface Sci. 129, 373–378.
Korb, J.P., 2018. Multiscale nuclear magnetic relaxation dispersion of complex liquids in bulk and confinement. Prog. Nucl. Magn. Reson. Spectrosc. 104, 12–55.
Kuila, U., McCarty, D.K., Derkowski, A., Fischer, T.B., Topor, T., Prasad, M., 2014. Nanoscale texture and porosity of organic matter and clay minerals in organic-rich mudrocks. Fuel 135, 359–373.
Lagaly, G., Dekany, I., 2013. Colloid clay science. In: Bergaya, F., Lagaly, G. (Eds.), Developments of Clay Science. Handbook of Clay Science, Part A Fundamentals. second ed. vol. 5. Elsevier, Amsterdam, pp. 243–346.
Laird, D.A., 1999. Layer charge influences on the hydration of expandable 2:1 phyllosilicates. Clays Clay Miner. 47, 630–636.
Lecloux, R., Pirard, J.P., 1979. Importance of standard isotherms in the analysis of adsorption-desorption isotherms for determining the porous texture of solids. J. Colloid Interface Sci. 70, 265–281.
Lee, S., Fischer, T.B., Stokes, M.R., Klingler, R.J., Ilavsky, J., McCarty, D.K., Wigand, M.O., Derkowski, A., Winans, R.E., 2014. Dehydration effect on the pore size, porosity, and fractal parameters of shale rocks: ultrasmall-angle X-ray scattering study. Energy Fuel 28, 6772–6779.
Leu, L., Georgiadis, A., Blunt, M.J., Busch, A., Bertier, P., Schweinar, K., Liebi, M., Menzel, A., Ott, H., 2016. Multiscale description of shale pore systems by scanning SAXS and WAXS microscopy. Energy Fuel 30, 10282–10297.
Michot, L.J., Villiéras, F., 2002. Assessment of surface energetic heterogeneity of synthetic Na-saponites. The role of layer charge. Clay Miner. 37, 39–57.
Michot, L.J., François, M., Cases, J.M., 1990. Surface heterogeneity studied by quasi-equilibrium gas adsorption procedure. Langmuir 6, 677–681.
Michot, L.J., Villiéras, F., Yvon, J., Fourty, G., 1993. Multistage wet grinding of talc. Relation between physico-chemical parameters of the filler and mechanical properties of filled polypropylenes. J. Mater. Sci. 28 (7), 1856–1866.
Michot, L.J., Villiéras, F., François, M., Yvon, J., Le Dred, R., Cases, J.M., 1994. The structural microscopic hydrophobicity of talc. Langmuir 10, 3765–3773.
Michot, L.J., Villiéras, F., Lambert, J.F., Bergaoui, L., Grillet, Y., Robert, J.L., 1998. Surface heterogeneity in micropores of pillared clays: the limits of classical pore-filling mechanisms. J. Phys. Chem. B 102, 3466–3476.
Michot, L.J., Bihannic, I., Pelletier, M., Rinnert, E., Robert, J.-L., 2005. Hydration and swelling of synthetic Na-saponites: influence of layer charge. Am. Mineral. 90, 166–172.

Michot, L.J., Bihannic, I., Maddi, S., Funari, S.S., Baravian, C., Levitz, P., Davidson, P., 2006. Liquid crystalline aqueous clay suspensions. PNAS 103, 16101–16104.
Michot, L.J., Bihannic, I., Maddi, S., Duval, J.F.L., Baravian, C., Moyne, C., Levitz, P., Davidson, P., 2009. Sol-Gel and isotropic/nematic transitions in aqueous suspensions of natural nontronite clay. Influence of particle anisotropy. Part 2. Gel structure and rheology. Langmuir 25, 127–139.
Michot, L.J., Bihannic, I., Thomas, F., Lartiges, B.S., Waldwogel, Y., Caillet, C., Thieme, J., Funari, S.S., Levitz, P., 2013. Coagulation of Na-montmorillonite by inorganic cations. A combined transmission X-ray microscopy, small angle and wide angle X-ray scattering study. Langmuir 29, 3500–3510.
Millan, H., Govea-Alcaide, E., Garcia-Fomaris, I., 2013. Truncated fractal modeling of H_2O-vapor adsorption isotherms. Geoderma 206, 14–23.
Montargès-Pelletier, E., Bogenez, S., Pelletier, M., Razafitianamaharavo, A., Ghanbajaa, J., Lartiges, B., Michot, L.J., 2005. Synthetic allophane-like particles: textural properties. Colloids Surf. A 255, 1–10.
Mooney, R.W., Keenan, A.G., Wood, L.A., 1952a. Adsorption of water vapour by montmorillonite. I. Heat of desorption and application of BET theory. J. Am. Chem. Soc. 74, 1367–1371.
Mooney, R.W., Keenan, A.G., Wood, L.A., 1952b. Adsorption of water vapour by montmorillonite. II. Effect of exchangeable ions and lattice swelling as measured by X-ray diffraction. J. Am. Chem. Soc. 74, 1371–1374.
Neumann, M.G., Gessner, F., Schmitt, C.C., Sartori, R., 2002. Influence of the layer charge and clay particle size on the interactions between the cationic dye methylene blue and clays in an aqueous suspension. J. Colloid Interface Sci. 255, 254–259.
Ohkubo, T., Ibaraki, M., Tachi, Y., Iwada, Y., 2016. Pore distribution of water-saturated compacted clay using NMR relaxometry and freezing temperature depression; effects of density and salt concentration. Appl. Clay Sci. 123, 148–155.
Paineau, E., Bihannic, I., Baravian, C., Philippe, A.-M., Davidson, P., Levitz, P., Funari, S.S., Rochas, C., Michot, L.J., 2011. Aqueous suspensions of natural swelling clay minerals. 1. Structure and electrostatic interactions. Langmuir 27, 5562–5573.
Patterson, A.L., 1939. The Scherrer formula for X-ray particle size determination. Phys. Rev. 56, 978.
Pedreira-Segade, U., Michot, L.J., Daniel, I., 2018. Effects of salinity on the adsorption of nucleotides onto phyllosilicates. Phys. Chem. Chem. Phys. 108, 1938–1952.
Perronnet, M., Villieras, F., Jullien, M., Razafitianamaharavo, A., Raynal, J., Bonnin, D., 2007. Towards a link between the energetic heterogeneities of the edge faces of smectites and their stability in the context of metallic corrosion. Geochim. Cosmochim. Acta 71, 1463–1479.
Rajec, P., Sucha, V., Eberl, D.D., Srodon, J., Elsass, F., 1999. Effect of illite particle shape on cesium sorption. Clays Clay Miner. 47, 755–760.
Rautureau, M., Tchoubar, C., 1976. Structural analysis of sepiolite by selected area electron diffraction. Relations with physicochemical properties. Clays Clay Miner. 24, 43–49.
Revil, A., Lu, N., 2013. Unified water isotherm for clayey porous materials. Water Resour. Res. 49, 5685–5699.
Rouquerol, J., 1970. L'analyse thermique à vitesse de décomposition constante. J. Thermal Anal. 2, 123–140.
Rouquerol, J., 1989. Controlled transformation rate thermal analysis: the hidden face of thermal analysis. Thermochim. Acta 144, 209–224.

Rouquerol, J., Bordère, S., Rouquerol, F., 1992. Controlled rate evolved gas analysis: recent experimental set-up and typical results. Thermochim. Acta 203, 193–202.

Rutherford, D.W., Chiou, C.T., Eberl, D.D., 1997. Effects of exchanged cation on the microporosity of montmorillonite. Clays Clay Miner. 45, 534–543.

Sayed Hassan, M., Villiéras, F., Razafitianamaharavo, A., Michot, L.J., 2005. Role of exchangeable cations on geometrical and energetic surface heterogeneity of kaolinites. Langmuir 21, 12283–12289.

Sayed Hassan, M., Villiéras, F., Gaboriaud, F., Razafitianamaharavo, A., 2006. AFM and low-pressure argon adsorption analysis of geometrical properties of phyllosilicates. J. Colloid Interface Sci. 296, 614–623.

Scherrer, P., 1912. Kolloidchemie Ein Lehrbuch. Springer, pp. 387–409.

Segad, M., Jönsson, B., Cabane, B., 2012a. Tactoid formation in montmorillonite. J. Phys. Chem. C 116, 25425–25433.

Segad, M., Anski, S., Olsson, U., Ruokolainen, J., Åkeson, T., Jönsson, B., 2012b. Microstructural and swelling properties of Ca and Na montmorillonite: (in situ) observations with Cryo-TEM and SAXS. J. Phys. Chem. C 116, 7596–7601.

Segad, M., Åkeson, T., Cabane, B., Jönsson, B., 2015. Nature of flocculation and tactoid formation in montmorillonite: the role of pH. Phys. Chem. Chem. Phys. 44, 29608–29615.

Shramm, L.L., Kwak, J.C.T., 1982. Influence of exchangeable cation composition on the size and shape of montmorillonite particles in dilute suspensions. Clays Clay Miner. 30, 40–48.

Sing, K.S.W., Everett, D.H., Haul, R.A.W., Moscou, L., Pierotti, R.A., Rouquerol, J., Siemieniewska, T., 1985. Reporting physisorption data for gas/solid systems with special reference to the determination of surface area and porosity. Pure Appl. Chem. 57, 603–619.

Suquet, H., Pézerat, H., 1987. Parameters influencing layer stacking types in saponite and vermiculite: a review. Clays Clay Miner. 35, 353–362.

Tan, B.W., Thomas, N.L., 2017. Tortuosity model to predict the combined effects of crystallinity and nano-sized clay mineral on the water vapour barrier properties of polylactic acid. Appl. Clay Sci. 141, 46–54.

Theng, B.K.G., 1995. On measuring the specific surface area of clays and soils by adsorption of para-nitrophenol: use and limitations. In: Churchman, G.J., Fitzpatrick, R.W., Eggleton, R.A. (Eds.), Clays Controlling the Environment. Proceedings of the 10th International Clay Conference, Adelaide, Australia. In: 1993. CSIRO Publishing, Melbourne, pp. 304–310.

Thomas Jr., J., Bohor, B.F., 1968. Surface area of montmorillonite from the dynamic sorption of nitrogen and carbon dioxide. Clays Clay Miner. 16, 83–91.

Tiller, K.G., Smith, L.H., 1990. Limitations of EGME retention to estimate the surface area of soils. Aust. J. Soil. Res. 28, 1–26.

Topór, T., Derkowski, A., Kuila, U., Fischer, T.B., McCarty, D.K., 2016. Dual liquid porosimetry: a porosity measurement technique for oil- and gas-bearing shales. Fuel 183, 537–549.

Tournassat, C., Neaman, A., Villiéras, F., Bosbach, D., Charlet, L., 2003. Nanomorphology of montmorillonite particles: estimation of the clay edge sorption site density by low-pressure gas adsorption and AFM observations. Am. Mineral. 88, 1989–1995.

Villiéras, F., Cases, J.M., François, M., Michot, L., Thomas, F., 1992. Texture and surface energetic heterogeneity of solids from modelling of low pressure gas adsorption isotherms. Langmuir 8, 1789–1795.

Villiéras, F., Michot, L.J., Bardot, F., Cases, J.M., François, M., Rudzinski, W., 1997a. An improved derivative isotherm summation method to study surface heterogeneity of clay minerals. Langmuir 13, 1104–1117.

Villiéras, F., Michot, L.J., Cases, J.M., Berend, I., Bardot, F., François, M., Gérard, G., Yvon, J., 1997b. Static and dynamic studies of the energetic surface heterogeneity of clay minerals. In: Rudzinski, W., Steele, W.A., Zgrablich, G. (Eds.), Equilibria and Dynamics of Gas Adsorption on Heterogeneous Solid Surfaces. Elsevier, Amsterdam, pp. 573–623.
Villiéras, F., Michot, L.J., Bardot, F., Chamerois, M., Eypert-Blaison, C., Gérard, G., Cases, J.M., 2002. Surface heterogeneity of minerals. C. R. Geosci. 334, 597–609.
Yukselen, Y., Kaya, A., 2008. Suitability of the methylene blue test for surface area, cation exchange capacity and swell potential determination of clayey soils. Eng. Geol. 102, 38–45.
Zhao, J.H., Jin, Z.J., Hu, Q.H., Barber, T.J., Zhang, Y.X., Bleuel, M., 2017. Integrating SANS and fluid-invasion methods to characterize pore structure of typical American shale oil reservoirs. Sci. Rep. 7. # 15413.

Quantum-chemical modelling of clay mineral surfaces and clay mineral–surface–adsorbate interactions

3

Sergey V. Churakov[*,†], *Xiandong Liu*[‡]
[*]Paul Scherrer Institute, Laboratory for Waste Management, Villigen, Switzerland
[†]University of Bern, Institute of Geological Sciences, Bern, Switzerland
[‡]State Key Laboratory for Mineral Deposits Research, School of Earth Sciences and Engineering, Nanjing University, Nanjing, PR China

3.1 Quantum mechanical description of interatomic interactions

The methods of atomistic modelling can predict structural and thermodynamics properties of chemical systems based on the accurate description of interatomic interaction. The interatomic interaction parameters can be derived as an empirical model fitted to the known experimental data or predicted from the first principles of quantum mechanics. The quantum mechanical (QM) description of a polyatomic system involves obtaining the distribution of electron density from which the interatomic forces between atoms and the interaction energy are calculated. According to the postulates of quantum mechanics, electron probability density (electron density) is equal to the square of the electron wave function (WF). The WF is the solution of the second order differential equation, known as the Schrödinger equation.

The exact analytical solution of the Schrödinger equation is only known for a single electron system. The solution for multielectron systems is obtained numerically using several approximations. These approximations determine the accuracy of the method, and set limits to the size of the system and the time scale accessible in the simulations. The most important contributions to chemical bonding and intermolecular interaction are related to Coulombic interaction and polarization effects. The electrostatic interaction depends on the electron density distribution, whereas the polarizability describes the readiness of the system to change electronic density distribution under an external static and dynamic electric field. The electron density distribution depends on the electrostatic interaction between the charged particles (electrons and nuclei), pair correlations between electrons due to their charge, and inherent magnetic moment (spin state). The electron–electron correlation describes mutual avoidance between electrons due to the repulsion between negatively charged particles. These correlations always result in a lower system energy compared with the uncorrelated

Developments in Clay Science, Vol. 9. https://doi.org/10.1016/B978-0-08-102432-4.00003-2

(averaged) electron density distributions. The so-called exchange energy arises due to the QM nature of electrons, which prevents electrons with parallel spins from occupying the same spatial orbital (Pauli exclusion principle).

3.1.1 Hartree–Fock method

Most commonly applied ab initio simulations are based on Hartree–Fock (HF) methods and density functional theory (DFT). The HF approach makes an assumption about the mathematical form of the total WF. In this method, the total WF of the N-electron system is represented as a determinant composed by N noninteracting electron orbitals (Slater determinant). The HF method provides an exact expression for the kinetic energy of a single electron wave function, as well as the Coulomb and exchange interaction of electrons with an average electrostatic potential created by all electrons in the system (so-called mean field potential). The pair-wise electron–electron correlation energy is entirely neglected in the HF method. Accordingly, the energy predicted by HF methods is always higher than the true energy of the system. The HF method has a tendency to overestimate the strength of chemical bonding. The predicted bond distances are, on average, ~1% too short, and the predicted vibrational frequencies are ~10% too high. The predicted electron density is more delocalised compared with the exact solution. The computational cost of solving the HF equation scales with the number of electrons is N^2—N^4 (depending on the implementation of numerical approximations), where N is the number of single electron WF. Accordingly, the HF methods have been traditionally limited to molecules and small molecular clusters.

Several methods have been proposed to account for electron–electron correlation on the basis of the HF equation. These methods express the total wave function as a linear combination of extended Slater determinants, which includes orbitals for excited electrons, in addition to the reference ground state solution of the HF equations. Such a multideterminant expression of the wave function allows, in principle, one to obtain an exact description of electron–electron correlations, taking the HF solution as the reference (e.g. *second* order Møller–Plesset perturbation theory (MP2)). The computational cost of such methods grows as N^6 and higher, which makes them impractical for large systems (Foresman and Frisch, 1996).

3.1.2 Density functional theory

The methods of DFT are based on the fundamental theorem formulated by Hohenberg and Kohn (1964), which proves that the ground state energy of the multielectron system is uniquely defined by the electron density distribution. This fact allows a drastic simplification of the numerical problem reducing the $3N$ dimensional solution of a full Schrödinger equation to the system of N 3-dimensional equations for the total electron density (Kohn and Sham, 1965). Despite the formal proof of the existence of a solution, the exact explicit equation for the system energy as a function of the electron density remains unknown.

Modern DFT approaches use electron orbitals to calculate the kinetic energy contributions and apply different semiempirical approximations for the exchange and the correlation energy. The major theoretical developments in the field of DFT are focused on formulating an accurate description of the exchange and the correlation functional. The exchange and correlation energies cannot be strictly separated within DFT. The first widely successful formulation of the exchange correlation functional (E_{xc}) was based on the so-called local density approximation (LDA). This approach uses the E_{xc} of homogeneous electron gas as a proxy to evaluate the properties of a real polyatomic system with heterogeneous (space dependent) electron density. The contribution to the E_{xc} of a chemical system in each point of space is obtained as the sum of contributions E_{xc} from a hypothetical electron gas with the same density. This approach is expected to perform well for systems with slowly varying electron density. Modern E_{xc} functionals take into account the density gradient (generalised gradient approximation, GGA) and second order derivatives of orbitals (kinetic energy of electron orbitals). These functionals are called local, because the calculation of E_{xc} in each point of space depends only on the density and their derivatives in this particular point. In contrast, the nonlocal E_{xc} functionals, which are usually more accurate, but computationally expensive, include integration in the local point and its proximity. The family of so-called hybrid potentials use a linear combination of nonlocal exchange interaction obtained in HF approximation and local gradient corrected exchange and correlation functional. The B3LYP and PBE0 hybrid potentials are currently two of the most commonly applied functionals to geochemical systems. Several popular exchange-correlation potentials are listed in Table 3.1.

A serious flaw of DFT functionals used in solid-state application is the underestimation of the band gap energy, which can be as high as 50%. In contrast, HF calculations tend to substantially overestimate the band gap. These errors are closely related to the prediction of the absolute energy for electron orbitals of highly localized states such as *3d* states in transition metals or *4f* and *5f* states in lanthanides and actinides (Furche and Perdew, 2006). The simultaneous presence of multivalence species (e.g. common presence of Fe^{2+} and Fe^{3+} ions) adds extra complexity. Correct partitioning (localization/de-localization) of electron density between ions of transition metals is very sensitive to the energy of *3d* electronic states. An efficient semiempirical correction can be obtained at a low computational overhead based on the Hubbard

Table 3.1 Popular exchange and correlation functionals used for the simulations of clay minerals

Functional acronym	GGA[a]	Hybrid[b]	Reference
PBE0	x	x	Adamo and Barone (1999)
BLYP	x		Becke (1988) and Lee et al. (1988)
B3LYP	x	x	Stephens et al. (1994)
PW91	x		Perdew and Wang (1992)
PBE	x		Perdew et al. (1996)

These functionals can also be augmented with van der Waals correction (Grimme et al., 2016).
[a] GGA—generalized gradient approximations.
[b] Hybrid functionals include fraction of the exact exchange energy calculated based on the HF theory.

model (Hubbard, 1963) (so-called DFT + U method). In this approach, an extra empirical potential is applied to obtain the correct energy of the localized electronic states (Anisimov et al., 1997). The model parameters have to be calibrated based on experimental data, however.

3.1.3 Dispersion correction

The description of van der Waals forces within the DFT approach needs special attention. Van der Waals forces arise due to instantaneous correlation between induced dipoles. Thus, this is a nonlocal property of the density, and therefore cannot be described by a local density functional. Two strategies are used to deal with the van der Waals forces in DFT. The approach developed by Grimme et al. (2016) applies dispersion correction in the form of a screened inverse power expansion to the local E_{xc}. The second approach pursues the development of a nonlocal density functional that includes the van der Waals interaction (Peverati and Truhlar, 2012). The approach of Grimme et al. (2016) has been widely applied in the simulation of systems, including water, because of its robustness and accuracy.

The van der Waals type interaction determines adsorption properties of the basal plane in noncharged clay minerals such as talc and pyrophyllite. The calculated interlayer distance is very sensitive to the description of the dispersion interaction, and is used as a benchmark to test the exchange correlation functional and dispersion correction schemas (Ugliengo et al., 2009; Tunega et al., 2012; Zhang et al., 2012). Systematic studies of structural parameters (Tunega et al., 2012) demonstrate that the LDA approach substantially underestimate lattice constants of clay minerals (Fig. 3.1). In contrast, the gradient corrected functionals systematically overestimate the lattice parameters. The error in the predicted lattice constant in the c-direction is as large as 5% on average. The inclusion of dispersion correction results in an excellent agreement between the calculations and the experiment (better than 1%). The dispersion correction also provides a better description of the distance-energy dependence of the interaction, resulting in accurate prediction of bulk moduli and the structural position of the interlayer cations (Voora et al., 2011; Zhang et al., 2012; Li et al., 2016). The performance of most popular exchange correlation functionals in predicting adsorption energies of small molecules on the surface of clay minerals was recently evaluated (Zen et al., 2016). The study concludes that proper consideration of dispersion correction results in a good agreement between calculated and experimentally measured relative adsorption energies of molecules. The predicted values of absolute adsorption energies still deviate from the experimental measurements beyond the measurements' uncertainties, and further improvement of the functionals is necessary to approach the experimental accuracy.

3.1.4 Basis set

Both in DFT and HF based approaches, the electron WF and the density have to be expressed in a convenient functional form, which is referred to as the basis set. Two types of basis functions are used. The origin of the atom centred WF coincides

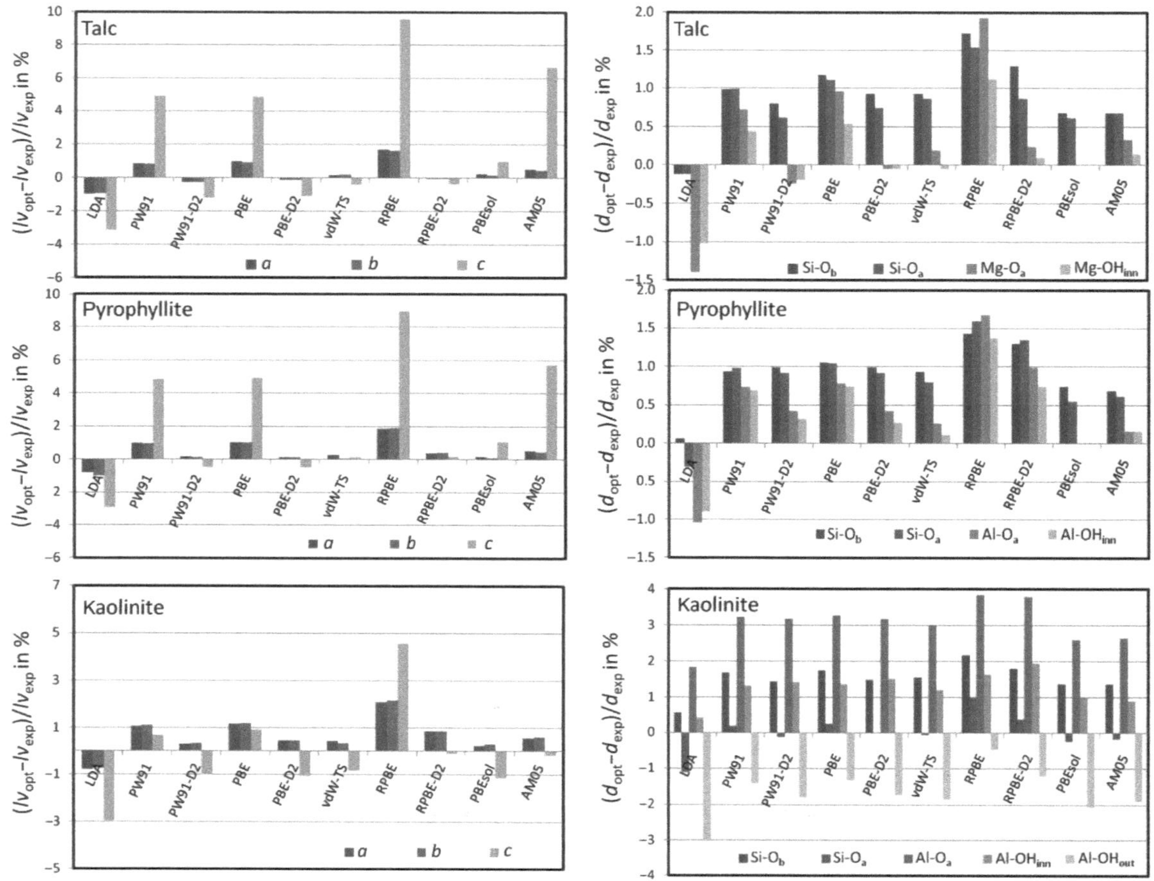

Fig. 3.1 Relative difference of main bond distances and crystallographic axes with respect to experimental data computed for talc, pyrophyllite, and kaolinite. Adapted with permission from Tunega, D., Bucko, T., Zaoui, A., 2012. Assessment of ten DFT methods in predicting structures of sheet silicates: importance of dispersion corrections. J. Chem. Phys. 137, 114105–114109. Copyright (2012) American Physical Society.

with the positions of the nuclei. Such atom centred orbitals reproduce the WF of an isolated atom expressed as a sum of Slater/Gaussian type functions. The major assumption behind this approach is that a linear combination of atomic orbitals can provide a good description of chemical bonding in molecules and solids.

Alternatives to the atom centred orbitals are the origin free, plane wave (PW) type basis sets. The PW describes the behaviour of a free electron in a periodic potential. This basis set is a natural choice for simulations with periodic boundary conditions. The main advantage of the PW basis set is that it guarantees a smooth convergence to the exact solution as the number of PW increases and provides an equally accurate representation of the WF in the entire simulation domain. Thus, the PW basis set is free from the basis set superposition error inherent to the atom centred basis sets. The numerical efficiency of DFT based calculations with a PW basis is achieved using the fast Fourier transform technique, which allows $n \log n$ scalability of the computational costs with the number of PW (n).

On the other hand, the PW basis set becomes less efficient if solid–vacuum interfaces or isolated clusters and molecules are investigated. In such systems, computational efforts are 'wasted' in representing the electron density of matter free domains (vacuum). Many modern QM simulation packages use the combination of atom centred and PW type basis sets to benefit from computational efficiency and accuracy of both types of basis sets. In such codes, the atom centred basis sets are used to represent the WF, whereas the density is mapped using the PW.

3.1.5 Effective core potentials

In a typical chemical environment, only the outermost electrons of an atom change their distribution upon formation of chemical bonds, or upon changes of the coordination environment. The core electrons remain unchanged, and can be accounted for implicitly by means of the so-called effective core potential (pseudopotential, PP). The PP represents the combined effect of the true electrostatic potential due to positively charged atomic nuclei, and the screening effect of the core electrons. The PP and corresponding pseudo WF are constructed based on an electronic structure calculation for an isolated atom in ground, and possibly in several excited states, see Dolg and Cao (2012) for an overview. The PP is constructed to satisfy a number of constraints. The resulting WF must reproduce true WF beyond a selected core cutoff distance and have smooth, node-free behaviour within this cutoff radius. Pseudo WF corresponding to the so-called norm conserving PP integrates into the same electron number density as the true WF (Troullier and Martins, 1991; Goedecker et al., 1996). The last condition is not required in the so-called ultrasoft PP, which allows having extrasmooth pseudo-WF at comparably small computational overhead required to retrieve the correct integral electron number density (Vanderbilt, 1990). The PP core cutoff is an element-specific parameter that determines the accuracy and the computational efficiency of the PP. Shorter cutoff means more accurate representation and a higher degree of transferability of the PP, but is usually associated with higher computational costs.

3.1.6 System size and boundary conditions

State-of-the-art simulation codes and high-performance-computing-resources currently available for researchers allow DFT-based simulations for systems of a few hundreds of atoms. This is very far from the size of real macroscopic systems. Two approaches can be applied to mimic the macroscopic system behaviour. One method uses periodic boundary conditions to simulate infinitely large periodic systems. A potential drawback of this approach is an artificial long-range periodicity, which may not be present in the real system. The second method is based on a cluster representation of the system. In this case, the structural relaxation at the surface may influence the simulation results. In both approaches, the effect of the system size and the boundary conditions has to be tested by systematically increasing the system size and monitoring the convergence of the simulation results. Due to the limitations of computational algorithms, the HF-based calculations are almost exclusively used for cluster system setup, whereas the DFT-based approach is equally applied to both periodic and cluster calculations.

3.1.7 Structural optimisation and molecular dynamics simulations

Two types of approaches are commonly applied to simulate structural and dynamic properties of solids. One, the structural optimization and quasi harmonic approximation, rely on the fact that at low temperatures (e.g., room temperatures) atoms in solids vibrate close to the position of energy minima. These lowest energy atomic positions can be obtained by minimizing the total energy of the system. The other properties of the system, such as vibrational spectra or elastic properties, are obtained by calculating the energy change of the system due to the small displacement of atoms from their equilibrium positions. These approximations hold very well for anhydrous solids, and have severe limitations when water containing clay minerals such as smectites are considered. A more general approach is based on molecular dynamics simulations (MD). In the MD method, the equilibrium motion of atoms in the system is obtained by solving the Newton equation of motion. The distinction of the ab initio MD (AIMD) from so-called classical MD simulations is related to the fact that in the first approach, the interatomic interaction forces are calculated based on a QM description. The AIMD is conducted either within the Born-Oppenheimer approximation (BOMD), or using the Car–Parrinello approach (CPMD). In the BOMD, the WF (electron density) is optimized after every displacement of the atomic coordinates. In this approach, the electrons are assumed to instantaneously adopt the ground state for every atomic configuration. In the CPMD approach, the electron WF are optimized in the beginning of the simulation only. The time evolution of the WF (electron density) and nuclei is obtained by solving a system of Newton-like equations for both nuclei and electrons. Both methods have been shown to give consistent results for structural and dynamic properties of solids and liquids (Kuo et al., 2004).

In the context of accuracy and efficiency, AIMD simulations are exclusively performed based on DFT. All uncertainties and limitations attributed to DFT calculations as discussed in previous sections are fully applicable to the DFT-based AIMD simulations. The major limitation of the AIMD simulations remains the system size and the length of the simulated trajectory. The longest time scale accessible for ab initio simulations of a realistic system is currently limited to a few hundred picoseconds. This can be sufficient to investigate diffusion and structural rearrangement of water, ions, and small molecules at clay mineral surfaces.

3.1.8 Ab initio spectroscopy

Various absorption and emission spectra of clay minerals are related to either transitions of electrons between different states, or to changes in the dynamic of nuclei. These experimentally observed properties can be evaluated directly from ab initio simulations. In particular, the IR and Raman spectra are related to the fluctuation of the total dipole moment, and polarizability of the system, respectively, caused by the thermal vibrations of atoms. The total dipole moment, for example, contains contributions from the displacement of positively charged atomic cores and the electron density. Both can be evaluated either by analyzing AIMD trajectories or evaluating the response of the charge density distribution to the external electric field relative to the equilibrium state (linear response theory). In a similar way, the polarizability reflects the ability of the system to change the polarization in response to the external electric field. The advantage of the MD approach is that it provides information about anharmonic effects. The spectra are obtained by Fourier transformation of the total dipole–dipole correlation function. The computational costs of obtaining sufficiently long AIMD are rather high. In contrast, the linear response theory-based calculations are essentially limited to harmonic effects. The experimentally measured and calculated spectra are compared in terms of position of the spectral lines and their amplitude. In general, DFT-based methods tend to underestimate the spectral line frequency, whereas the HF approach tends to overestimate frequencies in vibrational spectra. This shift is attributed to the systematic errors in the description of the electron exchange and correlation. Because the errors are systematic, the common practice is to apply a method-specific scaling factor to the calculated frequencies. Performing normal mode analysis, that is, decoupling the collective motion of atoms in the system into individual modes (stretching, bending, libration, etc.), the IR and Raman spectral lines can be assigned to specific functional groups. The quantitative prediction of the adsorption/emission intensity is much more difficult because the results depend not only on the equilibrium density distribution, but also on the changes, with respect to atomic displacements. The intensity further depends on the local symmetry of the IR or Raman active site.

Theoretical calculations of IR and Raman spectra have been successfully used for the assignment of the absorption bands in the experimentally measured spectra. Most of the studies are conducted based on DFT calculations using gradient corrected E_{XC}. In general, good agreement between calculated position of the absorption bands and measured spectra can be obtained. This information is particularly helpful for the identification of structural polymorphs in the experimentally measured data. The calculations can be further used to correlate the hydrogen bond lengths with the frequencies and the effect of isomorphous heterovalent structural substitutions. In most cases, a functional relationship between the frequencies of the spectral lines and clay mineral structure/compositions can be obtained. These relationships are particularly useful for understanding structural aspects of natural clay minerals, and allow quantitative conclusions about the structural properties of the clay minerals and the nature and concentration of isomorphous substitutions to be made.

3.2 Simulations of clay minerals structure

The basic structural element of clay minerals can be considered as a combination of a sheet with closely packed octahedrally coordinated cations (O-sheet) and a sheet of hexagonally ordered tetrahedra (T-sheet). The most common combinations of tetrahedral and octahedral sheets result in repeating TOT and TO arrangements. These clay minerals are referred to as 2:1 and 1:1 types, respectively.

In 2:1 clay minerals, the octahedral sheet is composed by two octahedrons with *cis*-and one octahedron with *trans*-orientation of OH groups. In trioctahedral clay minerals, all octahedral sites are occupied by cations. In dioctahedral clay minerals, only 2/3 of available octahedral positions are occupied. This results in the existence of *cis*- or *trans*-vacant dioctahedral clay minerals. The experimental observations suggest that the stability of *cis*- and *trans*-vacant structures, as well as the mechanism of the isomorphous substitution, depend on the local symmetry of the sites and the effective cation radius.

3.2.1 Structure of TO/TOT layer and isomorphous substitutions

The calculations indicate that the energy differences between dioctahedral smectites with vacancies in *cis*- or *trans*-OH sites are small, and that their relative energies vary with the chemical nature of the substitutions (Sainz-Diaz et al., 2002; Tunega et al., 2007b). Energies of most common isomorphous cation substitutions (e.g. Fe^{2+}/Fe^{3+} and Mg^{2+} substitution for Al^{3+}) are obtained by isodesmic reactions (e.g. reactants and products have the same number and type of bonds). The calculations confirm the preferred stability of Al^{3+}, Mg^{2+} mixed smectites relative to the stability of pyrophyllite. It was also found that Fe^{3+} more easily enters the structures containing Mg^{2+} compared with the systems containing only Al^{3+} or Al^{3+}/Fe^{3+} mixtures (Timon et al., 2003). In structures with divalent cations (e.g. Mg^{2+} and Fe^{2+}) substituting for Al^{3+} in octahedral sites, the arrangements with *trans*- and *cis*-OH coordination for the vacancy are energetically equivalent, and their stabilities are virtually identical (Keri et al., 2017). The *cis*-vacant structures were found to be more stable when the charge originated from substitution of Al^{3+} for Si^{4+} in the tetrahedral sheet. *trans*-OH vacant coordination for the vacancy was found to be more stable only with structures where a Fe^{3+} cation occupies an octahedral site. The structure of oxidised and reduced Fe-bearing smectites and the role of Fe substitutions was investigated by Car–Parrinello MD (Liu et al., 2010). Analysis of MD trajectories showed that the average Fe—O bond lengths are about 2.08 and 2.02 Å in the reduced and oxidised clay minerals, respectively. Fe^{3+} substitution has no obvious influence on hydroxyl orientation, but Fe^{2+} broadens the angle distribution of OH groups. Both Fe^{2+} and Fe^{3+} substitutions result in a red-shifted in-plane bending mode of hydroxyl.

Isomorphous substitutions in *cis*- and *trans*-positions of dioctahedral smectites were found to have a significant effect on structural orientation and vibrational properties of the OH groups (Botella et al., 2004; Hernandez-Haro et al., 2014). According to the results of theoretical calculations, different molar extinction coefficients should be assigned to OH groups associated with *cis*- and *trans*-positions. This would allow for a quantitative analysis of concentrations of cationic substitution in smectites using the Lambert–Beer law. A comparative study of electrostatic properties of *cis*- and *trans*-vacant dioctahedral smectites reveals that *trans*-vacant configuration shows a higher hydrophilic character, that is, stronger water surface interaction, than the *cis*-vacant configuration (Minisini and Tsobnang, 2005). Different illite models were built based on 1*M* and 2*M* mica structures in order to analyse the effect of Al^{3+} distribution in the tetrahedral layer and the various cations (K^+, $NH_4{}^+$, H_3O^+ and interlayer H_2O) in the interlayer space on the structural properties, and spectroscopic properties of illite (Geatches and Wilcox, 2014; Geatches et al., 2014). The obtained vibrational frequencies and the XRD-pattern agree well with the experimental observation and provide a basis for the more detailed interpretation of experimental measurements.

3.2.2 Structure of hydroxyl layer in 1:1 clay minerals

The surface properties of the octahedral and tetrahedral basal planes of 1:1 clay minerals are very different. Whereas the tetrahedral sheet resembles the properties of the basal plane of 2:1 clay minerals, the octahedral surface is covered with OH groups. Depending on the orientation, the surface OH groups can adsorb both positively and negatively charged ions and molecular complexes. The surface OH groups are also responsible for the stacking of adjacent TO layers. The variability in the structural arrangement of OH groups and relative displacement of TO layers leads to a large number of structural polymorphs of 1:1 clay minerals. Identification of such structural polymorphs with traditional X-diffraction techniques is particularly challenging. Ab initio simulations are widely used to assess the local structural environment and dynamics of OH groups (Hobbs et al., 1997; Balan et al., 2001; Benco et al., 2001b; Refson et al., 2003; Sato et al., 2004; Castro and Martins, 2005b; Tosoni et al., 2006; Mercier and Le Page, 2008, 2009; Ortega-Castro et al., 2008, 2009; Campos et al., 2009; Clausen et al., 2009b; White et al., 2009). In fact, 36 structurally distinct stacking polymorphs can be built considering translations and rotations consistent with the symmetry of the individual octahedral layers (Mercier and Le Page, 2008). Just three, namely, kaolinite (Fig. 3.2), dickite, and nacrite are common in nature. Ab initio calculation at zero Kelvin and ones considering thermal vibrations of atoms confirm the stability of kaolinite, dickite, and nacrite relative to other possible polymorphs. Analysis of the relative energies of different polymorphs suggests that interaction between the nonadjacent layers is nearly negligible. The stable phases are formed either by the repeated application of the same stacking of two layers, or an alternate application of a two-layer transformation, followed by that of the enantiomorphic transformation.

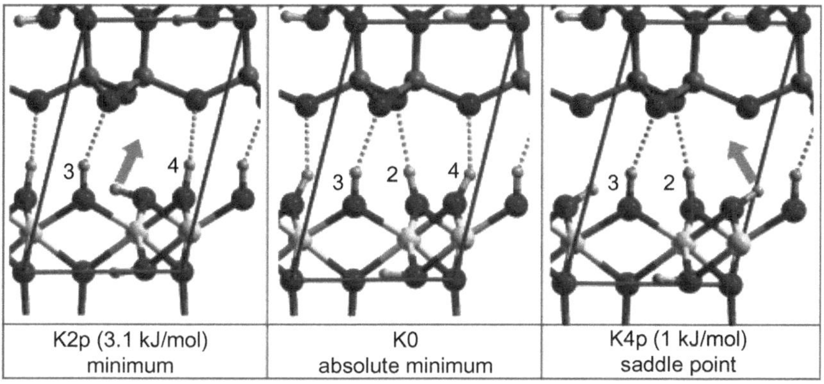

Fig. 3.2 Structural arrangement of OH1, OH2, OH3, and OH4 groups in kaolinite for the ground state (K0), local minima (K2p), and the transition state (K4p).
Reprinted (adapted) with permission from Tosoni, S., Doll, K., Ugliengo, P., 2006. Hydrogen bond in layered materials: structural and vibrational properties of kaolinite by a periodic B3LYP approach. Chem. Mater. 18, 2135–2143. Copyright (2006) American Chemical Society.

Periodic DFT studies of dickite reveal the existence of four structurally different OH groups. One (OH1) is located between the tetrahedral and octahedral sheet. Three others (OH2, OH3, OH4) are located on the external octahedral layer. These groups contribute to hydrogen bonds with the neighbouring tetrahedral sheet, and are responsible for the stacking of the TO layers. In contrast to the previous assignment, ab initio simulations could attribute OH2 and OH4 groups to the lowest energy spectral lines, whereas OH1 and OH3 forming weak hydrogen bonds are responsible for the high frequency absorption bands (Benco et al., 2001a).

A remarkable agreement in both position and intensity of the stretching bands of OH groups in kaolinite was obtained taking into account the size distribution of the kaolinite particles (Balan et al., 2001). A more detailed analysis of the simulation results suggests that the good agreement between the experimental data and vibrational frequencies of OH groups of kaolinite calculated in the harmonic approximation with the GGA functional may result from fortuitous error cancellation caused by overestimation of hydrogen bond length, resulting in higher frequencies and neglecting the bond anharmonicity, which compensates for the bond length effect, as shown in Table 3.2 (Benco et al., 2001a; Tosoni et al., 2006; Balan et al., 2007).

The arrangement of water molecules on the siloxane basal plane of kaolinite is dominated by hydrogen bonding between water molecules (Tunega et al., 2002a,c, 2004). In contrast, the OH groups of the aluminol surface of kaolinite interact strongly with water, and determine the orientation of the water film (Tunega et al., 2004). A monolayer I_h ice-like structure on the siloxane surface of kaolinite was found to have slightly hydrophobic properties, making further growth of multilayer ice energetically unfavourable (Hu and Michaelides, 2008).

Table 3.2 Harmonic (ω_h) and fundamental (ω_{01}) frequencies [cm^{-1}] calculated by decoupled anharmonic analysis (Benco et al., 2001b; Tosoni et al., 2006) are compared with measured FTIR and Raman frequencies

Mode assignment	(OH4)	(OH2)	(OH3)	(OH1)
FTIR[a]	3694	3668	3651	3620
Raman[a]	3699	3675	3656	3615
ω_h(K0)	3895	3859	3842	3809
ω_{01}(K0)	3714	3671	3657	3635
ω_{01}(K2p)	3608	3708	3617	3678
ω_{01}(K4p)	3791	3638	3623	3650
ω_{01}(slab)[b]	3780	3954	3960	3856

[a] Measurements (Johansson et al., 1998; Franco et al., 2004).
[b] Calculations (Benco et al., 2001b).

DFT studies of Pb^{2+}, PbO, $PbCl^+$, $Pb(OH)^-$, and $PbCl_2$ adsorption on kaolinite demonstrate that these species preferentially bind to the aluminol surface (Wang et al., 2014, 2015a,b). Moreover, PbO has the strongest interaction due to formation of covalent—O^{kaol}-PbO-Al^{kaol}—bonds (Wang et al., 2015a). Zn^{2+} forms mono- and bidentate complexes on the aluminol surface, whereas only weakly bound outer-sphere complexes are observed on the siloxane plane (Wang et al., 2017). Multivalent metal cations, such as Ca^{2+}, are used to stimulate the coagulation of a fine particle dispersion. The efficiency of coagulation depends on the ability of ions to bind to the clay mineral particle surface. Ca^{2+} is expected to hydrolyse to $Ca(OH)^+$ at pH greater than 10. The adsorption $Ca(OH)^+$ was found to be strongest on the edge surface of kaolinite (Peng et al., 2017).

3.2.3 Structure of the interlayer and basal plane in 2:1 clay minerals

The Al for Si substitution in the tetrahedral layer plays a key role in the stabilisation of the interlayer distance and the location of the interlayer cations in smectites. The interlayer cations Li^+, Na^+, K^+, Mg^{2+}, and Ca^{2+} were found to occupy the siloxane cavities on the basal plane, forming strong bonds to the oxygens of the Al-substituted tetrahedra (Shi et al., 2013). The smaller the radii and the higher the charge of the interlayer cation, the stronger the bonding is.

Counter ions in the interlayer space have a strong effect on the structural properties of the TOT layer. Interlayer counter ions with larger ionic radius result in a slight expansion of the tetrahedral sheet thickness and a compaction of the octahedral sheet (Li et al., 2016). The magnitude of the tetrahedral tilting and rotation is reduced with the increasing size of the interlayer cations. The smaller cations tend to sink in the ditrigonal cavities of the smectite layers, whereas larger cations are situated in the middle of the interlayer space.

Ab initio simulation was applied to simulate the dynamics of water and ions in the interlayer of smectites and brittle micas (Suter et al., 2008, 2012; Churakov and Kosakowski, 2010; Demontis et al., 2013; Peng et al., 2016). Comparative studies of hydration behaviour of Li^+, Na^+, and K^+ explain the role of interlayer cations and the distribution of heterovalent substitutions in tetrahedral and octahedral sheets of montmorillonite (Mt) (Mignon et al., 2010). K^+ was found to always bind to the basal plane of Mt in the presence and absence of water, acting as a swelling inhibitor. Li^+, in contrast, tends to hydrate easily and to detach from the surface, except for the clay mineral structure with adjacent heterovalent substitutions in tetrahedral and octahedral sites. Sodium forms inner-sphere complexes in the structure with the substitutions in the tetrahedral sheet. The tendency of Na^+ ions to be hydrated in the interlayer space was also confirmed by AIMD simulations (Boek and Sprik, 2003; Churakov and Kosakowski, 2010). An accurate description of the free energy landscape for different locations of interlayer cations and the degree of hydration at room temperature could be obtained from constrained AIMD simulations. The calculated free energies of Li^+ and Na^+ adsorption onto Mt and beidellite are consistent with the results obtained from static calculations, and explain the experimentally observed swelling behaviour of both minerals (Suter et al., 2008, 2012). At low water content, the interlayer cations have a stronger interaction with smectite surfaces having tetrahedral substitutions than ones having octahedral substitutions due to strong electrostatic attraction (Clausen et al., 2009a). Desorption energies of water at low coverage follow the trend of the cation hydration energies $Li^+ > Na^+ > K^+$ (Zhang et al., 2016a). Particularly difficult from the experimental point of view is the identification of the hydronium position in the interlayer space of clay minerals. The ab initio simulations resolve this difficulty and demonstrate that the complexation of the hydronium ion depends on the water content and the charge distribution in the TOT layer (Churakov and Kosakowski, 2010; Escamilla-Roa et al., 2016).

Investigations of hydrated Mt with divalent interlayer cations (Mg^{2+}/Ca^{2+}/Sr^{2+}/Ba^{2+}) indicate that the basal spacing is primarily controlled by the size and the degree of hydration of the countercations (Berghout et al., 2010). Ba^{2+}-Mt was found to have the largest interlayer space for any degree of hydration ('dry', one or two water layers). The basal spacing of 'dry' Mt exchanged with small cations, Mg^{2+} and Ca^{2+}, is very similar. In hydrated models, the interlayer expansion correlates with the ionic radius of the cations. The structures of Ca^{2+} hydrates in the interlayer space of Mt were studied by DFT with dispersion correction (Lou et al., 2014). It was found that at low humidity, Ca^{2+} is strongly adsorbed by the six-membered oxygen ring of the T-sheet. The predicted lattice constants of the Mt layers in a direction parallel to the basal place are typically in good agreement with the experimental observations. These parameters are not affected by cations and water in the interlayer space. The accurate prediction of basal spacing is more challenging. Static structural optimizations of hydrated structures do not take into account diffusive motion of the interlayer species. The time scale of available AIMD trajectories is typically too short to allow for the accurate sampling of all relevant configurations of the interlayer water molecules and cations.

Analysis of the relative nucleophilicity of interlayer cations obtained by Fukui functions formalism (Li and Evans, 1995) revealed that the activity order for the exchangeable cations $Ca^{2+} > Sr^{2+} > Mg^{2+} > Rb^{+} > Ba^{2+} > Na^{+} > Li^{+} > Cs^{+} > K^{+}$ correlate with experimental observations of the variation of the interlayer spacing in the monohydrated Mt (Chatterjee et al., 2004). The substitutions of Me^{2+} for Me^{3+} ions in the octahedral sheet were found to result in Mt with higher Brønsted acidity compared with the one with Al^{3+} for Si^{4+} in the tetrahedral sheet, and more readily release the structural proton (Briones-Jurado and Agacino-Valdes, 2009).

The stability of the vermiculite structure was found to correlate with the relative distribution of Al substitutions in the tetrahedral layer and hydrated Mg^{2+} cations in the interlayer space (Tunega and Lischka, 2003). The structure is stabilised by strong (about 1.5 Å long) hydrogen bonds between water hydrating the interlayer Mg^{2+} atoms and the basal surface, in which Mg^{2+} cation is located above the tetrahedral site. In contrast to electrostatic arguments suggesting the minimal cation–cation repulsion when Mg^{2+} is situated above Al tetrahedra, the most stable structures were found to form when Mg^{2+} is sandwiched in between Si tetrahedra of opposite layers. This configuration is stabilised due to a much stronger electron density localization of Si—O bonds, compared with Al—O bonds, and thus stronger Mg-—OH...O—Si hydrogen bonds. Formation of ice-like structures was also found in a super-cooled water film in the interlayer of Na^{+} exchanged vermiculite (Demontis et al., 2014). Using thermodynamic calculations, the temperature of subsequent layer by layer dehydration of smectites can be obtained. The good agreement with the experimental measurement could be obtained when a dispersion corrected functional was applied (Fonseca et al., 2017).

The interaction of Cs^{+} with basal planes of dioctahedral smectites, including Mt, beidellite, Fe—Mt and nontronite, was investigated by a combination of X-ray photoelectron spectroscopy (XPS) and DFT (Ebina et al., 1999). It was found that Cs^{+} ions enter the hexagonal cavities on the basal plane. Investigation of Cs^{+}/K^{+} exchange in the interlayer space of muscovite reveals nearly isoenergetic properties of Cs^{+} and K^{+} mica under constant pressure conditions (Rosso et al., 2001). Potassium in the interlayer space appears to screen the electrostatic repulsion better, and results in a decrease of the interlayer spacing with an increasing TOT layer charge, whereas the opposite trend was observed for Cs^{+}. It was identified that the exchange of Cs^{+} for K^{+} significantly reduces the migration barrier of neighbouring K^{+}, and thus increases the exchange rate (Ruiz Pestana et al., 2016). A concept of 'chemical–mechanical coupling' has been applied to explain the observed sharp exchange fronts that lead to interstratification.

3.3 Elastic properties of clay minerals

Mechanical properties of clay minerals are important for understanding their performance in technical installations such as backfill and drilling fluids, as well as the role of clay minerals as lubricants for tectonic deformations. The calculations help to explain the interplay between the clay mineral structure and the arrangement of the intercalating materials. Accordingly, the elastic tensors were derived for several clay minerals with different compositions based on DFT (Sato et al., 2005; Militzer et al., 2011). Good agreement between calculated and measured elastic constants of water-free clay minerals was obtained. Simulations show that the influence of

Table 3.3 Calculated bulk modulus (K), shear modulus (G), transverse wave velocity (V_s), longitudinal wave velocity (V_p) and Poisson's ratio (ν), of kaolinite aggregates

	DFT[a]	DFT-D2[a]	DFT[b]	Exp[c]
K_R[GPa]	59.21	74.25	23	44.0–59.7
K_V[GPa]	72.43	77.79		
G_R[GPa]	20.91	21.7	17	19.7–31.8
G_V[GPa]	40.66	41.48		
$V_{p,\ R}$[km s^{-1}]	5.75	6.26	7.34	5.488–6.23
$V_{p,\ V}$[km s^{-1}]	6.93	7.11		
$V_{s,\ R}$[km s^{-1}]	2.82	2.87		2.839–3.55
$V_{s,\ V}$[km s^{-1}]	3.93	3.97		
$V_{p,\ R}/V_{s,\ R}$	2.04	2.18	2.1	1.755–1.823
$V_{p,\ V}/V_{s,\ V}$	1.76	1.79		
ν_R	0.342	0.367	0.352	0.26–0.319
ν_V	0.264	0.274		

Subscripts R and V indicate lower and upper bound for the averaged elastic properties of crystalline aggregates calculated based on single crystal elastic moduli according to the method of Reuss and Voight (Hill, 1952). Range of most reliable experimental data is given for comparison.

[a] Weck et al. (2015).

[b] Sato et al. (2005).

[c] Range of experimental data summarized in table 5(d, e, h) (Weck et al., 2015).

cation disorder is minimal. On the other hand, stacking disorder was found to be of importance for clay minerals of the kaolin group. The kaolins exhibit a distinct type of seismic anisotropy, which can be related to hydrogen bonding (Weck et al., 2015). Predictions of the properties of polymineral rocks by the conventional averaging procedure, that is, Reuss, Voigt, and Hill approximations (Hill, 1952) are more challenging (Table 3.3). Calculated elastic properties display higher stiffness and lower P-wave anisotropy. The difference is likely to be due to the presence of oriented flattened pores in natural samples that are not considered in the simulations and the averaging procedure. Kaolinite is particularly interesting with respect to the viscosity of the nanoconfined fluid, because of the hydrophobic and hydrophilic properties of siloxane and aluminol surfaces, respectively. Recent ab initio simulation of shear viscosity of water films confined between kaolinite layers could not confirm the drop of viscosity reported experimentally (Feibelman, 2013). The discrepancies can be related to the system setup and the system size effect, and need to be further investigated.

3.4 Redox processes

Fe is the most common redox-sensitive element in clay minerals, and can participate in the redox reaction. The redox activity of Fe in clay mineral depends on the structural position of Fe and the kinetics of the electron transfer (ET) reaction, which in turn depends on the structure. The ratio between ferrous and ferric iron affects chemical

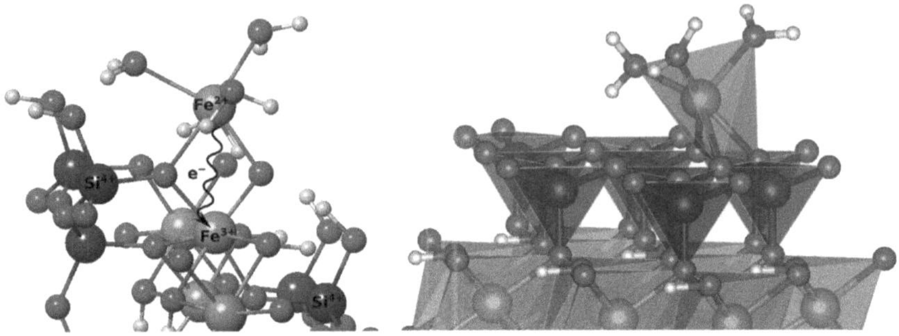

Fig. 3.3 *Left*: One of the lowest energy sixfold coordinated Fe^{2+} complex adsorbed on the (110) edge of nontronite. ET takes place across Fe^{2+}—O—Fe^{3+} bond chain. *Right*: One of possible Fe^{2+} complexes on the basal plane of nontronite. For any possible location of Fe^{2+} complex on the basal plane the ET is energetically unfavourable due to the large distance between Fe^{2+} complex on the basal plane acting as the electron donor and the electron acceptors (Fe^{3+} in the octahedral layer).
Adapted with permission from Alexandrov, V., Rosso, K.M., 2013. Insights into the mechanism of Fe(II) adsorption and oxidation at Fe-clay mineral surfaces from first-principles calculations. J. Phys. Chem. C 117, 22880–22886. Copyright (2013) American Chemical Society.

and physical properties of clay minerals, such as: specific surface area, cation exchange capacity, and the degree of swelling and surface reactivity with respect to specific functional groups of organic molecules (Srodon and McCarty, 2008). Most DFT studies focused on the Fe redox activity using nontronite as the structural prototype $\left(Fe^{(3+)}{}_2(Fe, Al, Si)_4O_{10}(OH)_2\right)$. Analysis of reorganisation energy in nontronite structures with different distribution of Fe in the octahedral and tetrahedral sheets and the calculations of adiabatic activation energy for ET suggest that electron hopping between neighbouring Fe^{2+}–Fe^{3+} octahedral sites occurs on the order of 10^5–$10^6\ s^{-1}$, whereas the contribution of Fe^{3+} located in the tetrahedral sites is negligible ($\sim 10^{-11}\ s^{-1}$). Adsorption of Fe^{2+} ions at basal and edge surfaces of nontronite and interfacial Fe^{2+}–Fe^{3+} ET were further studied by DFT calculations (Alexandrov and Rosso, 2013). Fe^{2+} can adsorb on different surface sites including ferrinol, silanol, and mixed sites, whereas the adsorption on ferrinol sites is the most favourable, and is coupled to proton transfer (Alexandrov et al., 2013). The comparison of the activation barrier for ET to the bulk from the Fe^{2+} adsorbed on the edge or on the basal plane (Fig. 3.3) suggests that the ET preferentially takes place at the edge, whereas the ET through the basal plane is highly unlikely (Alexandrov and Rosso, 2013).

3.5 Interaction of clay minerals with organic compounds

The interactions between clay mineral surfaces and organic matter are of great interest to diverse fields, including soil, environmental science, petroleum engineering, and development of functional materials.

3.5.1 Natural organic matter and environmental engineering

Adsorption of carboxylic acid on clay mineral surfaces plays a vital part in the interaction between natural organic matters and clay minerals. A combination of ATR-FTIR spectroscopy and molecular orbital calculations suggested that salicylic acid can form monodentate and bidentate complexes on clay mineral edges (Kubicki et al., 1997). Cluster models (Tunega et al., 2002d) and periodic calculations (Tunega et al., 2002b) suggest that acetic acid is strongly bound to the octahedral surfaces of kaolinite due to H-bonding. Taking acetate as a model molecule, the interaction of the carboxylate group with interlayer and edge surfaces of smectite was investigated with DFT calculations (Liu et al., 2008; Chen et al., 2017). It was found that H-bonding plays an important role in both inner- and outer-sphere complexes (Liu et al., 2008). Acetate forms stable monodentate complexes on edge surfaces and bidentate complexes in the interlayer space. Acetate complexation on edge surfaces is pH dependent, inner-sphere complexes on (110) surfaces is limited to a mildly acidic pH, while complexes on (010) surfaces can occur in the entire pH range relevant to environmental conditions (Liu et al., 2017). In another study involving carboxyl acid, the pathway for the decarboxylation of propionic acid into alkane and carbon dioxide catalysed by clay minerals was investigated using the ab initio transition state sampling method, with the aim of understanding the oil formation and biofuel production (Geatches et al., 2011). Comparative studies of adsorption energy of small molecules on the basal plane of smectites indicate the following trend in adsorption energies: $E(NH_3) > E(H_2O) > E(C_6H_6)$ (Benco and Tunega, 2009).

The mechanism of methylene blue (MB) adsorption on siloxane and hydroxyl basal planes of kaolinite has been studied combining DFT and classical molecular dynamics simulations (Greathouse et al., 2015). Neutral and cationic MB preferentially adsorb on the siloxane plane of kaolinite. At infinite dilution, aromatic rings are located above the siloxane ring parallel to the surface. At higher concentrations, MB molecules form stacks or one-dimensional chains.

Adsorption of small molar mass organic molecules attracts a lot of attention in the context of CO_2 sequestration, unconventional oil, and shale gas recovery. It has been suggested that maximal adsorption of dry supercritical CO_2 in the interlayer space of Mt is favoured up to monolayer coverage, and decreases at higher CO_2 concentrations (He et al., 2014; Schaef et al., 2014). In the presence of water, CO_2 does not wet the surface of smectites (Solling et al., 2015), and segregates in the centre of the interlayer space (Myshakin et al., 2013). This ensures that CO_2 can be efficiently trapped in clay rocks by the so-called snap-off mechanism. In this process, the CO_2 segregates in large pores, and the wetting phase (water or brine) blocks surrounding narrow pores' throats.

Sodium hexametaphosphate, a polymeric mixture of metaphosphates called Graham's salts, is a widely used dispersant in the flotation process. The ab initio simulations demonstrate that sodium hexametaphosphate can bind to the aluminol surface of kaolinite. Moreover, the deprotonated $-[PO_3]_m-$ species interact more strongly than the $-[HPO_3]_m-$ species. The negatively charged $-[PO_3]_m-$ species on the surface of kaolinite thus increase the electrostatic repulsion between particles (Han et al., 2016a). A similar effect was documented for the adsorption of negatively charged aqueous silicate species $SiO(OH)_3^-$ (Han et al., 2016b).

3.5.2 Organic contaminants

Due to their ubiquitous presence in the environment, clay minerals are considered as cheap and efficient geo-adsorbents for organic pollutants. Organic polymers are also used to modify mechanical, hydrophobic, and hydrophilic properties of clay mineral surfaces and to control their swelling behaviour. A number of ab initio studies systematically investigate interaction of different functional groups of organic polymers with clay minerals (Munoz-Santiburcio et al., 2009; Geatches et al., 2012; Scott et al., 2015). Aromatic compounds are infamous in the environment for their toxicity. Most of the ab initio adsorption studies are conducted without explicit consideration of water. These simulations are often used to test the performance of empirical force fields (Shapley et al., 2013).

The interaction between 1,3,5-trinitrobenzene and clay mineral surfaces was studied at different levels of theory (e.g. HF, MP2 and DFT approaches) using a cluster model (Pelmenschikov and Leszczynski, 1999; Gorb et al., 2006). It was shown that the attractive component of the adsorption energy includes both electrostatic and dispersive contributions. The interaction of *meta*- and *para*-substituted nitrobenzene molecules (—NH_2, —OH, —OCH_3, —CHO, —$OCOCH_3$, —$COCH_3$, —CF_3, —CN and —$COOCH_3$ radical) with smectite surfaces was studied by DFT using the cluster approach. Obtained interaction energies suggest that beidellite is a suitable adsorbent for the retention of various nitrobenzene molecules (Chatterjee et al., 2003). A combination of FTIR measurements and cluster-based DFT calculations showed that the formation of octachlorodibenzodioxin (OCDD) on Fe-smectite surfaces is initiated by the formation of pentachlorophenol (PCP) radical cations via single ET from pentachlorophenol to Fe^{3+} (Gu et al., 2011). Electrostatic interaction was found to be the main contribution to the adsorption of polychlorinated dibenzo-*p*-dioxins on pyrophyllite (Kat et al., 2008). Therefore, the chlorination of the molecule eventually strengthens the binding between dibenzo-*p*-dioxins and clay mineral surfaces. The same tendency was also confirmed with an empirical force field, and with periodic DFT calculations, taking into account dispersion effects at the DFT-D2 level of theory (Shapley et al., 2013). Periodic DFT simulations of trichloroethene adsorbed in the interlayer of nontronite showed that in the presence of water, the C—Cl bond at the end of the H—C—Cl chain is broken, thus leading to the dechlorination of trichloroethene (Teppen et al., 2002). The adsorption of 2,4-dinitrotoluene (DNT) and formamide (FA) on kaolinite was investigated using cluster model DFT calculations (Song et al., 2013). The study suggests that FA and DNT molecules preferentially adsorb on the octahedral surfaces by H-bonding. A cluster study of FA conformation on kaolinite reveals that the molecule interacts more strongly with the aluminol surface. Furthermore, the intercalation was found to be energetically more favourable than the adsorption on a single surface. The —NH_2 group in the FA molecule controls conformation on the basal surface of the tetrahedral sheet, whereas the carbonyl oxygen atom plays a crucial role in the formation of hydrogen bridges on the aluminol surface (Wang et al., 2013; Song et al., 2013). Extended studies of more complex molecules such as acetamide (AA), *cis-N*-methylformamide (NMFA1), *trans-N*-methylformamide (NMFA2), *cis-N*-methylacetamide (NMA1), and *trans-N*-methylacetamide (NMA2) confirms the observation obtained for FA; namely, the stronger interaction for aluminol surfaces and the tendency to intercalate. The stability of intercalated structures was found to decrease in the order of NMFA2 > FA > AA > NMFA1 > NMA2 > NMA1 (Michalkova et al., 2002; Song et al., 2014). Small differences were observed for different exchange correlation functionals.

Cluster studies of the adsorption of heteroaromatic hydrocarbons such as thiophene (TH), benzothiophene (BT), and dibenzothiophene (DBT) on the silanol surface of kaolinite based on DFT and perturbation theory suggest that C—H...O hydrogen bonding interactions between the heteroaromatic compounds and the tetrahedral surface are likely to control the adsorption mechanism. The sequence of stability for the cluster model of kaolinite complexed with the heteroaromatic compounds is found to be 3Si-O-DBT > 3Si-O-BT > 3Si-O-TH (Sainz-Diaz et al., 2010).

Reported predictions of benzene interaction with kaolinite are method- and basis-set dependent. Earlier studies of benzene adsorption on the hydroxyl layer of kaolinite performed with a small basis set at the HF level of theory suggest a tilted orientation of the molecule on the surface (Castro and Martins, 2005a). More recent DFT investigations with hybrid functionals predicted parallel orientation of the aromatic ring on both tetrahedral and octahedral sheets (Castro et al., 2012). These DFT results also agree with the predictions based on perturbation theory.

Phenoxyacid compounds are common pollutants in the environment belonging to the pesticide family. DFT studies of their adsorption on basal planes of muscovite in the gas phase indicate that ionized compounds interact more strongly than the neutral ones. In the neutral form, the OH group of the carboxylic acid forms hydrogen bonds to the basal plane, whereas in the ionized form, the oxygen of the carboxylic group bind to the K ions adsorbed on the surface (Ramalho et al., 2013). The carboxylate group of 2,4–dichlorophenoxyacetic acid forms a bidentate complex with Ca^{2+} in a dry interlayer region but a monodentate complex in a wet interlayer (Tunega et al., 2007a).

Interaction of complex polyaromatic compounds with surfaces depend on the acid–base properties of the functional groups, their protonation state, and the molecular dimensions (Sainz-Diaz et al., 2010; Carvalho et al., 2014; Bocquet et al., 2015; Pirillo et al., 2015). The presence of strongly interacting functional groups, that is, negatively charged sites, results in a tilted orientation of the molecules, eventually forming pillared structures, whereas weakly interacting groups favour flat orientation of the molecule on the surface. The methyl or phenyl radicals always bond upright to one of the outmost Si atoms of the clay mineral layers, while the H atom attaches to the inner Al—OH group, forming a water molecule (Bocquet et al., 2015). The covalent attachment of the radicals to the oxygen atoms of the clay mineral surface is always less favourable than to the silicon atoms. The adsorption results in formation of five coordinated Si surface sites with elongated Si—O bonds.

Fine kaolinite particles strongly aggregate with amine/ammonium salts, that is, dodecyl amine (DDA), *N*-methyldodecyl amine (MDA), *N*,*N*-dimethyldodecyl amine (DMDA), and dodecyl trimethyl ammonium chloride (DTAC). The aggregating abilities of the four amine/ammonium salts for kaolinite fall in the order DTAC > DDA > DMDA > MDA. These aggregates are stabilised by polyamine molecules forming strong hydrogen bonds N—H...O and weak C—H...O hydrogen bonds. The adsorption energy was found to be stronger on the siloxane surface than on the hydroxyl surface (Chen et al., 2017).

3.5.3 Pillared organo-clay nanocomposites

Organo-clay nanocomposites are of interest to the chemical and material community. Hybrid materials can be synthesised by intercalating organics into the interlayer space of clay minerals, thus modifying the surface properties. A large number of modelling studies employ classical MD simulations to investigate the interlayer structures and dynamics of pillared clay minerals and nanocomposites based on clay minerals, that is, as reviewed by Chen et al. (2008). Only a limited number of studies have been conducted using QM methods. Initial stages of the polymerization of methanol and ethylenediamine in the interlayer space of Mt was modelled using periodic DFT (Stackhouse et al., 2001). Nucleophilic attack of the amine on the aldehyde was only observed when the aldehyde was protonated or coordinated to a metal ion. No evidence was found for the dissociation of water in the hydration sphere of the sodium counterions. The effects of isomorphous substitutions on the binding energy of nylon 6 on clay mineral surfaces was studied by periodic DFT (Mrayed et al., 2009). The interaction strength grows with increasing surface charge. The binding energies were found to be higher when the substitution took place in the tetrahedral sheets. A combination of DFT calculation and IR spectroscopy revealed that *N*-butylammonium ($N=1$–4, *N*-BA) had a planar arrangement in the interlayer spaces of Mt (Scholtzová et al., 2016). H-bonding and electrostatic interactions were found to play an important part in stabilisation and orientation of these cations.

3.5.4 Interaction with petroleum molecules

Adsorption of methane and benzene on the illite basal surface was investigated with periodic DFT and DFT-D2 (Bocquet et al., 2015). The simulations showed that methane and benzene can be grafted to Si atoms. The process proceeds with low activation, if starting from the radicals of these molecules. This could represent the realistic kerogen phase. Using both DFT and a classical force field method, the adsorption of oil components, especially the representative resin molecule decahydro-2-napthoic acid (DHNA) and its deprotonated form, $DHNA^-$ on basal (Greathouse et al., 2017) and edge surfaces of clay minerals was investigated (Zeitler et al., 2017). The study explains the molecular mechanism of enhanced oil recovery using low salinity water flooding and highlights the role of cationic bridges in the adsorption of organic molecules.

3.5.5 Adsorption of biomolecules

Ab initio geometry optimizations of L-Alanine (L-Ala) and D-Alanine (D-Ala) on nontronite showed that chiral clay minerals' surfaces can have different effects on the chemical reactivity of enantiomers (Yu et al., 2001). Thus, D-Ala was found to form more stable surface complexes than L-Ala.

Adsorption of nucleobases on the surfaces of Na^+-Mt using periodic DFT was investigated in the presence and absence of Na^+ (Mignon et al., 2009). The face-to-face arrangement was found to be the favourable adsorption mode due to the dispersive forces between the nucleobases and the surfaces. However, this study did not include the effects of water, thus other adsorption modes may also be possible. The effect of solvation on the stabilisation of nucleobases thymine and uracil on octahedral and tetrahedral surfaces of kaolinite was investigated in cluster DFT (Michalkova et al., 2011). The simulations demonstrated that the presence of Na^+ leads to a significant stabilisation of nucleobases on both surfaces, and the stabilisation increases only slightly with the addition of water. The metal cation and nucleobases were not fully solvated in water due to the limited number of water molecules considered in the system setup. Therefore, further investigations are necessary to fully appreciate the solvation effects.

The adsorption mechanism of nitroaromatic compounds is the key factor controlling their bioavailability and biodegradability in a natural environment. Cluster-based DFT studies of 2,4-dinitrotoluene interaction with the kaolinite surface indicates that the molecule preferentially adsorbs by means of hydrogen bonding with N...H—O intermolecular interactions, providing the most important contribution to the interaction energy (Wang et al., 2013).

3.6 Acid–base properties of edge surfaces and cation complexation

Acid–base reactivity is central to the interfacial chemistry of clay minerals. The protonation states of clay mineral surfaces depends on pH conditions in the aqueous environments, and thus controls the interfacial reactivity, that is, retention of ions and catalytic properties. It is nontrivial to quantitatively characterise the protonation states of clay mineral surfaces experimentally. The potentiometric experiments measure a total proton adsorption capacity, but do not distinguish the individual contributions from different surface sites. The in situ spectroscopic techniques are still unable to identify the protons. Advanced ab initio simulations can predict the intrinsic pK_a of individual surface sites and surface complexation mechanism of the individual ions. Particularly powerful is the combination of spectroscopic studies and ab initio simulations. Using careful data analysis, the ab initio simulations can be used to interpret the experimentally measured EXAFS and XANES spectra and obtain information about concentration of several adsorbed species present in the natural samples (Kremleva et al., 2011, 2012, 2015a; Churakov and Daehn, 2012; Alvim and Miranda, 2015; Keri et al., 2017).

3.6.1 Edge surface structures and surface pKa

The schematic structures of edge surfaces on 1:1 and 2:1 clay minerals are shown in Figs. 3.4 and 3.5, respectively. It is generally accepted that the coordination of Si and Al at the edge of the tetrahedral sheets is fourfold. The anticipated sixfold coordination of (Al/Mg/Fe) edge atoms in the octahedral sheet (White and Zelazny, 1988) was confirmed by gas phase calculations (Bickmore et al., 2003; Churakov, 2006;

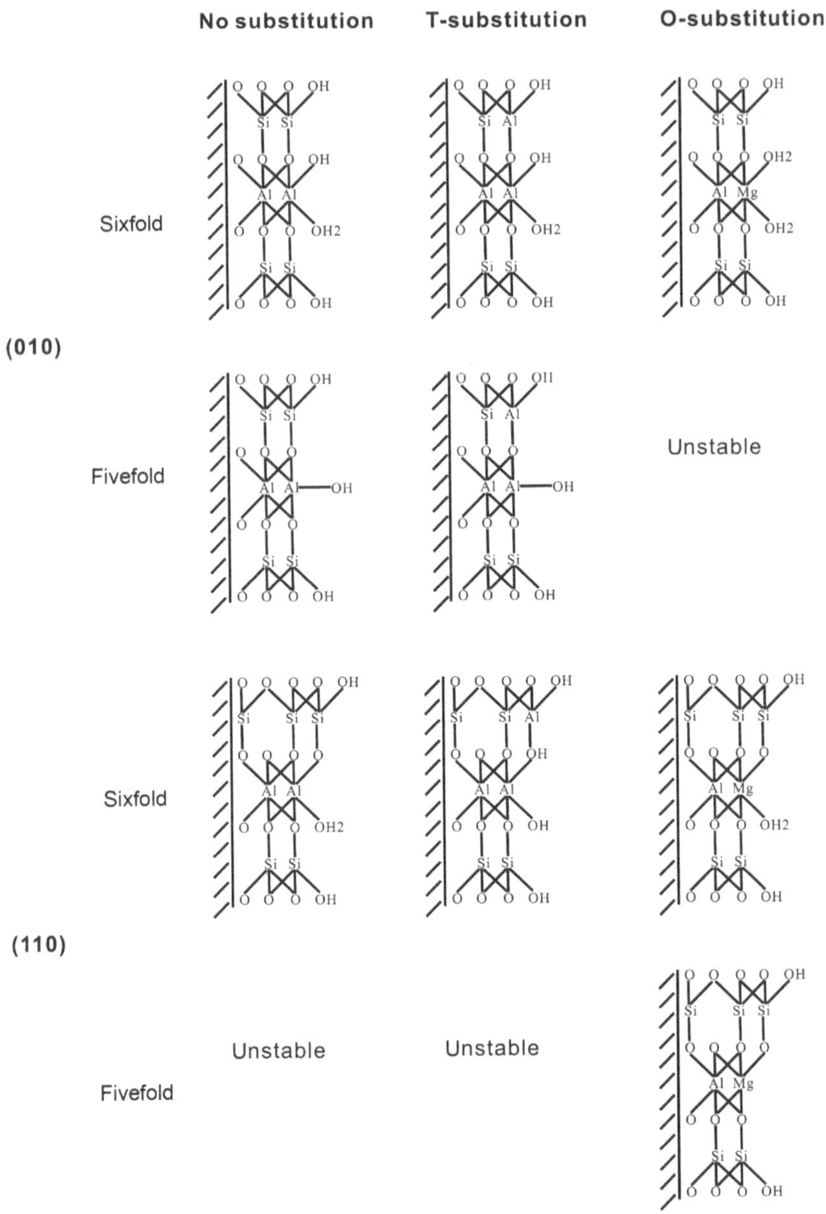

Fig. 3.4 Schematic representation of edges of 2:1-clay minerals.

Okumura et al., 2017) and AIMD simulations (Churakov, 2007; Kabalan et al., 2015). The calculated free energy of water adsorption indicates that sixfold coordinated surface sites coexist with fivefold coordinated ones (Liu et al., 2012b,c). It has also been shown that the coordination of the edge sites occupied by redox-sensitive elements

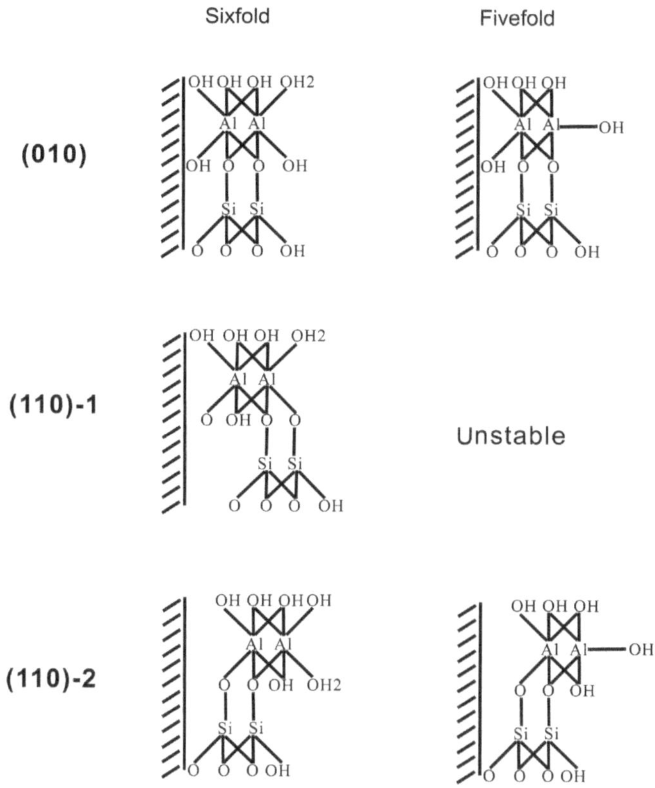

Fig. 3.5 Schematic representation of edges of 1:1-clay minerals.

(e.g. Fe) depends on the metal's oxidation state (Liu et al., 2015a). The predictions obtained at room temperature have been shown to hold at least up to 423 K (Liu et al., 2015b).

Earlier gas phase calculations performed on pyrophyllite were used to correlate the pK_a of the surface groups to their deprotonation enthalpy (Bickmore et al., 2003; Churakov, 2006). Recently, ab initio thermodynamic integration methods were used for direct estimation of pK_a constants of ≡OH surface groups at edge surfaces of clay minerals and to evaluate the effect of most common isomorphous substitutions (Liu et al., 2013, 2014, 2015a) (Table 3.4). The substitution of Al^{3+} for Me^{2+} in the octahedral sites ($\equiv Me_{octa}(OH_2)_2$) and of Si^{4+} for Al^{3+} in tetrahedral sites ($\equiv Al(OH)_{Tetra}$), substantially decreases the acidity of the OH at tetrahedral surface sites. Notably, the substitutions of Mg^{2+} and Fe^{2+} for Al^{3+} in octahedral sheets also decrease the acidity of the ≡SiOH in the tetrahedral layer by ~3–4 pK_a units.

According to the calculated pK_a values, only a few types of sites can participate in the protonation/deprotonation reactions and accommodate inner-sphere cationic complexes. For the (010)-edges, these sites include $\equiv SiO^-$, $\equiv Al/Fe^{3+}(OH_2)(OH)$ and $\equiv Al/Fe^{3+}(OH)_2$, $\equiv Fe^{2+}(OH)$. For (110)-edge, the ion complexing sites are $\equiv SiO^-$, $\equiv Al/Fe^{3+}(OH)$ and ≡Al/Mg/Fe—O—Si≡, ≡Mg—O—Si≡.

Table 3.4 pK_a of edge sites on individual surfaces of clay minerals

Surfaces	Sites	pKa
Kaolinite		
(010) Neutral	$\equiv Si(OH_2)/\equiv Si(OH)$	−6.7/6.9
	$\equiv Al(OH_2)(OH_2)/\equiv Al(OH_2)(OH)$	0.2/5.7
Pyrophyllite/Mt		
(110) Neutral	$\equiv Si(OH_2)/\equiv Si(OH)$	−6.8/8.3
	$\equiv AlSiOH$	1.7
	$\equiv Al(OH_2)$	5.5
(110) Mg-sub	$\equiv Si(OH_2)/\equiv Si(OH)$ (connecting Mg via O)	−2.5/11.0
	$\equiv MgSiOH$	4.2
(110) Al-sub	$\equiv Al(OH_2)_{Tetra}/\equiv Al(OH)_{Tetra}$	−2.4/17.5
	$\equiv AlAlOH$	10.2
	$\equiv Al(OH_2)_{Octa}$	12.7
(010) Neutral	$\equiv Si(OH_2)/\equiv Si(OH)$	−14.3/7.0
	$\equiv Al(OH_2)(OH_2)_{Octa}/\equiv Al(OH_2)(OH)_{Octa}$	3.1/8.3
(010) Mg-sub	$\equiv Si(OH_2)/\equiv Si(OH)$ (connecting Mg via O)	−10.9/10.8
	$\equiv Mg(OH_2)_2$	13.2
(010) Al-sub	$\equiv Al(OH_2)_{Tetra}/\equiv Al(OH)_{Tetra}$	−2.4/15.1
	$\equiv Al(OH_2)(OH_2)_{Octa}/\equiv Al(OH_2)(OH)_{Octa}$	4.9/8.5
(010) Fe^{3+}	$\equiv Si(OH)$	8.6
	$\equiv Fe(OH_2)(OH_2)/\equiv Fe(OH_2)(OH)$	1.2/5.1
(010) Fe^{2+}	$\equiv Si(OH)$	11.2
	$\equiv Fe(OH_2)$ [fivefold]/$\equiv Fe(OH_2)(OH_2)$ [sixfold]	6.6/10.2
Nontronite		
(010) Fe^{3+}	$\equiv Si(OH)$	8.7
	$\equiv Fe(OH_2)(OH_2)/\equiv Fe(OH_2)(OH)$	1.2/4.4
(010) Fe^{2+}	$\equiv Si(OH)$	9.9
	$\equiv Fe(OH_2)$ [fivefold]	8.4

Simulations of water between the edge sites provide unique information about the hydration mechanism of the broken surfaces of clay minerals and the solution structure at the interface (Kabalan et al., 2015). Particularly intriguing is the mechanism of proton transport, which was shown to involve cooperative motion of several protons along the hydrogen-bonded water molecules (Churakov, 2007).

Table 3.5 Acidity of water confined in the interlayer of smectites as obtained by ab initio simulations (Liu et al., 2011)

System		pK_a
Neutral 2:1 clay		12.6
Na^+-smectite Oct. Subs.	Cation bound/free water	11.5
Mg^{2+}-smectite Oct. Subs.	Cation bound water	10.1
	Free water	11.3
Mg^{2+}-smectite Tetr. Subs.	Cation bound water	10.4
	Free water	12.1

The effect of confinement on the dissociation constant of water in the interlayer space of smectites was investigated by constrained AIMD (Liu et al., 2011). The simulations show that the uncharged clay minerals (e.g. pyrophyllite) do not modify the dissociation constant of water, whereas all charged clay minerals substantially enhance the acidity of the interlayer water (Table 3.5). At the same time, the exact location of the layer charge does not have obvious influence on the acidity of the interlayer water.

3.6.2 Metal complexation at edge sites

Ab initio simulations have the potential to explain the molecular mechanism of the surface complexation of hazardous metals, which preferentially takes place at the edge surfaces of clay minerals. Due to the large computational costs and complexity, such studies are just emerging. First, such a study was applied to explain the uptake of Zn^{2+} on edge surfaces of Mt, and to interpret the measured K-edge EXAFS spectra of adsorbed Zn^{2+} (Churakov and Daehn, 2012). It was shown that the structurally incorporated Zn^{2+} preferentially substitutes for Al^{3+} in *trans*-symmetric octahedral sites. At low loading, Zn^{2+} is incorporated into the outermost *trans*-octahedra, and at medium loading, Zn^{2+} forms monodentate and bidentate complexes on the octahedral sheets (Fig. 3.6). Furthermore, the relative contribution of the (010) and (110) edge sites could be obtained by a linear combination fit to the experimental study (Fig. 3.7). A similar approach was successfully applied to discriminate Fe^{2+} and Fe^{3+} uptake by Mt (Keri et al., 2017). A possible structure of the inner-sphere complexes of Fe^{2+} (Liu et al., 2012a, 2013, 2014) and Cd^{2+}/Ni^{2+} (Zhang et al., 2016b, 2017) adsorbed on the edge surfaces of 2:1 clay minerals, including the complexes on silanol, aluminol, and octahedral vacancy sites was revealed by AIMD (Fig. 3.6). Cd^{2+}/Ni^{2+} cations form monodentate complexes on $\equiv SiO^-$ sites, bidentate complexes on $\equiv Al(OH)_2$ sites, and tetradentate complexes on vacancies. The calculated free energy of desorption indicate that the octahedral vacancy has

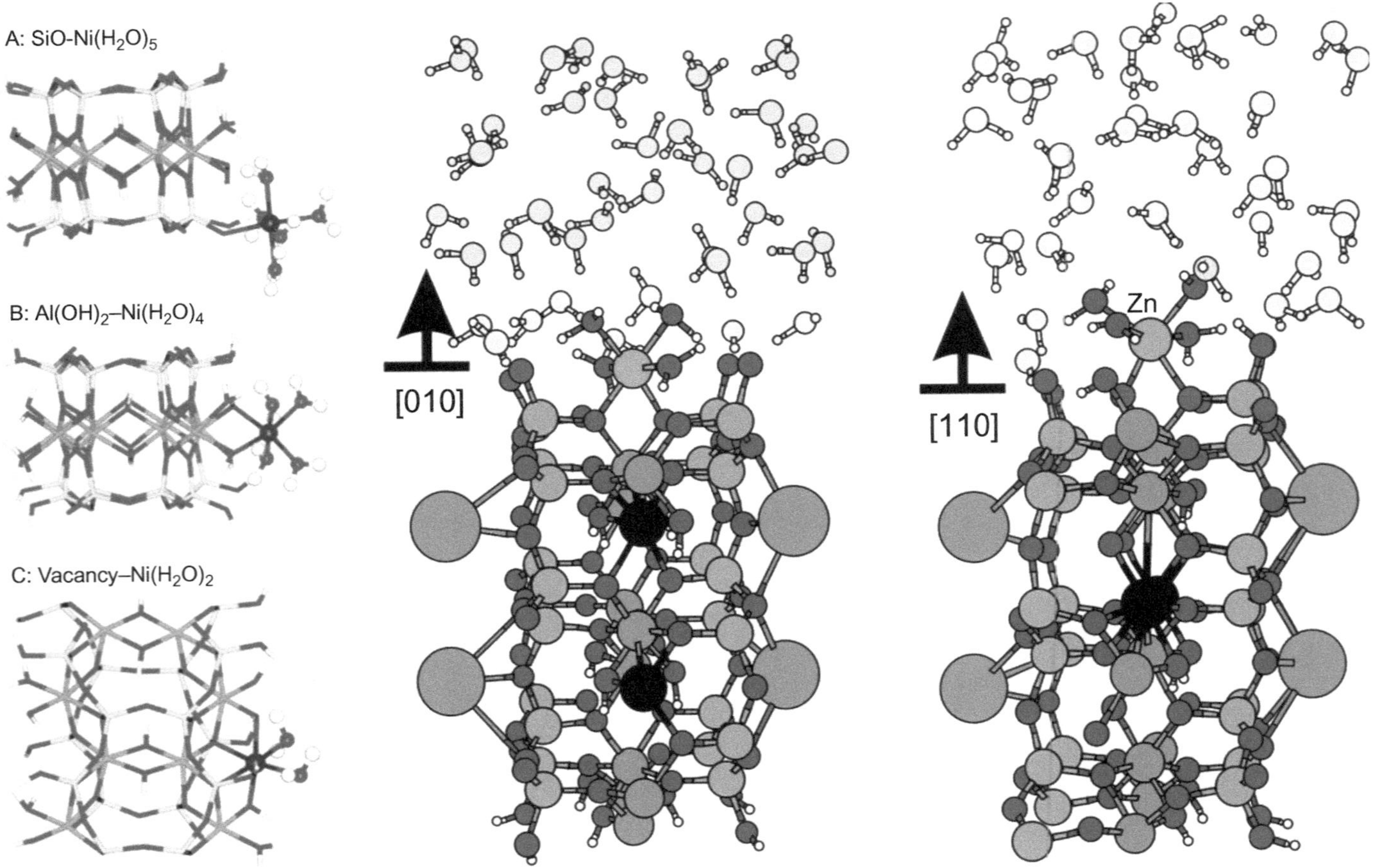

Fig. 3.6 See legend on opposite page.

significantly higher free energies than the other two sites. Other metal cations such as Co^{2+} and Cu^{2+} form similar complexes and demonstrate strong affinity to the vacancy sites (Zhang et al., 2017). The same complexation mechanism holds at elevated temperatures. The vacancy site can further accept polynuclear complexes that are stable at least up to 423 K (Zhang et al., 2018). According to this result, vacancy site and mono/bidentate sites on $\equiv SiO^-/\equiv Al(OH)_2$ sites were suggested to represent the strong site and weak site in the 2-site protolysis non-electrostatic surface complexation and cation exchange (2SPNE SC/CE) model (Bradbury and Baeyens, 1997), respectively. A similar conclusion was made earlier based on the comparative analysis of calculated and measured EXAFS spectra for Zn^{2+} (Churakov and Daehn, 2012).

Adsorption of actinides on basal and edge surfaces of clay minerals has been explored by several DFT studies (Kremleva et al., 2008, 2010, 2011, 2012, 2015a, b; Martorell et al., 2010; Kremleva and Krüger, 2016). The adsorption of uranyl on kaolinite basal surfaces was studied by static DFT calculations (Kremleva et al., 2008; Martorell et al., 2010). The structures of possible inner- and outer-sphere complexes were optimized, and the formation energies were calculated. Uranyl adsorption on edge surfaces was studied with regard to both kaolinite and 2:1 clay minerals (Kremleva et al., 2011, 2012, 2015a,b). It was found that the structure of the adsorption complexes is controlled by the availability of specific surface groups, and not by the types of clay minerals. Furthermore, various bidentate complexes were found to coexist on the same edge surfaces. Geometry optimization studies also showed that some complexes hydrolyse spontaneously. By comparison with spectroscopic experiments, it was concluded that $\equiv$AlOHOH and $\equiv$AlOHSiO sites of (010) surfaces and $\equiv AlO_mOH$, $\equiv SiOO_m$, and $\equiv$AlOHSiO of (110) surfaces have the largest affinity for surface complexation. The structure of corresponding surface complexes is illustrated in Fig. 3.8. Energy of different Np(V) monodentate and bidentate complexes on edge surfaces with different coordination numbers was explored by DFT (Kremleva and Krüger, 2016). It was suggested that the monodentate Np(V) complex with a coordination number of four is more stable, and these complexes show a lower tendency to form monohydroxides.

Fig. 3.6 *Left*: Structure of energetically favourable Ni^{2+} complexes adsorbed on (010) edge of pyrophyllite. *Right*: Structure of bidentate Zn^{2+} complexes on (010) and (110) edges of Mt. Some water molecules in the interlayer space are shaded or removed to improve visibility (Churakov and Daehn, 2012).
Left: After Zhang, C., Liu, X.D., Lu, X.C., He, M.J., Meijer, E.J., Wang, R.C., 2017. Surface complexation of heavy metal cations on clay edges: insights from first principles molecular dynamics simulation of Ni(II). Geochim. Cosmochim. Acta 203, 54–68, adopted with permission Copyright (2017) Elsevier. *Right*: Adapted with permission from Churakov, S.V., Daehn, R., 2012. Zinc adsorption on clays inferred from atomistic simulations and EXAFS spectroscopy. Environ. Sci. Technol. 46, 5713–5719. Copyright (2012) American Chemical Society.

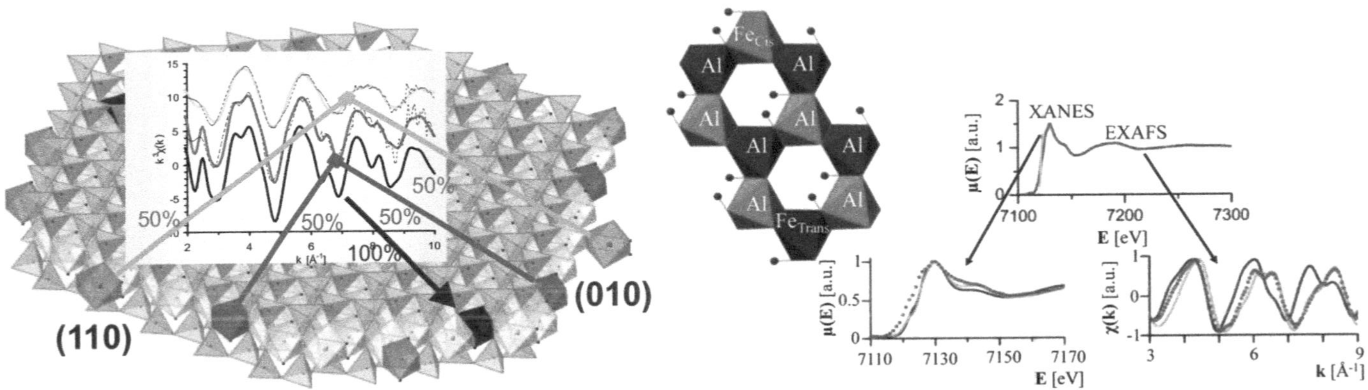

Fig. 3.7 Combined use of ab initio simulations and theoretical modelling of XAS spectra provide a link to molecular adsorption mechanism and experimental studies of metal uptake by clay minerals. *Left*: Simulations of inner-sphere complexes of Zn at (010) and (110) surface of Mt. *Right*: XANES and EXAFS spectra of structural Fe in Mt resolved from ab initio simulations.
Adapted with permission from Churakov, S.V., Daehn, R., 2012. Zinc adsorption on clays inferred from atomistic simulations and exafs spectroscopy. Environ. Sci. Technol. 46, 5713–5719 and Keri, A., Dahn, R., Krack, M., Churakov, S.V., 2017. Combined XAFS spectroscopy and ab initio study on the characterization of iron incorporation by Montmorillonite. Environ. Sci. Technol. 51, 10585–10594. Copyright (2017) American Chemical Society.

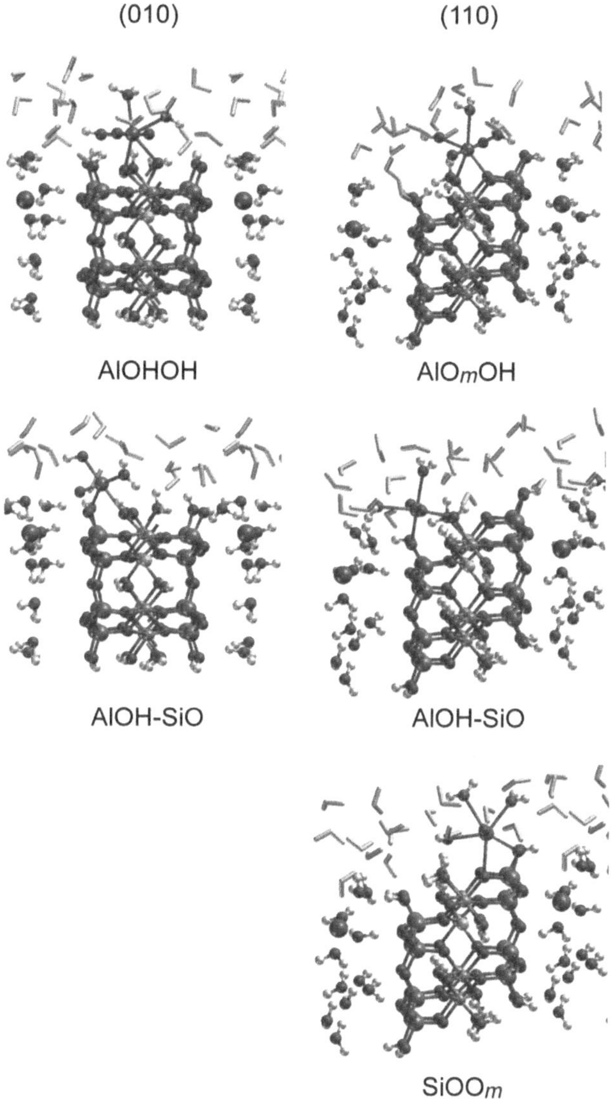

Fig. 3.8 Structure of energetically favourable UO_2^{2+} complexes on edges of Mt.
After Kremleva, A., Kruger, S., Rosch, N., 2015a. Uranyl adsorption at solvated edge surfaces of 2:1 smectites. A density functional study. Phys. Chem. Chem. Phys. 17, 13757–13768. Reprinted with permission. Copyright © 2015, Royal Society of Chemistry.

3.7 Outlook

The time scale and system size are two major factors limiting the application of ab initio methods to complex natural systems. Most recent simulations consider systems with up to 1000 atoms and efficient parallelisation algorithms allow obtaining continuous trajectories as long as several 100 ps. In combination with the advanced sampling techniques such as umbrella sampling and metadynamics, it has become possible to investigate thermodynamics of chemical reactions at the surface of clay minerals.

Recent developments of the numerical algorithms and codes made feasible simulations of periodic systems using hybrid density functionals (Guidon et al., 2009), as well as post-HF methods for electron correlation, such as MP2 (Del Ben et al., 2015), and random phase approximation (Wilhelm et al., 2016). A wider application of these techniques will further improve the accuracy and predictive power of the QM simulations. Use of these advanced levels of theory is essential for the description of activation energies of bond breaking/formation, and the ET reactions.

The advances in QM simulations will further stimulate application of the multiscale modelling and up-scaling to clay mineral systems. QM calculations provide a reference for classical force field developments. The results from classical MD simulations can be incorporated in the advanced multiscale models for ion transport (Churakov et al., 2014b). Further, the results of QM calculations can feed semiempirical thermodynamic adsorption models. For example, a macroscopic surface complexation model using ab initio derived pK_a could successfully reproduce the titration curves (Tournassat et al., 2016). More advanced multiscale models have already been applied to pore scale simulation of ion adsorption in clay minerals and hydrous calcium–aluminium silicates (Delhorme et al., 2010; Churakov et al., 2014a; Churakov and Labbez, 2017).

Rare-events sampling methods such as constrained MD and metadynamics have been widely employed to investigate the thermodynamics in clay mineral sciences. However, the kinetics have been rarely explored. Accurate estimations of activation barriers for surface reactions are essential for the mechanistic modelling of clay mineral dissolution and growth. So far, such studies have relied on the data from gas phase cluster calculations (Kurganskaya and Luttge, 2013). The transition path sampling and metadynamics simulations with high accuracy methods (e.g. hybrid functionals, post-HF methods) under consideration of water are extremely time consuming. Such studies will become increasingly feasible as computing power increases.

References

Adamo, C., Barone, V., 1999. Toward reliable density functional methods without adjustable parameters: the PBE0 model. J. Chem. Phys. 110, 6158.

Alexandrov, V., Rosso, K.M., 2013. Insights into the mechanism of Fe(II) adsorption and oxidation at Fe-clay mineral surfaces from first-principles calculations. J. Phys. Chem. C 117, 22880–22886.

Alexandrov, V., Neumann, A., Scherer, M.M., Rosso, K.M., 2013. Electron exchange and conduction in nontronite from first-principles. J. Phys. Chem. C 117, 2032–2040.

Alvim, R.S., Miranda, C.R., 2015. First principles characterization of silicate sites in clay surfaces. Phys. Chem. Chem. Phys. 17, 4952–4960.

Anisimov, V.I., Aryasetiawan, F., Lichtenstein, A.I., 1997. First-principles calculations of the electronic structure and spectra of strongly correlated systems: the LDA+U method. J. Phys. Condens. Matter 9, 767–808.

Balan, E., Saitta, A.M., Mauri, F., Calas, G., 2001. First-principles modeling of the infrared spectrum of kaolinite. Am. Mineral. 86, 1321–1330.

Balan, E., Lazzeri, M., Delattre, S., Meheut, M., Refson, K., Winkler, B., 2007. Anharmonicity of inner-OH stretching modes in hydrous phyllosilicates: assessment from first-principles frozen-phonon calculations. Phys. Chem. Miner. 34, 621–625.

Becke, A.D., 1988. Density-functional exchange-energy approximation with correct asymptotic behavior. Phys. Rev. A 38, 3098–3100.

Benco, L., Tunega, D., 2009. Adsorption of H2O, NH3 and C6H6 on alkali metal cations in internal surface of mordenite and in external surface of smectite: a DFT study. Phys. Chem. Miner. 36, 281–290.

Benco, L., Tunega, D., Hafner, J., Lischka, H., 2001a. Ab initio density functional theory applied to the structure and proton dynamics of clays. Chem. Phys. Lett. 333, 479–484.

Benco, L., Tunega, D., Hafner, J., Lischka, H., 2001b. Upper limit of the O-H···O hydrogen bond. Ab initio study of the kaolinite structure. J. Phys. Chem. B 105, 10812–10817.

Berghout, A., Tunega, D., Zaoui, A., 2010. Density functional theory (DFT) study of the hydration steps of Na+/Mg2+/Ca2+/Sr2+/Ba2+-exchanged montmorillonite. Clays Clay Miner. 58, 174–187.

Bickmore, B.R., Rosso, K.M., Nagy, K.L., Cygan, R.T., Tadanier, C.J., 2003. Ab initio determination of edge surface structures for dioctahedral 2:1 phyllosilicates: implications for acid-base reactivity. Clays Clay Miner. 51, 359–371.

Bocquet, M.L., Coasne, B., Pellenq, R., Ulm, F.J., 2015. Organic-clay interfacial chemical bonds probed by ab initio calculations. J. Phys. Chem. C 119, 6511–6517.

Boek, E.S., Sprik, M., 2003. Ab initio molecular dynamics study of the hydration of a sodium smectite clay. J. Phys. Chem. B 107, 3251–3256.

Botella, V., Timon, V., Escamilla-Roa, E., Hernandez-Languna, A., Sainz-Diaz, C.I., 2004. Hydrogen bonding and vibrational properties of hydroxy groups in the crystal lattice of dioctahedral clay minerals by means of first principles calculations. Phys. Chem. Miner. 31, 475–486.

Bradbury, M.H., Baeyens, B., 1997. A mechanistic description of Ni and Zn sorption on Na-montmorillonite part II: modelling. J. Contam. Hydrol. 27, 223–248.

Briones-Jurado, C., Agacino-Valdes, E., 2009. Bronsted sites on acid-treated montmorillonite: a theoretical study with probe molecules. J. Phys. Chem. A 113, 8994–9001.

Campos, R.B., Wypych, F., Martins, H.P., 2009. Theoretical estimates of the IR spectrum of water intercalated into kaolinite. Int. J. Quantum Chem. 109, 594–604.

Carvalho, A.J.P., Dordio, A.V., Ramalho, J.P.P., 2014. A DFT study on the adsorption of benzodiazepines to vermiculite surfaces. J. Mol. Model. 20, 2336–2344.

Castro, E.A.S., Martins, J.B.L., 2005a. Theoretical study of benzene interaction on kaolinite. J. Comput. Aided Mater. Des. 12, 121–129.

Castro, E.A.S., Martins, J.B.L., 2005b. Theoretical study of kaolinite. Int. J. Quantum Chem. 103, 550–556.

Castro, E.A.S., Gargano, R., Martins, J.B.L., 2012. Benzene-kaolinite interaction properties. Int. J. Quantum Chem. 112, 2828–2831.

Chatterjee, A., Ebina, T., Iwasaki, T., Mizukami, F., 2003. Intermolecular reactivity study to scale adsorption property of para- and meta-substituted nitrobenzene over 2:1 dioctahedral smectite. J. Chem. Phys. 118, 10212–10220.

Chatterjee, A., Ebina, T., Onodera, Y., Mizukami, F., 2004. Effect of exchangeable cation on the swelling property of 2:1 dioctahedral smectite—a periodic first principle study. J. Chem. Phys. 120, 3414–3424.

Chen, B., Evans, J.R.G., Greenwell, H.C., Boulet, P., Coveney, P.V., Bowden, A.A., Whiting, A., 2008. A critical appraisal of polymer-clay nanocomposites. Chem. Soc. Rev. 37, 568–594.

Chen, J., Min, F.F., Liu, L.Y., Liu, C.F., Lu, F.Q., 2017. Experimental investigation and DFT calculation of different amine/ammonium salts adsorption on kaolinite. Appl. Surf. Sci. 419, 241–251.

Churakov, S.V., 2006. Ab initio study of sorption on pyrophyllite: structure and acidity of the edge sites. J. Phys. Chem. B 110, 4135–4146.

Churakov, S.V., 2007. Structure and dynamics of the water films confined between edges of pyrophyllite: a first principle study. Geochim. Cosmochim. Acta 71, 1130–1144.

Churakov, S.V., Daehn, R., 2012. Zinc adsorption on clays inferred from atomistic simulations and EXAFS spectroscopy. Environ. Sci. Tech. 46, 5713–5719.

Churakov, S.V., Kosakowski, G., 2010. An ab initio molecular dynamics study of hydronium complexation in Na-montmorillonite. Philos. Mag. 90, 2459–2474.

Churakov, S.V., Labbez, C., 2017. Thermodynamics and molecular mechanism of Al incorporation in calcium silicate hydrates. J. Phys. Chem. C 121, 4412–4419.

Churakov, S.V., Labbez, C., Pegado, L., Sulpizi, M., 2014a. Intrinsic acidity of surface sites in calcium-silicate-hydrates and its implication to their electrokinetic properties. J. Phys. Chem. C 118, 11752–11762.

Churakov, S.V., Gimmi, T., Unruh, T., Loon, L.R.V., Juranyi, F., 2014b. Resolving diffusion in clay minerals at different time scales: combination of experimental and modeling approaches. Appl. Clay Sci. 96, 36–44.

Clausen, P., Andreoni, W., Curioni, A., Hughes, E., Plummer, C.J.G., 2009a. Adsorption of low-molecular-weight molecules on a dry clay surface: an ab initio study. J. Phys. Chem. C 113, 12293–12300.

Clausen, P., Andreoni, W., Curioni, A., Hughes, E., Plummer, C.J.G., 2009b. Water adsorption at a sodium smectite clay surface: an ab initio study of the first stage. J. Phys. Chem. C 113, 15218–15225.

Del Ben, M., Hutter, J., VandeVondele, J., 2015. Probing the structural and dynamical properties of liquid water with models including non-local electron correlation. J. Chem. Phys. 143, 054506. 054501-054512.

Delhorme, M., Labbez, C., Caillet, C., Thomas, F., 2010. Acid-base properties of 2:1 clays. I. Modeling the role of electrostatics. Langmuir 26, 9240–9249.

Demontis, P., Masia, M., Suffritti, G.B., 2013. Water nanoconfined in clays: the structure of Na vermiculite revisited by ab initio simulations. J. Phys. Chem. C 117, 15583–15592.

Demontis, P., Masia, M., Suffritti, G.B., 2014. Peculiar structure of water in slightly super-hydrated vermiculite clay studied by Car-Parrinello molecular dynamics simulations. J. Phys. Chem. C 118, 7923–7931.

Dolg, M., Cao, X., 2012. Relativistic pseudopotentials: their development and scope of applications. Chem. Rev. 112, 403–480.

Ebina, T., Iwasaki, T., Onodera, Y., Chatterjee, A., 1999. A comparative study of DFT and XPS with reference to the adsorption of caesium ions in smectites. Comput. Mater. Sci. 14, 254–260.

Escamilla-Roa, E., Nieto, F., Sainz-Diaz, C.I., 2016. Stability of hydronium cation in the structure of illite. Clays Clay Miner. 64, 413–424.
Feibelman, P.J., 2013. Viscosity of ultrathin water films confined between aluminol surfaces of kaolinite: ab initio simulations. J. Phys. Chem. C 117, 6088–6095.
Fonseca, C.G., Vaiss, V.S., Wypych, F., Diniz, R., Leitao, A.A., 2017. Structural and thermodynamic investigation of the hydration-dehydration process of Na (+)-Montmorillonite using DFT calculations. Appl. Clay Sci. 143, 212–219.
Foresman, J.B., Frisch, A., 1996. Exploring Chemistry With Electronic Structure Methods: A Guide to Using Gaussian. Gaussian Inc.
Franco, F., Perez-Maqueda, L.A., Perez-Rodriguez, J.L., 2004. The effect of ultrasound on the particle size and structural disorder of a well-ordered kaolinite. J. Colloid Interface Sci. 274, 107–117.
Furche, F., Perdew, J.P., 2006. The performance of semilocal and hybrid density functionals in 3d transition-metal chemistry. J. Chem. Phys. 124, 044103.
Geatches, D.L., Wilcox, J., 2014. Ab initio investigations of dioctahedral interlayer-deficient mica: modelling 1M polymorphs of illite found within gas shale. Eur. J. Mineral. 26, 127–144.
Geatches, D.L., Greenwell, H.C., Clark, S.J., 2011. Ab initio transition state searching in complex systems: fatty acid decarboxylation in minerals. Chem. A Eur. J. 115, 2658–2667.
Geatches, D.L., Jacquet, A., Clark, S.J., Greenwell, H.C., 2012. Monomer adsorption on kaolinite: modeling the essential ingredients. J. Phys. Chem. C 116, 22365–22374.
Geatches, D., McCarty, D., Wilcox, J., 2014. Ab initio investigations of dioctahedral interlayer-deficient mica: modeling particles of illite found within gas shale. Am. Mineral. 99, 1962–1972.
Goedecker, S., Teter, M., Hutter, J., 1996. Separable dual-space Gaussian pseudopotentials. Phys. Rev. B 54, 1703–1710.
Gorb, L., Lutchyn, R., Zub, Y., Leszczynska, D., Leszczynski, J., 2006. The origin of the interaction of 1,3,5-trinitrobenzene with siloxane surface of clay minerals. J. Mol. Struct. (THEOCHEM) 766, 151–157.
Greathouse, J.A., Geatches, D.L., Pike, D.Q., Greenwell, H.C., Johnston, C.T., Wilcox, J., Cygan, R.T., 2015. Methylene blue adsorption on the basal surfaces of kaolinite: structure and thermodynamics from quantum and classical molecular simulation. Clays Clay Miner. 63, 185–198.
Greathouse, J.A., Cygan, R.T., Fredrich, J.T., Jerauld, G.R., 2017. Adsorption of aqueous crude oil components on the basal surfaces of clay minerals: molecular simulations including salinity and temperature effects. J. Phys. Chem. C 121, 22773–22786.
Grimme, S., Hansen, A., Brandenburg, J.G., Bannwarth, C., 2016. Dispersion-corrected mean-field electronic structure methods. Chem. Rev. 116, 5105–5154.
Gu, C., Liu, C., Johnston, C.T., Teppen, B.J., Li, H., Boyd, S.A., 2011. Pentachlorophenol radical cations generated on Fe(III)-montmorillonite initiate octachlorodibenzo-p-dioxin formation in clays: density functional theory and Fourier transform infrared studies. Environ. Sci. Technol. 45, 1399–1406.
Guidon, M., Hutter, J., VandeVondele, J., 2009. Robust periodic Hartree-Fock exchange for large-scale simulations using Gaussian basis sets. J. Chem. Theory Comput. 5, 3010–3021.
Han, Y.H., Liu, W.L., Zhou, J., Chen, J.H., 2016a. Interactions between kaolinite Al-OH surface and sodium hexametaphosphate. Appl. Surf. Sci. 387, 759–765.
Han, Y.H., Liu, W.L., Chen, J.H., 2016b. DFT simulation of the adsorption of sodium silicate species on kaolinite surfaces. Appl. Surf. Sci. 370, 403–409.

He, M.C., Zhao, J., Li, Y., 2014. First principles ab initio study of CO2 adsorption on the kaolinite (001) surface. Clays Clay Miner. 62, 153–160.

Hernandez-Haro, N., Ortega-Castro, J., Pruneda, M., Sainz-Diaz, C.I., Hernandez-Laguna, A., 2014. Theoretical study on the influence of the Mg2+ and Al3+ octahedral cations on the vibrational spectra of the hydroxy groups of dioctahedral 2:1 phyllosilicate models. J. Mol. Model. 20, 2402–2410.

Hill, R., 1952. The elastic behaviour of a crystaline aggregate. Proc. Phys. Soc. Lond. 65A, 349–354.

Hobbs, J.D., Cygan, R.T., Nagy, K.L., Schultz, P.A., Sears, M.P., 1997. All-atom ab initio energy minimization of the kaolinite crystal structure. Am. Mineral. 82, 657–662.

Hohenberg, P., Kohn, W., 1964. Inhomogeneous electron gas. Phys. Rev. 136, 864–871.

Hu, X.L., Michaelides, A., 2008. Water on the hydroxylated (001) surface of kaolinite: from monomer adsorption to a flat 2D wetting layer. Surf. Sci. 602, 960–974.

Hubbard, J., 1963. Electron correlations in narrow energy bands. Proc. R. Soc. Lond. A Math. Phys. Sci. 285, 238–257.

Johansson, U., Frost, R.L., Forsling, W., Kloprogge, J.T., 1998. Raman spectroscopy of the kaolinite hydroxyls at 77 K. Appl. Spectrosc. 52, 1277–1282.

Kat, F.A., Toby, O.H.W., Arnaud, M., Steve, C.P., Emilio, A., Martin, T.D., 2008. Electrostatic versus polarization effects in the adsorption of aromatic molecules of varied polarity on an insulating hydrophobic surface. J. Phys. Condens. Matter. 20, 035215.

Keri, A., Dahn, R., Krack, M., Churakov, S.V., 2017. Combined XAFS spectroscopy and ab initio study on the characterization of iron incorporation by montmorillonite. Environ. Sci. Technol. 51, 10585–10594.

Kohn, W., Sham, L.J., 1965. Self-consistent equation including exchange and correlation effects. Phys. Rev. 140, A1133–A1138.

Kremleva, A., Krüger, S., 2016. Comparative computational study of Np (V) and U (VI) adsorption on (110) edge surfaces of montmorillonite. Clays Clay Miner. 64, 438–451.

Kremleva, A., Krüger, S., Rösch, N., 2008. Density functional model studies of uranyl adsorption on (001) surfaces of kaolinite. Langmuir 24, 9515–9524.

Kremleva, A., Krüger, S., Rösch, N., 2010. Quantum chemical modeling of uranyl adsorption on mineral surfaces. Radiochim. Acta 98, 635–646.

Kremleva, A., Kruger, S., Rosch, N., 2011. Uranyl adsorption at (010) edge surfaces of kaolinite: a density functional study. Geochim. Cosmochim. Acta 75, 706–718.

Kremleva, A., Martorell, B., Kruger, S., Rosch, N., 2012. Uranyl adsorption on solvated edge surfaces of pyrophyllite: a DFT model study. Phys. Chem. Chem. Phys. 14, 5815–5823.

Kremleva, A., Kruger, S., Rosch, N., 2015a. Uranyl adsorption at solvated edge surfaces of 2:1 smectites. A density functional study. Phys. Chem. Chem. Phys. 17, 13757–13768.

Kremleva, A., Krüger, S., Rösch, N., 2015b. Toward a reliable energetics of adsorption at solvated mineral surfaces: a computational study of uranyl (VI) on 2:1 clay minerals. J. Phys. Chem. C 120, 324–335.

Kubicki, J.D., Itoh, M.J., Schroeter, L.M., Apitz, S.E., 1997. Bonding mechanisms of salicylic acid adsorbed onto illite clay: an ATR–FTIR and molecular orbital study. Environ. Sci. Technol. 31, 1151–1156.

Kuo, I.-F.W., Mundy, C.J., McGrath, M.J., Siepmann, J.I., VandeVondele, J., Sprik, M., Hutter, J., Chen, B., Klein, M.L., Mohamed, F., Krack, M., Parrinello, M., 2004. Liquid water from first principles: investigation of different sampling approaches. J. Phys. Chem. B 108, 12990–12998.

Kurganskaya, I., Luttge, A., 2013. A comprehensive stochastic model of phyllosilicate dissolution: structure and kinematics of etch pits formed on muscovite basal face. Geochim. Cosmochim. Acta 120, 545–560.

Lee, C., Yang, W., Parr, R.G., 1988. Development of the Colle-Salvetti correlation-energy formula into a functional of the electron density. Phys. Rev. B 37, 785–789.

Li, Y., Evans, J.N.S., 1995. The Fukui function: a key concept linking frontier molecular orbital theory and the hard-soft-acid-base principle. J. Am. Chem. Soc. 117, 7756–7759.

Li, H.T., Kang, T.H., Zhang, B., Zhang, J.J., Ren, J., 2016. Influence of interlayer cations on structural properties of montmorillonites: a dispersion-corrected density functional theory study. Comput. Mater. Sci. 117, 33–39.

Liu, X., Lu, X., Wang, R., Zhou, H., Xu, S., 2008. Surface complexes of acetate on edge surfaces of 2:1 type phyllosilicate: insights from density functional theory calculation. Geochim. Cosmochim. Acta 72, 5896–5907.

Liu, X.D., Meijer, E.J., Lu, X.C., Wang, R.C., 2010. Ab initio molecular dynamics study of Fe-containing smectites. Clays Clay Miner. 58, 89–96.

Liu, X.D., Lu, X.C., Wang, R.C., Meijer, E.J., Zhou, H.Q., 2011. Acidities of confined water in interlayer space of clay minerals. Geochim. Cosmochim. Acta 75, 4978–4986.

Liu, X., Jan Meijer, E., Lu, X., Wang, R., 2012a. First-principles molecular dynamics insight into Fe2+ complexes adsorbed on edge surfaces of clay minerals. Clays Clay Miner. 60, 341–347.

Liu, X., Lu, X., Meijer, E.J., Wang, R., Zhou, H., 2012b. Atomic-scale structures of interfaces between phyllosilicate edges and water. Geochim. Cosmochim. Acta 81, 56–68.

Liu, X.D., Lu, X.C., Wang, R.C., Meijer, E.J., Zhou, H.Q., He, H.P., 2012c. Atomic scale structures of interfaces between kaolinite edges and water. Geochim. Cosmochim. Acta 92, 233–242.

Liu, X., Lu, X., Sprik, M., Cheng, J., Meijer, E.J., Wang, R., 2013. Acidity of edge surface sites of montmorillonite and kaolinite. Geochim. Cosmochim. Acta 117, 180–190.

Liu, X., Cheng, J., Sprik, M., Lu, X., Wang, R., 2014. Surface acidity of 2:1-type dioctahedral clay minerals from first principles molecular dynamics simulations. Geochim. Cosmochim. Acta 140, 410–417.

Liu, X., Cheng, J., Sprik, M., Lu, X., Wang, R., 2015a. Interfacial structures and acidity of edge surfaces of ferruginous smectites. Geochim. Cosmochim. Acta 168, 293–301.

Liu, X., Lu, X., Cheng, J., Sprik, M., Wang, R., 2015b. Temperature dependence of interfacial structures and acidity of clay edge surfaces. Geochim. Cosmochim. Acta 160, 91–99.

Liu, X., Lu, X., Zhang, Y., Zhang, C., Wang, R., 2017. Complexation of carboxylate on smectites surfaces. Phys. Chem. Chem. Phys. 19, 18400–18406.

Lou, Z., Liu, H., Zhang, Y., Meng, Y., Zeng, Q., Shi, J., Yang, M., 2014. A density functional theory study of the hydration of calcium ions confined in the interlayer space of montmorillonites. J. Theor. Comput. Chem. 13, 1450028.

Martorell, B., Kremleva, A., Krüger, S., Rösch, N., 2010. Density functional model study of uranyl adsorption on the solvated (001) surface of kaolinite. J. Phys. Chem. C 114, 13287–13294.

Mercier, P.H.J., Le Page, Y., 2008. Kaolin polytypes revisited ab initio. Acta Crystallogr. B 64, 131–143.

Mercier, P.H.J., Le Page, Y., 2009. Ab initio exploration of layer slipping transformations in kaolinite up to 60 GPa. Mater. Sci. Technol. 25, 437–442.

Michalkova, A., Tunega, D., Nagy, L.T., 2002. Theoretical study of interactions of dickite and kaolinite with small organic molecules. J. Mol. Struct. THEOCHEM 581, 37–49.

Michalkova, A., Robinson, T., Leszczynski, J., 2011. Adsorption of thymine and uracil on 1:1 clay mineral surfaces: comprehensive ab initio study on influence of sodium cation and water. Phys. Chem. Chem. Phys. 13, 7862–7881.

Mignon, P., Ugliengo, P., Sodupe, M., 2009. Theoretical study of the adsorption of RNA/DNA bases on the external surfaces of Na+-montmorillonite. J. Phys. Chem. C 113, 13741–13749.

Mignon, P., Ugliengo, P., Sodupe, M., Hernandez, E.R., 2010. Ab initio molecular dynamics study of the hydration of Li+, Na+ and K+ in a montmorillonite model. Influence of isomorphic substitution. Phys. Chem. Chem. Phys. 12, 688–697.

Militzer, B., Wenk, H.R., Stackhouse, S., Stixrude, L., 2011. First-principles calculation of the elastic moduli of sheet silicates and their application to shale anisotropy. Am. Mineral. 96, 125–137.

Minisini, B., Tsobnang, F., 2005. Ab initio comparative study of montmorillonite structural models. Appl. Surf. Sci. 242, 21–28.

Mrayed, S.M., Zeng, Q.H., Yu, A.B., 2009. Interfacial interactions in clay-based nylon 6 nanocomposites: a density functional theory study. Comput. Mater. Sci. 46, 942–949.

Munoz-Santiburcio, D., Ortega-Castro, J., Sainz-Diaz, C.I., Huertas, F.J., Hernandez-Laguna, A., 2009. Theoretical study of the adsorption of 2-nitro-1-propanol on smectite surface models. J. Mol. Struct. THEOCHEM 912, 95–104.

Myshakin, E.M., Saidi, W.A., Romanov, V.N., Cygan, R.T., Jordan, K.D., 2013. Molecular dynamics simulations of carbon dioxide intercalation in hydrated Na-montmorillonite. J. Phys. Chem. C 117, 11028–11039.

Okumura, M., Sassi, M., Rosso, K.M., Machida, M., 2017. Origin of 6-fold coordinated aluminum at (010)-type pyrophyllite edges. AIP Adv. 7, 055211–055219.

Ortega-Castro, J., Hernandez-Haro, N., Hernandez-Laguna, A., Sainz-Diaz, C.I., 2008. DFT calculation of crystallographic properties of dioctahedral 2:1 phyllosilicates. Clay Miner. 43, 351–361.

Ortega-Castro, J., Hernandez-Haro, N., Munoz-Santiburcio, D., Hernandez-Laguna, A., Sainz-Diaz, C.I., 2009. Crystal structure and hydroxyl group vibrational frequencies of phyllosilicates by DFT methods. J. Mol. Struct. THEOCHEM 912, 82–87.

Pelmenschikov, A., Leszczynski, J., 1999. Adsorption of 1, 3, 5-trinitrobenzene on the siloxane sites of clay minerals: ab initio calculations of molecular models. J. Phys. Chem. B. 103, 6886–6890.

Peng, C.L., Min, F.F., Liu, L.Y., Chen, J., 2016. Periodic DFT study of adsorption of water on sodium-montmorillonite (001) basal and (010) edge surface. Appl. Surf. Sci. 387, 308–316.

Peng, C.L., Min, F.F., Liu, L.Y., Jun, C., 2017. The adsorption of CaOH+ on (001) basal and (010) edge surface of Na-montmorillonite: a DFT study. Surf. Interface Anal. 49, 267–277.

Perdew, J.P., Wang, Y., 1992. Accurate and simple analytic representation of the electron-gas correlation energy. Phys. Rev. B 45, 13244–13249.

Perdew, J.P., Burke, K., Ernzerhof, M., 1996. Generalized gradient approximation made simple. Phys. Rev. Lett. 77, 3865–3868.

Peverati, R., Truhlar, D.G., 2012. Quest for a universal density functional: the accuracy of density functionals across a broad spectrum of databases in chemistry and physics. Philos. Trans. R. Soc. A 372, 20120476.

Pirillo, S., Luna, C.R., Lopez-Corral, I., Juan, A., Avena, M.J., 2015. Geometrical and electronic properties of hydrated sodium montmorillonite and tetracycline montmorillonite from DFT calculations. J. Phys. Chem. C 119, 16082–16088.

Ramalho, J.P.P., Dordio, A.V., Carvalho, A.J.P., 2013. Adsorption of two phenoxyacid compounds on a clay surface: a theoretical study. Adsorption 19, 937–944.
Refson, K., Park, S.H., Sposito, G., 2003. Ab initio computational crystallography of 2:1 clay minerals: 1. Pyrophyllite-1Tc. J. Phys. Chem. B 107, 13376–13383.
Rosso, K.M., Rustad, J.R., Bylaska, E.J., 2001. The Cs/K exchange in muscovite interlayers: an ab initio treatment. Clays Clay Miner. 49, 500–513.
Ruiz Pestana, L., Kolluri, K., Head-Gordon, T., Lammers, L.N., 2016. Direct exchange mechanism for interlayer ions in non-swelling clays. Environ. Sci. Technol. 51, 393–400.
Sainz-Diaz, C.I., Timon, V., Botella, V., Artacho, E., Hernandez-Laguna, A., 2002. Quantum mechanical calculations of dioctahedral 2:1 phyllosilicates: effect of octahedral cation distributions in pyrophyllite, illite, and smectite. Am. Mineral. 87, 958–965.
Sainz-Diaz, C.I., Francisco-Marquez, M., Vivier-Bunge, A., 2010. Molecular structure and spectroscopic properties of polyaromatic heterocycles by first principle calculations: spectroscopic shifts with the adsorption of thiophene on phyllosilicate surface. Theor. Chem. Acc. 125, 83–95.
Sato, H., Ono, K., Johnston, C.T., Yamagishi, A., 2004. First-principle study of polytype structures of 1:1 dioctahedral phyllosilicates. Am. Mineral. 89, 1581–1585.
Sato, H., Ono, K., Johnston, C.T., Yamagishi, A., 2005. First-principles studies on the elastic constants of a 1:1 layered kaolinite mineral. Am. Mineral. 90, 1824–1826.
Schaef, H.T., Glezakou, V.A., Owen, A.T., Ramprasad, S., Martin, P.F., McGrail, B.P., 2014. Surface condensation of CO2 onto kaolinite. Environ. Sci. Technol. Lett. 1, 142–145.
Scholtzová, E., Madejová, J., Jankovič, L.u., Tunega, D., 2016. Structural and spectroscopic characterization of montmorillonite intercalated with N-butylammonium cations (N = 1–4)—modeling and experimental study. Clays Clay Miner. 64, 401.
Scott, A.M., Burns, E.A., Lafferty, B.J., Hill, F.C., 2015. Theoretical predictions of thermodynamic parameters of adsorption of nitrogen containing environmental contaminants on kaolinite. J. Mol. Model. 21, 21.
Shapley, T.V., Molinari, M., Zhu, R.L., Parker, S.C., 2013. Atomistic modeling of the sorption free energy of dioxins at clay-water interfaces. J. Phys. Chem. C 117, 24975–24984.
Shi, J., Liu, H.B., Lou, Z.Y., Zhang, Y., Meng, Y.F., Zeng, Q., Yang, M.L., 2013. Effect of interlayer counterions on the structures of dry montmorillonites with Si4+/Al3+ substitution. Comput. Mater. Sci. 69, 95–99.
Solling, T.I., Budi, A., Mogensen, K., 2015. The competition between H2O and CO2 adhesion at reservoir conditions: a DFT study of simple mineral models and the entropy, ZPE, dispersion and T,P variations. Comput. Theor. Chem. 1073, 55–60.
Song, K.-h., Wang, X., Qian, P., Zhang, C., Zhang, Q., 2013. Theoretical study of interaction of formamide with kaolinite. Comput. Theor. Chem. 1020, 72–80.
Song, K.H., Zhong, M.J., Wang, L., Li, Y., Qian, P., 2014. Theoretical study of interaction of amide molecules with kaolinite. Comput. Theor. Chem. 1050, 58–67.
Srodon, J., McCarty, D.K., 2008. Surface area and layer charge of smectite from CEC and EGME/H2O-retention measurements. Clays Clay Miner. 56, 155–174.
Stackhouse, S., Coveney, P.V., Sandré, E., 2001. Plane-wave density functional theoretic study of formation of clay-polymer nanocomposite materials by self-catalyzed in situ intercalative polymerization. J. Am. Chem. Soc. 123, 11764–11774.
Stephens, P.J., Devlin, F.J., Chabalowski, C.F., Frisch, M.J., 1994. Ab-initio calculation of vibrational absorption and circular-dichroism spectra using density-functional force-fields. J. Phys. Chem. 98, 11623–11627.
Suter, J.L., Boek, E.S., Sprik, M., 2008. Adsorption of a sodium ion on a smectite clay from constrained ab initio molecular dynamics simulations. J. Phys. Chem. C 112, 18832–18839.

Suter, J.L., Sprik, M., Boek, E.S., 2012. Free energies of absorption of alkali ions onto beidellite and montmorillonite surfaces from constrained molecular dynamics simulations. Geochim. Cosmochim. Acta 91, 109–119.
Suter, J.L., Kabalan, L., Khader, M., Coveney, P.V., 2015. Ab initio molecular dynamics study of the interlayer and micropore structure of aqueous montmorillonite clays. Geochim. Cosmochim. Acta 169, 17–29.
Teppen, B.J., Yu, C.-H., Newton, S.Q., Miller, D.M., Schäfer, L., 2002. Quantum molecular dynamics simulations regarding the dechlorination of trichloro ethene in the interlayer space of the 2:1 clay mineral nontronite. Chem. A Eur. J. 106, 5498–5503.
Timon, V., Sainz-Diaz, G.I., Botella, V., Hernandez-Laguna, A., 2003. Isomorphous cation substitution in dioctahedral phyllosilicates by means of ab initio quantum mechanical calculations on clusters. Am. Mineral. 88, 1788–1795.
Tosoni, S., Doll, K., Ugliengo, P., 2006. Hydrogen bond in layered materials: structural and vibrational properties of kaolinite by a periodic B3LYP approach. Chem. Mater. 18, 2135–2143.
Tournassat, C., Davis, J.A., Chiaberge, C., Grangeon, S., Bourg, I.C., 2016. Modeling the acid-base properties of montmorillonite edge surfaces. Environ. Sci. Technol. 50, 13436–13445.
Troullier, N., Martins, J.L., 1991. Efficient pseudopotentials for plane-wave calculations. Phys. Rev. B 43, 1993–2006.
Tunega, D., Lischka, H., 2003. Effect of the Si/Al ordering on structural parameters and the energetic stabilization of vermiculites—a theoretical study. Phys. Chem. Miner. 30, 517–522.
Tunega, D., Benco, L., Haberhauer, G., Gerzabek, M.H., Lischka, H., 2002a. Ab initio molecular dynamics study of adsorption sites on the (001) surfaces of 1:1 dioctahedral clay minerals. J. Phys. Chem. B 106, 11515–11525.
Tunega, D., Haberhauer, G., Gerzabek, M.H., Lischka, H., 2002c. Theoretical study of adsorption sites on the (001) surfaces of 1:1 clay minerals. Langmuir 18, 139–147.
Tunega, D., Gerzabek, M.H., Lischka, H., 2004. Ab initio molecular dynamics study of a monomolecular water layer on octahedral and tetrahedral kaolinite surfaces. J. Phys. Chem. B 108, 5930–5936.
Tunega, D., Gerzabek, M.H., Haberhauer, G., Lischka, H., 2007a. Formation of 2,4-D complexes on montmorillonites—an ab initio molecular dynamics study. Eur. J. Soil Sci. 58, 680–691.
Tunega, D., Goodman, B.A., Haberhauer, G., Reichenauer, T.G., Gerzabek, M.H., Lischka, H., 2007b. Ab initio calculations of relative stabilities of different structural arrangements in dioctahedral phyllosilicates. Clays Clay Miner. 55, 220–232.
Tunega, D., Bucko, T., Zaoui, A., 2012. Assessment of ten DFT methods in predicting structures of sheet silicates: importance of dispersion corrections. J. Chem. Phys. 137, 114105–114109.
Ugliengo, P., Zicovich-Wilson, C.M., Tosoni, S., Civalleri, B., 2009. Role of dispersive interactions in layered materials: a periodic B3LYP and B3LYP-D* study of Mg(OH)(2), Ca(OH)(2) and kaolinite. J. Mater. Chem. 19, 2564–2572.
Vanderbilt, D., 1990. Soft self-consistent pseudopotentials in a generalized eigenvalue formalism. Phys. Rev. B 41, 7892–7895.
Voora, V.K., Al-Saidi, W.A., Jordan, K.D., 2011. Density functional theory study of pyrophyllite and M-Montmorillonites (M = Li, Na, K, Mg, and Ca): role of dispersion interactions. J. Phys. Chem. A 115, 9695–9703.
Wang, X., Qian, P., Song, K.H., Zhang, C., Dong, J., 2013. The DFT study of adsorption of 2,4-dinitrotoluene on kaolinite surfaces. Comput. Theor. Chem. 1025, 16–23.

Wang, J., Xia, S.W., Yu, L.M., 2014. Adsorption mechanism of hydrated Pb(OH)(+) on the kaolinite (001) surface. Acta Phys. Chim. Sin. 30, 829–835.

Wang, X.Y., Huang, Y.J., Pan, Z.G., Wang, Y.X., Liu, C.Q., 2015a. Theoretical investigation of lead vapor adsorption on kaolinite surfaces with DFT calculations. J. Hazard. Mater. 295, 43–54.

Wang, J., Xia, S.W., Yu, L.M., 2015b. Structure and bonding nature of PbCl (+) adsorption on the kaolinite(001) surface in aqueous system. Appl. Surf. Sci. 330, 411–417.

Wang, Q., Kong, X.P., Zhang, B.H., Wang, J., 2017. Adsorption of Zn(II) on the kaolinite(001) surfaces in aqueous environment: a combined DFT and molecular dynamics study. Appl. Surf. Sci. 414, 405–412.

Weck, P.F., Kim, E., Jove-Colon, C.F., 2015. Relationship between crystal structure and thermo-mechanical properties of kaolinite clay: beyond standard density functional theory. Dalton Trans. 44, 12550–12560.

White, G.N., Zelazny, L.W., 1988. Analysis and implications of the edge structure of dioctahedral phyllosilicates. Clays Clay Miner. 36, 141–146.

White, C.E., Provis, J.L., Riley, D.P., Kearley, G.J., van Deventer, J.S.J., 2009. What is the structure of kaolinite? Reconciling theory and experiment. J. Phys. Chem. B 113, 6756–6765.

Wilhelm, J., Seewald, P., Del Ben, M., Hutter, J., 2016. Large-scale cubic-scaling random phase approximation correlation energy calculations using a Gaussian basis. J. Chem. Theory Comput. 12, 5851–5859.

Yu, C.-H., Newton, S.Q., Miller, D.M., Teppen, B.J., Schäfer, L., 2001. Ab initio study of the nonequivalence of adsorption of D- and L-peptides on clay mineral surfaces. Struct. Chem. 12, 393–398.

Zeitler, T.R., Greathouse, J.A., Cygan, R.T., Fredrich, J.T., Jerauld, G.R., 2017. Molecular dynamics simulation of resin adsorption at kaolinite edge sites: effect of surface deprotonation on interfacial structure. J. Phys. Chem. C 121, 22787–22796.

Zen, A., Roch, L.M., Cox, S.J., Hu, X.L., Sorella, S., Alfe, D., Michaelides, A., 2016. Toward accurate adsorption energetics on clay surfaces. J. Phys. Chem. C 120, 26402–26413.

Zhang, G.Z., Al-Saidi, W.A., Myshakin, E.M., Jordan, K.D., 2012. Dispersion-corrected density functional theory and classical force field calculations of water loading on a pyrophyllite(001) surface. J. Phys. Chem. C 116, 17134–17141.

Zhang, Y., Meng, Y.F., Liu, H.B., Yang, M.L., 2016a. First-principles study of water desorption from montmorillonite surface. J. Mol. Model. 22, 105–109.

Zhang, C., Liu, X., Lu, X., Meijer, E.J., Wang, K., He, M., Wang, R., 2016b. Cadmium (ii) complexes adsorbed on clay edge surfaces: insight from first principles molecular dynamics simulation. Clays Clay Miner. 64, 337–347.

Zhang, C., Liu, X.D., Lu, X.C., He, M.J., Meijer, E.J., Wang, R.C., 2017. Surface complexation of heavy metal cations on clay edges: insights from first principles molecular dynamics simulation of Ni(II). Geochim. Cosmochim. Acta 203, 54–68.

Zhang, C., Liu, X., Lu, X., He, M., 2018. Complexation of heavy metal cations on clay edges at elevated temperatures. Chem. Geol. 479, 36–46.

Clay mineral–water interactions

4

Cliff T. Johnston
Departments of Earth, Atmospheric and Planetary Sciences and Agronomy, Purdue University, West Lafayette, IN, United States

4.1 Introduction

Water is comprised of two of the three most abundant elements in the Universe, and is the most abundant molecule found on the surface of the Earth to a depth of several thousand metres (Fraser et al., 2002; Mottl et al., 2007). The 'Blue Planet' is defined and controlled by the presence of water. It is the key compound for sustaining life on Earth. In some way, shape, or form, water is involved in the great majority of all geological, biological, and chemical processes. In geology, for example, the continual interaction of water with exposed mineral surfaces at the rock–water interface has produced a broad array of weathering products, including clay minerals (Hazen and Ferry, 2010). Thus, the interaction of water with 'high surface area to volume' clay minerals is critically important to many different disciplines, including all aspects of geology, soil, and water science, geotechnical engineering, geomorphology, seismology, geophysics, petroleum engineering, atmospheric science, and astrobiology.

In addition to being the most abundant molecule on the Earth's surface, it is also the most anomalous. The anomalous macroscopic properties of water are familiar and are clearly demonstrated in terms of its isobaric temperature dependence of density, thermal expansion coefficient, compressibility, and heat capacity (Debenedetti, 2003; Gallo et al., 2016). As shown in Fig. 4.1, the density of water reaches a maximum at 4°C, and the solid phase has lower density than that of the liquid (i.e. ice floats). Both the isothermal compressibility and heat capacity of 'simple liquids' increase with increasing temperature. In the case of water, however, these quantities increase with decreasing temperature. The isothermal compressibility and specific heat of water reach minima at 46°C and 35°C, respectively (Fig. 4.1). Although these unusual properties of water are well known, the underlying static and dynamic properties of water remain a current area of research (Pettersson et al., 2016). These properties are attributed to entropy and volume fluctuations in the structure of water that increase upon cooling, in the case of water. These have been described as a 'cascade of anomalies' which, in turn, lead to the unusual response of the thermodynamic properties (Gallo et al., 2016).

The interaction of water with ions is central to understanding clay mineral–water interactions. From a thermodynamic perspective, it is useful to consider the stepwise interaction of H_2O with selected alkali metal cations to illustrate the concept of the enthalpy of hydration. As shown in Table 4.1, the change in enthalpy for one H_2O molecule associating with Li^+ is -142 kJ/mol, the second H_2O molecule contributes another -100 kJ/mol. Negative enthalpy values mean that heat is released during the

Developments in Clay Science, Vol. 9. https://doi.org/10.1016/B978-0-08-102432-4.00004-4

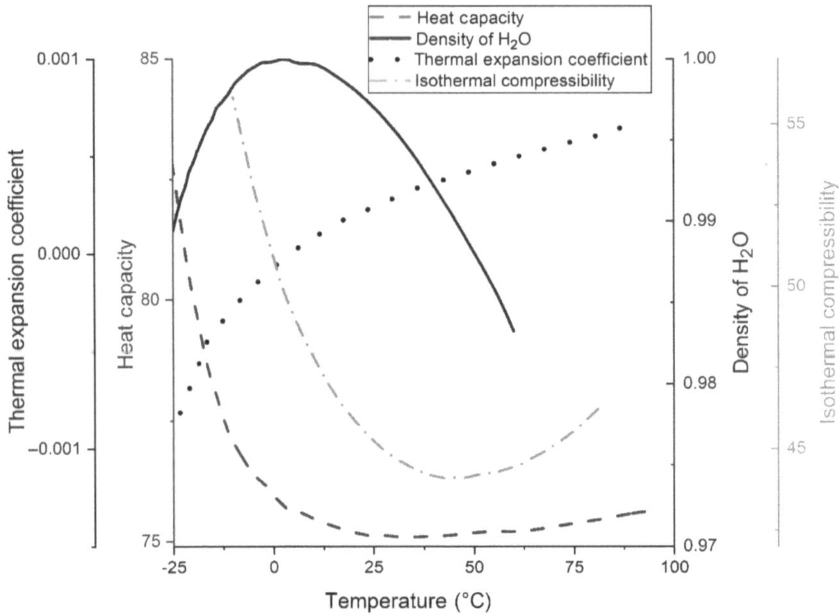

Fig. 4.1 Anomalous behaviour of water as a function of temperature: thermal expansion coefficient (α_p units of K^{-1} shown in *black*; *dotted line* in print version), specific heat capacity (C_p units J K^{-1} mol^{-1} shown in *red*; *dashed line* in print version), density (ρ units of g cm^{-3} shown in *blue*; *solid line* in print version), isothermal compressibility (K_T units of bar^{-1} shown in *green*; *dash-dot line* in print version).
Adapted from Angell, C.A., Oguni, M., Sichina, W.J., 1982. Heat capacity of water at extremes of supercooling and superheating. *Journal of Physical Chemistry*, 86, 998–1002; Debenedetti, P.G., 2003. Supercooled and glassy water. J. Phys. Condens. Matter, 15, R1669–R1726.

reaction. The cumulative change in enthalpy after complexation by six H_2O molecules is −515 kJ/mol, which is close to the 'infinite' hydration, or the Standard Molar Enthalpy of Hydration ($-\Delta_h H_I^\infty$), value of −531 kJ/mol (Marcus, 1987, 2015a). Stepwise hydration data are also presented for Na^+ and K^+ in Table 4.1. Standard molar enthalpies, entropies, and free energies for a broader suite of ions are listed in Table 4.2, and the following trends can be observed. For both the alkali and alkaline earth metal cations, the $-\Delta_h H_I^\infty$ values become less negative with increasing atomic numbers within a given row. The cations with the smaller ionic radii (e.g. Li^+ or Mg^{2+}) have larger negative enthalpies of hydration compared with their larger counterparts (e.g. Cs^+ and Ba^{2+}), meaning these smaller cations associate with water more strongly. Second, the $-\Delta_h H_I^\infty$ values become more negative with increasing valence. For example, the $-\Delta_h H_I^\infty$ values for Na^+, Mg^{2+}, and Al^{3+} for the Period 3 elements are −416, −1949, and −4715 kJ/mol, respectively.

Also included in Table 4.2 are the molar structural entropies for some of the ions. A molar structural entropy change is defined that measures the change in the structural

Table 4.1 The standard molar enthalpies of stepwise clustering of water molecules around ions in the ideal gas phase, $-\Delta_{n-1,n}H^{\circ}$/kJ/mol, at $T^{\circ} = 298.15$ K

	Li^+		Na^+		K^+	
	Stepwise hydration enthalpy (kJ/mol)	Cumulative hydration enthalpy (kJ/mol)	Stepwise hydration enthalpy (kJ/mol)	Cumulative hydration enthalpy (kJ/mol)	Stepwise hydration enthalpy (kJ/mol)	Cumulative hydration enthalpy (kJ/mol)
$M^+_{(g)} + H_2O = M(H_2O)^+_{(g)}$	−142		−100		−75	
$M(H_2O)^+_{(g)} + H_2O \rightarrow M(H_2O)_2{}^+_{(g)}$	−108	−250	−83	−183	−67	−142
$M(H_2O)_2{}^+_{(g)} + H_2O \rightarrow M(H_2O)_3{}^+_{(g)}$	−87	−337	−66	−249	−55	−197
$M(H_2O)_3{}^+_{(g)} + H_2O \rightarrow M(H_2O)_4{}^+_{(g)}$	−69	−406	−58	−307	−49	−246
$M(H_2O)_4{}^+_{(g)} + H_2O \rightarrow M(H_2O)_5{}^+_{(g)}$	−58	−464	−51	−358	−45	−291
$M(H_2O)_5{}^+_{(g)} + H_2O \rightarrow M(H_2O)_6{}^+_{(g)}$	−51	−515	−45	−403	−42	−333
Standard molar enthalpy of hydration (kJ/mol) $\Delta_h H_I^{\infty}$	Li^+	−531	Na^+	−416	K^+	−334

Data from Dzidic, I., Kebarle, P., 1970. Hydration of the alkali ions in the gas phase. Enthalpies and entropies of reactions M+(H2O) n-1 + H2O = M+(H2O)n. J. Phys. Chem. 74, 1466–1474; Marcus, Y., 2015. Ions and their properties. In: Ions in Solution and Their Solvation. John Wiley & Sons, Inc.

Table 4.2 Standard molar enthalpies, entropies, Gibb energies of hydration structural entropy, and *B* NMR and *B* viscosity terms

Ion	$-\Delta_h H_I^{\infty}$	$-\Delta_h S_I^{\infty}$	$-\Delta_h G_I^{\infty}$	$\Delta_{str}S_I$	B_{NMR}	B_{η}
	kJ/mol			J/mol	$dm^3\ mol^{-1}$	
Li^+	531	141.8	489	−52	0.14	0.146
Na^+	416	111.2	383	−14	0.06	0.085
K^+	334	74.3	312	47	−0.01	−0.009
Rb^+	308	65.1	289	52	−0.04	−0.033
Cs^+	283	58.6	266	68	−0.05	−0.047
Mg^{2+}	1949	331.2	1837	−113	0.5	0.385
Ca^{2+}	1602	252.4	1527	−59	0.27	0.284
Sr^{2+}	1470	241.7	1398	−53	0.23	0.261
Ba^{2+}	1332	208.2	1270	−18	0.18	0.216
Cr^{2+}	1933	308.1	1841			
Mn^{2+}	1874	291.8	1787			
Fe^{2+}	1972	362.4	−1864			
Al^{3+}	4715	538.5	4554			
Fe^{3+}	4462	556.1	−4296			

Data from Marcus, Y., 2015. Ions and their properties. In: Ions in Solution and Their Solvation. John Wiley & Sons, Inc.

order of an infinitely dilute solution caused by the introduction of the ion into the solvent (water, in this case). This value is obtained from the standard molar entropy of solvation of the ion, after certain unrelated quantities are subtracted (Marcus, 2009, 2015b). It is related to the qualitative terms 'kosmotropes' and 'chaotropes' (Collins, 1997, 2004; Collins et al., 2007) used to describe 'structure making' and 'structure breaking' ions. Kosmotropes are small ions of high charge density with the ability to bind strongly to water molecules. In contrast, chaotropes are larger ions of low charge density that have lower affinities for water (Collins, 1997). Similarly, Burgess introduced the term 'standard partial molar entropy,' which is based on the assumption of zero for the hydrated proton (Burgess, 1999). Conceptually, these terms emphasise the competition between water–water interactions and the perturbation introduced by the presence of the ion. Note that the hydration structural entropies ($\Delta strS_I$) are much smaller, with values that range from -113 for Mg^{2+} to +68 J/mol for Cs^+ (J/mol not kJ/mol). When small cations with high charge densities (e.g. Mg^{2+}) are added into water, it results in a large number of electrostricted water molecules (water molecules that are strongly associated with the cation), causing a decrease in entropy (-113 J/mol for Mg^{2+}). These cations are called 'structure making' ions. On the other hand, addition in water of ions having a large size and low charge density results in an increase in entropy. For example, Cs^+ introduced in water results in a $\Delta strS_I$ change of +68 J/mol, causing an increase in entropy, and thus Cs^+ is classified as a 'structure breaking' ion. A list of structure making and structure breaking ions (Marcus, 2009) is presented in Table 4.3.

Two additional terms are listed in Table 4.2 that provide insight about how a particular ion influences the structure of water.

Table 4.3 A partial list of structure making and structure breaking ions (Marcus, 2009)

Structure breaking ions
I^-, ClO_4^-, MnO_4^-, TcO_4^-, $Cr_2O_7^{2-}$, Br^-, SCN^-, Cl^-
K^+, Rb^+, Cs^+, Tl^+, CN^-, NO_2^-, NO_3^-, $Al(OH)_4^-$, S^{2-}, $(CH_3)_4N^+$,
Ra^{2+}, HCO_2^-, HSO_4^-, SeO_4^{2-}, CrO_4^{2-}, NH_4^+, $B(OH)_4^-$, SO_4^{2-}, MoO_4^{2-}, WO_4^{2-}, $C_2O_4^{2-}$
Borderline ions
Na^+, Ag^+, Ba^{2+}, Pb^{2+}, F^-, HCO_3^-, $H_2PO_4^-$
Structure making ions
Li^+, Sr^{2+}, Sn^{2+}, Al^{3+}, Cr^{3+}, Bi^{3+}, OH^-, Ca^{2+}, Eu^{2+}, Hg_2^{2+}, Sc^{3+}, Co^{3+}, Pu^{4+}, HPO_4^{2-}
V^{2+}, Cr^{2+}, Mn^{2+}, Cu^{2+}, Cd^{2+}, Fe^{3+}, AsO_4^{3-}
Mg^{2+}, Co^{2+}, Ni^{2+}, Zn^{2+}, Y^{3+}, La^{3+} to Eu^{3+}, $PO4^{3-}$, Fe^{2+}, UO_2^{2+}

(i) The first is the viscosity B_η coefficient that is part of the Jones–Dole expression to calculate the relative viscosity changes of water in the presence of a particular ion:

$$\left[\left(\frac{\eta}{\eta^*}\right) - 1\right] = A_\eta c^{1/2} + B_\eta c + \ldots$$

Different types of ions can increase or decrease the viscosity of water, and this concept is captured in the B_η coefficient (dm^3/mol). Li^+, for example, causes the viscosity to increase, whereas Cs^+ results in a decrease in viscosity.

(ii) Analogous to the viscosity B_η coefficient, is the NMR longitudinal proton relaxation time B_{NMR} (dm^3 /mol). NMR relaxation times are longer from structure-making ions and shorter from structure-breaking ions. Proton (^{1}H) NMR longitudinal relaxation times are restricted to diamagnetic ions. The alkali metal cations, that is, Li^+, Na^+, K^+, Rb^+, and Cs^+ all are diamagnetic due to the absence of unpaired electrons. For paramagnetic transition metal cations, ^{17}O NMR spin–lattice relaxation of D_2O can be used to look at water reorientation.

When considering molecular properties of water and hydrated ions, it is important to consider the timescale associated with a particular experimental method (Sposito and Prost, 1982; Guven, 1992). Vibrational transitions, for example, occur in the timescale from 10^{-15} to 10^{-12}s and NMR transitions are considerably slower (10^{-11} to 10^{-3}s). Thus, a molecular motion may appear to be static using vibrational spectroscopy, whereas a motion-averaged structure may be observed in NMR, EPR, or neutron scattering.

The ratio of reorientation times of hydration water molecules in 1 M salt solutions to that of pure water (Table 4.4) is determined using dielectric relaxation (Giese et al., 1970; Marcus, 2009). The reorientation time increases with decreasing cation size. The structure of ionic hydration shells has been explored using a diverse suite of

Table 4.4 Ratio of H_2O reorientation times of ion in 1 M salt solution (τ_{IM}) to that in pure H_2O (τ_W)

Ion	τ_{IM}/τ_W	Ion	τ_{IM}/τ_W
Li^+	2.41	F^-	2.61
Na^+	1.53	Cl^-	0.90
K^+	0.90	Br^-	0.73
Rb^+	0.78	I^-	0.41
Cs^+	0.68	OH^-	2.44
H_3O^+	1.62	NO_3^-	0.73
NH_4^+	0.72		
Me_4N^+	1.59		
Et_4N^+	1.96		
Pr_4N^+	2.37		
Bu_4N^+	2.80		

Data from Marcus, Y., 2009. Effect of ions on the structure of water: structure making and breaking. Chem. Rev. 109, 1346–1370.

methods, including X-ray and neutron diffraction, elastic and inelastic neutron scattering, vibrational, EXAFS, NMR, and dielectric relaxation spectroscopies, and molecular simulation (Ohtaki and Radnai, 1993; Marcus, 2009, 2015a; Van der Vegt et al., 2016).

Because water is a polar molecule, it interacts favourably with ions and other polar solutes in solution through Coulombic interactions. Neutron diffraction studies revealed that water molecules coordinated to metal ions through orientation of the oxygen atom towards the cation. $NiCl_2$, for example, revealed a Ni—O distance of 0.207 nm. The diameter of a water molecule has a radius of 0.140 nm, the corresponding ionic radius of the Ni^{2+} ion is 0.067 nm (Soper et al., 1977; Marcus, 1988; Ohtaki and Radnai, 1993). For water molecules coordinating to anions, the opposite orientation was found, as illustrated in Fig. 4.2.

The structure and dynamics of the complex hydrogen-bond (HB) network formed by water molecules is an area of active study using both static and advanced time resolved vibrational spectroscopies (Perakis et al., 2016). Vibrational spectroscopy provides unique information about the HB network and its dynamics. Linear FTIR and Raman spectra of neat and isotopically dilute water (e.g. HOD in D_2O) reveal insight about HB strength and the complex coupling of vibrational modes. In addition, ultra-fast time-resolved vibrational methods (pump-probe, photon-echo,

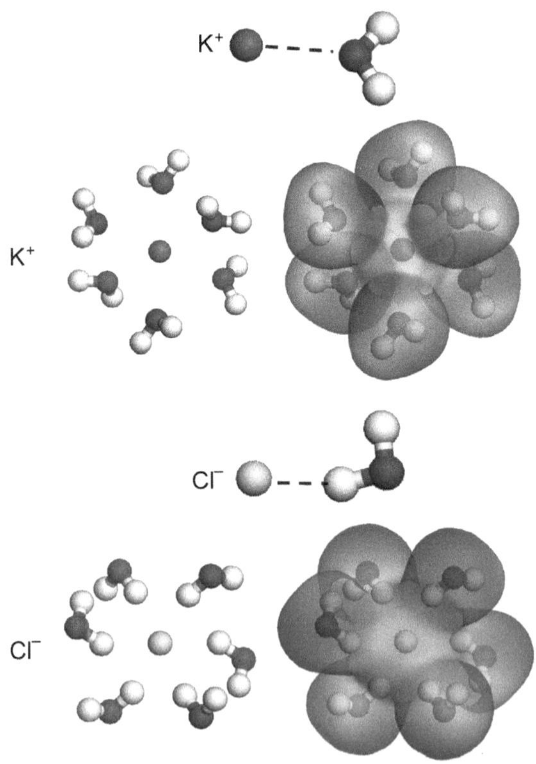

Fig. 4.2 Representation of H_2O molecules clustered around K^+ and Cl^- ion. Note orientation of the oxygen atoms of H_2O towards the K^+ ion and opposite orientation for H_2O molecules clustered around Cl^-. Images on the *right* correspond to an electrostatic potential map of the structures shown on the *left* where *blue* (*light grey* in the print version) shows partial negative charge and *red* (*dark grey* in the print version) represents partial positive charge.

and two-dimensional infrared (2D-IR)) can resolve transitions ranging from femtoseconds to nanoseconds and provide insights into the dynamics of the HB network and vibrational energy redistribution (Auer et al., 2007; Auer and Skinner, 2008; Li and Skinner, 2010b; Tainter et al., 2013; Ni and Skinner, 2015; Perakis et al., 2016; Götte et al., 2017). In addition, interface-specific techniques such as sum frequency generation (SFG) can probe the molecular structure of water and ions at both liquid–air and solid–liquid interfaces (Du et al., 1994; Hua et al., 2014; Ni and Skinner, 2015; DeWalt-Kerian et al., 2017; Götte et al., 2017). These experimental advances are helping to develop accurate molecular models of water and ions and solutes dissolved in water, and the behaviour of water at interfaces (Auer and Skinner, 2008; Li and Skinner, 2010a,b).

4.2 Water interactions with 'neutral clay mineral surfaces'

4.2.1 *Talc and pyrophyllite*

The interaction of water with clay minerals is complex and is influenced by the nature of the clay mineral surface, extent and location of isomorphous substitution, type of exchangeable cations, pH, ionic composition of the aqueous phase, particle size and shape, and the overall pore size distribution. The simplest systems to consider are talc and pyrophyllite, the 2:1 clay minerals that do not possess permanent structural charge, and have hydrophobic basal surfaces. Understanding how water interacts with these surfaces is important because these surface features are also present on with other clay minerals (e.g. kaolin group minerals) and on the neutral surfaces between isomorphous heterovalent substitution (IHS) sites of permanently charged clay minerals (e.g. smectites).

The structures of talc ($Si_4Mg_3O_{10}(OH)_2$) and pyrophyllite ($Si_4Al_2O_{10}(OH)_2$) are shown in Fig. 4.3 (refined structures based respectively on Lee and Guggenheim, 1981; Perdikatsis and Burzlaff, 1981). Talc is a trioctahedral mineral, meaning that

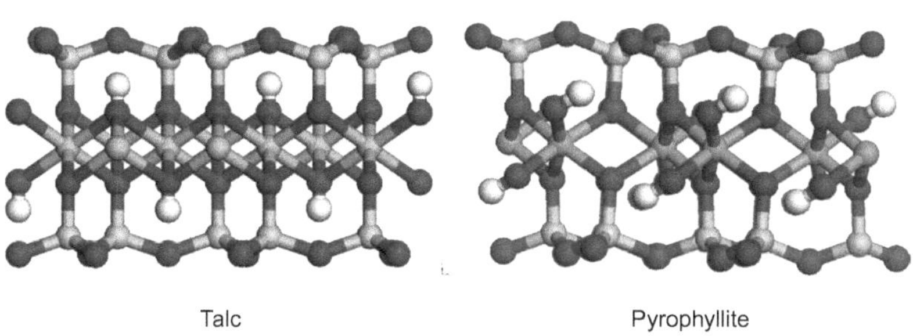

Fig. 4.3 Crystal structures of talc and pyrophyllite projected along [100]. The octahedral cation in talc is Mg^{2+} and that in pyrophyllite is Al^{3+}.
Structures refined from Lee, J.H., Guggenheim, S., 1981. Single-crystal X-ray refinement of pyrophyllite-1TC. Am. Mineral. 66, 350–357; Perdikatsis, B., Burzlaff, H., 1981. Structural refinement of talc $Mg_3(OH)_2Si_4O_{10}$. Z. Kristallogr. 156, 177–186.

all of the octahedral sites are occupied by a divalent cation (i.e. Mg^{2+}); the orientation of the OH is perpendicular to the clay mineral surface (Fig. 4.3). In contrast, only two out of the three possible octahedral sites in pyrophyllite are occupied by a trivalent ion, and the hydroxyl group is tilted towards the vacant site (Fig. 4.3).

Interaction of water with talc and pyrophyllite has been investigated using water adsorption isotherms (Michot et al., 1994; Malandrini et al., 1997; Douillard et al., 2002; Tarasevich, 2007), immersion calorimetry (Michot et al., 1994; Malandrini et al., 1997; Douillard et al., 2002), contact angle measurements (Schrader and Yariv, 1990; Giese et al., 1991; Malandrini et al., 1997; Douillard et al., 2002), atomic force microscopy (AFM) (Yan et al., 2013), Raman and IR spectroscopy (Rashchenko et al., 2016), and thermal analysis with evolved gas analysis (Michot et al., 1994). Water vapour adsorption isotherms on talc (and kaolinite) are shown in Fig. 4.4 (Douillard et al., 2002; Tarasevich, 2007), showing that water has an overall low affinity for talc, but does exhibit some hydrophilic character at low relative vapour pressure of H_2O. These results are supported by recent molecular modelling studies of H_2O-talc/pyrophyllite interactions (Wang et al., 2009; Rotenberg et al., 2011; Zeitler et al., 2012; Tarasevich and Aksenenko, 2014; Zarzycki and Gilbert, 2016). Both talc and pyrophyllite have low N_2 specific surface areas (2–18 m^2/g), low H_2O monolayer adsorption values (0.017–0.055 mmol H_2O/g), moderate H_2O cross-sectional areas (surface area occupied by one H_2O molecule nm^2/molecule), intermediate enthalpy of immersion values (mJ/m^2), and experimentally measured advancing contact angles ranging from 29 to 83 degrees (Schrader and Yariv, 1990; Giese et al., 1991; Michot et al., 1994; Malandrini et al., 1997; Douillard et al., 2002; Tarasevich, 2007).

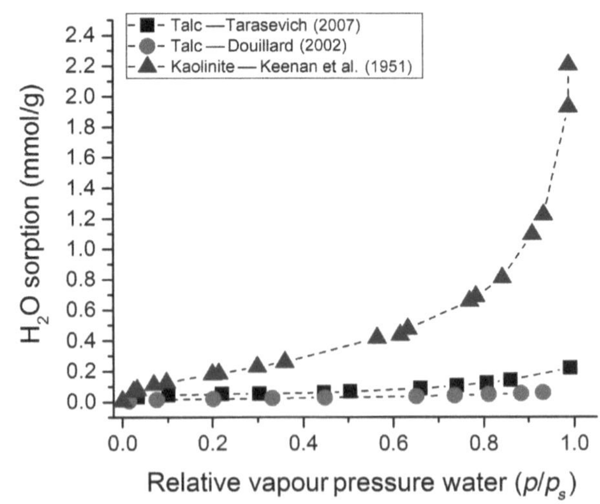

Fig. 4.4 Water vapour adsorption isotherms on talc and kaolinite.
Data read from plots in Keenan, A.G., Mooney, R.W., Wood, L.A., 1951. The relation between exchangeable ions and water adsorption on kaolinite. J. Phys. Colloid Chem. 55, 1462–1474; Douillard, J.M., Zajac, J., Malandrini, H., Clauss, F., 2002. Contact angle and film pressure: study of a talc surface. J. Colloid Interface Sci. 255, 341–351; Tarasevich, Y.I., 2007. The surface energy of hydrophilic and hydrophobic adsorbents. Colloid J. 69, 212–220.

There is general agreement that the basal surfaces of talc and pyrophyllite are hydrophobic, and the edges are hydrophilic. Due to their intrinsically small particle size, measurement of contact angles on a single crystal surface of talc and pyrophyllite is not possible. However, wicking and thin film measurements have yielded consistent results with values of 79–80 degrees (hydrophobic range) (Giese et al., 1991). Lower values have been reported in the literature for some samples, and this has been ascribed to the varying contribution of hydrophilic edge sites (Douillard et al., 1994; Tarasevich, 2007). For strongly dehydrated talc samples, both experimental and molecular modelling studies have shown an affinity of H_2O for the siloxane surface (Michot et al., 1994; Rotenberg et al., 2011). In the absence of water (e.g. strongly desiccated surface), talc behaves as a hydrophilic mineral (Michot et al., 1994). Quantitative thermal desorption measurements show that water is strongly retained by the surface, and immersion calorimetry measurements show a greater enthalpy of immersion for strongly dehydrated surfaces (Michot et al., 1994). These results are further supported by recent molecular modelling results, showing that H_2O has a high affinity of the basal surface at low surface coverage (Wang et al., 2009; Rotenberg et al., 2011). At low surface coverage, adhesive (water-surface) interactions out-compete cohesive (water–water interactions) and surface functions as a hydrophilic surface (Fig. 4.4). Classical density functional theory calculations have shown that H_2O adsorbed on pyrophyllite is oriented with the water dipole extending from the oxygen atom to the centre of mass of the hydrogen atoms, pointing away from the surface (Jeanmairet et al., 2014). At higher surface coverage of H_2O, cohesion (water–water) forces between water molecules out-compete the surface-water interactions, resulting in a strongly hydrophobic surface at higher water coverage, and calculated contact angles of 96 and 105 degrees for talc and pyrophyllite, respectively (Rotenberg et al., 2011). These results are consistent with the observed water vapour adsorption isotherms on talc (Fig. 4.4) that show an overall low affinity of talc for H_2O but with high affinity at low surface coverage (Michot et al., 1994; Douillard et al., 2002; Tarasevich, 2007). In addition, the cross-sectional surface areas derived from these water adsorption isotherms reveal values between 0.15 and 0.18 nm^2/H_2O molecules at monolayer coverage. For strongly hydrophobic surfaces, much larger values are generally observed.

Additional insight about the nature of the basal and edge surfaces of talc comes from force measurements on talc (and muscovite) particles using AFM combined with DLVO theory (Yan et al., 2013). DLVO is a dispersion stabilisation theory named after Derjaguin, Landau, Verwey, and Overbeek (see Derjaguin and Landau, 1941; Verwey and Overbeek, 1948). AFM-derived Stern potential measurements of the basal surface of both talc and muscovite were negative for all three ionic strengths, with the surface of muscovite showing a more negative charge (Fig. 4.5). The edge sites were less negative in distilled H_2O, and charge reversal was observed for Ca^{2+} and Mg^{2+} at higher ionic strength (Fig. 4.5).

Although talc is considered as a hydrophobic mineral under ambient conditions, talc–water interactions become important at elevated temperature and pressure. Under these conditions, talc reacts with H_2O to form a 10 Å phase (abbr. 'TAP'

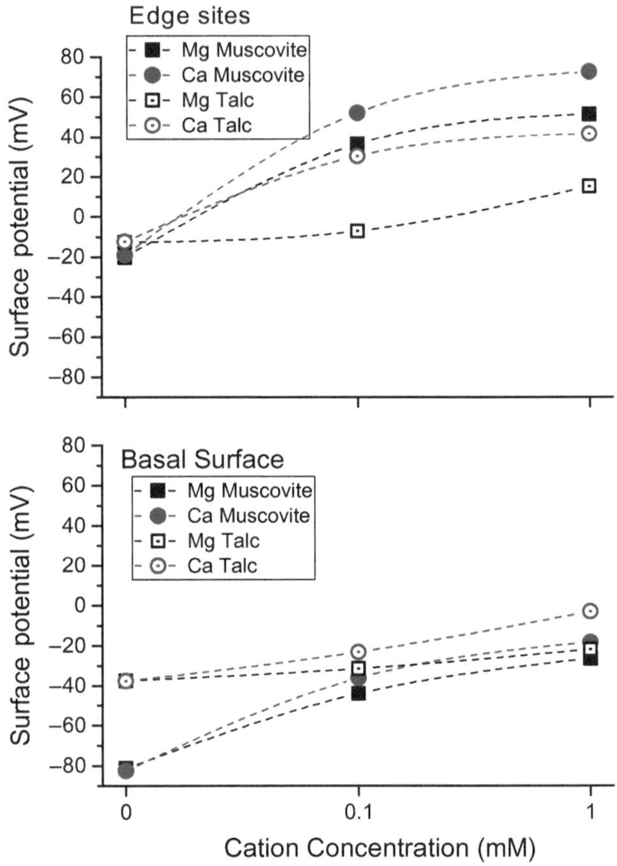

Fig. 4.5 AFM and DLVO-derived surface potentials of muscovite and talc basal planes and edges surfaces. Data read from plot in Yan, L.J., Masliyah, J.H., Xu, Z.H., 2013. Interaction of divalent cations with basal planes and edge surfaces of phyllosilicate minerals: muscovite and talc. J. Colloid Interface Sci. 404, 183–191.

for Ten Ångstrom Phase) with implications to serve as a host of water in subduction zones (Chinnery et al., 1999; Parry et al., 2007; Pawley and Welch, 2014; Rashchenko et al., 2016). A combination of high pressure Raman, XRD, and IR methods have revealed the transformation of talc to TAP. For example, the Raman spectrum of talc in the ν(OH) region is characterised by a sharp, well-resolved O—H stretching band at 3681 cm^{-1} corresponding to the ν(OH) mode of the structural OH groups. In the presence of H_2O at elevated temperature (500°C) and pressure (8 GPa), the 3681 cm^{-1} band is replaced by the two TAP ν(OH) bands at 3590 and 3631 cm^{-1} (Rashchenko et al., 2016). Similar spectral perturbations are observed using IR spectroscopy and XRD (Parry et al., 2007; Pawley and Welch, 2014).

4.2.2 Kaolin group mineral

Similar to talc and pyrophyllite, the 1:1 clay minerals do not possess significant structural charge, that is, little, if any, isomorphous substitution (Brigatti et al., 2013). This section will focus on the dioctahedral 1:1 structures of the kaolin group. The structure of kaolin group minerals is based on the different arrangements of the structural unit of $Si_2Al_2O_5(OH)_4$ in the form of three main polytypes: kaolinite (most common), dickite, and nacrite (least common). In addition, halloysite is a hydrated form of kaolinite where one H_2O layer resides between the layers.

4.2.2.1.1 Kaolinite

The crystal structure of kaolinite (Kaol) is shown in Fig. 4.6. Unlike talc and pyrophyllite, Kaol is terminated by two different types of surfaces that produce an overall electrostatic dipole (Hu and Michaelides, 2010). As shown in Fig. 4.6, the top surface is a silica tetrahedral surface, similar to that found on pyrophyllite. The bottom surface is terminated by a 'gibbsite layer' of exposed Al–OH groups, similar to that of gibbsite.

The surfaces of the clay minerals and their interaction with H_2O have been investigated using an ever-increasing array of methods. Like most clay minerals, measuring the contact angle of water on Kaol is influenced by surface roughness, particle size, surface preparation, and pretreatment, temperature, and humidity (Yin and Miller, 2012). These values are generally not reliable. In this case, further complexity is introduced because Kaol has three types of surfaces. In addition to edge sites, the basal surface on the siloxane side of Kaol is hydrophobic, and the gibbsite-like surface is hydrophilic. Contact angle measurements vary from 15 to 46 degrees, reflecting the hydrophilic character of the gibbsite surface (Shang et al., 2010; Yin and Miller, 2012; Yin et al., 2012). This is also reflected by a wide variation in immersion

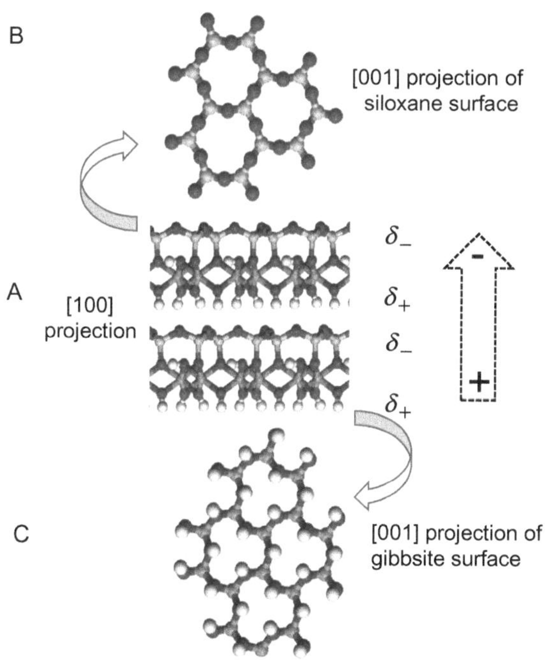

Fig. 4.6 Crystal structure of kaolinite (2 layers) projected along [100] (A). [001] projections showing the siloxane surface (B), and $[00\overline{1}]$ projection showing the gibbsite surface. Also shown are the partial positive charges on the gibbsite surface and partial negative charges of the siloxane surface resulting in an overall electrostatic dipole moment. Crystal structure from Bish, D.L., 1993. Rietveld refinement of the kaolinite structure at 1.5 K. Clays Clay Miner. 41, 738–744.

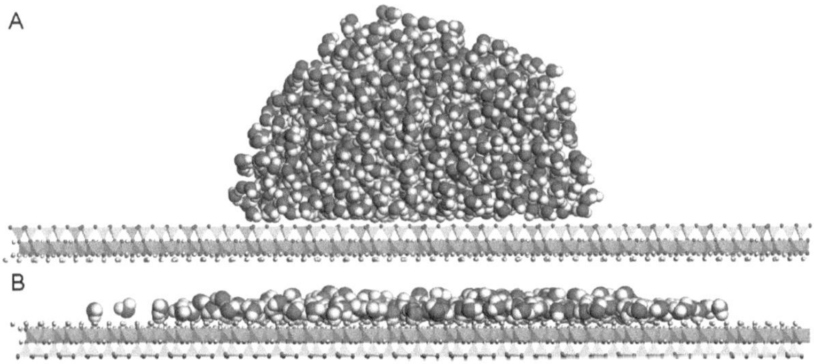

Fig. 4.7 Molecular dynamics simulation of a nanodrop of water in contact with the siloxane surface (A), and with the gibbsite surface (B) of kaolinite.
Illustration adapted from Solc, R., Gerzabek, M.H., Lischka, H., Tunega, D., 2011. Wettability of kaolinite (001) surfaces—molecular dynamic study. Geoderma 169, 47–54.

enthalpies ranging from -196 to $-308\,mJ/m^2$ (Partyka and Douillard, 1995; Douillard and Medout-Marere, 2000). Molecular dynamics simulations of nanodrops of water on the two surfaces of Kaol (Fig. 4.7) showed that the gibbsite surface was 'water wet' with a contact angle of $\sim 0°$ in contrast to the siloxane surface with a calculated contact angle of 105 degrees (Solc et al., 2011). Ions can also be introduced into nanopores to observe surface interactions (Vasconcelos et al., 2007). A simulated solution of CsCl was introduced to a nanopore within Kaol, and after 600ps the ions had clearly segregated with the Cs^+ ions migrating to the siloxane surface, and the Cl^- ions to the opposing hydroxyl surface (Vasconcelos et al., 2007). High resolution AFM has provided further insight about the surface charge and charge distribution across particles of Kaol deposited on both sapphire and mica surfaces, exposing both silica and gibbsite surfaces, respectively (Kumar et al., 2016). It follows that the affinity of H_2O for Kaol is modest, but higher than that for talc (Fig. 4.4) (Keenan et al., 1951).

This dipolar nature of Kaol has led to recent interest in their role as Janus particles that can stabilise emulsions (Hirsemann et al., 2012; Weiss et al., 2013; Kirillova et al., 2014, 2016). Kaol has been successfully used to create Janus particles. These particles take advantage of the surface polarity differential between the siloxane and gibbsite surfaces. This allows each surface to be functionalised with different chemistries that can interact favourably with a hydrophobic phase on one side (e.g. oil) and an aqueous phase on the other.

4.2.2.1.2 Halloysite

The previous discussion focused on clay minerals that have minimal affinities for H_2O and whose overall water contents are generally less than 50mg of H_2O/g of clay mineral. In the kaolin group minerals, halloysite (Hal) is a hydrated form of Kaol with 2 H_2O molecules per half unit cell intercalated between the siloxane surface of one layer and the gibbsite surface of the opposing layer (Fig. 4.8) with a structural formula

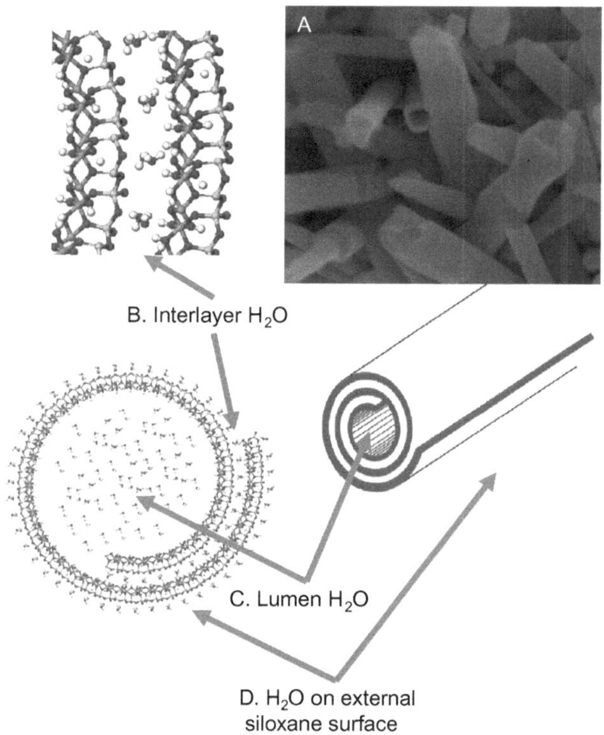

Fig. 4.8 Cryo-SEM micrograph and crystal structure of tubular halloysite showing three types of water. The cryo-SEM image (A) was obtained from a hydrated halloysite from the Gardner Mine Ridge deposit in Southern Indiana, USA (Ambers and Murray, 1995). Based on TEM analysis, the average inner diameter was ~15 nm and the outer diameter was 45–20 nm. A schematic of the tubular structure is shown in the *upper right portion*. The crystal structure of halloysite is from Ferrante et al. (2015). Three types of H_2O molecules are present, consisting of interlayer H_2O (B), H_2O present in the tubular structure or lumen (C), and H_2O external to the halloysite structure (D) (H_2O molecules not shown).
Data were adapted from Santagata, M.C., Premachandra, G.S., Johnston, C.T., 2009. Nanoconfined water in clay minerals. In: Mico et Nano Scientiae Mare Magnum. Castellaneta Marina (TA), Italy, AIPEA.

of $Si_2Al_2O_5(OH)_4(H_2O)_2$. Tubular morphology of Hal is the most common form (shown in Fig. 4.8); however, pseudo-spherical and spheroidal, platy or tubular, fibre, prismatic, and lath morphologies have been observed (Joussein et al., 2005). In addition to the H_2O molecules between the layers, water molecules are also found within the lumen (Fig. 4.8). Thermogravimetric analysis of Hal that is both fully hydrated and has never been dried has a mass loss of about 11.1 mmol H_2O/g of clay mineral. Loss of the two interlayer H_2O molecules (per $Si_2Al_2O_5(OH)_4$ unit) accounts for about 7.8 mmol H_2O/g of clay mineral. In addition, a loss of an additional 2.9 mmol H_2O/g of clay mineral was assigned to 'pore/lumen' water (Cruz et al., 1978).

Dehydration of Hal is irreversible due to the loss of the interlayer H_2O molecules (Bordallo et al., 2008). Although water will rehydrate the lumen, loss of interlayer H_2O is irreversible (Joussein et al., 2005).

Although Hal has been characterised extensively using IR and Raman spectroscopy (Janik and Keeling, 1993; Frost, 1995; Frost and Kristof, 1997; Frost and Shurvell, 1997; Kristof et al., 1997; Fialips et al., 2000; Adamo et al., 2001; Cuadros and Dudek, 2006; Zsirka et al., 2017), most of these studies examined dry samples that did not contain interlayer or lumen H_2O. A few studies have examined Hal samples containing both interlayer and lumen water (Tarasevi and Gribina, 1972; Yariv and Shoval, 1975; Johnston, 2017), which revealed that the interlayer and lumen water were somewhat more hydrogen bonded than in bulk water. Loss of interlayer water occurs rapidly, and a key point is that water should be considered as an integral part of the Hal structure (Johnston, 2017). In addition, the nanoconfined geometry of water in Hal is distinct from other strongly hydrated clay minerals in that interlayer cations are absent. The interlayer water molecules in Hal are interacting only with the siloxane and gibbsite surfaces, and with each other (Bordallo et al., 2008).

Neutron studies of Hal have revealed insight about the mobility of H_2O on the timescale of neutron scattering measurements. Common to many NMR and neutron studies of water mobility near clay mineral surfaces, the residence time between jumps, τ_0, of water in fully hydrated Hal was longer (7.1 ps) compared with that of bulk water (1.57 ps), and the corresponding self-diffusion coefficient was lower (0.2–$0.8 \times s10^{-9}$ m^2/s) compared with that of bulk water (2.49×10^{-9} m^2/s) (Bordallo et al., 2008). However, earlier studies showed that the specific heat of water in Hal was close to that of ice at temperatures <150 K, and was similar to that of liquid water at temperatures >250 K. Corresponding NMR measurements revealed that interlayer H_2O molecules rapidly rotated about their C_2 axis with a slower tumbling motion perpendicular to the C_2 axis (Cruz et al., 1978). Because different types of water molecules are present, the NMR and neutron data reflect contributions from both interlayer and lumen water. At present, there is significant interest in using Hal in a wide range of applications, taking advantage of its tubular structure and mesoporosity for applications in catalysis, drug delivery, and clay mineral polymer nanocomposites (Shchukin et al., 2005; Veerabadran et al., 2007; Lvov et al., 2008; Polansky et al., 2017; Zsirka et al., 2017).

4.3 Water interactions with 'charged clay mineral surfaces' (ion–dipole)

4.3.1 Smectites

Smectites are ubiquitous clay minerals that occur in soils, sediments, and rocks and control many important subsurface processes (Bergaya and Lagaly, 2013a,b). These clay minerals are strongly hydrated, and water is part of their structures. For example, attempts to remove all of the water from a smectite will generally result in the irreversible collapse of the clay mineral structure, and alteration of its surface properties.

The structure and properties of smectites have been reviewed extensively (Bergaya and Lagaly, 2013a; Brigatti et al., 2013; Emmerich, 2013; Schoonheydt and Johnston, 2013), and this section will focus on the interaction of water with smectites.

Smectites represent a diverse group of dioctahedral and trioctahedral planar 2:1 clay minerals with intermediate layer charge (χ) values ranging from 0.2 to 0.6 charges per formula unit. Isomorphous heterovalent substitution (IHS) that occurs in the crystal lattice produces a charge deficit within the crystal structure. Examples of IHS in smectites are shown in Fig. 4.9, illustrating Al^{3+} for Si^{4+} substitution in the tetrahedral sheet (common in saponite and beidellite) and Mg^{2+} for Al^{3+} substitution in the octahedral sheet (common in montmorillonite). The extent of IHS, also referred to as

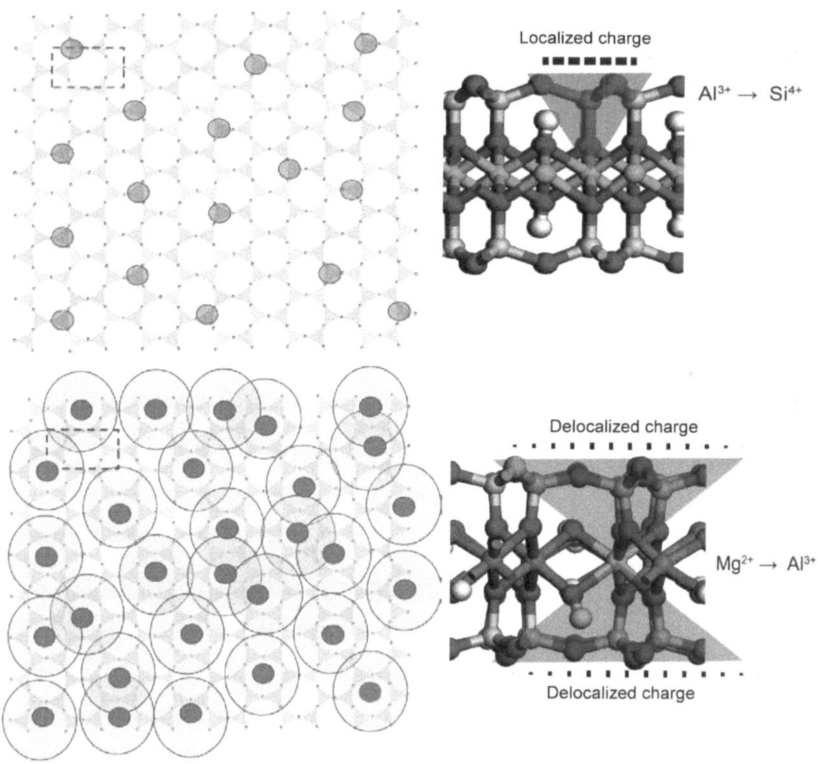

Fig. 4.9 Illustration of isomorphous heterovalent substitution (IHS) in the tetrahedral sheet ($Al^{3+} \rightarrow Si^{4+}$, *top right illustration*) and octahedral sheet ($Mg^{2+} \rightarrow Al^{3+}$, *lower right illustration*). The negative charge from IHS in the tetrahedral sheet is localised and is mainly manifest on one surface (*top surface* in this illustration). Looking down onto a 5 nm × 5 nm section of the basal surface, the distribution of the localised charges are conceptually illustrated by the *orange circles* (*grey* in the print version). In contrast, the charge from IHS in the octahedral sheet is delocalised over a larger number of oxygen atoms on both the *top* and *bottom surfaces* of layer. Because the charge is manifest on both surfaces, essentially the entire basal surface has some charged character.

structural charge, in a particular clay mineral is represented by χ, which corresponds to the number of IHSs in a half unit cell formula. Smectites are defined by the extent and type of IHS (Brigatti et al., 2013; Emmerich, 2013). In comparison with other types of clay minerals (kaolin group minerals, talc and pyrophyllite all have χ values of $\sim$0), vermiculites have higher layer charge ($0.6<\chi<0.9$), mica $\chi=1$, and brittle micas have $\chi=2$.

If the charge deficit created by IHS were point charges, the distance between the charge sites would vary between 0.6 and 1.2 nm for an isolated particle, depending on the surface charge density and type of IHS (Schoonheydt and Johnston, 2013; Johnston, 2017). However, because these substitutions occur within the crystal structure, the resultant negative charge will be distributed over a number of surface oxygen atoms on the siloxane surface. For tetrahedral substitution, the charge will be highly localised over a few oxygen atoms on 'one side' of the structure (Fig. 4.9). For IHS in the octahedral sheet, the charge will be distributed over a larger number of surface oxygen atoms on 'both sides' of the clay mineral structure (Fig. 4.9). Because the charge is more localised in smectites with tetrahedral substitution and the substitution is only manifest on one surface (as opposed to substitution in the octahedral sheet), a significantly greater proportion of the smectite surface will be 'neutral' and have similar properties to those of the pyrophyllite and talc surfaces reviewed earlier. This is illustrated in Fig. 4.9, comparing charge distribution in saponite and hectorite based on crude approximations of how the charge will be distributed over the surface oxygen atoms. Quantum chemical studies are beginning to provide additional information about how partial atomics charges are distributed on the surface oxygen atoms (Mignon et al., 2010; Lavikainen et al., 2015).

The negative structural charge deficit resulting from IHS is balanced by the presence of positively charged exchangeable cations in close proximity to the clay mineral surface illustrated in Fig. 4.10. The most common cations found in natural clay minerals are Ca^{2+}, Na^{+}, Mg^{2+} and K^{+}; however, the charge can be balanced by any positively charged ion, including H_3O^{+}, Al^{3+}, $Fe^{3+,2+}$, Cs^{+}, Sr^{2+}, or positively charged organic species. A distinguishing feature of the 'common' cations is that all have large negative enthalpies of hydration (Table 4.2), meaning that H_2O molecules are attracted to these interlayer cations, as shown in Fig. 4.10, resulting in the intrinsically high affinity of H_2O for smectites.

Water adsorption on smectites is generally high, although strongly dependent on the nature of the exchangeable cation, type, and extent of IHS. Key features of smectite–water are illustrated in Figs 4.11–4.13, comparing water adsorption isotherms for Na^{+}- and Ca^{2+}-montmorillonite (Mt) at different water contents (Prost et al., 1998). The lower figure plots the amount of water adsorbed in mmol H_2O/g vs the activity of water (a_w) (vapour pressure of H_2O/saturated vapour pressure of H_2O) ranging from 0 to 0.8 (p/p_s). The enthalpy of hydration of Ca^{2+} of -1669 kJ/mol is significantly higher than that of Na^{+} (-444 kJ/mol), resulting in more H_2O adsorbed to the Ca^{2+}-exchanged clay mineral in the crystalline swelling region (Fig. 4.11). The *d*-value of the Ca^{2+}-Mt at two different water contents is indicated. At an a_w value of 0.1, the *d*-value is 1.25 nm, corresponding to a one-layer hydrate, and at a_w value of 0.5, the *d*-value is 1.52 nm, corresponding to a two-layer hydrate. At a_w values >0.98 p/p_s (osmotic

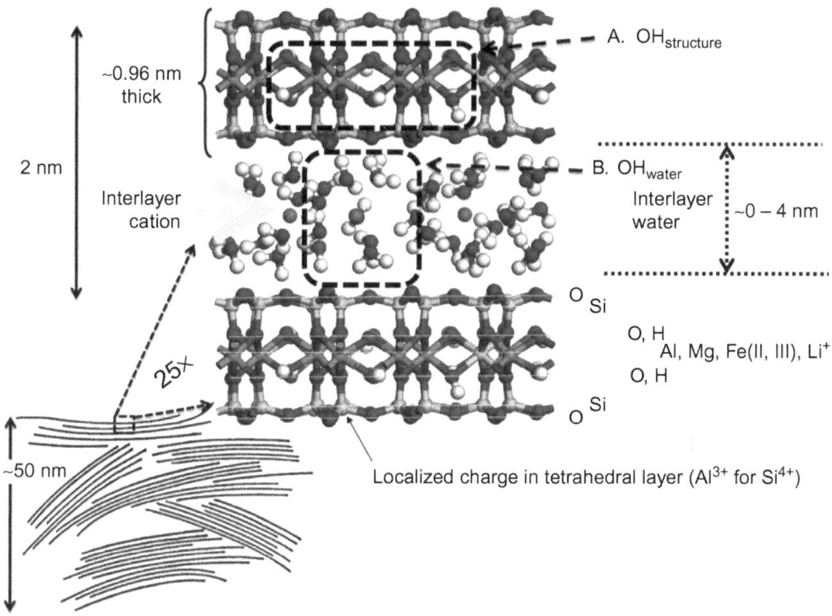

Fig. 4.10 Two montmorillonite particles with octahedral ISH separated by approximately three layers of H_2O. Interlayer H_2O molecules are clustered around exchangeable cations that balance the negative charge created by IHS. Also shown in the *lower left* is a conceptual view the morphology of montmorillonite particles stacked on top of each in the form of aggregates. Used with permission from Johnston, C.T., 2010. Probing the nanoscale architecture of clay minerals. Clay Miner. 45, 245–279.

swelling region), the situation is reversed, where the Na^+-exchanged clay mineral particles can be separated by larger distances, and more H_2O is associated with the Na^+-Mt at high a_w values (Prost et al., 1998). This is shown in Fig. 4.12 using $\log(\log(1/a_w))$ transformed x-axis to show the behaviour at high a_w values (see top axis in Fig. 4.12). In the high a_w range (>0.98), capillary condensation of H_2O occurs and the Na^+-exchanged clay mineral layers can swell to have d-values >5 nm (Norrish, 1954; Gilbert et al., 2015; Tester et al., 2016). Gilbert and coworkers have recently captured osmotic hydrates of Na^+-SWy-2 Mt using cryo-TEM and SAXS showing well-defined separations of particles as high as 60 nm (Gilbert et al., 2015; Tester et al., 2016).

The influence of the exchangeable cation on controlling water adsorption on smectites is shown in Fig. 4.13. Water desorption and adsorption isotherms for Na^+-, Ca^{2+}-, and Mg^{2+}-exchanged SAz-1 Mt in the region of crystalline swelling are shown in Fig. 4.13 (Johnston et al., 1992; Xu et al., 2000). SAz-1 Cheto Mt is a high surface charge density Mt with a χ value 0.47 and substitution almost exclusively in the octahedral sheet (Costanzo and Guggenheim, 2001; Srodon and McCarty, 2008). These isotherms reveal three important features of water adsorption to smectites: (i) the amount of water adsorbed is strongly influenced by the enthalpy of hydration of

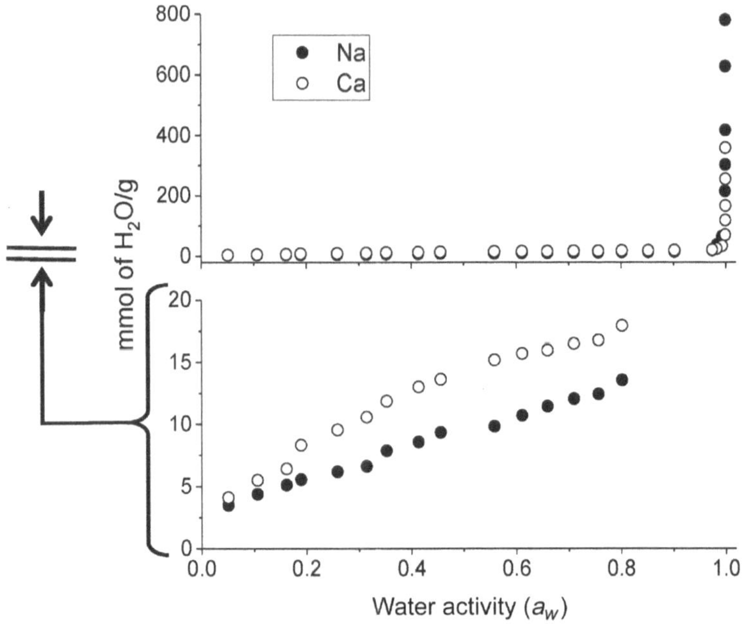

Fig. 4.11 Water adsorption isotherms of Na^+ and Ca^{2+} on montmorillonite. *Top figure* shows full range of relative vapour pressure of H_2O and water content, *lower plot* shows expanded range of 'crystalline swelling' in the $0 < a_w < 0.8$.
Data were read from plot in Prost, R., Koutit, T., Benchara, A., Huard, E., 1998. State and location of water adsorbed on clay minerals: consequences of the hydration and swelling-shrinkage phenomena. Clays Clay Miner. 46, 117–131.

the exchangeable cations with $Mg^{2+} > Ca^{2+} > Na^+$ (Table 4.2) in the domain of crystalline swelling (i.e. $0 < a_w < 0.9$); (ii) the desorption leg of the isotherm shows greater adsorption values related to the fact that it takes more energy to remove H_2O from the surface of the clay mineral than it does to adsorb it; (iii) the extent of hysteresis between the adsorption and desorption legs is related to the nature of the exchangeable cations. In the case of strongly hydrated cations, such as Mg^{2+}, the cations remain fully hydrated, and water has greater access to the interlayer spaces. In the case of less hydrated monovalent cations such as Na^+, the clay mineral layers can collapse to 1 W, or even 0 W hydrates. These clay minerals exhibit significant hysteresis because energy is required to rehydrate and separate the clay mineral layers. In a classic paper on smectite–water hysteresis, Mooney et al. (1952a,b) showed that water adsorption isotherms were not reliable because of this collapse-induced hysteresis, and that only desorption isotherms were reproducible. There have been numerous experimental studies of water adsorption to various smectites; examples include (Mooney et al., 1952a,b; Norrish, 1954; Hall and Astill, 1989; Cases et al., 1992, 1997; Johnston et al., 1992; Berend et al., 1996; Cancela et al., 1997; Xu et al., 2000; Schuttlefield et al., 2007a; Schnetzer et al., 2017). Water adsorption isotherms continue to play an important role in validating and benchmarking molecular simulation models.

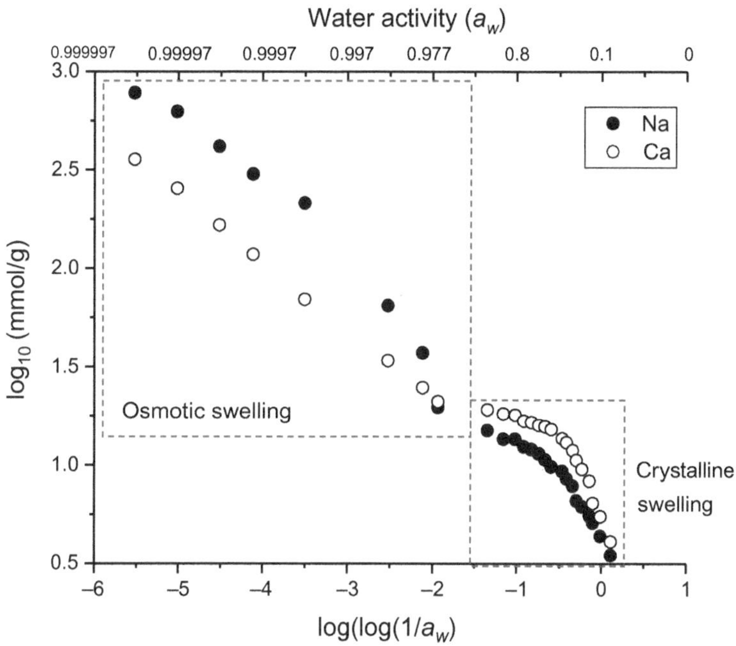

Fig. 4.12 Water adsorption isotherms of Na^+ and Ca^{2+} on montmorillonite using a log(log(1/a_w)) transformed x-axis. Actual relative vapour pressures of H_2O are shown on the top x-axis. Data were read from plot in Prost, R., Koutit, T., Benchara, A., Huard, E., 1998. State and location of water adsorbed on clay minerals: consequences of the hydration and swelling-shrinkage phenomena. Clays Clay Miner. 46, 117–131.

4.4 Molecular probe and reporter group studies of smectite–water interactions

Since the first reported IR study of water adsorbed to clay minerals more than 100 years ago (Coblentz, 1910, 1911), spectroscopic and related advanced characterisation methods have provided insight about how water interacts with clay mineral surfaces (Sposito and Prost, 1982; Newman, 1987; Guven, 1992; Marry et al., 2008; Johnston, 2010, 2017). A comprehensive review of smectite–water interactions at the molecular level is beyond the scope of this chapter; rather, a framework for understanding these interactions will be introduced by categorising studies as molecular probe or reporter group studies (Johnston et al., 1993; Zecchina and Arean, 1996; Johnston, 2010, 2017). Both molecular probe and reporter group studies rely on one or more 'diagnostic properties' that are sensitive to changes in their local environment. Molecular probes are molecules that are used to interrogate a surface (e.g. H_2O), whereas reporter groups are part of the crystal structure of the clay minerals itself (e.g. structural OH group of a clay mineral). For example, consider the ^2H NMR

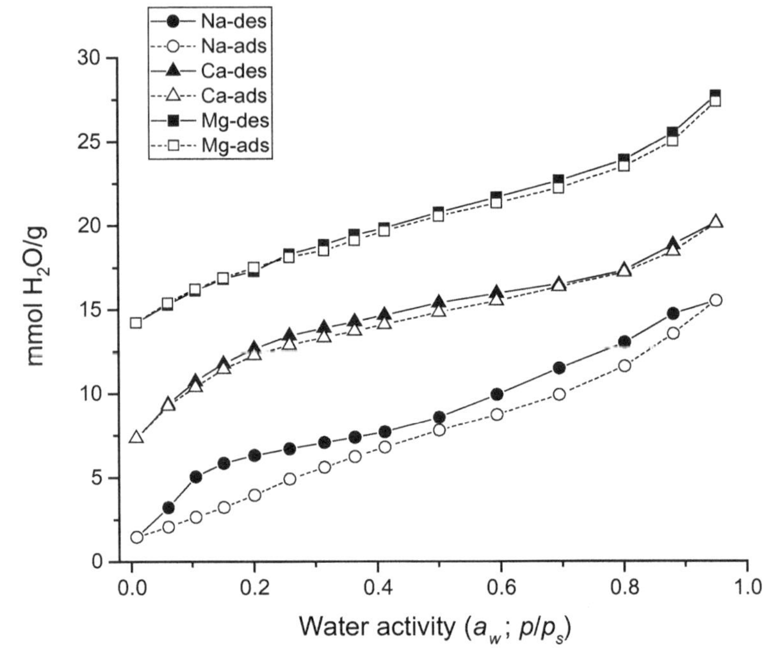

Fig. 4.13 Water desorption and adsorption isotherms of Na^{+}, Ca^{2+} and Mg^{2+} SAz-1 montmorillonite.
Data were used with permission Xu, W., Johnston, C.T., Parker, P., Agnew, S.F., 2000. Infrared study of water sorption on Na-, Li-, Ca- and Mg-exchanged (SWy-1 and SAz-1) montmorillonite. Clays Clay Miner. 48, 120-131.

behaviour of $^{2}H_2O$ (i.e. deuterium oxide or deuterated water). In this case, the quadrupolar coupling constant (QCC) is the diagnostic property of interest that changes with experimental conditions, and $^{2}H_2O$ functions as a molecular probe because it can be introduced or removed from the surface. In clay mineral studies, the QCC of $^{2}H_2O$ of bulk water ($^{2}H_2O$) can be compared to QCC of interlayer water as a function of temperature, water content, and exchangeable cation (Reddy et al., 2016). An ever-expanding group of diagnostic properties are being reported for clay mineral–water interactions that can be observed using vibrational, NMR, EPR, neutron scattering, X-reflectivity, and dielectric relaxation. A partial listing of NMR and vibrational molecular probes and reported groups is illustrated in Fig. 4.14.

Characterisation of smectite–water interactions is strongly dependent on the water activity, nature of the exchangeable cation, clay mineral properties, and environmental variables (e.g. temperature and pressure). Thus, it is important to establish a clear connection between the spectroscopic/structural information with these variables (e.g. water activity, temperature, pressure). This can be achieved using in situ methods that combine techniques such as XRD, IR, or NMR spectroscopy, with methods that can quantify water adsorption (e.g. quartz crystal microbalance). Recent examples of

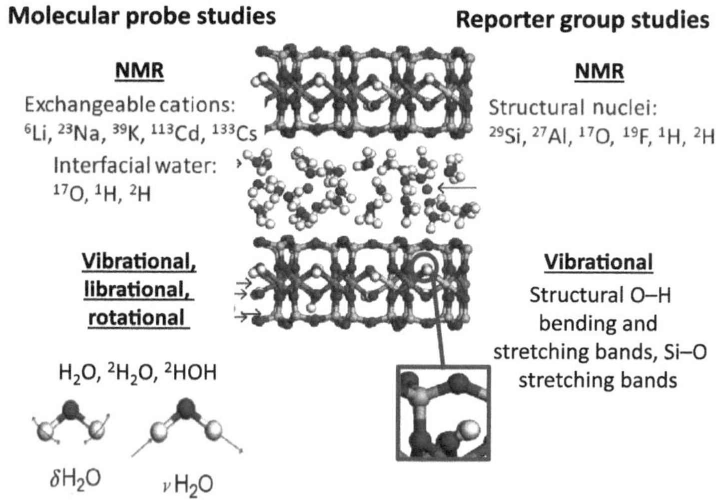

Fig. 4.14 Common molecular probes and reporter groups used to study clay-water interactions.

in situ studies include IR (Schuttlefield et al., 2007a,b; Loring et al., 2014; Thompson et al., 2014; Tang et al., 2016; Schnetzer et al., 2017), NMR (Bowers et al., 2014a,b; Kirkpatrick et al., 2015; Reddy et al., 2016), and XRD (Schaef et al., 2015, 2017). In situ studies provide a direct connection between the points, or a clearly defined region, on the water adsorption isotherm to the structural/spectroscopic properties of interest. A good example is the recent study of Schnetzer et al. (2017) where the concomitant shifts in the vibrational frequency of the H_2O bending band and the structural OH bending bands could be quantitatively related to the number of H_2O molecules coordinated to the interlayer Na^+ ions (Fig. 4.15).

4.4.1 Vibrational studies of smectite–water interactions

The vibrational properties of smectite–water interactions have been used in both molecular probe and reporter group studies (Johnston, 2010, 2017). Water is an ideal molecular probe, as its vibrational modes are sensitive to short-range interactions. In smectite–water studies, the stretching, bending, and librational vibrational modes of H_2O (and ^{2}HOH and $^{2}H_2O$) have been used to study both ion–dipole and hydrogen-bonding interactions (Russell and Farmer, 1964; Poinsignon et al., 1978; Johnston et al., 1992; Xu et al., 2000; Johnston and Premachandra, 2001; Madejova et al., 2002; Dontsova et al., 2004; Schuttlefield et al., 2007b; Kuligiewicz et al., 2015a, b; Schnetzer et al., 2016; Tang et al., 2016; Kuligiewicz and Derkowski, 2017). Vibrational studies have shown that the water molecules clustered around the exchangeable cations are strongly polarised by the cation. The remaining water molecules have similar vibrational properties to those of bulk water. Interestingly, water molecules coordinated to cations in clay mineral interlayers at low water content are less hydrogen

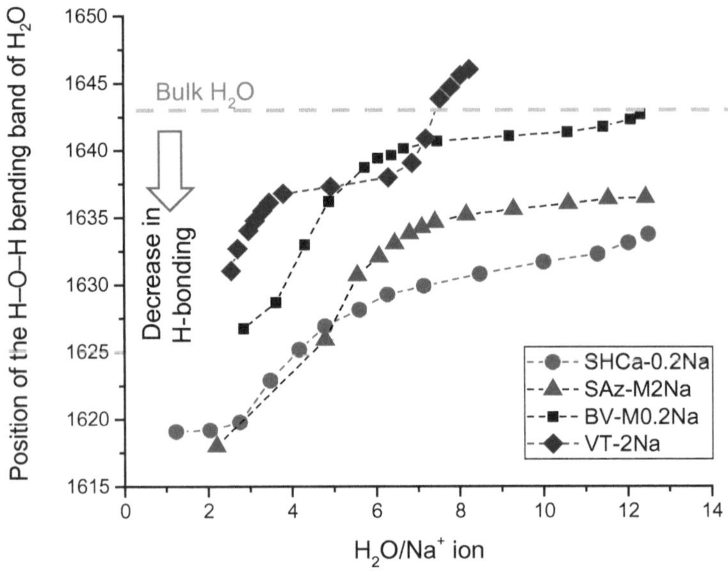

Fig. 4.15 Position of the H–O–H bending band of H_2O as a function of water content expressed as the number of water molecules per Na^+ ion for hectorite (SHCa), two montmorillonite samples (SAz-1 Cheto montmorillonite, and Volclay), and vermiculite (VT).
Data were used with permission from Schnetzer, F., Johnston, C.T., Premachandra, G.S., Giraudo, N., Schuhmann, R., Thissen, P., Emmerich, K., 2017. Impact of intrinsic structural properties on the hydration of 2:1 layer silicates. ACS Earth Space Chem. 1, 608–620.

bonded than in bulk water. Evidence comes from in situ FTIR studies of H_2O adsorbed to different smectites where the shift in the H–O–H bending band of water reflects the overall extent of hydrogen bonding between water molecules (Xu et al., 2000; Schnetzer et al., 2017). A shift of the H–O–H bending band to lower frequencies compared with bulk water indicates an overall decrease in hydrogen bonding (Fig. 4.15). For example, the position of the H–O–H bending band of water in the IR spectra of matrix-isolated or gas phase H_2O is 1594–1597 cm^{-1} and that of liquid H_2O is 1644 cm^{-1} (Venyaminov and Prendergast, 1997). Depending on the water content of the clay, the H–O–H bending can range from $\sim$1647 cm^{-1} at high water content to values $<$1620 cm^{-1} at low water content (Schnetzer et al., 2017), corresponding to increased and decreased hydrogen bonding, respectively. In addition, the ν(OH) band of adsorbed ^2H_2O and ^2HOH has been used as a probe of smectite surface properties (e.g. charge density) (Kuligiewicz et al., 2015a,b; Szczerba et al., 2016; Kuligiewicz and Derkowski, 2017).

In addition to water, the vibrational features of the clay mineral structure itself provide a type of reporter group. In some cases, these bands (e.g. the structural O—H bending or deformation bands) are sensitive to the presence of water molecules in the interlayer space. Examples of reporter group vibrational studies include (Sposito et al., 1983; Johnston et al., 1992; Xu et al., 2000; Schnetzer et al., 2017).

The structural OH deformation bands, in particular, are sensitive to both the nature of the exchangeable cation and to changes in water content (Xu et al., 2000; Schnetzer et al., 2017).

4.4.2 NMR and EPR studies of smectite–water interactions

NMR spectroscopy provides a comprehensive suite of both molecular probes and reporter groups for studying clay mineral–water interactions (Fig. 4.14). Advantages of NMR spectroscopy for clay mineral–water studies include access to element specific probes, sensitive to both local structure and dynamics over a wide range of timescales from milli- to nanoseconds, and the ability to study longer-range interactions (Kirkpatrick et al., 2015). NMR molecular probes include water (nuclei = ^{2}H, ^{1}H and ^{17}O) (Reddy et al., 2016) and exchangeable cations (e.g. $^{7}Li^{+}$, $^{23}Na^{+}$, $^{39}K^{+}$, $^{43}Ca^{2+}$, $^{87}Sr^{2+}$, $^{113}Cd^{2+}$, $^{133}Cs^{+}$) (Bank et al., 1989; Laperche et al., 1990; Weiss et al., 1990; Alba et al., 1998; Bowers et al., 2014a; Steudel et al., 2015). Structural nuclei used as reporter groups include ^{1}H, ^{2}H, ^{17}O, ^{19}F, ^{27}Al, and ^{29}Si (Sanz et al., 2015; Steudel et al., 2015). Some of these molecular probe and reporter group nuclei are shown in Fig. 4.14. NMR methods can be used to study the coordination and mobility of exchangeable cations, the residence time of interlayer H_2O molecules bonding to other interlayer H_2O molecules, or the residence time of interlayer water bonding to an exchangeable cation (Reddy et al., 2016). Because H_2O in nanoconfined geometries 'freezes' at lower temperatures than in bulk H_2O (e.g. approx. −40°C for Hal), controlled temperature NMR methods are useful to study the transition of liquid H_2O to ice-like structures (Bowers et al., 2014b). The concept of 'freezing' becomes less relevant as water content decreases to values of <10 H_2O molecules per exchangeable cation. In addition, NMR methods can measure longer-range interactions of interest in the mesoscale (2–50 nm, Schoonheydt and Bergaya, 2011) through T_1 and T_2 relaxation rates at timescales extending from frequencies of >1 GHz to <1 kHz (Kirkpatrick et al., 2015). Because these frequencies are much lower than those of vibrational frequencies (THz to PHz), complementary information is obtained using the two methods. The presence of Fe or other paramagnetic impurities strongly interfere with NMR measurements, restricting NMR to clay minerals containing little or no paramagnetic impurities.

Similar to the use of NMR methods, which look at nuclear spin transitions, electron paramagnetic resonance (also known at electron spin resonance, EPR, or ESR) measures electron spin transitions. Similar to NMR, EPR methods have been used to study cationic molecular probes (Cu^{2+}, Fe^{3+}, Mn^{2+}, Cr^{3+}, VO^{2+}) and structural nuclei (Fe^{3+} and Mn^{2+}). EPR spectra of self-supporting clay mineral films oriented parallel and perpendicular to the applied magnetic field provided the first detailed studies of hydrated cation orientation in the interlayer spaces (McBride et al., 1975a,b). In addition, the spin probes (e.g. Cu^{2+} and Fe^{3+}) were used to study single-electron transfer reactions of aromatic hydrocarbons in the interlayer spaces that occurred at low water content. EPR measurements showed both the loss of the spin-probe signal (signifying the metal cation was reduced) and the concomitant formation of a free radical from the organic species (Mortland and Pinnavaia, 1971; Pinnavaia and Mortland, 1971; Fenn et al., 1973; Johnston et al., 1991; Hinedi et al., 1993).

4.4.3 Inelastic and quasielastic neutron scattering of smectite–water interactions

Neutron scattering methods have contributed to the understanding of clay mineral–water interactions for more than four decades (Olejnik et al., 1970; Hunter et al., 1971; Johnston et al., 2000; Bordallo et al., 2008; Martins et al., 2014; Cygan et al., 2015; Jimenez-Ruiz et al., 2017). One of the unique attributes of neutron scattering methods is that these methods are particularly sensitive to the motions of hydrogen atoms. Inelastic neutron scattering (INS) of clay mineral–water interactions in the mid-infrared region can be compared directly with other vibrational spectroscopies, and the spectra can be calculated using molecular simulation (Johnston et al., 2000; Ockwig et al., 2009; Cygan et al., 2015; Jimenez-Ruiz et al., 2017). Similar to earlier EPR and IR studies, oriented clay mineral films can be used to obtain polarisation and orientation information (Michot et al., 2016; Jimenez-Ruiz et al., 2017). Furthermore, advanced molecular simulation methods, such as ab initio molecular dynamics, molecular dynamics (MD), and density functional theory, are now able to simulate the INS spectra at different levels of approximation, and this is an exciting area where experiment and theory can be meaningfully combined (Cygan et al., 2015; Jimenez-Ruiz et al., 2017). The ability to both measure and model water dynamics in confined geometries at geologically relevant temperature and pressures has immediate relevance (Cygan et al., 2015).

In comparison with INS, quasielastic neutron scattering (QENS) examines small energy transfer compared with the incident energy of the scattered particles between neutrons and the sample on 'both sides' (i.e. Stokes and anti-Stokes) of the elastic line (i.e. the energy of incident neutrons). The main contributions to QENS come from the translational and rotational motions of H_2O molecules that have characteristic timescales of 10^{-9}–10^{-10} s, and 10^{-11}–10^{-12} s, respectively (Martins et al., 2014). Early studies compared the self-diffusion coefficient of water (2.14×10^{-9} m^2/s) to that of interlayer H_2O and showed that diffusion of H_2O was significantly hindered by the clay mineral (0.6–2.5×10^{-10} m^2/s) depending on the type of clay mineral (vermiculite or Mt), exchangeable cation and overall water content (Olejnik et al., 1970; Hunter et al., 1971; Cebula et al., 1981). Additional insight was provided by Swenson et al. (2000) in a QENS study of oriented one-H_2O and two-H_2O layer vermiculites. For a two-H_2O layer vermiculite, the translational diffusion of water was about two to three times slower than that of bulk H_2O. More recently, QENS data were combined with molecular simulations to study both the parallel and perpendicular translations of H_2O in a synthetic fluoro-hectorite where the perpendicular component was found to be nonnegligible. In this case, the contribution from structural OH groups was eliminated by replacing OH^- with F^- (Marry et al., 2013). This method has been further extended by combining in situ XRD and QENS with controlled relative humidity (Gates et al., 2017). These authors provided molecular insight about the dynamics of H_2O associated with clay mineral hysteresis and clay mineral dehydration (Gates et al., 2017). In a related study of H_2O in Mt and Hal, the self-diffusion coefficient, average residence time between jumps (τ_0) on interlayer-intercalated H_2O molecules were compared with bulk water and 'surface water.' For both Hal and Mt, the self-diffusion coefficient of interlayer H_2O molecules was about three times slower than that of bulk water, the average residence time between jumps (τ_o) was 5 to 16ps compared with 1.5ps for bulk water (Bordallo et al., 2008).

4.4.4 Dielectric relaxation spectroscopy

Dielectric relaxation spectroscopy (DRS) measures the interaction of a sample with an oscillating electric field. The field-induced polarisation can be measured in the time domain or as a function of the field frequency (Buchner and Hefter, 2009). Compared with other methods, the range of timescales is quite large, extending from far-IR transitions (femtoseconds) to relaxation processes that can take hours (Schmuttenmaer, 2004). In the case of clay mineral–water interactions, DRS is sensitive to a wide range of dynamical processes involving water (Calvet, 1975; Rotenberg et al., 2005; Cadene et al., 2006; Rahman et al., 2015). Different relaxation processes occur over the wide range of timescales, and one of the challenges of DRS is to derive clear mechanistic insight from these processes. In a study of smectite–water interactions, five distinct relaxations were observed, ranging in frequency from 3kHz to 3GHz (Cadene et al., 2006). In order to resolve these features, spectral deconvolution methods are required, which can introduce some ambiguity. Of particular interest are the DRS-derived mobilities of the exchangeable cations that can be compared with MD results (Rotenberg et al., 2005). In the terahertz frequency range (also known as far-IR), the librational motions of water, cations, and related species can be resolved (Diaz et al., 2000, 2002a,b; Zich et al., 2013).

4.5 Probing the hydrophobic/hydrophilic character of clay mineral surfaces

The interaction of water with clay mineral surfaces provides important information about the overall hydrophilic character of clay minerals. Although 'wettability' of well-defined macroscopic surfaces can easily be measured by traditional methods (e.g. advancing contact angles), these methods are irrelevant for submicron-sized clay mineral particles due to surface roughness, and contribution of basal vs edge sites. Water adsorption, molecular spectroscopy, structural and imaging methods, and molecular simulation of water on clay mineral surfaces continue to provide mechanistic insight about the fundamental interactions in clay mineral–water systems.

In addition, however, adsorption of organic molecules on clay mineral surfaces provides complementary insight about the overall hydrophobic/hydrophilic nature of a particular surface, and three prototypical examples are given here as examples of hydrophobic/semipolar probes. First, consider the case of metal-uncharged ligand complexes, where in the case of the bis-Cu^{2+}-ethylenediamine complex ($Cu(en)_2^{2+}$), the overall stability constant of this complex exceeds the bulk solution value by three orders of magnitude (Maes et al., 1978; Maes and Cremers, 1979, 1981; Maes et al., 1980). In subsequent spectroscopic work, the authors concluded that the interlayer space of clay minerals behaves as a 'rigid, anionic solvent' towards $Cu(en)_2^{2+}$ and stabilised the complex. $[Cu(en)_2(H_2O)_2]^{2+}$ loses its two axially coordinated water molecules in the interlayer space. The clay mineral surface acts then as axial ligands, which are much weaker than water (Maes et al., 1980). The second example is from an important study that examined the adsorption of atrazine, a weak organic base, with a range of 14 Ca-exchanged smectites from aqueous solution. Adsorption ranged from essentially complete removal (Freundlich adsorption coefficient (K_f) >1300 for a beidellite), to essentially no adsorption ($K_f = 0.01$) for Otay Mt (Laird et al., 1992). What is interesting from this study was that adsorption varied by three orders of magnitude, but all of the adsorbents were Ca-smectites. Adsorption was related to differences in IHS and surface charge density effects. Based on Laird's work, a number of related compounds, including nitroaromatics and carbamates, have been investigated (Boyd et al., 2011). The last example is from a combined adsorption and spectroscopic study of a strongly hydrophobic compound, dibenzo-*p*-dioxin, with smectites exchanged with different exchangeable cations. In general, one would expect that smectites, which are considered to function as hydrophilic materials, would have little, if any, affinity for dibenzo-*p*-dioxin. For strongly hydrated monovalent and divalent cations, this is indeed the case. However, when the exchangeable cations are K^+ and especially Cs^+, surprisingly high adsorption is observed, and the overall amount of adsorbed dioxin is 10,000mg/kg (1wt%) (Liu et al., 2009, 2012; Rana et al., 2009). These studies show that clay mineral surfaces play an important role in stabilising organic compounds; however, a comprehensive understanding about how clay mineral surfaces interact with both H_2O and organic molecules is not yet well understood.

4.6 Conclusions

Because clay minerals are the most abundant nanoparticles and water is the only abundant inorganic liquid on Earth, it follows that clay mineral–water interactions are central to *many* important chemical, physical, or biological process in the biosphere (Robertson, 1986; Hochella et al., 2008). In soil and subsurface environments, the behaviour of water near the surfaces of clay minerals plays essential roles in regulation of most of the interactions occurring in the 'critical zone' (Brantley et al., 2007; Lin, 2010). The world is increasingly dependent upon hydrocarbon recovery from shale oil and gas reservoirs, the energy-water nexus, where clay mineral-water interactions control flow through porous media. As materials, clay minerals are continually finding new applications in materials science and clay-polymer nanocomposites.

Water is intimately associated with the structures of many clay minerals, and is an integral structural component for a number of clay minerals. From a mechanistic perspective, many clay minerals (and their counterions) interact strongly with interfacial water molecules through ion–dipole, hydrogen bonding, and related dipole–dipole interactions. For these clay minerals, the properties of interfacial water are strongly influenced by the surface. For example, the basal siloxane surface of clay minerals can be both hydrophobic in the absence of IHS, or hydrophilic, depending on the extent of charge substitution and the nature of the exchangeable cation. In reality, both hydrophobic and hydrophilic sites are present on a given clay mineral. In the case of smectites, the overall hydrophobic/hydrophilic character of clay mineral can be controlled through ion exchange of cations with large or small enthalpies of hydration. Even when water molecules are excluded from hydrophobic clay mineral surfaces, these clay minerals play an important role in stabilising organic species. For clay minerals that possess an exposed aluminium or magnesium octahedral sheet surface (e.g. kaolin group minerals), these surfaces are hydrophilic where adhesion (water–surface) interactions are greater than cohesion (water–water) interactions. Kaolinite is an interesting example because it possesses both a hydrophobic siloxane surface and a hydrophilic aluminium hydroxyl surface. After many years of studying clay mineral–water interactions, a broad spectrum of experimental and theoretical insights has been achieved. The great majority of these relate to 'short range' molecular interactions that occur at distances from 0 to 2 nm. As increasingly powerful experimental methods are combined with more sophisticated and powerful molecular simulation methods capable of simulating larger numbers of atoms and electrons, it will be exciting to see how clay mineral-water interaction governs interfacial behaviour at progressively larger scales of >2 nm.

References

Adamo, P., Violante, P., Wilson, M.J., 2001. Tubular and spheroidal halloysite in pyroclastic deposits in the area of the Roccamonfina volcano (Southern Italy). Geoderma 99, 295–316.

Alba, M.D., Alvero, R., Becerro, A.I., Castro, M.A., Trillo, J.M., 1998. Chemical behavior of lithium ions in reexpanded Li-montmorillonites. J. Phys. Chem. B 102, 2207–2213.

Ambers, C.P., Murray, H.H., 1995. The role of carbonate bedrock in the formation of Indianaite halloysitic clays. Indiana Geol. Survey Bull. 65, 1–29.

Auer, B.M., Skinner, J.L., 2008. IR and Raman spectra of liquid water: theory and interpretation. J. Chem. Phys. 128, 12.

Auer, B., Kumar, R., Schmidt, J.R., Skinner, J.L., 2007. Hydrogen bonding and Raman, IR, and 2D-IR spectroscopy of dilute HOD in liquid D2O. Proc. Natl. Acad. Sci. U.S.A. 104, 14215–14220.

Bank, S., Bank, J.F., Ellis, P.D., 1989. Solid-state 113Cd nuclear magnetic resonance study of exchanged montmorillonites. J. Phys. Chem. 93, 4847–4855.

Berend, I., Cases, J.-M., Francois, M., Uriot, J.P., Michot, L., Masion, A., Thomas, F., 1996. Mechanism of adsorption and desorption of water vapor by homoionic montmorillonites: 2. The Li^+ Na^+, K^+, Rb^+ and Cs^+-exchanged forms. Clays Clay Miner. 43, 324–336.

Bergaya, F., Lagaly, G. (Eds.), 2013a. Handbook of Clay Science. Part A: Fundamentals. second ed. Elsevier, Amsterdam, NL.

Bergaya, F., Lagaly, G. (Eds.), 2013b. Handbook of Clay Science. Part B: Techniques and Applications. second ed. Elsevier, Amsterdam, NL.

Bordallo, H.N., Aldridge, L.P., Churchman, G.J., Gates, W.P., Telling, M.T.F., Kiefer, K., Fouquet, P., Seydel, T., Kimber, S.a.J., 2008. Quasi-elastic neutron scattering studies on clay interlayer-space highlighting the effect of the cation in confined water dynamics. J. Phys. Chem. C 112, 13982–13991.

Bowers, G.M., Hoyt, D.W., Burton, S.D., Ferguson, B.O., Varga, T., Kirkpatrick, R.J., 2014a. In situ C^{-13} and Na^{-23} magic angle spinning NMR investigation of supercritical CO_2 incorporation in smectite-natural organic matter composites. J. Phys. Chem. C 118, 3564–3573.

Bowers, G.M., Singer, J.W., Bish, D.L., Kirkpatrick, J., 2014b. Structural and dynamical relationships of Ca^{2+} and H_2O in smectite/$(H_2O)^{-}H^{-2}$ systems. Am. Mineral. 99, 318–331.
Boyd, S.A., Johnston, C.T., Laird, D.A., Teppen, B.J., Li, H., 2011. Comprehensive study of organic contaminant adsorption by clays: methodologies, mechanisms, and environmental implications. In: Xing, B., Senesi, N., Huang, P.M. (Eds.), Biophysico-Chemical Processes of Anthropogenic Organic Compounds in Environmental Systems. John Wiley & Sons, Inc.
Brantley, S.L., Goldhaber, M.B., Ragnarsdottir, K.V., 2007. Crossing disciplines and scales to understand the critical zone. Elements 3, 307–314.
Brigatti, M.F., Galán, E., Theng, B.K.G., 2013. Chapter 2—structure and mineralogy of clay minerals. In: Bergaya, F., Lagaly, G. (Eds.), Developments in Clay Science. Elsevier.
Buchner, R., Hefter, G., 2009. Interactions and dynamics in electrolyte solutions by dielectric spectroscopy. Phys. Chem. Chem. Phys. 11, 8984–8999.
Burgess, J., 1999. 4—Ion–Solvent Interactions. Ions in Solution, second ed. Woodhead Publishing.
Cadene, A., Rotenberg, B., Durand-Vidal, S., Badot, J.C., Turq, P., 2006. Dielectric spectroscopy as a probe for dynamic properties of compacted smectites. Phys. Chem. Earth 31, 505–510.
Calvet, R., 1975. Dielectric properties of montmorillonites saturated by bivalent cations. Clays Clay Miner. 23, 257–265.
Cancela, G.D., Huertas, F.J., Taboada, E.R., Sanchezrasero, F., Laguna, A.H., 1997. Adsorption of water vapor by homoionic montmorillonites. Heats of adsorption and desorption. J. Colloid Interface Sci. 185, 343–354.
Cases, J.M., Berend, I., Besson, G., Francois, M., Uriot, J.P., Thomas, F., Poirier, J.E., 1992. Mechanism of adsorption and desorption of water vapor by homoionic montmorillonite. 1. The sodium-exchanged form. Langmuir 8, 2730–2739.
Cases, J.M., Berend, I., Francois, M., Uriot, J.P., Michot, L.J., Thomas, F., 1997. Mechanism of adsorption and desorption of water vapor by homoionic montmorillonite. 3. The Mg^{2+}, Ca^{2+}, Sr^{2+} and Ba^{2+} exchanged forms. Clays Clay Miner. 45, 8–22.
Cebula, D.J., Thomas, R.K., White, J.W., 1981. Diffusion of water in Li-montmorillonite studied by quasielastic neutron scattering. Clays Clay Miner. 29, 241–248.
Chinnery, N.J., Pawley, A.R., Clark, S.M., 1999. In situ observation of the formation of 10 angstrom phase from talc plus H_2O at mantle pressures and temperatures. Science 286, 940–942.
Coblentz, W.W., 1910. Note on water of crystallization. Phys. Rev. 30, 322–327.
Coblentz, W.W., 1911. The role of water in minerals. J. Franklin Inst. 172, 309–335.
Collins, K.D., 1997. Charge density-dependent strength of hydration and biological structure. Biophys. J. 72, 65–76.
Collins, K.D., 2004. Ions from the Hofmeister series and osmolytes: effects on proteins in solution and in the crystallization process. Methods 34, 300–311.
Collins, K.D., Neilson, G.W., Enderby, J.E., 2007. Ions in water: characterizing the forces that control chemical processes and biological structure. Biophys. Chem. 128, 95–104.
Costanzo, P.M., Guggenheim, S., 2001. Baseline studies of the clay minerals society source clays. Clays Clay Miner. 49, 371–452.
Cruz, M.I., Letellier, M., Fripiat, J.J., 1978. NMR study of adsorbed water. II. Molecular motions in the monolayer hydrate of halloysite. J. Chem. Phys. 69, 2018–2027.
Cuadros, J., Dudek, T., 2006. FTIR investigation of the evolution of the octahedral sheet of kaolinite-smectite with progressive kaolinization. Clays Clay Miner. 54, 1–11.

Cygan, R.T., Daemen, L.L., Ilgen, A.G., Krumhansl, J.L., Nenoff, T.M., 2015. Inelastic neutron scattering and molecular simulation of the dynamics of inter layer water in smectite clay minerals. J. Phys. Chem. C 119, 28005–28019.

Debenedetti, P.G., 2003. Supercooled and glassy water. J. Phys. Condens. Matter 15, R1669–R1726.

Derjaguin, B., Landau, L., 1941. Theory of the stability of strongly charged lyophobic sols and of the adhesion of strongly charged particles in solutions of electrolytes. Acta Physicochim. URSS 14, 633.

Dewalt-Kerian, E.L., Kim, S., Azam, M.S., Zeng, H.B., Liu, Q.X., Gibbs, J.M., 2017. pH-dependent inversion of Hofmeister trends in the water structure of the electrical double layer. J. Phys. Chem. Lett. 8, 2855–2861.

Diaz, M., Farmer, V.C., Prost, R., 2000. Characterization and assignment of far infrared absorption bands of K^+ in muscovite. Clays Clay Miner. 48, 433–438.

Diaz, M., Huard, E., Prost, R., 2002a. Far infrared analysis of the structural environment of interlayer K^+, NH_4^+, Rb^+ and Cs^+ selectively retained by vermiculite. Clays Clay Miner. 50, 284–293.

Diaz, M., Laperche, V., Harsh, J., Prost, R., 2002b. Far infrared spectra of K^+ in dioctahedral and trioctahedral mixed-layer minerals. Am. Mineral. 87, 1207–1214.

Dontsova, K.M., Norton, L.D., Johnston, C.T., Bigham, J.M., 2004. Influence of exchangeable cation on water adsorption by soil clays. Soil Sci. Soc. Am. J. 68, 1218–1227.

Douillard, J.M., Medout-Marere, V., 2000. Surface Energy and Acid-Base Properties of Solids Studied by Immersion Calorimetry. Vsp Bv-C/O Brill Academic Publishers, Leiden.

Douillard, J.M., Malandrini, H., Zoungrana, T., Clauss, F., Partyka, S., 1994. Surface-tension of talc and talc-chlorite mixtures. J. Thermal Anal. 41, 1205–1210.

Douillard, J.M., Zajac, J., Malandrini, H., Clauss, F., 2002. Contact angle and film pressure: study of a talc surface. J. Colloid Interface Sci. 255, 341–351.

Du, Q., Freysz, E., Shen, Y.R., 1994. Vibrational spectra of water molecules at quartz-water interfaces. Phys. Rev. Let. 72, 238–241.

Emmerich, K., 2013. Chapter 2.13—full characterization of smectites. In: Bergaya, F., Lagaly, G. (Eds.), Developments in Clay Science. Elsevier.

Fenn, D.B., Mortland, M.M., Pinnavaia, T.J., 1973. The chemisorption of anisole on Cu(II)-hectorite. Clays Clay Miner. 21, 315–322.

Ferrante, F., Armata, N., Lazzara, G., 2015. Modeling of the halloysite spiral nanotube. J. Phys. Chem. C 119, 16700–16707.

Fialips, C.I., Petit, S., Decarreau, A., 2000. Hydrothermal formation of kaolinite from various metakaolins. Clay Miner. 35, 559–572.

Fraser, H.J., Mccoustra, M.R.S., Williams, D.A., 2002. The molecular universe. Astron. Geophys. 43, 10–18.

Frost, R.L., 1995. Fourier transform Raman spectroscopy of kaolinite, dickite and halloysite. Clays Clay Miner. 43, 191–195.

Frost, R.L., Kristof, J., 1997. Intercalation of halloysite: a Raman spectroscopic study. Clays Clay Miner. 45, 551–563.

Frost, R.L., Shurvell, H.F., 1997. Raman microprobe spectroscopy of halloysite. Clays Clay Miner. 45, 68–72.

Gallo, P., Arnann-Winkel, K., Angell, C.A., Anisimov, M.A., Caupin, F., Chakravarty, C., Lascaris, E., Loerting, T., Panagiotopoulos, A.Z., Russo, J., Sellberg, J.A., Stanley, H.E., Tanaka, H., Vega, C., Xu, L.M., Pettersson, L.G.M., 2016. Water: a tale of two liquids. Chem. Rev. 116, 7463–7500.

Gates, W.P., Aldridge, L.P., Carnero-Guzman, G.G., Mole, R.A., Yu, D.H., Iles, G.N., Klapproth, A., Bordallo, H.N., 2017. Water desorption and absorption isotherms of sodium montmorillonite: a QENS study. Appl. Clay Sci. 147, 97–104.

Giese, K., Kaatze, U., Pottel, R., 1970. Permittivity and dielectric and proton magnetic relaxation of aqueous solutions of alkali halides. J. Phys. Chem. 74, 3718.

Giese, R.F., Costanzo, P.M., Vanoss, C.J., 1991. The surface free energies of talc and pyrophyllite. Phys. Chem. Miner. 17, 611–616.

Gilbert, B., Comolli, L.R., Tinnacher, R.M., Kunz, M., Banfield, J.F., 2015. Formation and restacking of disordered smectite osmotic hydrates. Clays Clay Miner. 63, 432–442.

Götte, L., Parry, K.M., Hua, W., Verreault, D., Allen, H.C., Tobias, D.J., 2017. Solvent-shared ion pairs at the air–solution interface of magnesium chloride and sulfate solutions revealed by sum frequency spectroscopy and molecular dynamics simulations. J. Phys. Chem. A 121, 6450–6459.

Guven, N., 1992. Molecular aspects of clay/water interactions. In: Guven, N., Pollastro, R.M. (Eds.), Clay-Water Interface and Its Rheological Implications. The Clay Minerals Society, Boulder, CO.

Hall, P.L., Astill, D.M., 1989. Adsorption of water by homoionic exchange forms of Wyoming montmorillonite (SWy-1). Clays Clay Miner. 37, 355–363.

Hazen, R.M., Ferry, J.M., 2010. Mineral evolution: mineralogy in the fourth dimension. Elements 6, 9–12.

Hinedi, Z.R., Johnston, C.T., Erickson, C., 1993. Chemisorption of benzene on Cu-montmorillonite as characterized by FTIR and ^{13}C MAS NMR. Clays Clay Miner. 41, 87–94.

Hirsemann, D., Shylesh, S., De Souza, R.A., Diar-Bakerly, B., Biersack, B., Mueller, D.N., Martin, M., Schobert, R., Breu, J., 2012. Large-scale, low-cost fabrication of Janus-type emulsifiers by selective decoration of natural kaolinite platelets. Angew. Chem. Int. Ed. 51, 1348–1352.

Hochella, M.F., Lower, S.K., Maurice, P.A., Penn, R.L., Sahai, N., Sparks, D.L., Twining, B.S., 2008. Nanominerals, mineral nanoparticles, and Earth systems. Science 319, 1631–1635.

Hu, X.L., Michaelides, A., 2010. The kaolinite (001) polar basal plane. Surf. Sci. 604, 111–117.

Hua, W., Verreault, D., Huang, Z.S., Adams, E.M., Allen, H.C., 2014. Cation effects on interfacial water organization of aqueous chloride solutions. I. Monovalent cations: Li^+, Na^+, K^+, and NH_4^+. J. Phys. Chem. B 118, 8433–8440.

Hunter, R.J., Stirling, G.C., White, J.W., 1971. Water dynamics in clays by neutron spectroscopy. Nat. Phys. Sci. 230, 192.

Janik, L.J., Keeling, J.L., 1993. FT-Ir partial least-squares analysis of tubular halloysite in kaolin samples from the mount hope kaolin deposit. Clay Miner. 28, 365–378.

Jeanmairet, G., Marry, V., Levesque, M., Rotenberg, B., Borgis, D., 2014. Hydration of clays at the molecular scale: the promising perspective of classical density functional theory. Mol. Phys. 112, 1320–1329.

Jimenez-Ruiz, M., Ferrage, E., Blanchard, M., Fernandez-Castanon, J., Delville, A., Johnson, M.R., Michot, L.J., 2017. Combination of inelastic neutron scattering experiments and ab initio quantum calculations for the study of the hydration properties of oriented saponites. J. Phys. Chem. C 121, 5029–5040.

Johnston, C.T., 2010. Probing the nanoscale architecture of clay minerals. Clay Miner. 45, 245–279.

Johnston, C.T., 2017. Infrared studies of clay mineral-water interactions. In: Gates, W.P., Kloprogge, J.T., Madejova, J., Bergaya, F. (Eds.), Infrared and Raman Spectroscopies of Clay Minerals. Elsevier, Amsterdam.

Johnston, C.T., Premachandra, G.S., 2001. Polarized ATR-FTIR study of smectite in aqueous suspension. Langmuir 17, 3712–3718.
Johnston, C.T., Tipton, T., Stone, D.A., Erickson, C., Trabue, S.L., 1991. Chemisorption of p-dimethoxybenzene on Cu-montmorillonite. Langmuir 7, 289–296.
Johnston, C.T., Sposito, G., Erickson, C., 1992. Vibrational probe studies of water interactions with montmorillonite. Clays Clay Miner. 40, 722–730.
Johnston, C.T., Sposito, G., Earl, W.L., 1993. Surface spectroscopy of environmental particles by Fourier transform infrared and nuclear magnetic resonance spectroscopy. In: Buffle, J., Van Leeuwen, H.P. (Eds.), Environmental Particles. In: The Environmental Analytical and Physical Chemistry Series, vol. 2. Lewis Publ, Boca Raton.
Johnston, C.T., Bish, D.L., Eckert, J., Brown, L.A., 2000. Infrared and inelastic neutron scattering study of the 1.03-and 0.95-nm kaolinite-hydrazine intercalation complexes. J. Phys. Chem. B 104, 8080–8088.
Joussein, E., Petit, S., Churchman, J., Theng, B., Righi, D., Delvaux, B., 2005. Halloysite clay minerals—a review. Clay Miner. 40, 383–426.
Keenan, A.G., Mooney, R.W., Wood, L.A., 1951. The relation between exchangeable ions and water adsorption on kaolinite. J. Phys. Colloid Chem. 55, 1462–1474.
Kirillova, A., Stoychev, G., Ionov, L., Eichhorn, K.J., Malanin, M., Synytska, A., 2014. Platelet Janus particles with hairy polymer shells for multifunctional materials. ACS Appl. Mater. Interfaces 6, 13106–13114.
Kirillova, A., Marschelke, C., Friedrichs, J., Werner, C., Synytska, A., 2016. Hybrid hairy Janus particles as building blocks for antibiofouling surfaces. ACS Appl. Mater. Interfaces 8, 32591–32603.
Kirkpatrick, R.J., Kalinichev, A.G., Bowers, G.M., Yazaydin, A.O., Krishnan, M., Saharay, M., Morrow, C.P., 2015. NMR and computational molecular modeling studies of mineral surfaces and interlayer galleries: a review. Am. Mineral. 100, 1341–1354.
Kristof, J., Toth, M., Gabor, M., Szabo, P., Frost, R.L., 1997. Study of the structure and thermal behaviour of intercalated kaolinites. J. Thermal Anal. 49, 1441–1448.
Kuligiewicz, A., Derkowski, A., 2017. Tightly bound water in smectites. Am. Mineral. 102, 1073–1090.
Kuligiewicz, A., Derkowski, A., Emmerich, K., Christidis, G.E., Tsiantos, C., Gionis, V., Chryssikos, G.D., 2015a. Measuring the layer charge of dioctahedral smectite by O-D vibrational spectroscopy. Clays Clay Miner. 63, 443–456.
Kuligiewicz, A., Derkowski, A., Szczerba, M., Gionis, V., Chryssikos, G.D., 2015b. Revisiting the infrared spectrum of the water-smectite interface. Clays Clay Miner. 63, 15–29.
Kumar, N., Zhao, C.L., Klaassen, A., Van Den Ende, D., Mugele, F., Siretanu, I., 2016. Characterization of the surface charge distribution on kaolinite particles using high resolution atomic force microscopy. Geochim. Cosmochim. Acta 175, 100–112.
Laird, D.A., Barriuso, E., Dowdy, R.H., Koskinen, W.C., 1992. Adsorption of atrazine on smectites. Soil Sci. Soc. Am. J. 56, 62–67.
Laperche, V., Lambert, J.F., Prost, R., Fripiat, J.J., 1990. High-resolution solid-state NMR of exchangeable cations in the interlayer surface of a swelling mica: ^{23}NA, ^{111}Cd, and ^{133}Cs vermiculites. J. Phys. Chem. 94, 8821–8831.
Lavikainen, L.P., Tanskanen, J.T., Schatz, T., Kasa, S., Pakkanen, T.A., 2015. Montmorillonite interlayer surface chemistry: effect of magnesium ion substitution on cation adsorption. Theor. Chem. Accounts 134, 7.
Lee, J.H., Guggenheim, S., 1981. Single-crystal X-ray refinement of pyrophyllite-1TC. Am. Mineral. 66, 350–357.

Li, F., Skinner, J.L., 2010a. Infrared and Raman line shapes for ice Ih. I. Dilute HOD in H_2O and D_2O. J. Chem. Phys. 132, 204505.
Li, F., Skinner, J.L., 2010b. Infrared and Raman line shapes for ice Ih. II. H_2O and D_2O. J. Chem. Phys. 133, 244504.
Lin, H., 2010. Earth's critical zone and hydropedology: concepts, characteristics, and advances. Hydrol. Earth Syst. Sci. 14, 25–45.
Liu, C., Li, H., Teppen, B.J., Johnston, C.T., Boyd, S.A., 2009. Mechanisms associated with the high adsorption of dibenzo-p-dioxin from water by smectite clays. Environ. Sci. Technol. 43, 2777–2783.
Liu, C., Li, H., Johnston, C.T., Boyd, S.A., Teppen, B.J., 2012. Relating clay structural factors to dioxin adsorption by smectites: molecular dynamics simulations. Soil Sci. Soc. Am. J. 76, 110–120.
Loring, J.S., Ilton, E.S., Chen, J., Thompson, C.J., Martin, P.F., Benezeth, P., Rosso, K.M., Felmy, A.R., Schaef, H.T., 2014. In situ study of CO_2 and H_2O partitioning between Na-montmorillonite and variably wet supercritical carbon dioxide. Langmuir 30, 6120–6128.
Lvov, Y.M., Shchukin, D.G., Mohwald, H., Price, R.R., 2008. Halloysite clay nanotubes for controlled release of protective agents. ACS Nano 2, 814–820.
Madejova, J., Janek, M., Komadel, P., Herbert, H.J., Moog, H.C., 2002. FTIR analyses of water in MX-80 bentonite compacted from high salinary salt solution systems. Appl. Clay Sci. 20, 255–271.
Maes, A., Cremers, A., 1979. Stability of metal uncharged ligand complexes in ion exchangers. Part 4. Hydration effects and stability changes of copper-ethylenediamine complexes in montmorillonite. J. Chem. Soc. Faraday Trans. 1 75, 513–524.
Maes, A., Cremers, A., 1981. Influence of charge-density on the acid-base-equilibrium of ethylenediamine in montmorillonites. J. Chem. Soc., Faraday Trans.1 77, 1553–1559.
Maes, A., Peigneur, P., Cremers, A., 1978. Stability of metal uncharged ligand complexes in ion exchangers. Part 2.—the copper + ethylenediamiine complex in montmorillonite and sulphonic acid resin. J. Chem. Soc. Faraday Trans. 1 74, 182–189.
Maes, A., Schoonheydt, R.A., Cremers, A., Uytterhoeven, J.B., 1980. Spectroscopy of Cu(en)2 2^+ on clay surfaces. Surface and charge-density effects. J. Phys. Chem. 84, 2795–2799.
Malandrini, H., Clauss, F., Partyka, S., Douillard, J.M., 1997. Interactions between talc particles and water and organic solvents. J. Colloid Interface Sci. 194, 183–193.
Marcus, Y., 1987. The thermodynamics of solvations of ions. 2. The enthalpy of hydration at 298.15 K. J. Chem. Soc., Faraday Trans.1 83, 339–349.
Marcus, Y., 1988. Ionic radii in aqueous-solutions. Chem. Rev. 88, 1475–1498.
Marcus, Y., 2009. Effect of ions on the structure of water: structure making and breaking. Chem. Rev. 109, 1346–1370.
Marcus, Y., 2015a. Ion solvation in neat solvents. In: Ions in Solution and Their Solvation. John Wiley & Sons, Inc.
Marcus, Y., 2015b. Mutual effects of ions and solvents. In: Ions in Solution and Their Solvation. John Wiley & Sons, Inc.
Marry, V., Rotenberg, B., Turq, P., 2008. Structure and dynamics of water at a clay surface from molecular dynamics simulation. Phys. Chem. Chem. Phys. 10, 4802–4813.
Marry, V., Dubois, E., Malikova, N., Breu, J., Haussler, W., 2013. Anisotropy of water dynamics in clays: insights from molecular simulations for experimental QENS analysis. J. Phys. Chem. C 117, 15106–15115.
Martins, M.L., Gates, W.P., Michot, L., Ferrage, E., Marry, V., Bordallo, H.N., 2014. Neutron scattering, a powerful tool to study clay minerals. Appl. Clay Sci. 96, 22–35.

Mcbride, M.B., Pinnavaia, T.J., Mortland, M.M., 1975a. Electron spin relaxation and the mobility of manganese(II) exchange ions in smectites. Am. Mineral. 60, 66–72.
Mcbride, M.B., Pinnavaia, T.J., Mortland, M.M., 1975b. Electron spin resonance studies of cation orientation in restricted water layers on phyllosilicate (smectite) surfaces. J. Phys. Chem. 79, 2430–2435.
Michot, L.J., Villieras, F., Francois, M., Yvon, J., Ledred, R., Cases, J.M., 1994. The structural microscopic hydrophilicity of talc. Langmuir 10, 3765–3773.
Michot, L.J., Ferrage, E., Delville, Jimenez-Ruiz, M. 2016. Influence of layer charge, hydration state and cation nature on the collective dynamics of interlayer water in synthetic swelling clay minerals. Appl. Clay Sci., 119, 375–384.
Mignon, P., Ugliengo, P., Sodupe, M., Hernandez, E.R., 2010. Ab initio molecular dynamics study of the hydration of Li^+, Na^+ and K^+ in a montmorillonite model. Influence of isomorphic substitution. Phys. Chem. Chem. Phys. 12, 688–697.
Mooney, R.W., Keenan, A.G., Wood, L.A., 1952a. Adsorption of water vapor by montmorillonite. I. Heat of desorption and application of BET theory. J. Am. Chem. Soc. 74, 1367–1374.
Mooney, R.W., Keenan, A.G., Wood, L.A., 1952b. Adsorption of water vapor by montmorillonite. II. Effect of exchangeable ions and lattice swelling as measured by X-ray diffraction. J. Am. Chem. Soc. 74, 1371–1374.
Mortland, M.M., Pinnavaia, T.J., 1971. Formation of copper(II) arene complexes on the interlamellar surfaces of montmorillonite. Nature 229, 75–77.
Mottl, M.J., Glazer, B.T., Kaiser, R.I., Meech, K.J., 2007. Water and astrobiology. Chem. Erde-Geochem. 67, 253–282.
Newman, A.C.D., 1987. The interaction of water with clay mineral surfaces. In: Newman, A.C.D. (Ed.), Chemistry of Clays and Clay Minerals. Mineralogical Society, London. Monograph No. 6.
Ni, Y.C., Skinner, J.L., 2015. IR and SFG vibrational spectroscopy of the water bend in the bulk liquid and at the liquid-vapor interface, respectively. J. Chem. Phys. 143, 12.
Norrish, K., 1954. The swelling of montmorillonite. Discuss. Faraday Soc. 18, 120–134.
Ockwig, N.W., Greathouse, J.A., Durkin, J.S., Cygan, R.T., Daemen, L.L., Nenoff, T.M., 2009. Nanoconfined water in magnesium-rich 2:1 phyllosilicates. J. Am. Chem. Soc. 131, 8155–8162.
Ohtaki, H., Radnai, T., 1993. Structure and dynamics of hydrated ions. Chem. Rev. 93, 1157–1204.
Olejnik, S., Stirling, G.C., White, J.W., 1970. Neutron scattering studies of hydrated layer silicates. Spec. Discuss. Faraday Soc. 1, 194–201.
Parry, S.A., Pawley, A.R., Jones, R.L., Clark, S.M., 2007. An infrared spectroscopic study of the OH stretching frequencies of talc and 10-A phase to 10 GPa. Am. Mineral. 92, 525–531.
Partyka, S., Douillard, J.M., 1995. Nature of interactions between organic pure liquids and model rocks—a calorimetric investigation. J. Petrol. Sci. Eng. 13, 95–102.
Pawley, A.R., Welch, M.D., 2014. Further complexities of the 10 angstrom phase revealed by infrared spectroscopy and X-ray diffraction. Am. Mineral. 99, 712–719.
Perakis, F., De Marco, L., Shalit, A., Tang, F.J., Kann, Z.R., Kuhne, T.D., Torre, R., Bonn, M., Nagata, Y., 2016. Vibrational spectroscopy and dynamics of water. Chem. Rev. 116, 7590–7607.
Perdikatsis, B., Burzlaff, H., 1981. Structural refinement of talc $Mg_3(OH)2Si_4O_{10}$. Z. Kristallogr. 156, 177–186.
Pettersson, L.G.M., Henchman, R.H., Nilsson, A., 2016. Water—the most anomalous liquid. Chem. Rev. 116, 7459–7462.

Pinnavaia, T.J., Mortland, M.M., 1971. Interlamellar metal complexes on layer silicates. I. Copper(II)-arene complexes on montmorillonite. J. Phys. Chem. 75 (26), 3957–3962.

Poinsignon, C., Cases, J.M., Fripiat, J.J., 1978. Electrical-polarization of water molecules adsorbed by smectites. An infrared study. J. Phys. Chem. 82, 1855–1860.

Polansky, R., Kadlec, P., Kolska, Z., Svorcik, V., 2017. Influence of dehydration on the dielectric and structural properties of organically modified montmorillonite and halloysite nanotubes. Appl. Clay Sci. 147, 19–27.

Prost, R., Koutit, T., Benchara, A., Huard, E., 1998. State and location of water adsorbed on clay minerals: consequences of the hydration and swelling-shrinkage phenomena. Clays Clay Miner. 46, 117–131.

Rahman, R., Mccarty, D.K., Prasad, M., 2015. Sub-THz complex dielectric constants of smectite clay thin samples with $Na+/Ca^{++}$ ions. J. Geophys. Res. Solid Earth 120, 6219–6225.

Rana, K., Boyd, S.A., Teppen, B.J., Li, H., Liu, C., Johnston, C.T., 2009. Probing the microscopic hydrophobicity of smectite surfaces. A vibrational spectroscopic study of dibenzo-p-dioxin sorption to smectite. Phys. Chem. Chem. Phys. 11, 2976–2985.

Rashchenko, S.V., Likhacheva, A.Y., Goryainov, S.V., Krylov, A.S., Litasov, K.D., 2016. In situ spectroscopic study of water intercalation into talc: new features of 10 angstrom phase formation. Am. Mineral. 101, 431–436.

Reddy, U.V., Bowers, G.M., Loganathan, N., Bowden, M., Yazaydin, A.O., Kirkpatrick, R.J., 2016. Water structure and dynamics in smectites: X-ray diffraction and ^2H NMR spectroscopy of Mg–, Ca–, Sr–, Na–, Cs–, and Pb–hectorite. J. Phys. Chem. C 120, 8863–8876.

Robertson, R.H.S., 1986. Fuller's Earth: A History of Calcium Montmorillonite. Volturna Press, Hythe, Kent, UK.

Rotenberg, B., Cadene, A., Dufreche, J.F., Durand-Vidal, S., Badot, J.C., Turq, P., 2005. An analytical model for probing ion dynamics in clays with broadband dielectric spectroscopy. J. Phys. Chem. B 109, 15548–15557.

Rotenberg, B., Patel, A.J., Chandler, D., 2011. Molecular explanation for why talc surfaces can be both hydrophilic and hydrophobic. J. Am. Chem. Soc. 133, 20521–20527.

Russell, J.D., Farmer, V.C., 1964. Infra-red spectroscopic study of the dehydration of montmorillonite and saponite. Clay Miner. Bull. 5, 443–464.

Sanz, J., Sobrados, I., Robert, J.L., 2015. Influence of hydration on Na-23, Al-27, and Si-29 MAS-NMR spectra of sodium saponites and sodium micas. Am. Mineral. 100, 1076–1083.

Schaef, H.T., Loring, J.S., Glezakou, V.A., Miller, Q.R.S., Chen, J., Owen, A.T., Lee, M.S., Ilton, E.S., Felmy, A.R., Mcgrail, B.P., Thompson, C.J., 2015. Competitive sorption of CO2 and H2O in 2:1 layer phyllosilicates. Geochim. Cosmochim. Acta 161, 248–257.

Schaef, H.T., Loganathan, N., Bowers, G.M., Kirkpatrick, R.J., Yazaydin, A.O., Burton, S.D., Hoyt, D.W., Thanthiriwatte, K.S., Dixon, D.A., Mcgrail, B.P., Rosso, K.M., Ilton, E.S., Loring, J.S., 2017. Tipping point for expansion of layered aluminosilicates in weakly polar solvents: supercritical CO2. ACS Appl. Mater. Interfaces 9, 36783–36791.

Schmuttenmaer, C.A., 2004. Exploring dynamics in the far-infrared with terahertz spectroscopy. Chem. Rev. 104, 1759–1779.

Schnetzer, F., Thissen, P., Giraudo, N., Emmerich, K., 2016. Unraveling the coupled processes of (de)hydration and structural changes in Na^+-saturated montmorillonite. J. Phys. Chem. C 120, 15282–15287.

Schnetzer, F., Johnston, C.T., Premachandra, G.S., Giraudo, N., Schuhmann, R., Thissen, P., Emmerich, K., 2017. Impact of intrinsic structural properties on the hydration of 2:1 layer silicates. ACS Earth and Space Chem. 1, 608–620.

Schoonheydt, R.A., Bergaya, F., 2011. Industrial clay minerals as nanomaterials, Chapter 10. In: Christidis, G. (Ed.), Advances in the Characterization of Industrial Minerals. EMU Notes in Mineralogy, vol. 9. Miner. Soc. Publishers, pp. 415–440.

Schoonheydt, R.A., Johnston, C.T., 2013. Surface and interface chemistry of clay minerals. In: Bergaya, F., Theng, B.K.G., Lagaly, G. (Eds.), Handbook of Clay Science, second ed. Part A. Fundamentals, Elsevier, Amsterdam.

Schrader, M.E., Yariv, S., 1990. Wettability of clay minerals. J. Colloid Interface Sci. 136, 85–94.

Schuttlefield, J., Al-Hosney, H., Zachariah, A., Grassian, V.H., 2007a. Attenuated total reflection Fourier transform infrared spectroscopy to investigate water uptake and phase transitions in atmospherically relevant particles. Appl. Spectrosc. 61, 283–292.

Schuttlefield, J.D., Cox, D., Grassian, V.H., 2007b. An investigation of water uptake on clays minerals using ATR-FTIR spectroscopy coupled with quartz crystal microbalance measurements. J. Geophys. Res. Atmos. 112, D21303, https://doi.or/10.1029/2007JD008973.

Shang, J.Y., Flury, M., Harsh, J.B., Zollars, R.L., 2010. Contact angles of aluminosilicate clays as affected by relative humidity and exchangeable cations. Colloids Surf. A Physicochem. Eng. Asp. 353, 1–9.

Shchukin, D.G., Sukhorukov, G.B., Price, R.R., Lvov, Y.M., 2005. Halloysite nanotubes as biomimetic nanoreactors. Small 1, 510–513.

Solc, R., Gerzabek, M.H., Lischka, H., Tunega, D., 2011. Wettability of kaolinite (001) surfaces—molecular dynamic study. Geoderma 169, 47–54.

Soper, A.K., Neilson, G.W., Enderby, J.E., Howe, R.A., 1977. Neutron diffraction study of hydration effects in aqueous solutions. J. Phys. C Solid State Phys. 10, 1793–1801.

Sposito, G., Prost, R., 1982. Structure of water adsorbed on smectites. Chem. Rev. 82, 553–573.

Sposito, G., Prost, R., Gaultier, J.P., 1983. Infrared spectroscopic study of adsorbed water on reduced-charge Na/Li montmorillonites. Clays Clay Miner. 31, 9–16.

Srodon, J., Mccarty, D.K., 2008. Surface area and layer charge of smectite from CEC and EGME/H_2O-retention measurements. Clays Clay Miner. 56, 155–174.

Steudel, A., Heinzmann, R., Indris, S., Emmerich, K., 2015. CEC and 7Li MAS NMR study of interlayer Li^+ in the montmorillonite-beidellite series at room temperature and after heating. Clays Clay Miner. 63, 337–350.

Swenson, J., Bergman, R., Howells, W.S., 2000. Quasielastic neutron scattering of two-dimensional water in a vermiculite clay. J. Chem. Phys. 113, 2873–2879.

Szczerba, M., Kuligiewicz, A., Derkowski, A., Gionis, V., Chryssikos, G.D., Kalinichev, A.G., 2016. Structure and dynamics of water-smectite interfaces: hydrogen bonding and the origin of the sharp O-D-W/O-H-W infrared band from molecular simulations. Clays Clay Miner. 64, 452–471.

Tainter, C.J., Ni, Y., Shi, L., Skinner, J.L., 2013. Hydrogen bonding and OH^- stretch spectroscopy in water: hexamer (cage), liquid surface, liquid, and ice. J. Phys. Chem. Lett. 4, 12–17.

Tang, M.J., Cziczo, D.J., Grassian, V.H., 2016. Interactions of water with mineral dust aerosol: water adsorption, hygroscopicity, cloud condensation, and ice nucleation. Chem. Rev. 116, 4205–4259.

Tarasevi, Y.I., Gribina, I.A., 1972. Infrared spectroscopic study of state of water in Halloysite. Colloid J. Russ. Acad. of Sci. 34, 346–350.

Tarasevich, Y.I., 2007. The surface energy of hydrophilic and hydrophobic adsorbents. Colloid J. 69, 212–220.

Tarasevich, Y.I., Aksenenko, E.V., 2014. Hydrophobicity of talc basal surface. Colloid J. 76, 483–489.

Tester, C.C., Aloni, S., Gilbert, B., Banfield, J.F., 2016. Short- and long-range attractive forces that influence the structure of montmorillonite osmotic hydrates. Langmuir 32, 12039–12046.

Thompson, C.J., Martin, P.F., Chen, J., Benezeth, P., Schaef, H.T., Rosso, K.M., Felmy, A.R., Loring, J.S., 2014. Automated high-pressure titration system with in situ infrared spectroscopic detection. Rev. Sci. Instrum. 85, 11.

Van Der Vegt, N.F.A., Haldrup, K., Roke, S., Zheng, J., Lund, M., Bakker, H.J., 2016. Water-mediated ion pairing: occurrence and relevance. Chem. Rev. 116, 7626–7641.

Vasconcelos, I.F., Bunker, B.A., Cygan, R.T., 2007. Molecular dynamics modeling of ion adsorption to the basal surfaces of kaolinite. J. Phys. Chem. C 111, 6753–6762.

Veerabadran, N.G., Price, R.R., Lvov, Y.M., 2007. Clay nanotubes for encapsulation and sustained release of drugs. Nano 2, 115–120.

Venyaminov, S.Y., Prendergast, F.G., 1997. Water (H2O and D2O) molar absorptivity in the 1000-4000 cm-1 range and quantitative infrared spectroscopy of aqueous solutions. Anal. Biochem. 248, 234–245.

Verwey, E.J.W., Overbeek, J.T.G., 1948. Theory of the Stability of Lyophobic Colloids. Elsevier, Amsterdam.

Wang, J.W., Kalinichev, A.G., Kirkpatrick, R.J., 2009. Asymmetric hydrogen bonding and orientational ordering of water at hydrophobic and hydrophilic surfaces: a comparison of water/vapor, water/talc, and water/mica interfaces. J. Phys. Chem. C 113, 11077–11085.

Weiss, C.A., Kirkpatrick, R.J., Altaner, S.P., 1990. The structural environments of cations adsorbed onto clays: cesium-133 variable temperature MAS NMR spectroscopy of hectorite. Geochim. Cosmochim. Acta 54, 1655–1669.

Weiss, S., Hirsemann, D., Biersack, B., Ziadeh, M., Muller, A.H.E., Breu, J., 2013. Hybrid Janus particles based on polymer-modified kaolinite. Polymer 54, 1388–1396.

Xu, W., Johnston, C.T., Parker, P., Agnew, S.F., 2000. Infrared study of water sorption on Na-, Li-, Ca- and Mg-exchanged (SWy-1 and SAz-1) montmorillonite. Clays Clay Miner. 48, 120–131.

Yan, L.J., Masliyah, J.H., Xu, Z.H., 2013. Interaction of divalent cations with basal planes and edge surfaces of phyllosilicate minerals: muscovite and talc. J. Colloid Interface Sci. 404, 183–191.

Yariv, S., Shoval, S., 1975. Nature of interaction between water-molecules and kaolin-like layers in hydrated halloysite. Clays Clay Miner. 23, 473–474.

Yin, X., Miller, J.D., 2012. Wettability of kaolinite basal planes based on surface force measurements using atomic force microscopy. Miner. Metall. Process. 29, 13–19.

Yin, X.H., Gupta, V., Du, H., Wang, X.M., Miller, J.D., 2012. Surface charge and wetting characteristics of layered silicate minerals. Adv. Colloid Interface Sci. 179, 43–50.

Zarzycki, P., Gilbert, B., 2016. Long-range interactions restrict water transport in pyrophyllite interlayers. Sci. Rep. 6, 5.

Zecchina, A., Arean, C.O., 1996. Diatomic molecular probes for mid-IR studies of zeolites. Chem. Soc. Rev. 25, 187.

Zeitler, T.R., Greathouse, J.A., Cygan, R.T., 2012. Effects of thermodynamic ensembles and mineral surfaces on interfacial water structure. Phys. Chem. Chem. Phys. 14, 1728–1734.

Zich, D., Zacher, T., Darmo, J., Szocs, V., Lorenc, D., Janek, M., 2013. Far-infrared investigation of kaolinite and halloysite intercalates using terahertz time-domain spectroscopy. Vib. Spectrosc. 69, 1–7.

Zsirka, B., Taborosi, A., Szabo, P., Szilagyi, R.K., Horvath, E., Juzsakova, T., Fertig, D., Kristof, J., 2017. Surface characterization of mechanochemically modified exfoliated halloysite nanoscrolls. Langmuir 33, 3534–3547.

Adsorption of heavy metals including radionuclides

5

B. Baeyens, M. Marques Fernandes
Laboratory for Waste Management, Paul Scherrer Institute, Villigen, Switzerland

5.1 Clay mineral adsorption mechanisms and modelling

Montmorillonites (Mt) and illites are among the most commonly occurring dioctahedral 2:1 clay minerals. Their high surface charge, originating from isomorphous substitution in the crystal lattice and from the presence of pH dependent surface edge or broken bond sites, in conjunction with their morphological properties, that is, small particle size (<2 μm) and high surface area, confer upon them their substantial and characteristic capacity to adsorb cations.

Depending on the chemical composition of the tetrahedral and octahedral sheets, clay minerals will have a permanent electrostatic negative charge at the basal surfaces, arising from isomorphous substitution of lattice cations by cations of a lower valence. These developed charges are not pH-dependent; the number of O- or OH-ions in the tetrahedral or octahedral layers is fixed. Mt has the isomorphous substitution (Fe^{2+}, Mg^{2+} for Al^{3+}) in the octahedral sheet, but little or none in the tetrahedral sheet (Srodon and McCarty, 2008; Bergaya et al., 2013). In illite, the substitution (Al^{3+} for Si^{4+}) occurs essentially in the tetrahedral layer. The permanent negative charge on the basal surfaces is neutralised by weak electrostatic interactions with no direct chemical bonding to the siloxane surface (i.e. outer-sphere complexation) of positive charged interlayer cations such as Na^+, K^+, Mg^{2+}, Ca^{2+}. The maximum amount of cations that can be retained/exchanged on these sites determines the cation exchange capacity (CEC) of the given clay mineral. Mt has a CEC of approx. 70–120 cmol (+)/kg, whereas for illites the CEC varies between 10 and 40 cmol(+)/kg (Grim, 1953; Środoń and MaCarty, 2008; Bergaya et al., 2013).

The surface charge that develops along the clay mineral edges results from the protonation and deprotonation of amphoteric (i.e. that can act as either an acid or a base) surface hydroxyl functional groups. Unlike permanent charges, pH-dependent edge charges are variable. With increasing pH, edge sites on the clay minerals become more negatively charged, and control cation adsorption because of the stronger, covalent chemical interactions (i.e. inner-sphere surface complexation). A typical pH-dependent metal adsorption behaviour occurs similarly, as observed on (hydr)oxide minerals (Dzombak and Morel, 1990). A good overview of the surface properties of clay minerals is given by Tournassat et al. (2015).

Up to now, the thermodynamic adsorption models used to describe reactions occurring at the clay mineral-water interface have been essentially mass action law-based.

Developments in Clay Science, Vol. 9. https://doi.org/10.1016/B978-0-08-102432-4.00005-6

5.1.1 Permanent charge: Cation exchange

Cation exchange reactions occurring on the planar surface sites are often expressed in terms of a selectivity coefficient. There are different approaches for describing cation exchange equilibria with the two most common being the Gaines and Thomas (K_c) (Gaines and Thomas, 1953) and the Vanselow (K_v) (Vanselow, 1932) convention. The former one defines the fractional ion occupancies on the planar sites of clay minerals by an equivalent concentration scale, whereas the latter uses the molar scale in the mass action equation.

The cation exchange reaction of a metal B, of valence z_B, exchanging with a metal A, of valence z_A, on a clay mineral in the A-form, can be written as:

$$z_B A^{z_A} - \text{clay} + z_A B^{z_B} \Leftrightarrow z_A B^{z_B} - \text{clay} + z_B A^{z_A} \tag{5.1}$$

Following the Gaines and Thomas convention, a selectivity coefficient, ${}_A^B K_c$, for reaction (5.1) can be defined by the following mass action law:

$${}_A^B K_c = \frac{(N_B)^{z_A}}{(N_B)^{z_B}} \cdot \frac{[A]^{z_B}}{[B]^{z_A}} \cdot \frac{(\gamma_A)^{z_B}}{(\gamma_B)^{z_A}} \tag{5.2}$$

N_A and N_B = equivalent fractional occupancies, defined as the equivalents of A (or B) adsorbed per kg of clay mineral divided by the CEC [cmol(+)/kg]. A and B are aqueous concentrations [M]. γ_A and γ_B are single ion aqueous phase activity coefficients.

An extensive literature review of cation exchange on clay minerals has been compiled by Bruggenwert and Kamphorst (1982). Fletcher and Sposito (1989) developed an approach to model the interactions of Mt with a range of metals by transforming classical ion-exchange constants into sets of formation constants that can be used in computer-based predictions. A tabulation and evaluation of ion exchange data in the form of equilibrium constants on smectites has been given by Benson (1982). The data are presented as log K values, and reflect very well the ranges found for some common ion equilibria on smectite. A more specific modelling approach is the adsorption of Cs on illite. Brouwer et al. (1983) have developed a three-site cation exchange model of Cs adsorption on illite. This model has been extended to a generalised Cs adsorption model (Bradbury and Baeyens, 2000) to predict Cs adsorption isotherms on argillaceous rocks in the framework of the safety analysis of radioactive waste repositories in deep geological formations.

5.1.2 Variable charge: Amphoteric edge sites

5.1.2.1 Acid–base reactions

The surface hydroxyl groups (≡SOH) situated along the edges of the clay mineral layers (Sposito, 1984) undergo protonation/deprotonation (protolysis) reactions that generate a surface charge in which the value varies with pH, but has a maximum value equal to the number of binding sites. Acid–base titration experiments carried out on

Mt and illite suggest that both clay minerals have an edge site capacity of 80 mmol/kg (Bradbury and Baeyens, 1997, 2009a) (details are given in Section 5.1.3).

This amphoteric behaviour of the surface hydroxyl sites is usually described by the following protonation and deprotonation reactions with their corresponding mass action relationship:

$$\equiv SOH + H^+ \Leftrightarrow \equiv SOH_2{}^+ \quad K^+ = \frac{[\equiv SOH_2{}^+]}{[\equiv SOH] \cdot \{H^+\}} \tag{5.3}$$

$$\equiv SOH \Leftrightarrow \equiv SO^- + H^+ \quad K^- = \frac{[\equiv SO^-] \cdot \{H^+\}}{[\equiv SOH]} \tag{5.4}$$

where [ ] represents molar concentration and { } activities.

Note that the preceding mass action equations are written without an electrostatic term. In electrostatic models (see Table 5.1), a Coulombic term ($\exp(-F\Phi/RT)$) is included in the equations (depending on reaction stoichiometry) where F is the Faraday's constant (96,485 Coulombs per mole), Φ is the electrical potential at the surface, R is the gas constant, and T is temperature in Kelvin (K).

There are at least two comprehensive reviews on this subject. The first paper by Duc et al. (2005) gives a substantial overview of the experimental difficulties/constraints regarding titration measurements and the influence they can have on modelling. The second review is by Bourg et al. (2007), giving an overview of available acid–base models. This compilation contains pK_a data, site densities, and specific edge surface areas from 12 different studies on acid–base titration models of Na^+-Mt. The site types, their density, and the acid–base constants can vary in the various model approaches from one single site, up to 27 sites (Tournassat et al., 2004). In general,

Table 5.1 Acid–base titration data and models on Mt and illite

Clay	Model	Edge sites	References
Mt	CCM	1	Wanner et al. (1994)
	NEM	2	Bradbury and Baeyens (1997)
	DDLM	1	Kraepiel et al. (1999)
	CCM	1	Avena and De Pauli (1998)
	MUSIC	4	Tournassat et al. (2004, 2016)
	DDLM	2	Tombácz and Szekeres (2004)
Illite	No model		Beene et al. (1991)
	GEMS	2	Kulik et al. (2000)
	NEM	2	Bradbury and Baeyens (2009a)

CCM, constant capacitance model (Stumm et al., 1970); *DDLM*, diffuse double layer model (Huang and Stumm, 1973); *GEMS*, Gibbs energy minimisation selector; *MUSIC*, multisite complexation model (Hiemstra and Van Riemsdijk, 1996); *NEM*, nonelectrostatic model.

two sites are considered, and are associated with the tetrahedral silanol sites and the octahedral aluminol sites. Table 5.1 lists selected model approaches on the acid/base behaviour of clay minerals. The models that are isolated from metal adsorption data can only be used for describing the proton buffer capacity of clay minerals.

5.1.2.2 Surface complexation of cations on clay minerals

With the exception of the alkaline and alkaline-earth metals, most metal ions such as transition metals, lanthanides, and actinides form surface complexes at the edge sites of clay minerals. Surface complexation reactions on generic edge sites can be written as:

$$\equiv S^sOH + Me^z + yH_2O \Leftrightarrow \equiv S^sOMe(OH)_y^{z-(y+1)} + (y+1)H^+ \qquad (5.5)$$

where Me is a metal with valence z, y is an integer, and the corresponding surface stability constant is expressed as ${}^SK_y$. For $y=0$ the surface complex is $\equiv S^SOMe^{(z-1)}$.

In a nonelectrostatic model, the corresponding surface complexation constant, ${}^SK_y$ can be expressed as:

$${}^sK_y = \frac{\left[\equiv S^sOMe(OH)_y^{z-(y+1)}\right]}{[\equiv S^sOH]} \cdot \frac{f_{\equiv S^sOMe(OH)_y^{z-(y+1)}}}{f_{\equiv S^sOH}} \frac{\{H\}^{(y+1)}}{\{Me^z\}} \qquad (5.6)$$

where [ ] are concentrations, f are surface activity coefficients, and { } are aqueous activities. In most adsorption models, the surface activity coefficient is assumed to be unity (e.g. Dzombak and Morel, 1990). The chemical equilibrium is very often solved using a system of simultaneous equations and nonlinear least squares optimisation routines (Westall, 1982). The model results (best fit parameters) are then compared with the experimental data plotted as solid/liquid distribution coefficient (R_d, L/kg) as a function of pH or as Me adsorbed (mol/kg) as a function of Me equilibrium concentration (Me_{eql}, M). Table 5.2 gives examples of selected adsorption models of heavy metals on clay minerals described in the open literature.

5.1.3 The 2 site protolysis nonelectrostatic surface complexation and cation exchange model

All the experimental adsorption data on the clay minerals Wyoming Mt (SWy), Texas Mt (STx), synthetic iron free Mt (IFM), and illite du Puy (IdP) presented in this chapter have been modelled with a quasi-mechanistic 2 site protolysis nonelectrostatic surface complexation and cation exchange (2SPNE SC/CE) model (Bradbury and Baeyens, 1997, 2009b). The general approach to provide the parameter values required for the surface complexation component of the adsorption model are illustrated for the case of Ni adsorption on Mt. Three different types of experimental data are required: (i) acid/base titrations, (ii) Ni adsorption measured at trace concentrations as a function of pH at a fixed ionic strength (adsorption edge), and (iii) Ni adsorption measured as a function of metal concentration at constant pH and ionic strength (adsorption isotherm).

Table 5.2 An overview of adsorption of heavy metals on clay minerals by surface complexation

Clay	Model		Site types	Heavy metals	References
Smectite	NEM	2 pK_a	2 sites	Cd	Zachara and Smith (1994)
Mt	NEM	2 pK_a	Strong/weak	Ni, Zn, Eu, Sn, Am, Np, U	Bradbury and Baeyens (2005a)
Mt	EM	2 pK_a	Strong/weak	Ni, Zn	Kraepiel et al. (1999)
Mt	EM	4 pK_a	≡AlOH ≡SiOH	U	McKinley et al. (1995)
Mt	EM	2 pK_a	1 site K_{SOM}	Pb, Cd	Barbier et al. (2000)
Mt	DLM	3 pK_a (two ≡AlOH, ≡SiOH)	≡AlOH ≡SiOH	Np	Turner et al. (1998)
Mt	EM	4 pK_a	B-type surface; Si_T–Al_{Oc}–Si_T edge sites (≡Site H_3)	U	Tournassat et al. (2018)
Illite	2SPNE SC/CE	2 pK_a	Strong/weak	Ni, Eu, Am, Sn, Pa, Np, U	Bradbury and Baeyens (2009a,b)
Mt mica alumina	ECCM	3 pK_a (K_{A1}, K_{si1} and K_{si2})	≡AlOH and ≡SiOH	U	Kowal-Fouchard et al. (2004)
Mt	CCM	2 pK_a	2 sites	Cu	Stadler and Schindler (1993)
Beidellite	TLM		2 sites (≡AlOH and ≡SiOH)	U	Turner et al. (1996)

EM, electrostatic model; *DLM*, diffuse layer model (Dzombak and Morel, 1990); *ECCM*, (extended) constant capacitance model (Nilsson et al., 1996); *NEM*, nonelectrostatic model; *TLM*, triple-layer model (Hayes et al., 1991).

In Fig. 5.1A, the results of acid/base titration measurements in 0.01 M $NaClO_4$, 0.1 M $NaClO_4$, and 0.5 M $NaClO_4$ for Mt dispersions via a batch technique with back titrations of the supernatant solutions are illustrated (Baeyens and Bradbury, 1997). The titration data were fitted in terms of protonation and deprotonation reactions (Eqs. 5.3 and 5.4) on two types of weak sites ($\equiv S^{W1}OH$ and $\equiv S^{W2}OH$), described in terms of stability constants and mass action relationships without an electrostatic term. The fitted site capacities and protonation/deprotonation constants for Mt are summarised in Table 5.3. (A similar procedure was carried out for illite, and these data are also included in Table 5.3.).

In Fig. 5.1B, adsorption edges for Ni in 0.1 M $NaClO_4$, 0.03 M $NaClO_4$, and 0.01 M $NaClO_4$ on homoionic Na^+-SWy are shown (Baeyens and Bradbury, 1997). The fitting of these data sets yields surface complexation constants on strong sites ($\equiv S^SOH$). Note that the protolysis constants for $\equiv S^SOH$ sites are assumed to be

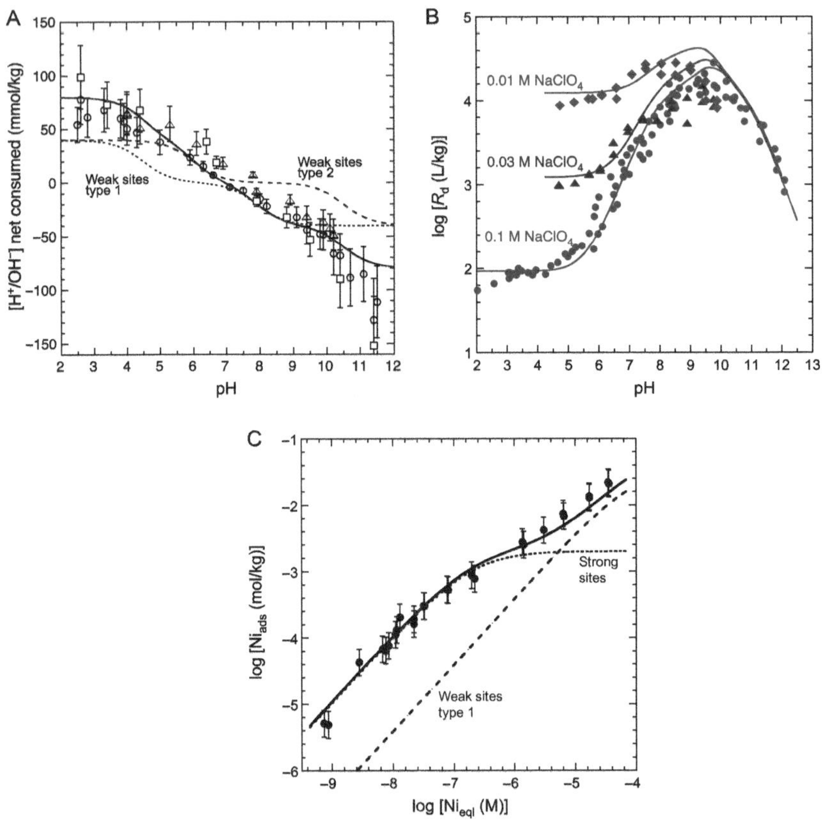

Fig. 5.1 (A) Titration data for Na^+-SWy in 0.01 M $NaClO_4$ (△), 0.1 M $NaClO_4$ (O), and 0.5 M $NaClO_4$ (□). The continuous lines are calculated with the 2SPNE model and the parameters given in Table 5.3. The contribution of each weak site type to the overall H^+/OH^- net consumption is indicated by the *dotted and dashed lines.* (B) Adsorption edges for Ni on Na^+-SWy in 0.1 M $NaClO_4$, 0.03 M $NaClO_4$, and 0.01 M $NaClO_4$. (C) Adsorption isotherm for Ni on Na^+-SWy in 0.1 M $NaClO_4$ at pH 8.2. The *continuous lines* are calculated with the 2SPNE SC/CE model and the parameters given in Tables 5.3, 5.4, and 5.6. The contribution of the Ni adsorption on the strong sites and on the weak sites in the Ni adsorption isotherm are shown by the *dotted and dashed lines*, respectively, (Bradbury and Baeyens, 1997).

Table 5.3 Site type and capacity values and protolysis reactions and constants for Na^+-Mt (Bradbury and Baeyens, 1997) and Na^+-illite (Bradbury and Baeyens, 2009a) used in the 2SPNE SC/CE adsorption model

Amphoteric edge sites		**Na^+-Mt**		**Na^+-Illite**
Site type		**Capacity (mol/kg)**		
$\equiv S^{S}OH$		2.0×10^{-3}		2.0×10^{-3}
$\equiv S^{W1}OH$		4.0×10^{-2}		4.0×10^{-2}
$\equiv S^{W2}OH$		4.0×10^{-2}		4.0×10^{-2}
		$\log K$		
Protolysis reactions	**$\equiv S^{S/W1}OH$**	**$\equiv S^{W2}OH$**	**$\equiv S^{S/W1}OH$**	**$\equiv S^{W2}OH$**
$\equiv SOH + H^+ \Leftrightarrow \equiv SOH_2{}^+$	4.5	6.0	4.0	8.5
$\equiv SOH \Leftrightarrow \equiv SO + H^+$	−7.9	−10.5	−6.2	−10.5

the same as for the $\equiv S^{W1}OH$ sites. The data in Fig. 5.1B illustrate clearly the ionic strength dependency of Ni adsorption at low pH (cation exchange), and the pH dependency of the Ni adsorption at pH above 5 (surface complexation).

Fig. 5.1C shows an adsorption isotherm for Ni on Na^+-SWy in 0.1 M $NaClO_4$ at pH 8.2 (Baeyens and Bradbury, 1997). From this data, the strong site surface capacity and the surface complexation constant for the weak sites can be obtained. The overall modelling approach to these experimental data is iterative. In a first step, the titration data are analysed to yield the weak site capacities and the protolysis constants. With these parameters fixed, the adsorption edge data are fitted to fix the surface complexation constants on the strong sites. In a third step, the isotherm is analysed to derive the strong site capacity and the surface complexation constants for the weak sites. A series of iterations follow until all parameters satisfactorily describe all the data sets. The results of this modelling approach are illustrated by the continuous curves in Fig. 5.1.

5.2 Adsorption of heavy metals and radionuclides on 2:1 clay minerals

In this section, a compilation of adsorption data of heavy metals and radionuclides on Mt and illite is presented. All the experimental data sets have been consistently modelled with the quasi mechanistic 2SPNE SC/CE adsorption model (Bradbury and Baeyens, 1997, 2009a). All data are "in house" measurements that have been published over the past two decades. The main motivation for the development of this comprehensive data/modelling investigation is that it forms the basis for the derivation of adsorption databases to be used in the safety analysis of deep geological radioactive waste repositories in argillaceous rocks (Nagra, 2002).

5.2.1 Adsorption by cation exchange

5.2.1.1 Heavy metals and radionuclides

In the quantitative description of the adsorption of metals on clay minerals, the cation exchange component is always present, and needs to be included in the adsorption model. For most of the heavy metals described in Section 5.2, the uptake on clay minerals at pH < 4–5 at trace concentration is governed by cation exchange. The solid–liquid distribution ratio, ${}^{B}R_{CE}$, for the adsorption by cation exchange of cation B can be defined as

$$ {}^{B}R_{CE} = \frac{B \text{ adsorbed by cation exchange (mol/kg)}}{\text{Aqueous concentration of B (mol/L)}} \quad (5.7) $$

For a bi-ionic system where one of the cations, B, is present at trace level, that is, $N_A \sim 1$, Eq. (5.2) can be used together with Eq. (5.7) to relate the selectivity coefficient to the adsorption of cation B by

$$ {}_{A}^{B}K_c = (z_B)^{Z_A} \cdot \left({}^{B}R_{CE}\right)^{Z_A} \cdot \frac{1}{(\text{CEC})^{Z_A}} \cdot \frac{(\gamma_A)^{Z_B}}{(\gamma_B)^{Z_A}} \cdot (A)^{Z_B} \quad (5.8) $$

where all parameters have been defined previously (see Eq. 5.2). If the values on the right hand side of Eq. (5.8) are known, the K_c can be readily calculated (Bradbury and Baeyens, 1994). Table 5.4 lists the K_c values for cation exchange equilibria involving heavy metals treated in this chapter.

5.2.1.2 Rubidium, cesium and thallium

The alkali metals K^+, Rb^+, and Cs^+, as well as $NH_4{}^+$ and Tl^+, exhibit high selective adsorption behaviour toward illite, especially at trace concentrations. A comparative study of Cs^+, Rb^+, and Tl^+ ion equilibria in illite saturated either with Na^+, K^+, and Ca^{2+} is presented in this section.

Cs (together with Sr) is the most abundant fission product produced in commercial nuclear power reactors. The fission of one hundred ${}^{235}U$ nuclides in a thermal reactor produces six ${}^{137}Cs$ ($t_{1/2} = 30.2$ years) and six ${}^{90}Sr$ ($t_{1/2} = 28.8$ years) nuclides. In addition, radiocaesium prevails in the environment because of nuclear weapon programmes, and from fallout caused by severe nuclear accidents. It is well known that Cs is preferentially taken up by illite and other micaceous minerals (Tamura and Jacobs, 1960; McKinley et al., 2001; Zachara et al., 2002), and multisite adsorption models have been developed on the uptake of Cs on illite (Brouwer et al., 1983; Bradbury and Baeyens, 2000).

Tl is a highly toxic trace element, and occurs in the environment predominantly as monovalent Tl^I, but can be oxidised to Tl^{III}. Anthropogenic Tl contamination can result from the production of cement (Xiao et al., 2012), metal mining and smelting

Table 5.4 Overview of Gaines and Thomas selectivity coefficients for cation exchange on the planar sites of Mt and illite used to describe the data sets given in this chapter

	Mt		**Illite**	
A-Clay + B	K_c	**References**	K_c	**References**
Na-clay + Fe^{2+}	6.3	Soltermann et al. (2014a)	—	—
Na-clay + Co^{2+}	3.7	Maes et al. (1975)	12.7	Bradbury and Baeyens (2009a)
Na-clay + Ni^{2+}	3.1	Baeyens and Bradbury (1997)	12.6	Bradbury and Baeyens (2009a)
Na-clay + Zn^{2+}	3.9	Baeyens and Bradbury (1997)	4.0	Montoya et al. (2018)
Na-clay + Pb^{2+}	3.7	Marques Fernandes and Baeyens (2018)	17.9	Marques Fernandes and Baeyens (2018)
Na-clay + Eu^{3+}	30	Bradbury and Baeyens (2002)	76	Bradbury and Baeyens (2009a)
Na-clay + Am^{3+}	29	Bradbury and Baeyens (2006)	40	Bradbury and Baeyens (2009b)
Na-clay + NpO_2^+	1.1	Bradbury and Baeyens (2006)	—	—
Na-clay + UO_2^{2+}	0.7	Bradbury and Baeyens (2005a)	4.5	Bradbury and Baeyens (2009b)

(Kersten et al., 2014), or coal burning (Peter and Viraraghavan, 2005). High geogenic Tl contents may occur in soils formed from rocks hosting Tl-rich mineralisations (Voegelin et al., 2015). The affinity of Tl^I to illite has long been assumed to be the key adsorbent for Tl in soils (Vaněk et al., 2011), and was recently confirmed by a spectroscopy study on the speciation of Tl in a geogenically Tl-rich soil (Voegelin et al., 2015).

The adsorption of Cs and Rb on illite has been extensively studied by Brouwer et al. (1983), and a three-site cation exchange model consisting of high affinity sites (site I), intermediate sites (site II), and planar sites (site III) was developed. In a recent study by Wick et al. (2018), the uptake of Tl^+ on illite has been thoroughly studied, and resembles the adsorption behaviour of the alkali metals Cs^+ and Rb^+ on illite. The experimental and modelled data of Tl^+ exchange on homoionic Na^+-, K^+-, and Ca^{2+}-illites obtained in Wick et al. (2018) are compared with the original Cs^+ and Rb^+ data from Brouwer et al. (1983) in Fig. 5.2, and the best fitted parameters are summarised in Table 5.5.

The adsorption of Cs^+, Tl^+ and Rb^+ at trace concentration ($<10^{-7}$M) is mainly governed by the so-called frayed edge sites, FES (site I) and the affinity of Tl^+ fits well between Cs^+ and Rb^+ on this site. The driving force for the adsorption on the

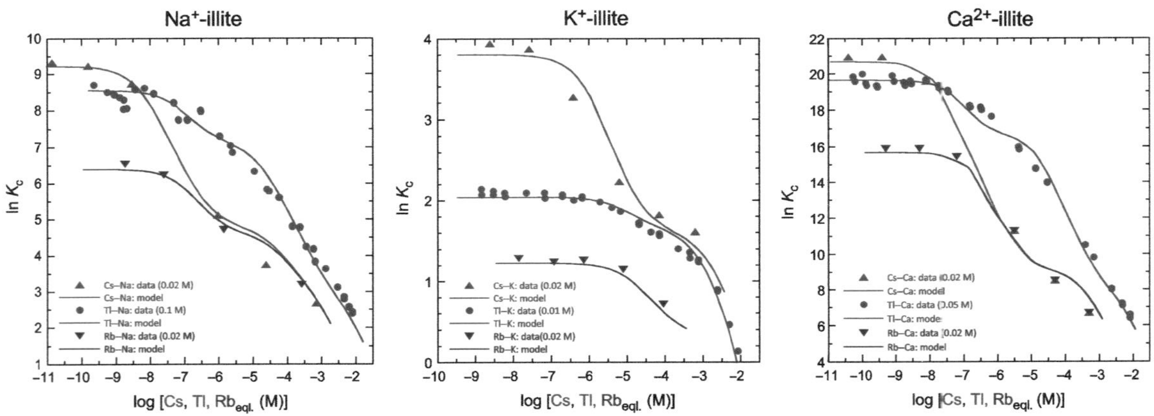

Fig. 5.2 Data and modelling of Cs^+, Tl^+ and Rb^+ on Na^+-, K^+- and Ca^{2+}-illites.
Data from Brouwer, E., Baeyens, B., Maes, A., Cremers, A., 1983. Cesium and rubidium ion equilibriums in illite clay. J. Phys. Chem. 87, 1213–1219; Wick, S., Baeyens, B., Marques Fernandes, M., Voegelin, A., 2018. Thallium adsorption onto illite. Environ. Sci. Technol. 52, 571–580.

Table 5.5 $\ln K_c$ values for the exchanges (25°C) on the three site types of illite for Cs^+, Tl^+ and Rb^+ on Na^+-, K^+- and Ca^{2+}-illite

Site type	Site I	Site II	Site III
Site capacity (cmol(+)/kg)	**0.05**	**4**	**16**
Exchange equilibrium	$\ln K_c$		
$Cs^+ + Na^+$-illite ⇔ Cs^+-illite + Na^+	15.2	6.2	3.2
$Tl^+ + Na^+$-illite ⇔ Tl^+-illite + Na^+	14.5	9.0	3.2
$Rb^+ + Na^+$-illite ⇔ Rb^+-illite + Na^+	12.2	6.0	3.2
$Cs^+ + K^+$-illite ⇔ Cs^+-illite + K^+	9.7	2.5	1.2
$Tl^+ + K^+$-illite ⇔ Tl^+-illite + K^+	6.9	2.3	1.4
$Rb^+ + K^+$-illite ⇔ Rb^+-illite + K^+	6.7	1.2	0.0
2 $Cs^+ + Ca^{2+}$-illite ⇔ 2 Cs^+-illite + Ca^{2+}	32.2	11.7	1.2
2 $TL^+ + Ca^{2+}$-illite ⇔ 2 TL^+-illite + Ca^{2+}	28.6	17.3	5.3
2 $Rb^+ + Ca^{2+}$-illite ⇔ 2 Rb^+-illite + Ca^{2+}	27.2	11.8	1.2

high affinity FES can be ascribed to the difference in free energies of hydration between, for example, Cs^+ and Na^+, or Ca^{2+} (Brouwer et al., 1983). This is also applicable for Tl^+, which has similar dehydration energy and ionic radius as Cs^+ and Rb^+. At a higher concentration, Tl^+ behaves in a similar way to Cs^+ and Rb^+ on K^+-illite; whereas Tl^+ appears to be more selective on the less energetic adsorption sites (site II and planar sites) on Na^+- and Ca^{2+}-illites. This modelling study is important to evaluate the transport behaviour in soils and sediments, and to describe the speciation and solubility of these highly (radio-)toxic elements.

5.2.2 Adsorption by surface complexation

5.2.2.1 Cobalt, nickel and zinc

The occurrence of Co, Ni, and Zn in the environment stems mainly from industrial activities, such as in the production of alloys and in paints, varnishes, enamels, and inks (Co), production of steels and alloys, batteries and electrical contacts (Ni) and metal coatings, alloys, and galvanised pipes (Zn) (de Haan and Zwerman, 1976).

Co, Ni, and Zn adsorption on Mt (SWy) and illite (IdP) was thoroughly investigated with experimental and modelling studies (Baeyens and Bradbury, 1997; Bradbury and Baeyens, 1997, 1999, 2005a, 2009a; Marques Fernandes et al., 2015; Montoya et al., 2018). A selected overview of the adsorption behaviour of these three transition metals is presented in Fig. 5.3, where for each metal, the pH dependent adsorption at trace concentration and an adsorption isotherm at pH ~ 7 in 0.1 M $NaClO_4$ on Na^+-SWy and Na^+-IdP are illustrated together with the modelling curves. The overall picture that emerges from the experimental data clearly demonstrates that the

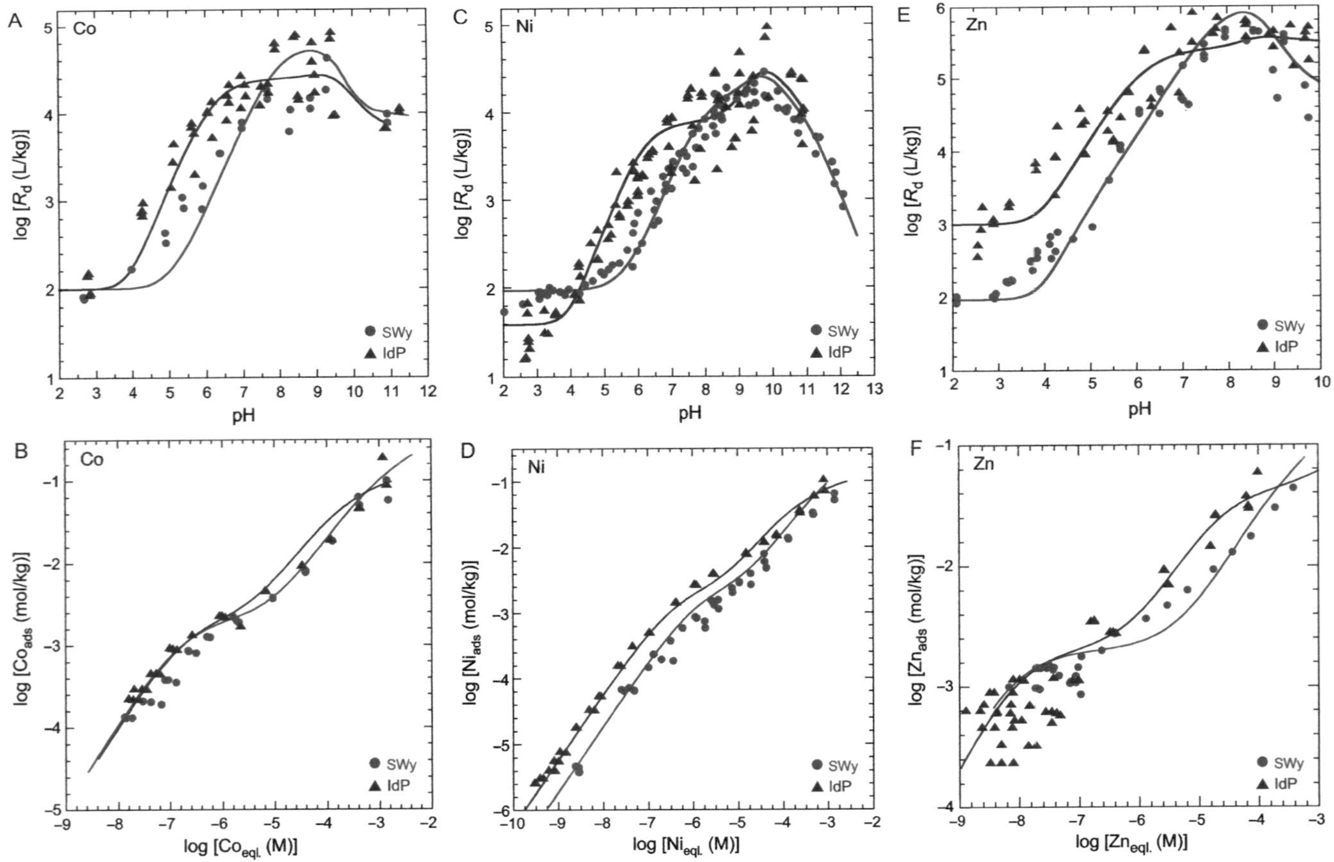

Fig. 5.3 See legend on next page.

adsorption behaviour of transition metals is very similar on Mt and illite, with a tendency that the affinity toward illite is higher compared with Mt. The parameters (selectivity coefficients, hydrolysis, and surface complexation constants) used to model the data sets in Fig. 5.3 are summarised in Tables 5.3 and 5.6.

In the 2SPNE SC/CE adsorption model, there are two broad categories of edge adsorption sites, the so-called strong ($\equiv S^{S}OH$) and weak ($\equiv S^{W1}OH$) sites (see Section 5.1.3). Because of their different adsorption characteristics, it was expected that the coordination environments of the surface complexes on the two site types would be different.

The complementary application of polarised extended X-ray absorption fine structure (P-EXAFS) spectroscopy (Dähn et al., 2011) and ab initio molecular dynamics (MD) calculations (Churakov and Dähn, 2012) allowed demonstration of the existence and structural characteristics of different edge site types, in agreement with the strong–weak site concept in the 2SPNE SC/CE adsorption model. Dähn et al. (2011) used the Zn isotherm data on Na^{+}-SWy presented in Fig. 5.3F to define the most favourable experimental conditions under which Zn adsorption was either dominated by the strong or by the weak sites. P-EXAFS measurements were performed on highly oriented self-supporting iron-poor Mt films (Milos and STx) to reduce the contribution of structural iron in the fluorescence signal, particularly at low Zn loadings. The structural parameters (Table 5.7) obtained from the fitting of the different polarisation angle-dependent spectra allowed to clearly discriminate between three types of Zn complexes, that is, Zn incorporated in the bulk structure and Zn adsorbed on two different types of edge sites. The structural parameters for the intrinsic Zn were obtained from the measurements on the as-received Mt from Milos (no Zn added), which contains ~1 mmol intrinsic Zn per kg clay mineral. The distances and coordination numbers fitted are characteristic of edge-sharing linkages between Al octahedra, and of corner-sharing linkages between Al octahedra and Si tetrahedra, and indicating that Zn^{2+} substitutes for Al^{3+} in the dioctahedral aluminium sheet. The spectra of Zn adsorbed on the "strong sites" exhibited a pronounced angular dependency and formed surface complexes in the continuity of the Al-octahedral sheets at the Mt edges. In contrast, the Zn "weak site" spectra showed comparatively decreased angular dependency, and a significantly longer Zn–Si_2 bond length, indicating that the

Fig. 5.3 Adsorption of the transition metals Co^{2+}, Ni^{2+} and Zn^{2+} on Mt and illite. Experimental data taken from Bradbury, M.H., Baeyens, B., 1997. A mechanistic description of Ni and Zn sorption on Na- montmorillonite. 2. Modelling. J. Contam. Hydrol. 27, 223–248; Bradbury, M.H., Baeyens, B., 2009. Sorption modelling on illite. Part I: titration measurements and the sorption of Ni, Co, Eu and Sn. Geochim. Cosmochim. Acta 73, 990–1003; Marques Fernandes, M., Vér, N., Baeyens, B., 2015. Predicting the uptake of Cs, Co, Ni, Eu, Th and U on argillaceous rocks using sorption models for illite. Appl. Geochem. 59, 189–199; Montoya, V., Baeyens, B., Glaus, M.A., Kupcik, T., Marques Fernandes, M., Van Laer, L., Bruggeman, C., Maes, N., Schäfer, T., 2018. Sorption of Sr, Co and Zn on illite: batch experiments and modelling including Co in-diffusion measurements on compacted samples. Geochim. Cosmochim. Acta 223, 1–20.

Table 5.6 Hydrolysis (Baes and Mesmer, 1976) and surface complexation reactions and corresponding constants to describe the adsorption data presented in Fig. 5.3

Hydrolysis	**Co**		**Ni**		**Zn**	
	log ^{OH}K					
$Me^{2+} + H_2O \Leftrightarrow MeOH^+ + H^+$	−9.65		−9.86		−8.96	
$Me^{2+} + 2H_2O \Leftrightarrow Me(OH)_2 + 2H^+$	−18.8		−19		−16.9	
$Me^{2+} + 3H_2O \Leftrightarrow Me(OH)_3^- + 3H^+$	−31.5		−30		−28.4	
Surface complexation formation reactions	**log K_{SC}**					
	Mt	**Illite**	**Mt**	**Illite**	**Mt**	**Illite**
$\equiv S^SOH + Me^{2+} + H_2O \Leftrightarrow \equiv S^SOMe^+ + H^+$	0.0	1.3	−0.6	0.7	1.6	2.1
$\equiv S^SOH + Me^{2+} + 2H_2O \Leftrightarrow \equiv S^SOMeOH + 2H^+$	−10.5	−8.5	−10	−8.2	−7.4	−6.4
$\equiv S^SOH + Me^{2+} + 3H_2O \Leftrightarrow \equiv S^SOMe(OH)_2^- + 3H^+$	−20	−18.3	−20	−17.3	−17	−15
$\equiv S^{W1}OH + Me^{2+} + H_2O \Leftrightarrow \equiv S^{W1}OMe^+ + H^+$	−3.3	−2.0	−3.3	−1.8	−2.7	−1.0

Table 5.7 Main structural parameters obtained by fitting the P-EXAFS data ($\alpha = 35$ degrees) of different Zn loaded Mt samples

	Zn—O		Zn—Al		Zn—Si_1		Zn—Si_2	
Samples	**CN**	***R* (Å)**	**CN**	***R* (Å)**	**CN**	***R* (Å)**	**CN**	***R* (Å)**
Na^+-Milos: incorporated Zn	6.4(4)	2.07(1)	3.3(6)	3.02(2)			4.0(6)	3.22(2)
Na^+-Milos: Zn adsorbed on strong sites	6.0(3)	2.05(1)	1.5(4)	3.02(2)	1.2(3)	3.10(3)	1.5(3)	3.26(2)
Na^+-STx: Zn adsorbed on strong sites	5.1(4)	2.05(1)	1.1(3)	3.02(2)	1.3(2)	3.06(3)	2.1(3)	3.26(2)
Na^+-STx: Zn adsorbed on weak sites	5.4(4)	2.05(1)	0.7(4)	3.04(2)			1.0(5)	3.33(3)

Taken from Dähn, R., Baeyens, B., Bradbury, M.H., 2011. Investigation of the different binding edge sites for Zn on montmorillonite using P-EXAFS—The strong/weak site concept in the 2SPNE SC/CE sorption model. Geochim. Cosmochim. Acta 75, 5154–5168.

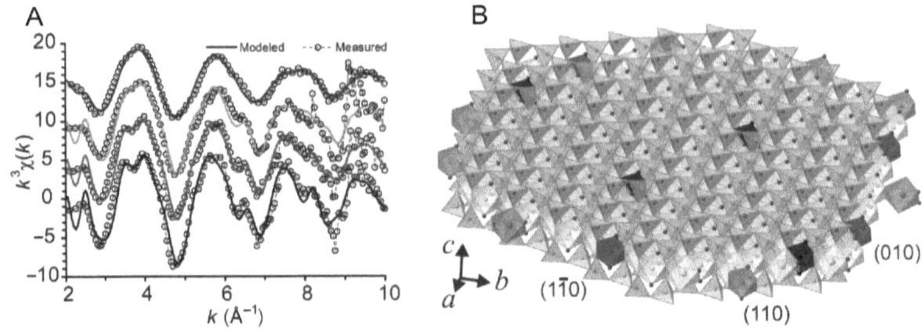

Fig. 5.4 (A) Best fit of the measured EXAFS spectra, from *top* to *bottom*: STx medium loading (*blue line*), STx low loading (*green line*; *grey* in the print version), Milos low loading (*red line*; *dark grey* in the print version), and Milos incorporated (*black line*). (B) Schematic view of adsorption sites on (010) and (110) edge facets of Mt at low (*red octahedra*; *dark grey* in the print version) and medium Zn loadings (*green octahedra*; *grey* in the print version). Structural incorporated Zn is illustrated by the (*black polyhedra*). *Trans*- and *cis*-Al octahedral are shown in *blue* (*grey* in the print version) and *yellow* (*white* in the print version) *colours*, respectively. Silica tetrahedral are *orange* (*light grey* in the print version).
Taken from Churakov, S.V., Dähn, R., 2012. Zinc adsorption on clays inferred from atomistic simulations and EXAFS spectroscopy. Environ. Sci. Technol. 46, 5713–5719, reprint with permission.

adsorption onto edge sites is not occurring as in the case for low-Zn loadings. However, the fully resolved structural configurations of the different surface sites for strong sites or for weak sites could not be obtained from this study due to the averaged nature of the EXAFS signal.

An unambiguous distinction was obtained by ab initio molecular dynamics calculations (Churakov and Dähn, 2012). The results of the atomistic modelling are depicted in Fig. 5.4 and are in line with the interpretation of the EXAFS data. The atomistic configurations derived evidence that structurally incorporated Zn preferentially substitutes for Al in the *trans*-symmetric sites of the octahedral layer. Adsorbed Zn is located at the outermost amphoteric *trans*-octahedra site on (010) and (110) edges when present at low loadings, whereas at medium loading, Zn^{2+} forms mono- and bidentate inner-sphere surface complexes attached to the octahedral layer of (010) and (110) edge sites.

5.2.2.2 Tin and lead

Sn and Pb are heavy metals belonging to the Group 6 elements of the periodic table with predominant oxidation states of +IV and +II, respectively. Radioisotopes of both nuclides are present in radioactive waste, and their migration from deep geological repositories through the geosphere and biosphere needs to be considered. Pb is further an important heavy metal occurring in the environment from industrial nonactive waste (automobile industry, batteries production, metal production).

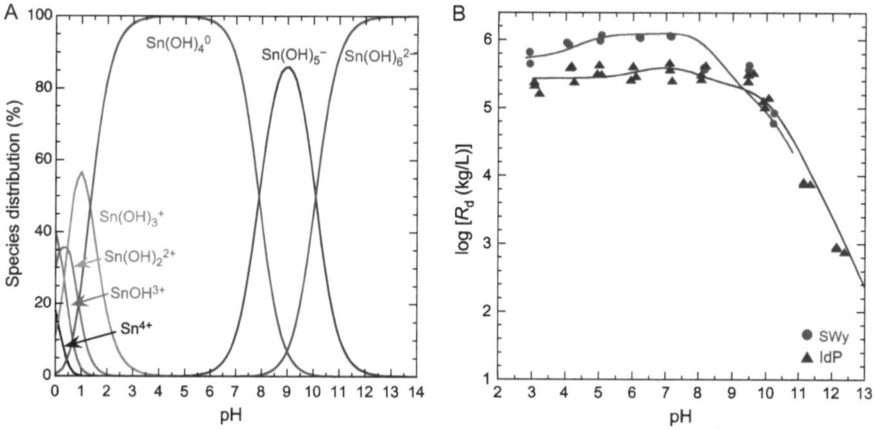

Fig. 5.5 (A) Distribution of Sn^{IV} (10^{-8} M) hydrolysis species at $I = 0.1$ M (ClO_4^-) and 25°C. (B) Adsorption edges of stanate on Mt and illite.
Experimental data taken from Bradbury, M.H., Baeyens, B., 2005a. Modelling the sorption of Mn(II), Co(II), Ni(II), Zn(II), Cd(II), Eu(III), Am(III), Sn(IV), Th(IV), Np(V) and U(VI) on montmorillonite: linear free energy relationships and estimates of surface binding constants for some selected heavy metals and actinides. Geochim. Cosmochim. Acta 69, 875–892; Bradbury, M.H., Baeyens, B., 2009a. Sorption modelling on illite. Part I: titration measurements and the sorption of Ni, Co, Eu and Sn. Geochim. Cosmochim. Acta 73, 990–1003.

The hydrolysis behaviour of stanate (Sn^{IV}) (Baes and Mesmer, 1976) is extremely strong, as shown in Fig. 5.5A. In the pH range 2–8, the neutral tetrahydroxy complex is dominant, whereas under alkaline conditions, the negatively charged penta- and hexahydroxystanate complexes dominate. This strong hydrolysis behaviour is reflected in the very strong adsorption of stanate as shown in Fig. 5.5B. Compared with the divalent transition metals, there is no characteristic pH dependency in the pH region 2–8. The adsorption is very high and constant in the pH region 3 to ~9, after which the adsorption decreases due to the formation of the negatively charged aqueous hydroxyl complexes (Fig. 5.5B). The model parameters for the stanate uptake in Mt and illite are summarised in Table 5.8. Because of the very low solubility (Amaya et al., 1997), adsorption of stanate will never exceed the strong site capacities for both Mt and illite, and no isotherms can be determined under pH neutral conditions.

The hydrolysis behaviour of Pb^{2+} is, although slightly stronger, similar to the transition metals, and is shown in Fig. 5.6A. Whereas Sn^{4+} is fully hydrolysed, Pb^{2+} is still present as free divalent aqua cation up to pH ~7, and its uptake behaviour on Mt and illite is characterised by a typical strong pH dependency, as illustrated in Fig. 5.6B. Because the solubility of PbO in a simple 1:1 background electrolyte is high, the concentration dependent adsorption can be determined, and the isotherms measured and modelled at pH 7 are shown in Fig. 5.6C. Similar to the case of the transition metals, the strong–weak-site concept is required to describe quantitatively the

Table 5.8 Hydrolysis (Baes and Mesmer, 1976) and surface complexation reactions and corresponding constants to describe the adsorption data presented in Fig. 5.5

Hydrolysis	**log ^{OH}K**	
$Sn^{4+} + 2H_2O \Leftrightarrow Sn(OH)_2^{2+} + 2H^+$	1.7	
$Sn^{4+} + 3H_2O \Leftrightarrow Sn(OH)_3^{+} + 3H^+$	1.6	
$Sn^{4+} + 4H_2O \Leftrightarrow Sn(OH)_4^{0} + 4H^+$	0.4	
$Sn^{4+} + 5H_2O \Leftrightarrow Sn(OH)_5^{-} + 5H^+$	−7.7	
$Sn^{4+} + 6H_2O \Leftrightarrow Sn(OH)_6^{-} + 6H^+$	−18.1	
Surface complexation formation reactions	**log K_{SC}**	
	Mt	**Illite**
$\equiv S^SOH + Sn^{4+} + 2H_2O \Leftrightarrow \equiv S^SOSn(OH)_2^{+} + 3H^+$	13.3	12.5
$\equiv S^SOH + Sn^{4+} + 3H_2O \Leftrightarrow \equiv S^SOSn(OH)_3^{0} + 4H^+$	9.2	8.5
$\equiv S^SOH + Sn^{4+} + 4H_2O \Leftrightarrow \equiv S^SOSn(OH)_4^{-} + 5H^+$	1.6	2.5
$\equiv S^SOH + Sn^{4+} + 5H_2O \Leftrightarrow \equiv S^SOSn(OH)_5^{2-} + 6H^+$	−7.6	−5.7

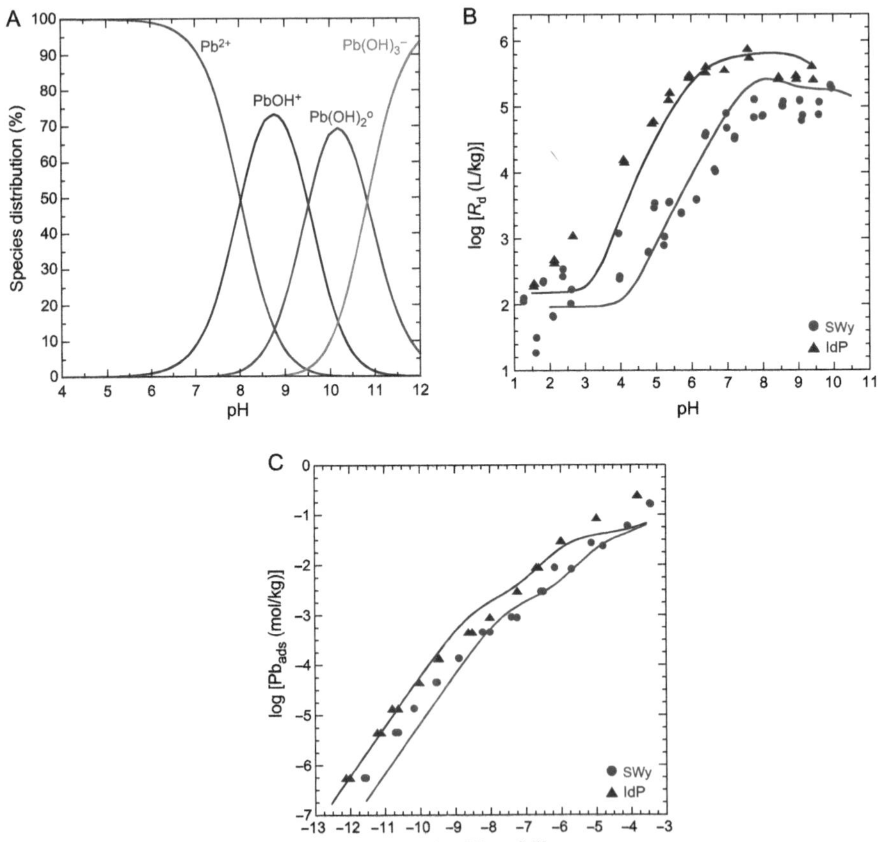

Fig. 5.6 (A) Distribution of Pb^{2+} (10^{-8} M) hydrolysis species at $I = 0.1$ M (ClO_4^-) and 25°C. (B) Adsorption edges and (C) adsorption isotherms at pH = 7 of lead on Mt and illite. Taken from Marques Fernandes, M., Baeyens, B., 2018. Cation exchange and surface complexation studies of lead on montmorillonite and illite including competitive adsorption effects. Appl. Geochem. (submitted for publication).

Table 5.9 Hydrolysis (Baes and Mesmer, 1976) and surface complexation reactions and corresponding constants to describe the adsorption data presented in Fig. 5.6

Hydrolysis	**log ^{OH}K**	
$Pb^{2+} + H_2O \Leftrightarrow PbOH^+ + H^+$	−7.71	
$Pb^{2+} + 2H_2O \Leftrightarrow Pb(OH)_2{}^0 + 2H^+$	−17.12	
$Pb^{2+} + 3H_2O \Leftrightarrow Pb(OH)_3{}^- + 3H^+$	−28.06	
Surface complexation formation reactions	**log K_{SC}**	
	Mt	**Illite**
$\equiv S^SOH + Pb^{2+} \Leftrightarrow \equiv S^SOPb^+ + H^+$	1.0	2.7
$\equiv S^SOH + Pb^{2+} + H_2O \Leftrightarrow \equiv S^SOPbOH^0 + 2H^+$	−7.6	−5.3
$\equiv S^{W1}OH + Pb^{2+} \Leftrightarrow \equiv S^{W1}OPb^+ + H^+$	−1.5	0.3

Pb uptake in the concentration range from trace to the millimolar scale. At the highest Pb concentration, the uptake on the clay mineral surfaces may be influenced by precipitation of lead (hydr)oxide phases.

The Sn^{4+} and Pb^{2+} model parameters given in Tables 5.3, 5.8, and 5.9 allow one to quantify the retention of both elements in a bentonite near-field or an argillaceous rock far-field in the framework of performance assessment of high-level waste repositories.

5.2.3 Adsorption of radionuclides

Large amounts of radioactivity have and will be released in the environment due to nuclear activities such as nuclear test explosions, nuclear operational reactor wastes, spent fuel reprocessing, accidents, and the geological disposal of high-level radioactive waste (Choppin and Rydberg, 1980). Chemical speciation, oxidation state, and adsorption characteristics are necessary in describing solubility of the different radionuclides and their migration behaviour in aqueous and terrestrial environments. The adsorption behaviour on 2:1 type clay minerals of trivalent lanthanides (Eu) and actinides (Am, Cm), tetravalent thorium, pentavalent protactinium, and neptunium and hexavalent uranium is described in this section. Chemical analogy can be invoked for elements that have variable oxidation states, and for which no experimental data are available. Th exhibits only the +IV oxidation state, and is a valuable chemical analogue for the +IV states of U, Np, and Pu when they prevail under the reducing "in situ" conditions of a deep geological waste repository. A similar approach is valid for penta- and hexavalent actinides. Np^V can act as a chemical analogue for Pu^V, and, U^{VI} for Np^{VI} and Pu^{VI} (Choppin, 2007).

5.2.3.1 Europium

The ionic radii for trivalent lanthanides and Am^{3+} are very similar (Shannon, 1976), and Eu is often used as a chemical analogue in adsorption studies. The radioactivity levels of analogue actinides are too high for practical application in laboratory

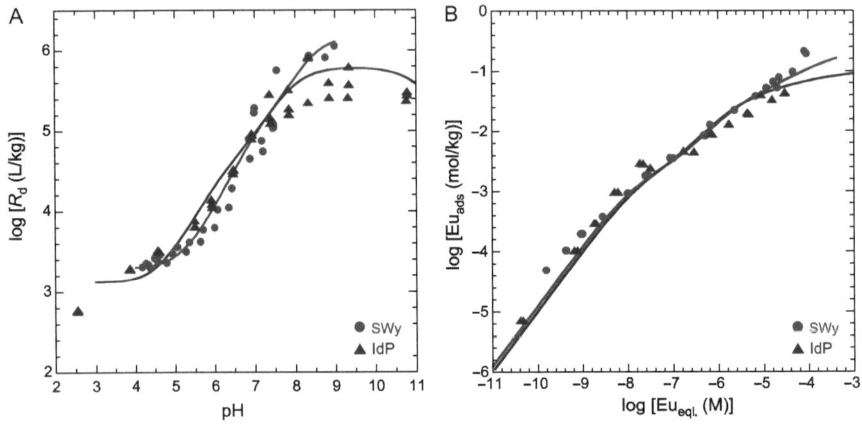

Fig. 5.7 (A) Adsorption edges for Eu^{3+} in 0.1 M $NaClO_4$ and (B) adsorption isotherms for Eu^{3+} in 0.1 M $NaClO_4$ on Na^+-Mt and Na^+-IdP at pH 7.
Experimental data taken from Bradbury, M.H., Baeyens, B., 2002. Sorption of Eu on Na- and Ca-montmorillonites: experimental investigations and modelling with cation exchange and surface. Geochim. Cosmochim. Acta 66, 2325–2334; Bradbury, M.H., Baeyens, B., 2009. Sorption modelling on illite. Part I: titration measurements and the sorption of Ni, Co, Eu and Sn. Geochim. Cosmochim. Acta 73, 990–1003; Marques Fernandes, M., Vér, N., Baeyens, B., 2015. Predicting the uptake of Cs, Co, Ni, Eu, Th and U on argillaceous rocks using sorption models for illite. Appl. Geochem. 59, 189–199.

experiments. Eu adsorption on Mt and illite was thoroughly investigated by experimental and modelling studies (Bradbury and Baeyens, 2002, 2005a; Bradbury et al., 2005; Marques Fernandes et al., 2008b, 2015), and spectroscopic studies (Stumpf et al., 2001, 2004; Rabung et al., 2005; Schnurr et al., 2015; Marques Fernandes et al., 2016). A pH dependent adsorption edge at trace concentration and adsorption isotherms at pH 7 in 0.1 M $NaClO_4$ on SWy and IdP are illustrated together with the modelling curves in Fig. 5.7. In contrast to the transition metals, the adsorption behaviour of Eu^{3+} on both clay minerals is very similar. The model parameters are summarised in Table 5.10. Similar to the metals described previously, the 2SPNE SC/CE model is capable of quantitatively describing the uptake of Eu^{3+} over a wide pH and concentration range.

5.2.3.2 Americium and curium

A complementary approach to the validation of the adsorption model was obtained in a time-resolved laser fluorescence spectroscopy (TRLFS) study of Cm^{3+} interaction with Ca^{2+}-Mt and Na^+-illite carried out by Rabung et al. (2005). The adsorption of trivalent actinides is very similar to trivalent lanthanides as shown in Fig. 5.8A, where the Eu^{3+} adsorption on illite is compared with Am^{3+} and Cm^{3+}. Adsorption of Cm and Eu trace concentrations onto Ca^{2+}-SWy and Na^+-IdP were investigated under anoxic conditions by TRLFS to determine their surface speciation on both clay minerals under varying pH conditions. Changes in the emission spectra of TRLFS (shift to

Table 5.10 Hydrolysis (Guillaumont et al., 2003) and surface complexation reactions and corresponding constants to describe the adsorption data presented in Fig. 5.7

Hydrolysis[a]	$\log {}^{OH}K$	
$Am^{3+} + H_2O \Leftrightarrow AmOH^+ + H^+$	−7.2	
$Am^{3+} + 2H_2O \Leftrightarrow Am(OH)_2^0 + 2H^+$	−15.1	
$Am^{3+} + 3H_2O \Leftrightarrow Am(OH)_3^- + 3H^+$	−26.2	
Surface complexation formation reactions[b]	$\log K_{SC}$	
	Mt	**Illite**
$\equiv S^SOH + Me^{3+} \Leftrightarrow \equiv S^SOMe^{2+} + H^+$	1.6	1.9
$\equiv S^SOH + Me^{3+} + H_2O \Leftrightarrow \equiv S^SOMeOH^+ + 2H^+$	−6.4	−4.7
$\equiv S^SOH + Me^{3+} + 2H_2O \Leftrightarrow \equiv S^SOMe(OH)_2^0 + 3H^+$	−14	−12.7
$\equiv S^{W1}OH + Me^{3+} \Leftrightarrow \equiv S^{W1}OMe^{2+} + H^+$	−0.5	0.3
$\equiv S^{W1}OH + Me^{3+} + H_2O \Leftrightarrow \equiv S^{W1}OMeOH^+ + 2H^+$	–	−6.2

[a] The hydrolysis constants for Eu^{III} and Cm^{III} are assumed to be the same as for Am^{III}.
[b] $Me^{3+} = Eu^{3+}$, Am^{3+} or Cm^{3+}.

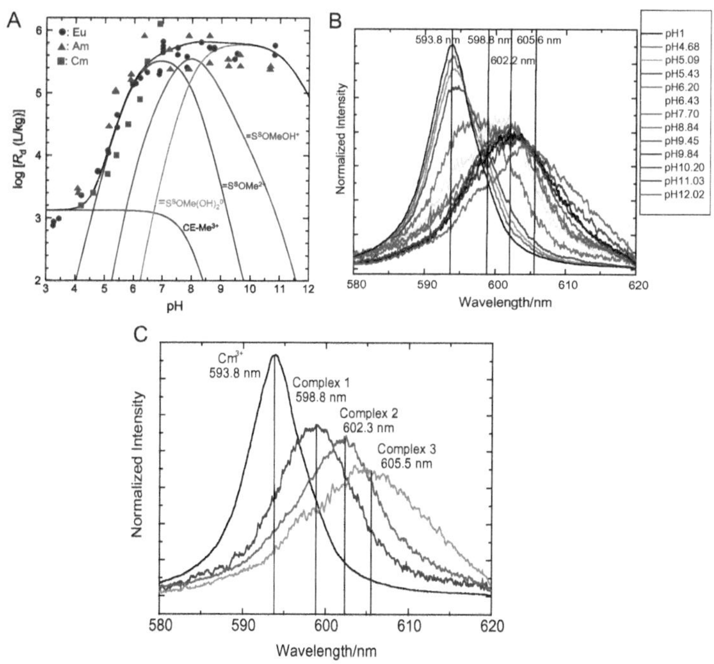

Fig. 5.8 (A) Eu^{3+} (Bradbury and Baeyens, 2009a), Am^{3+} (Bradbury and Baeyens, 2009b) and Cm^{3+} (Rabung et al., 2005) adsorption edge measurements onto Na^+-IdP. The experimental data (*symbols*) were modelled using the parameters given in Tables 5.3 and 5.10. (B) Cm^{3+} fluorescence spectra normalised to same peak area and (C) spectra of individual surface complexes.
Derived from peak deconvolution: Rabung, T., Pierret, M.C., Bauer, A., Geckeis, H., Bradbury, M.H., Baeyens, B., 2005. Sorption of Eu(III)/Cm(III) on Ca-montmorillonite and Na-illite. Part 1: batch sorption and time-resolved laser fluorescence spectroscopy experiments. Geochim. Cosmochim. Acta 69, 5393–5402 (print with permission).

higher wavelengths, shape) are due to variations in the first coordination sphere of the fluorescent metal ion and are indicative for inner-sphere complexation processes. This is illustrated in Fig. 5.8B where the evolution of TRLFS spectra obtained for the Cm^{3+} adsorption on Na^+-IdP in 0.1 M $NaClO_4$ in the pH range 4–12 is shown. Three different surface adsorbed Cm^{3+} species could be distinguished with peak maxima at 598.8, 602.3, and 605.5 nm (Fig. 5.8B). In Fig. 5.8A, the surface complexation modelling illustrates that three inner-sphere surface complexes are required to describe the adsorption edge in the pH range 4–11. The spectroscopic surface species as obtained from the deconvolution of the TRLFS spectra shown in Fig. 5.8C are in good agreement with the surface complexation modelling. Analysis of the fluorescence lifetime (τ) allows determining the number of fluorescence quenching entities. The fluorescence lifetime is mainly influenced by the number of OH^- ions or H_2O units in the first coordination sphere, which act as quenchers due to energy transfer to OH^- vibronic states. Replacing these entities by other ligands generally increases the fluorescence lifetime. The Cm aqua ion has a fluorescence lifetime $\tau = 68\,\mu s$. In the Cm^{3+}–Na^+-illite system, the fluorescence lifetime of the different surface species is $116 \pm 4\,\mu s$, which is indicative for inner-sphere surface complexes where four H_2O molecules have been removed from the first coordination sphere. The postulated surface complex has the form of $\equiv S{-}O{-}Cm(OH)_x^{(2-x)}(H_2O)_{5-x}$.

5.2.3.3 *Thorium*

Thorium is a naturally occurring, very long-lived ($t_{1/2} = 1.4 \times 10^{10}$ years) radioactive element that is predominately present as ^{232}Th. It is widely distributed in small amounts in the Earth's crust, being about as abundant as Pb and Mo, and three times as abundant as U (Weast et al., 1983). Soils commonly contain an average of about 6 ppm Th. The principal mineral of thorium is monazite, a complex phosphate containing rare earth elements and uranium. Thorium occurs widely in other minerals as well (silicates, e.g., thorite, huttonite). Large amounts of waste arising from mining, oil, and gas extraction contain naturally occurring thorium (Paschoa, 1998), or fly ashes from lignite power stations are often contaminated with Th (Savanonda et al., 1985). The most critical environmental aspect concerning Th is its presence in radioactive waste. The disposal of radioactive waste in geological formations can potentially result in the release in the near- and far-fields of such repositories, and ultimately into the biosphere. An understanding of the adsorption mechanism of tetravalent Th and it extension to other tetravalent actinides onto clay minerals is of great importance for the safety of radioactive waste repositories.

Th occurs in a single +IV redox state, and is characterised by strong hydrolysis, as shown in Fig. 5.9A. As is the case for all previously presented heavy metals, the adsorption of Th on illite and Mt is very similar. Because of the strong hydrolysis, no typical pH edge is observed. Up to pH 11, adsorption remains constant and high, due to the absence of the formation of negatively charged hydrolysis products (as is the case for tetravalent Sn). The results of the experimental adsorption and the modelling are shown in Fig. 5.9B. The hydrolysis constants and surface complexation reactions and corresponding constants are given in Table 5.11.

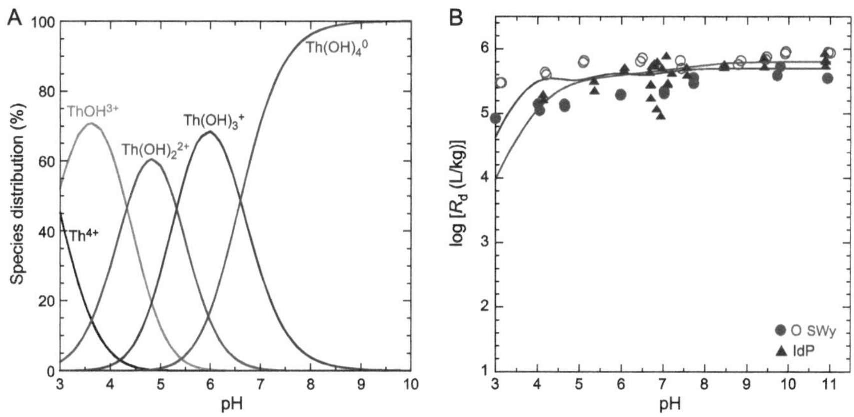

Fig. 5.9 (A) Distribution of Th^{4+} (10^{-8} M) hydrolysis species at $I=0.1$ M (ClO_4^-) and 25°C. (B) pH dependent adsorption of trace Th^{4+} on Mt and illite at $I=0.1$ M $NaClO_4$ (*closed symbols*) and 1 M $NaClO_4$ (*open symbols*).
Experimental data taken from Bradbury, M.H., Baeyens, B., 2005a. Modelling the sorption of Mn(II), Co(II), Ni(II), Zn(II), Cd(II), Eu(III), Am(III), Sn(IV), Th(IV), Np(V) and U(VI) on montmorillonite: linear free energy relationships and estimates of surface binding constants for some selected heavy metals and actinides. Geochim. Cosmochim. Acta 69, 875–892; Bradbury, M.H., Baeyens, B., 2009. Sorption modelling on illite. Part II: actinide sorption and linear free energy relationships. Geochim. Cosmochim. Acta 73, 1004–1013.

Table 5.11 Hydrolysis (Neck and Kim, 2001) and surface complexation reactions and corresponding constants to describe the adsorption data presented in Fig. 5.9

Hydrolysis	**log ^{OH}K**	
$Th^{4+} + H_2O \Leftrightarrow ThOH^{3+} + H^+$	−2.2	
$Th^{4+} + 2H_2O \Leftrightarrow Th(OH)_2^{2+} + 2H^+$	−6.0	
$Th^{4+} + 3H_2O \Leftrightarrow Th(OH)_3^{+} + 3H^+$	−11.0	
$Th^{4+} + 4H_2O \Leftrightarrow Th(OH)_4^{0} + 4H^+$	−17.5	
Surface complexation formation reactions	**log K_{SC}**	
	Mt	**Illite**
$\equiv S^SOH + Th^{4+} \Leftrightarrow \equiv S^SOTh^{3+} + H^+$	7.2	7.4
$\equiv S^SOH + Th^{4+} + H_2O \Leftrightarrow \equiv S^SOTh(OH)^{2+} + 2H^+$	2.7	2.3
$\equiv S^SOH + Th^{4+} + 2H_2O \Leftrightarrow \equiv S^SOTh(OH)_2^{+} + 3H^+$	−2.6	−2.4
$\equiv S^SOH + Th^{4+} + 3H_2O \Leftrightarrow \equiv S^SOTh(OH)_3^{0} + 4H^+$	−9.1	−8.8
$\equiv S^SOH + Th^{4+} + 4H_2O \Leftrightarrow \equiv S^SOTh(OH)_4^{-} + 5H^+$	−16.9	−15.3

5.2.3.4 Protactinium

Protactinium (Pa), the third element of the actinide group, occurs naturally in association with uranium and thorium minerals. Pa exhibits the same radioactive level (secular equilibrium), but due to its much shorter half-life, in very small mass concentration. On the contrary, the inventory of Pa in radioactive waste is much more important, because U is present in large concentrations.

Adsorption measurements for Pa are very sparse in the open literature. Bradbury and Baeyens (2006, 2009b) have measured the pH dependent adsorption of Pa^V on Na^+-illite and Na^+-Mt. For the modelling of the adsorption edges, the hydrolysis constants of Trubert et al. (2003) have been used. The reactions and corresponding hydrolysis constants are given in Table 5.12, and Fig. 5.10A illustrates the hydrolysis behaviour. Only three species are calculated, each with a well-defined dominant stability region within the pH field. In strongly acidic solutions, the $PaOOH^{2+}$ species is dominant. In the pH range from ~2 to ~7, $PaO(OH)_2^+$ is the most stable aqueous species, and above pH ~7, $Pa(OH)_5^0$. The speciation plot shown in Fig. 5.10A has been used to guide the choice of surface complexes included in the modelling of pH-dependent adsorption of Pa on illite and Mt.

The adsorption of ^{233}Pa on illite and Mt shown in Fig. 5.10B is very similar. The very high adsorption plateau exhibited by the uptake of Pa^V on Na^+-SWy and Na^+-IdP at $4 < pH \leq 11$ are reproduced well in the modelling. Table 5.12 gives the parameters used to model the data. Surface complexation reactions were written with respect to the $PaO(OH)^{2+}$ species, and no cation exchange-like behaviour is discernible from the adsorption edge profile because of the strong adsorption, due to surface complexation already apparent at low pH values. At pH below <3, the data deviated from the model predictions. The reason for this discrepancy is not known. In order to model the adsorption in the region below pH ~4 using the current adsorption model, a further surface species would be required that would in turn imply the presence of an additional aqueous species at very low pH.

Table 5.12 Hydrolysis (Trubert et al., 2003) and surface complexation reactions and corresponding constants to describe the adsorption data presented in Fig. 5.10

Hydrolysis	**log ^{OH}K**	
$PaOOH^{2+} + H_2O \Leftrightarrow PaO(OH)_2^+ + H^+$	−1.24	
$PaOOH^{2+} + 3H_2O \Leftrightarrow PaO(OH)_5^0 + 2H^+$	−8.27	
Surface complexation formation reactions	**log K_{SC}**	
	Mt	**Illite**
$\equiv S^SOH + PaOOH^{2+} \Leftrightarrow \equiv S^SOPaOOH^+ + H^+$	6.7	6.4
$\equiv S^SOH + PaOOH^{2+} \Leftrightarrow \equiv S^SOPaO_2^0 + 2H^+$	−0.5	0.4
$\equiv S^SOH + PaOOH^{2+} + H_2O \Leftrightarrow \equiv S^SOPaO_2OH^- + 3H^+$	−8.6	−6.7

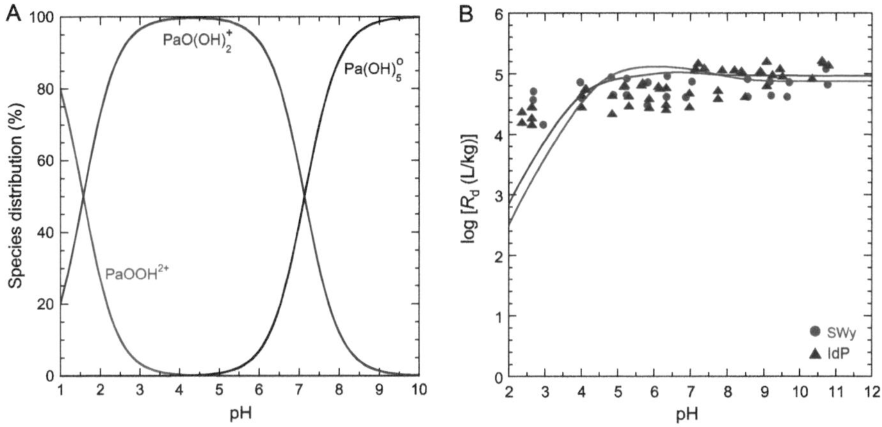

Fig. 5.10 (A) Distribution of Pa^V (10^{-8} M) hydrolysis species at $I=0.1$ M (ClO_4^-) and 25°C. (B) pH dependent adsorption of trace ^{233}Pa on Na^+-Mt and Na^+-illite in 0.1 M $NaClO_4$. Experimental data taken from Bradbury, M.H., Baeyens, B., 2005a. Modelling the sorption of Mn(II), Co(II), Ni(II), Zn(II), Cd(II), Eu(III), Am(III), Sn(IV), Th(IV), Np(V) and U(VI) on montmorillonite: linear free energy relationships and estimates of surface binding constants for some selected heavy metals and actinides. Geochim. Cosmochim. Acta 69, 875–892; Bradbury, M.H., Baeyens, B., 2009. Sorption modelling on illite. Part II: actinide sorption and linear free energy relationships. Geochim. Cosmochim. Acta 73, 1004–1013.

5.2.3.5 Neptunium

The long-lived neptunium isotope ^{237}Np ($t_{1/2}=10^5$ years) is created by irradiating ^{235}U or ^{238}U in a nuclear reactor, but also during the storage from the slow decay of ^{235}Am, and consequently will dominate the radiotoxicity of nuclear waste. Once released from a deep geological waste repository, neptunyl is quite soluble, poorly adsorbed, and readily mobile under environmental conditions (Choppin, 2007).

The adsorption behaviour of neptunyl (NpO_2^+) at trace concentration on Mt was studied by Bradbury and Baeyens (2006) in 0.01 M and 0.1 M $NaClO_4$ as function of pH in the absence of inorganic carbon. The hydrolysis of NpO_2^+ clearly indicates that the predominant NpO_2^+ species up to pH ~ 10 is the free neptunyl ion (Fig. 5.11A). In Fig. 5.11B, the pH-dependent adsorption of NpO_2^+ is shown. A clear effect of the Na^+ background concentration is seen at low pH. Adsorption of NpO_2^+ increases by one log unit if the Na^+ concentration decreases by one order of magnitude. Above pH ~ 6, NpO_2^+ uptake by Mt increases as a function of pH, and the ionic strength dependency on adsorption disappears. The uptake curves of NpO_2^+ on Na^+-SWy calculated using the 2SPNE SC/CE model and the parameters given in Tables 5.3 and 5.13 are shown by the curves in Fig. 5.11B. A NpO_2^+–Na^+ exchange reaction with a selectivity coefficient ($K_c=1.1$) was included in the modelling (see Table 5.4).

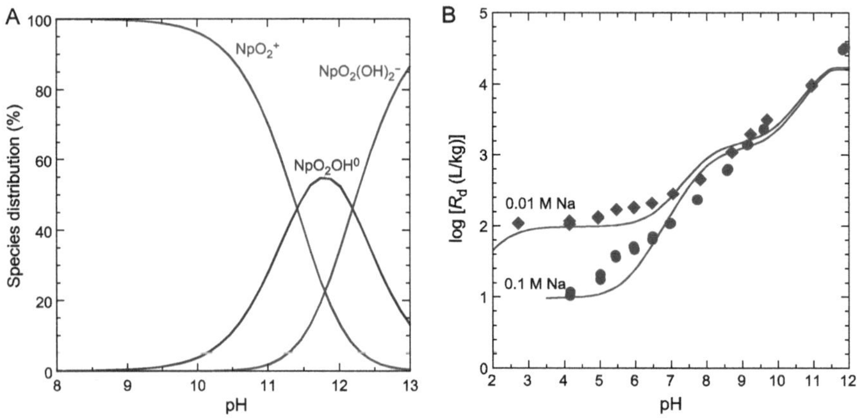

Fig. 5.11 (A) Distribution of Np^V (10^{-8} M) hydrolysis species at $I = 0.1$ M (ClO_4^-) and 25°C. (B) pH dependent adsorption of trace $^{237}NpO_2^+$ on 0.01 M and 0.1 M Na^+-SWy. Experimental data taken from Bradbury, M.H., Baeyens, B., 2006. Modelling sorption data for the actinides Am(III), Np(V) and Pa(V) on montmorillonite. Radiochim. Acta 94, 619–625.

Table 5.13 Hydrolysis (Guillaumont et al., 2003), surface complexation reactions and constants used to model the adsorption of NpO_2^+ on Na^+-Mt

Hydrolysis	**log ^{OH}K**
$NpO_2^+ + H_2O \Leftrightarrow NpO_2(OH)^0 + H^+$	−11.3
$NpO_2^+ + 2H_2O \Leftrightarrow NpO_2(OH)_2^- + 2H^+$	−23.6
Surface complexation formation reactions	**log K_{SC} Mt**
$\equiv S^SOH + NpO_2^+ \Leftrightarrow \equiv S^SONpO_2^0 + H^+$	−2.2
$\equiv S^SOH + NpO_2^+ + H_2O \Leftrightarrow \equiv S^SO\ NpO_2OH^- + 2H^+$	−12

5.2.3.6 Uranium

The presence of uranium in the environment is due to its natural occurrence in soil, rock and water; but also from activities such as mining, nuclear industry, and fertilisers (Burns and Finch, 1999; Merkel and Hasche-Berger, 2006). The migration of high levels of dissolved U from mill tailings has often resulted in contamination of soils and aquifers (Bond et al., 2008). Large quantities of vitrified high-level waste and spent fuel will ultimately be disposed of in deep geological repositories (Nagra, 2002). A detailed understanding of the adsorption processes over a wide range of conditions is essential for the development of predictive transport models.

Fig. 5.12A shows the aqueous hydrolysis of UO_2^{2+} in 0.1 M $NaClO_4$ at trace concentration (no polynuclear species are formed). The results of adsorption measurements of UO_2^{2+} on Mt and illite are shown in Fig. 5.12B–D (Bradbury and Baeyens, 2005b, 2009b; Marques Fernandes et al., 2012). The adsorption edge and

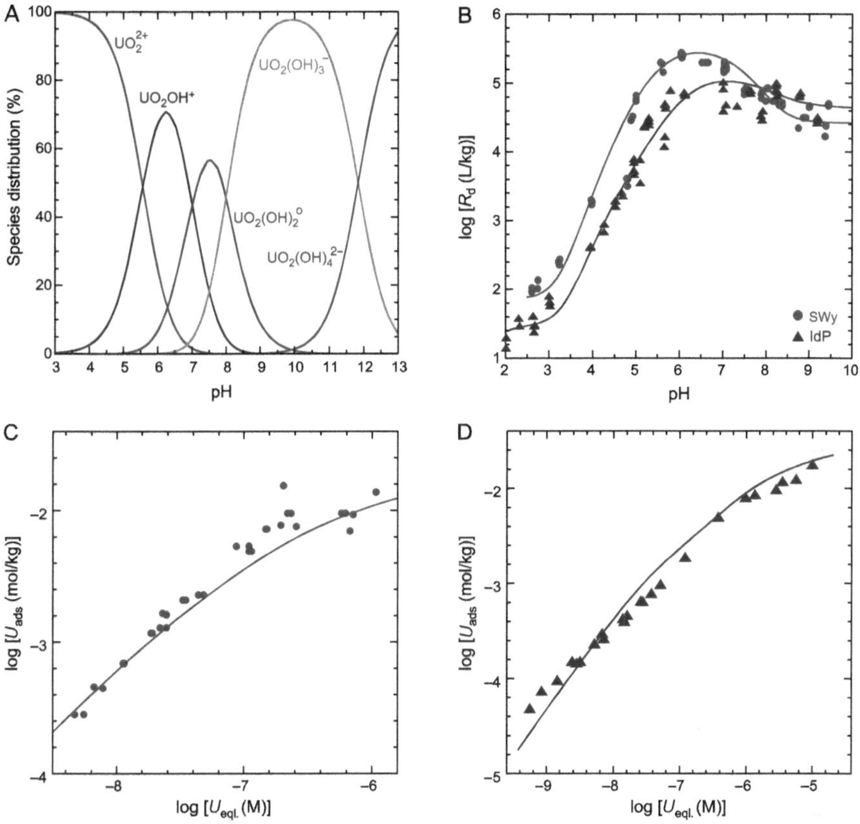

Fig. 5.12 (A) Distribution of UO_2^{2+} (10^{-8} M) hydrolysis species at $I = 0.1$ M (ClO_4^-) and 25°C. (B) UO_2^{2+} adsorption edges on Na^+-Mt and Na^+-illite in 0.1 M $NaClO_4$. UO_2^{2+} adsorption isotherms at (C) pH 8 on Na^+-SWy and (D) pH 6.2 on Na^+-IdP in 0.1 M $NaClO_4$. The continuous *curves* are the best fits obtained with the 2SPNE SC/CE adsorption model Experimental data taken from Marques Fernandes, M., Baeyens, B., Daehn, R., Scheinost, A.C., Bradbury, M.H., 2012. U(VI) sorption on montmorillonite in the absence and presence of carbonate: a macroscopic and microscopic study. Geochim. Cosmochim. Acta 93, 262–277; Bradbury M.H. and Baeyens B., 2009b. Sorption modelling on illite. Part II: actinide sorption and linear free energy relationships, Geochim. Cosmochim. Acta **73**, 1004–1013; Bradbury, M.H. and Baeyens, B. 2005b. Experimental and modelling investigations on Na-illite: acid-base behaviour and the sorption of strontium, nickel, europium and uranyl. PSI Report Nr. 05-02. Paul Scherrer Institut, Villigen, Switzerland.

isotherms were modelled using the 2SPNE SC/CE adsorption model. The aqueous uranyl hydrolysis constants were selected from the NEA TDB (Guillaumont et al., 2003).

The pH dependent adsorption data were modelled in terms of cation exchange on the planar sites with selectivity coefficients given in Table 5.4 for Mt and illite at low pH, and in terms of surface complexation at the amphoteric surface hydroxyl groups with a series of four surface complexation reactions on the strong sites of 2:1

Table 5.14 Hydrolysis (Guillaumont et al., 2003) and surface complexation constants used to model the adsorption of UO_2^{2+} on Na^+-Mt shown in Fig. 5.12 and on Na^+-IdP

Hydrolysis	**log ^{OH}K**	
$UO_2^{2+} + H_2O \Leftrightarrow UO_2OH^+ + H^+$	5.25	
$UO_2^{2+} + 2H_2O \Leftrightarrow UO_2(OH)_2^0 + 2H^+$	−12.15	
$UO_2^{2+} + 3H_2O \Leftrightarrow UO_2(OH)_3^- + 3H^+$	−20.25	
$UO_2^{2+} + 4H_2O \Leftrightarrow UO_2(OH)_4^{2-} + 4H^+$	−32.4	
Surface complexation formation reactions	**log K_{SC}**	
	Mt	**Illite**
$\equiv S^SOH + UO_2^{2+} \Leftrightarrow \equiv S^SOUO_2^+ + H^+$	3.1	2.0
$\equiv S^SOH + UO_2^{2+} + H_2O \Leftrightarrow \equiv S^SOUO_2OH^0 + 2H^+$	−4.6	−3.5
$\equiv S^SOH + UO_2^{2+} + 2H_2O \Leftrightarrow \equiv S^SOUO_2(OH)_2^- + 3H^+$	−12.6	−10.6
$\equiv S^SOH + UO_2^{2+} + 3H_2O \Leftrightarrow \equiv S^SOUO_2(OH)_3^{2-} + 4H^+$	−20.9	−19
$\equiv S^{W1}OH + UO_2^{2+} \Leftrightarrow \equiv S^{W1}OUO_2^+ + H^+$	0.5	0.1
$\equiv S^{W1}OH + UO_2^{2+} + H_2O \Leftrightarrow \equiv S^{W1}OUO_2OH^0 + 2H^+$	−5.7	−5.3

clay minerals. Two additional surface complexation reactions on the weak sites were required to model the adsorption isotherm on Na^+-Mt at pH 8 (Fig. 5.12C) and on Na^+-IdP at pH 5.8 (Fig. 5.12.D). In an iterative procedure, all the adsorption data were successfully modelled with the same set of parameters. The adsorption isotherm of UO_2^{2+} on Na^+-IdP was modelled in Marques Fernandes et al. (2015). The model parameters obtained for UO_2^{2+} on both clay minerals are summarised in Table 5.14.

5.2.4 *Influence of carbonate on the adsorption of Eu, Np, and U*

The omnipresence of inorganic carbonate in soils and sediments will largely influence the adsorption behaviour of trivalent lanthanides and actinides, pentavalent Np, and hexavalent uranium. These metals can form very stable and highly soluble aqueous carbonato complexes (Guillaumont et al., 2003). The formation of such complexes can potentially lead to a decrease in adsorption, and thereby an increase in the migration rates of these elements. The aqueous speciation of Eu^{3+}, NpO_2^+ and UO_2^{2+} in a system that is in equilibrium with air pCO_2 is illustrated in Fig. 5.13A, B and C, respectively. The thermodynamic constants for the formation of the metal-carbonato complexes are taken from Guillaumont et al. (2003) (Table 5.15). Clearly, for all elements, the carbonato complexes are dominant in the pH ranges from 7 to 10 for Eu, from 8.5 to 10 for NpO_2^+, and from 5 to 10 for UO_2^{2+} (Fig. 5.12).

Adsorption edge measurements of Eu^{3+} (Marques Fernandes et al., 2008a), NpO_2^+ (Wendt, 2008) and UO_2^{2+} (Marques Fernandes et al., 2012) on Na^+-SWy in the presence of dissolved carbonate have been carried out, and the experimental results are

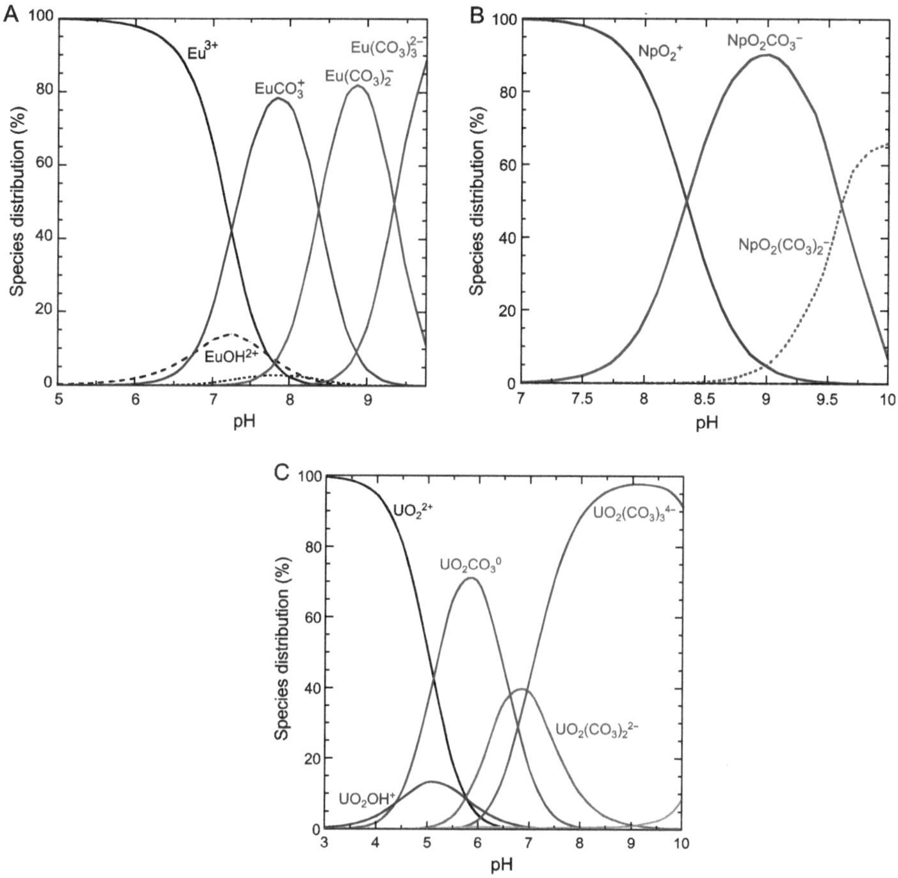

Fig. 5.13 Aqueous speciation of (A) Eu^{3+}, (B) $NpO_2{}^+$, and (C) $UO_2{}^{2+}$ in 0.1 M $NaClO_4$ in equilibrium with air pCO_2 and metal concentration of 10^{-8} M.

shown in Fig. 5.14 (closed symbols). The data are compared with the measurements in the absence of carbonate (open symbols). The adsorption behaviour for all three elements is similar. A pronounced decrease in adsorption is observed, with increasing pH and carbonate concentration.

By modelling the data under the assumption that no additional Eu^{3+}, $NpO_2{}^+$ and $UO_2{}^{2+}$ surface complexation reactions are taking place (i.e. aqueous carbonate complexes are treated as being non adsorbing) a large under prediction is observed at high pH for all three elements (broken lines in Fig. 5.14). An iterative modelling procedure was applied to derive the surface complexation constants in which the aqueous speciation for Eu^{3+}, $NpO_2{}^+$, and $UO_2{}^{2+}$ (Fig. 5.14) was used to guide the choice of the potential surface complexes. The adsorption reactions involving ternary surface complexes are summarised in Table 5.15, and the modelling results are shown in Fig. 5.14 (solid lines).

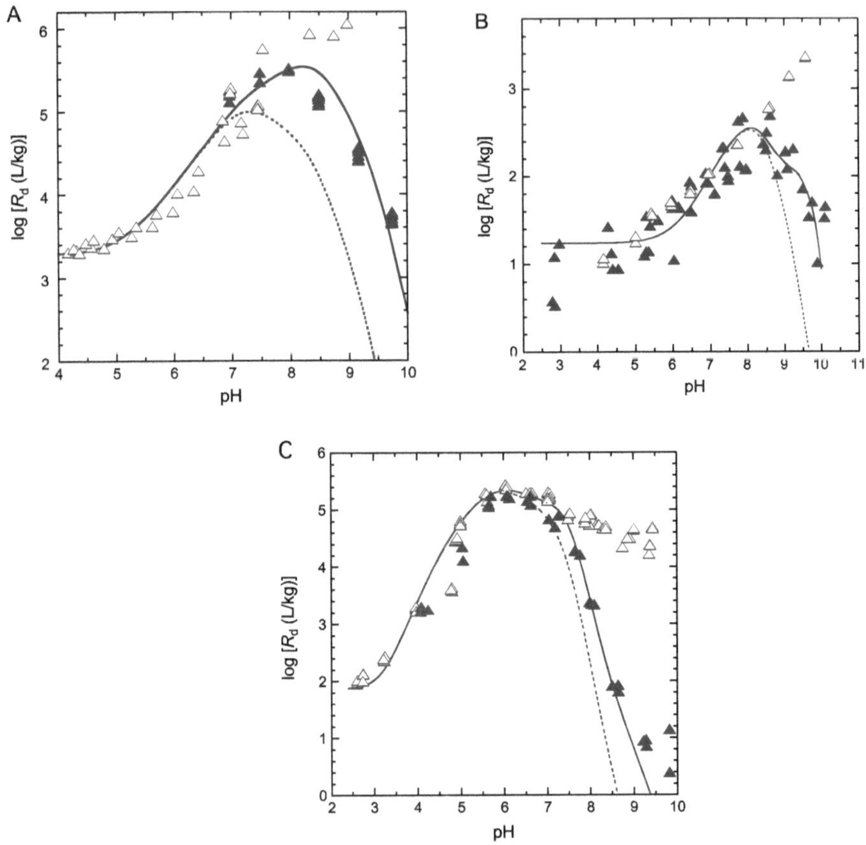

Fig. 5.14 Adsorption edge measurements on Na^+-SWy for (A) Eu^{3+}, (B) NpO_2^+ and (C) UO_2^{2+} in the absence (*open symbols*) and the presence (*closed symbols*) of air pCO_2.
Panel (A): Data taken from Marques Fernandes, M., Baeyens, B., Bradbury, M.H., 2008. The influence of carbonate complexation on lanthanide/actinide sorption on montmorillonite. Radiochim. Acta 96, 691–698. Panel (B): Data taken from Wendt, S., 2008. Sorption and direct speciation of neptunium(V) on aluminium oxide and montmorillonite. PhD thesis. Johannes Gutenberg University, Mainz, Germany. Panel (C): Data taken from Marques Fernandes, M., Baeyens, B., Daehn, R., Scheinost, A.C., Bradbury, M.H., 2012. U(VI) sorption on montmorillonite in the absence and presence of carbonate: a macroscopic and microscopic study. Geochim. Cosmochim. Acta 93, 262–277.

The inclusion of ternary metal-carbonate complexes on the strong sites in the adsorption modelling allowed the uptake of Eu^{3+}, NpO_2^+ and UO_2^{2+} on Na^+-SWy to be quantitatively described over a wide range of conditions.

Recently, Tournassat et al. (2018) have carried out very similar experiments to the one presented in Fig. 5.14C, and the experimental data are in excellent agreement. However, contrary to the modelling proposed by Marques Fernandes et al. (2015), this

Table 5.15 Aqueous carbonate complexation constants and ternary carbonato surface complexation reactions and associated surface complexation constants on Na+-Mt for Eu^{3+}, NpO_2^{+} and UO_2^{2+}

Aqueous carbonate complexation reactions	**log ^{CO3}K**
$An^{3+} + CO_3^{2-} \Leftrightarrow AnCO_3^{+}$	8.0
$An^{3+} + 2CO_3^{2-} \Leftrightarrow An(CO_3)_2^{-}$	12.9
$An^{3+} + 3CO_3^{2-} \Leftrightarrow An(CO_3)_3^{3-}$	15.0
$NpO_2^{+} + CO_3^{2-} \Leftrightarrow NpO_2CO_3^{-}$	4.96
$NpO_2^{+} + 2CO_3^{2-} \Leftrightarrow NpO_2(CO_3)_2^{3-}$	6.53
$NpO_2^{+} + 3CO_3^{2-} \Leftrightarrow NpO_2(CO_3)_3^{5-}$	5.5
$UO_2^{2+} + CO_3^{2-} \Leftrightarrow UO_2CO_3^{0}$	9.94
$UO_2^{2+} + 2CO_3^{2-} \Leftrightarrow UO_2(CO_3)_2^{2-}$	16.61
$UO_2^{2+} + 3CO_3^{2-} \Leftrightarrow UO_2(CO_3)_3^{4-}$	21.84
$2UO_2^{2+} + CO_3^{2-} + 3H_2O \Leftrightarrow (UO_2)_2CO_3(OH)_3^{3-}$	−0.86
$3UO_2^{2+} + CO_3^{2-} + 3H_2O \Leftrightarrow (UO_2)_3CO_3(OH)_3^{3-}$	0.66
$3UO_2^{2+} + 6CO_3^{2-} \Leftrightarrow (UO_2)_3(CO_3)_6^{6-}$	54.0
Ternary surface complexation formation reactions	**log K_{SC}**
$\equiv S^SOH + Eu^{3+} + CO_3^{2-} \Leftrightarrow \equiv S^SOEuCO_3^{0} + H^+$	8.3
$\equiv S^SOH + Eu^{3+} + CO_3^{2-} + H_2O \Leftrightarrow \equiv S^SOEuOHCO_3^{-} + 2H^+$	−0.25
$\equiv S^SOH + NpO_2^{+} + CO_3^{2-} \Leftrightarrow \equiv S^SONpO_2CO_3^{2-} + H^+$	2.0
$\equiv S^SOH + UO_2^{2+} + CO_3^{2-} \Leftrightarrow \equiv S^SOUO_2CO_3^{-} + H^+$	9.8
$\equiv S^SOH + UO_2^{2+} + 2CO_3^{2-} \Leftrightarrow \equiv S^SOUO_2(CO_3)_2^{3-} + H^+$	15.5

recent study did not require the formation of ternary U^{VI}-carbonato surface complexes. Tournassat et al. (2018) applied a new surface complexation model developed by Tournassat et al. (2016), including an electrostatic term that specifically accounts for the electrostatic surface potential of planar sites influencing the surface potential of neighbouring edge sites. This example illustrates nicely that similar adsorption data sets can be quantitatively described equally well by different surface complexation models.

5.3 Adsorption of iron on Mt

Iron (Fe) is the fourth most abundant metal in the Earth's crust, and is omnipresent in natural environments, occurring essentially in two oxidation states, the ferrous form (Fe^{2+}) and the ferric form (Fe^{3+}). Owing to its important redox reactivity, it is involved in many environmental geochemical processes. Its speciation, and consequently, its geochemical fate in aqueous environments is essentially governed by the prevailing pH and E_h conditions, because both oxidation states exhibit very different thermodynamic properties (e.g. solubility, aqueous complexation). The relatively soluble ferrous iron dominates in anaerobic and reducing environments. In the particular case of radioactive waste repositories, it is foreseen that through the corrosion process of steel canisters, large amounts of Fe^{2+} will be released into the bentonite buffer system. These high Fe^{2+} concentrations might impact the retention of other safety-relevant radionuclides in terms of competitive adsorption for available surface sites, but might also be involved in interfacial electron transfer processes, affecting more particularly the retention of redox sensitive radionuclides. In order to predict reliably the influence of Fe^{2+} on the adsorption behaviour of other metal ions in natural anoxic and reducing environments, it is essential to understand the uptake of Fe^{2+} on clay minerals and to develop robust models covering a large range of relevant geochemical conditions.

It should be emphasised that due to its important redox reactivity, investigating systems involving ferrous iron (in particularly at low Fe^{2+} concentrations ($\leq 10^{-7}$ M)), is quite challenging. To minimise the risk of oxidation, it is indispensable to work under strictly anoxic conditions. Even in inert atmosphere glove boxes, small amounts of O_2 are nearly unavoidable. Furthermore, adsorbed Fe^{2+} is much more susceptible to oxidation by O_2 than ferrous iron in the aqueous phase. Electrochemical reduction can be applied at higher Fe^{2+} concentrations; however, trace amounts of Fe^{3+} might still remain. Moreover, the aqueous quantification of Fe^{2+} without alteration of the chemical Fe^{2+}/Fe^{3+} equilibrium is an analytical challenge. Colorimetric methods using, for example, phenanthroline or ferrozine, allow an accurate determination of ferrous iron; however, yield a higher detection limit ($\geq 10^{-7}$ M) compared with common analytical techniques for elemental analysis, such as, for instance, ICP-MS (Verschoor and Molot, 2013).

Only a few studies explicitly addressed the adsorption of Fe^{2+} on clay minerals (i.e. in terms of adsorption quantification, thermodynamic modelling; e.g. Singhal et al., 1975; Tournassat, 2003; Schultz and Grundl, 2004; Charlet and Tournassat, 2005; Géhin et al., 2007; Jaisi et al., 2008; Soltermann et al., 2013b, 2014a,b).

An extensive study was carried out by Tournassat (2003) who investigated the interaction of Fe^{2+} with Mt specifically and systematically by batch adsorption experiments under anoxic conditions, adsorption modelling, and Mössbauer spectroscopy. The major outcomes of this study were: (i) Fe^{2+} and Ca^{2+} exhibit a similar cation exchange affinity (ii) the measured Fe^{2+} adsorption on clay mineral edge surfaces is considerably higher compared with Zn^{2+}; (iii) Fe^{2+} adsorption is controlled by several different reactions, that is, effective competitive adsorption with replacement of previously adsorbed or structural cations; co-adsorption with silica, suggesting a possible surface precipitation of an Fe—Si phase, and finally, oxidation of adsorbed Fe^{2+} on the edge sites to Fe^{3+}. In the case of adsorption on the planar sites (i.e. cation exchange) no oxidation was observed.

Géhin et al. (2007) investigated the pH-dependent adsorption of Fe^{2+} onto a synthetic iron free Ca^{2+}-saturated Mt under anoxic conditions. Under pH conditions where Fe^{2+} surface complexation is taking place, Mössbauer spectroscopic measurements showed that, with increasing pH, up to 7% of the total surface-bound

Fe is present as Fe^{3+}. The oxidation of Fe appeared to be fully reversible, suggesting that oxygen impurities were not the cause. Based on the adsorption data, and Mössbauer hyperfine parameters, an adsorption model considering cation exchange, surface complexation, and electron transfer was developed. The authors considered that oxidation only occurs on the strong sites, and proposed that the adsorbed Fe^{2+} acquires a steric configuration that stabilises Fe^{3+} relative to Fe^{2+}, thereby inducing a fully reversible Fe to clay mineral particle electron transfer under the formation of H_2.

The Mössbauer study of Schaefer et al. (2011) turned out to be of great importance in the understanding and description of adsorption of Fe^{2+}, because it yielded conclusive spectroscopic evidence for the reduction of structural Fe^{3+} to Fe^{2+} in ferruginous smectites by adsorbed Fe^{2+}.

Soltermann (2014) focused on the adsorption of ferrous iron on Mt with varying structural iron contents (SWy (3.4 wt.%), STx (0.7 wt.%) and IFM (0 wt.%)). A multidisciplinary approach was followed, whereby batch adsorption experiments under anoxic and electrochemically controlled reducing conditions, surface complexation modelling, Mössbauer, and EXAFS spectroscopy were combined. The following section will concentrate mostly on the work of this author, because it is the most up-to-date and complete study.

5.3.1 Fe^{2+} adsorption on Mt with no or low structural iron

5.3.1.1 Experimental adsorption data and modelling

Adsorption edges and isotherms of Fe^{2+} on Na^+-IFM and Na^+-STx shown in Fig. 5.15 were measured under anoxic, respectively, reducing conditions (Soltermann et al., 2013a, 2014a). The adsorption edges showed a strong pH dependency, which is typical for surface complexation of metal cations at the clay mineral edge sites and the nonlinear isotherms were consistent with adsorption on more than one type of surface sites. The similarity of the adsorption of Fe^{2+} with Zn^{2+} on both clay minerals (Fig. 5.15A) and the unchanged adsorption under reducing conditions (−0.640 V vs Standard Hydrogen Electrode, SHE) (Fig. 5.15B) suggested ferrous iron being the predominant surface complex (Soltermann et al., 2014a). The experimental data were fitted using the 2SPNE SC/CE adsorption model and the derived reactions and constants for Fe^{2+} on Na^+-IFM and Na^+-STx are given in Table 5.16. Note that the Fe^{2+}–Na^+ selectivity coefficient and the surface complexation constants for Fe^{2+} on the strong sites given in Table 5.16 slightly differ from the published data (Soltermann et al., 2013a, 2014a).

5.3.1.2 Spectroscopic studies

The predominance of adsorbed ferrous iron was demonstrated by Mössbauer spectroscopy on IFM and STx samples with different Fe loadings (1.2 and 10.3 mmol/kg on IFM and 10 and 40 mmol/kg on STx), supporting the model assumptions (Soltermann et al., 2013a,b, 2014a). However, the presence of ferric iron was also detected (up to 1.3 mmol/kg on IFM and ~5 mmol/kg on STx). Mössbauer

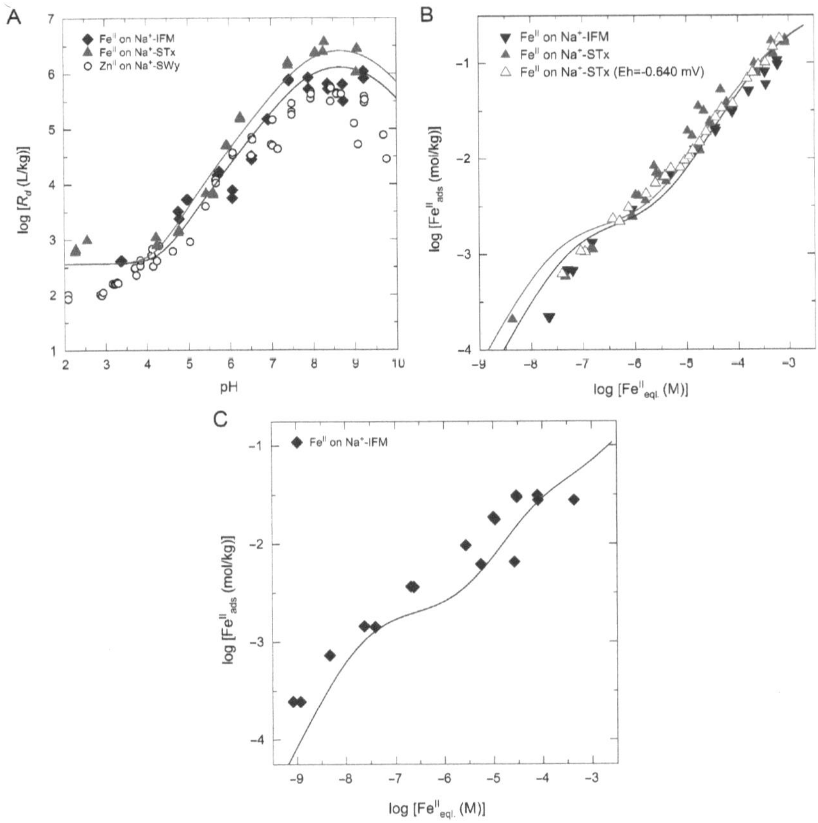

Fig. 5.15 (A) Fe^{2+} adsorption edges in 0.1 M $NaClO_4$ on Na^+-IFM and Na^+-STx and Zn^{2+} adsorption edge in 0.1 M $NaClO_4$ on Na^+-SWy. (B) Fe^{2+} adsorption isotherms at pH 6.2 in 0.1 M $NaClO_4$ on Na^+-IFM and Na^+-STx and under reducing conditions on Na^+-STx. (C) Adsorption isotherm at pH 6.7 in 0.3 M $NaClO_4$ on Na^+-IFM. *Symbols* represent the experimental data; *Continuous lines* represent the modelled curve using the 2SPNE SC/CE model with the parameters given in Tables 5.3 and 5.16.
Experimental data from Soltermann, D., Fernandes, M.M., Baeyens, B., Dähn, R., Miehé-Brendlé, J., Wehrli, B., Bradbury, M.H., 2013. Fe(II) sorption on a synthetic montmorillonite. A combined macroscopic and spectroscopic study. Environ. Sci. Technol. 47, 6978–6986; Soltermann, D., Baeyens, B., Bradbury, M.H., Fernandes, M.M., 2014. Fe(II) uptake on natural montmorillonites. II. Surface complexation modeling. Environ. Sci. Technol. 48, 8698–8705.

spectroscopy could be performed at much lower iron loading on IFM due to the absence of structural iron. The presence of oxidised iron on the IFM surface confirmed the previous observations by Géhin et al. (2007) on the same synthetic clay mineral. The reason for this oxidation could not yet be elucidated.

Table 5.16 Hydrolysis (Baes and Mesmer, 1976) and surface complexation and cation exchange reactions and the corresponding constants used to model the adsorption of Fe^{II} on IFM, STx, and SWy (Soltermann et al., 2013a,b, 2014a, this study)

Fe^{II} hydrolysis			log ^{OH}K
$Fe^{2+} + H_2O \Leftrightarrow Fe(OH)^+ + H^+$			−9.1
$Fe^{2+} + 2H_2O \Leftrightarrow Fe(OH)_2{}^0 + H^+$			20.6
$Fe^{2+} + 3H_2O \Leftrightarrow Fe(OH)_3{}^- + H^+$			31.0
	log $^{S,W1}K_{SC}$		
Surface complexation reactions	**IFM**	**STx**	**SWy**
$\equiv S^SOH + Fe^{2+} \Leftrightarrow \equiv S^SOFe^+ + H^+$	1.5	1.8	1.8
$\equiv S^{W1}OH + Fe^{2+} \Leftrightarrow \equiv S^{W1}OFe^+ + H^+$	−2.0	−1.7	−1.7
$\equiv S^SOH + Fe^{2+} \Leftrightarrow \equiv S^SOFe^{2+} + H^+ + e^-$	—	—	−1.7
$\equiv S^{W1}OH + Fe^{2+} \Leftrightarrow \equiv S^{W1}OFe^{2+} + H^+ + e^-$	—	—	−3.8
Cation exchange reaction	log K_c		
$2Na^+\text{-clay} + Fe^{2+} \Leftrightarrow Fe^{2+}\text{-clay} + 2Na^+$	1.1	1.1	1.1

Bulk and polarised EXAFS were successfully employed in other studies to unravel the structural atomic scale environment of iron in different smectites (Manceau et al., 2000a,b; Gates et al., 2002; Vantelon et al., 2003; Finck et al., 2015). Recently, a few studies addressed more specifically the identification of the surface products after reacting substantial Fe^{2+} concentrations (mM range) with different clay minerals. Zhu and Elzinga (2014) observed onto a synthetic mica-Mt the formation of Fe^{2+}–Al^{3+} layered double hydroxides (LDH) at near neutral pH, whereas at pH 8 the formation of trioctahedral Fe^{2+} phyllosilicates were formed. Jones et al. (2017), however, reported under similar conditions (pH 7.8), though on nontronite and Mt, the quantitative transformation of the added Fe^{2+} into chloride-green rust (Cl-GR), an Fe^{2+}–Fe^{3+}LDH.

Soltermann et al. (2013a) tried to unravel the structure of adsorbed Fe^{2+} on Na^+-IFM at much lower iron loadings than in the aforementioned studies using EXAFS spectroscopy. IFM was used instead of natural STx to avoid the spectral interference of the structural iron, enabling to probe solely the surface species. EXAFS samples were prepared under strictly anoxic conditions, and the iron loaded IFM pastes were immediately shock frozen in liquid N_2 to avoid oxidation during transport to the synchrotron facility. On site, the measurements were carried out under constant N_2 flow. Based on the wet chemical results, the reproducibility of the fitted structural parameters (i.e. invariance of Fe—O distance (2.01 Å) in all samples) and their similarity with Zn (Dähn et al., 2011), the authors concluded that Fe^{2+} forms, at low loadings, surface complexes in the continuity of the Al-octahedral sheets on well-defined

Fig. 5.16 (A) Fe^{2+} adsorption on Na^{+}-IFM at pH 7 in 0.1 M $NaClO_4$: *symbols* represent the EXAFS samples and *lines* the blind prediction using the 2SPNE SC/CE model with the surface complexation and selectivity constants on Na^{+}-IFM given in Table 5.16. Fe–K edge EXAFS spectra of the Fe loaded IFM samples: (B) XANES spectra and (C) Fourier transforms magnitudes.

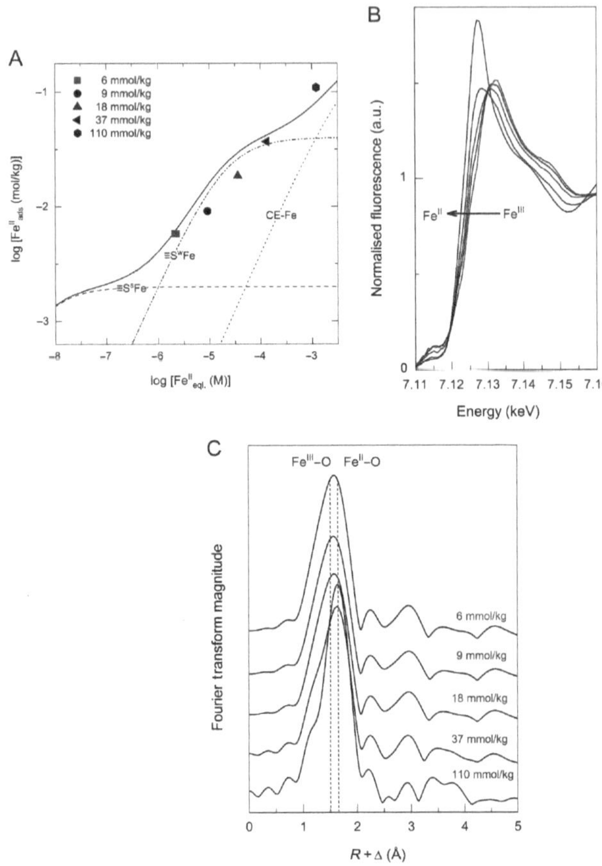

crystallographic binding sites at the Mt edges. At higher iron loadings, adsorption is characterised by a less crystallographic and more disordered structural environment (see also Section 5.2.2). However, it could also be argued that the Fe—O distance of 2.01 Å resembles the Fe^{3+}—O distance determined for structural iron in clay minerals (Vantelon et al., 2003; Finck et al., 2015), and not to Fe^{2+}—O distances of 2.10 Å as observed for instance in reduced nontronites (Manceau et al., 2000a), thus leaving a doubt on the oxidation state of the iron surface species. Recently, EXAFS spectra on newly prepared IFM samples at pH 7 with increasing Fe loadings ranging from 6 to 110 mmol/kg were recorded (Fig. 5.16A). The measurements were performed at the Fe K edge in fluorescence mode under cryogenic conditions (15 K) using a closed-cycle He-cryostat at the ROBL Beamline, ESRF in Grenoble. XANES and EXAFS data evaluation (one and two O-shell fit) (Fig. 5.15B and C and Table 5.17) suggests that iron is present in a mixed Fe^{2+}–Fe^{3+} oxidation state in all samples, with the average Fe—O distance obtained by one O-shell fit increasing from

Table 5.17 Main structural parameters obtained by fitting the EXAFS data (one and two O-shell fit) of the Fe loaded IFM samples

	One O shell fit		Two O shell fit				
Samples	**CN—O[a]**	**R (Å)**	**CN—O_1[c]**	**R (Å)[b]**	**CN—O_2[c]**	**R (Å)**	**% Fe^{II}**
6 mmol/kg	6	2.04	3.9	2.01	2.1	2.12	35
9 mmol/kg	6	2.05	3.1	2.01	2.9	2.12	48
18 mmol/kg	6	2.05	2.4	2.01	3.6	2.12	60
37 mmol/kg	6	2.08	2.0	2.01	4.0	2.12	66
110 mmol/kg	6	2.11	0.2	2.01	5.8	2.12	96

2.04 to 2.11 Å and the two O-shell fit yielding Fe—O distances of 2.01 and 2.12 Å with increasing total Fe loading, reflecting the increasing Fe^{2+}/Fe_{total} ratio. These results further suggest that iron in the IFM samples measured by Soltermann et al. (2013a) was mainly oxidised. Photochemical oxidation by the synchrotron beam could be a possible explanation, particularly because the samples were "only" maintained under N_2 flow during the measurements; however, there is no clear evidence at present. It is known that performing EXAFS measurements under cryogenic conditions prevents major oxidation of redox-sensitive nuclides. However, it was not possible to prepare samples with low and medium Fe loadings (i.e. below saturation of the edges sites), showing only Fe^{2+}. Iron in the highest-loaded IFM samples (110 mmol/kg) appeared to be mainly in its reduced state; however, under these conditions, 50% of the adsorbed iron is expected to be located on the planar sites, (Fig. 5.16A) which might be less prone to oxidation.

5.3.2 Fe^{2+} adsorption on Mt with moderate structural iron

As mentioned earlier, Soltermann et al. (2014a,b) also investigated the adsorption of Fe^{2+} on Na^+-SWy with 3.4 wt.% structural Fe^{3+}. The pH dependent adsorption of trace Fe^{2+} and an isotherm at pH 6.2 under anoxic conditions are shown in Fig. 5.17. Similar data sets measured on STx are shown for comparison. Under anoxic conditions, Fe^{2+} appeared to adsorb much stronger on Na^+-SWy than on the low iron content STx (Fig. 5.17). The rise in the adsorption edge starts at considerably lower pH values, and the R_d values are approximately 2 orders of magnitude higher than the corresponding adsorption edge for STx, clearly suggesting that the uptake of Fe^{2+} on SWy is distinctly different from that on STx (and also on IFM). This increased adsorption was attributed to the oxidation of the Fe^{2+} adsorbed onto the edge sites of SWy, resulting in a pronounced contribution of Fe^{3+} surface species, as already suggested by Tournassat (2003). Few data points along the adsorption isotherm at pH 6.2 were also measured under electrochemically reducing conditions (Fig. 5.17B), showing essentially the same

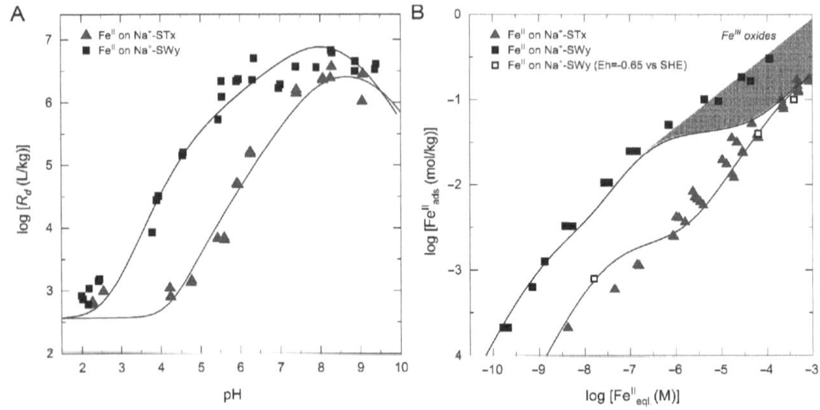

Fig. 5.17 Fe^{2+} adsorption in 0.1 M $NaClO_4$ on Na^+-SWy and on Na^+-STx (A) pH dependency and (B) adsorption isotherm at pH 6.2. *Continuous lines*: modelled curve using the 2SPNE SC/CE model with the parameters given in Tables 5.2 and 5.16. *Symbols*: experimental data. Experimental data from Soltermann, D., 2014. Ferrous Iron Uptake Mechanisms at the Montmorillonite-Water Interface under Anoxic and Electrochemically Reduced Conditions. PhD Thesis. ETH-Zürich.; Soltermann, D., Baeyens, B., Bradbury, M.H., Fernandes, M.M., 2014. Fe(II) uptake on natural montmorillonites. II. Surface complexation modeling. Environ. Sci. Technol. 48, 8698–8705.

behaviour as on Na^+-STx, and consequently validating the surface complexation constants derived for ferrous iron.

Both the oxidation of the surface species (up to 129 mmol/kg) and the associated reduction of the structural Fe^{3+} could be evidenced by using selectively the Mössbauer specificity of ^{56}Fe and ^{57}Fe (Soltermann, 2014). Based on this knowledge, Soltermann (2014) considered the formation of ferric surface complexes ($\equiv S^{S,W1}OFe^{2+}$), along with the previously postulated ferrous surface complexes ($\equiv S^{S,W1}OFe^{+}$) to model the experimental data on SWy. Given that the electron transfer mechanism and the nature of the Fe^{III} surface complexes are yet not fully mechanistically understood, the simplest reaction reflecting the oxidation of adsorbed Fe^{II} on the edge sites was used (see Table 5.16). It was assumed that in a first step, Fe^{2+} adsorbs on the strong and weak sites, and subsequently oxidises to Fe^{3+}. This redox-specific modelling step required the consideration of the electron activity, $p_e = -\log\{e^-\}$ in the aqueous phase. To model the isotherm, the redox potential (E_h) in the solution was determined at Fe^{2+} equilibrium concentrations $>10^{-5}$ M and the measured E_h values were converted to electron activities, $\{e^-\}$.To model the pH-dependent adsorption of trace Fe^{II}, an E_h–pH relation was included in the adsorption model. This relationship was obtained by keeping the surface complexation constants derived from the adsorption isotherm modelling fixed, and by adjusting for each pH the $\{e^-\}$ in order to reproduce the experimental data. The developed model is able to quantitatively describe the adsorption of Fe^{2+} on different Mt over a wider range of pH and E_h conditions; however, it does not provide a realistic mechanistic description, because it does not account for the intrinsic redox properties of the clay mineral itself.

Most studies in which the interaction of ferrous iron with clay minerals was addressed suggested that adsorbed Fe^{2+} is to some degree oxidised to Fe^{3+} at the clay mineral surface, even on IFM. In the case of iron containing clay minerals, this oxidation reaction is assumed to occur through electron transfer to the structural Fe^{3+}. The accurate mechanistic description of surface-induced electron transfer processes accounting for coupled redox changes (i.e. in the bulk structure and at the clay mineral surface) and its implementation in adsorption models requires a sound understanding of the redox properties of iron containing clay minerals (e.g. redox reactivity of structural Fe, conducting properties (efficiency, electron transfer pathway)). Investigating the redox properties and the associated electron transfer mechanisms in clay minerals is a challenging task. Fortunately, the number of studies dealing with these aspects, and additionally, applying different approaches (e.g. Fe^{2+} adsorption experiments, Mössbauer spectroscopy, mediated electrochemical reduction/oxidation, atomistic calculations) has increased in past years (e.g. Schaefer et al., 2011; Gorski et al., 2012a,b, 2013; Alexandrov and Rosso, 2013; Neumann et al., 2013, 2015; Latta et al., 2017).

A promising and successful approach to directly measure the redox properties of structural Fe in clay minerals is the mediated electrochemical reduction (MER) and oxidation (MEO) developed by Gorski et al. (2012a,b, 2013). The use of one-electron-transfer mediating compounds facilitates electron transfer between structural Fe in clay minerals, and the electrochemical set up (glassy carbon, working and reference electrode) enabling the direct measurement of reductive and oxidative currents. The excellent analytical capabilities of this mediated approach have allowed us to characterise and quantify the fraction of redox-active structural Fe (Fe^{2+}/Fe_{total}) in natural Mt with different structural iron content (native, reduced and reoxidised SWy-2, SWa-1), and in nontronites (NAu-1 and NAu-2) over a wide range of applied E_h-values (-0.6 V to $+0.6$ V). An Fe^{2+}/Fe_{total}–E_h relationship was deduced for each of the four clay minerals that correlate well with both bulk and molecular-scale properties of the smectites.

5.3.3 Influence of adsorbed iron on the uptake of redox-sensitive heavy metals

The mobility in geochemical environments of redox-sensitive heavy metals such as the actinides U, Np, Pu, or the fission product Tc, highly depends on their oxidation state. In their oxidised form, that is, UO_2^{2+}, NpO_2^{+}, TcO_4^{-} these radionuclides are highly mobile, whereas their transport is strongly decreased due to higher adsorption and lower solubility when reduced to the oxidation state +IV. The majority of adsorption studies on clay minerals so far have focused on the uptake of their oxidised states (see Section 5.2). Studies addressing the retention of the reduced forms are nearly non-existent, essentially due to their very low solubilities (Guillaumont et al., 2003) and the difficulty of maintaining reducing conditions during the experiments.

With respect to reduction reactions of inorganic and organic compounds, Fe^{2+} bound to oxide/clay minerals has been observed to be much more reactive than dissolved Fe^{II} (i.e. faster reaction kinetics, Fe^{2+}–Fe^{3+} associated with the clay mineral/oxide has a much lower redox potential than solute iron (Wehrli, 1990; Stumm and Sulzberger, 1992; Liger et al., 1999)). Depending on the Fe location (adsorbed on the clay mineral as a surface complex or by cation exchange, or present in the bulk structure), its reactivity can change by orders of magnitude. Iron associated with clay minerals may oxidise or reduce radionuclides at their surface (by acting either as an electron donor or acceptor), thus reducing/enhancing their mobility and/or (bio-)availability.

In the past two decades, an increasing number of studies have addressed the influence of coupled adsorption and interface charge transfer reactions on the retention of redox-sensitive radionuclides such as, for example, U, Np, Tc, or Pu, showing that these elements can be abiotically reduced by, e.g., Fe-bearing minerals/soils, Fe-oxides, and by Fe in clay minerals (e.g. Liger et al., 1999; O'Loughlin et al., 2003; Ilton et al., 2004, 2010; Behrends and Van Cappellen, 2005; Jeon et al., 2005; Jang et al., 2008; Peretyazhko et al., 2008; Jaisi et al., 2009; Boland et al., 2011; Kirsch et al., 2011; Latta et al., 2012; Peretyazhko et al., 2012; Singer et al., 2012; Fox et al., 2013; Brookshaw et al., 2015; Scheinost et al., 2016; Yalçıntaş et al., 2016). Studies focusing specifically on the influence of Fe^{II} adsorbed on clay minerals are, however, still sparse.

Chakraborty et al. (2010) observed the formation of an adsorbed molecular form of U^{IV} after abiotic reduction reaction of U^{VI} by Fe^{2+} adsorbed on Ca^{2+}-IFM. In this study three IFM samples with increasing Fe^{2+} surface loadings (20, 60 and 146 mmol/kg) prepared at pH 6.1, 7.5 and 8.5 and further equilibrated with U^{VI} were studied. Analysis of the XANES spectra suggested an increasing reduction of U^{VI} to U^{IV} with pH (14% U^{IV} at pH 6.1 to 96% U^{IV} at pH 8.5) in agreement with the pH dependent increasing Fe^{2+} adsorption. The reduction kinetics followed by XPS during 24 h at pH 7.5 confirmed the presence of a mixture of U^{VI} and U^{IV} surface species. Based on the measured redox potentials, the authors suggested that the Fe^{2+}–Fe^{3+}-IFM suspensions were controlled by the Fe^{2+}/hydrous ferric oxide couple at pH 6.2 and by the Fe^{2+}/lepidocrocite couple at pH 7.5. The exact structural environment of the U surface complexes could not be elucidated, since fitting of the EXAFS spectra only provided U–O distances.

In a more recent study, Tsarev et al. (2016) studied the retention of U^{VI} on three smectites with different structural iron contents (two nontronites with 25.2 and 26.5 wt.% and one smectite with 2.1 wt.%) in the absence and presence of significant aqueous Fe^{2+} concentrations (1 and 2 mM) in equilibrium with CO_2(g) under anoxic conditions as a function of pH. In the presence of 1 and 2 mM Fe^{2+}, U^{VI} uptake by the three smectites was characterised by a significant increase above pH 6 compared to the Fe^{2+} free U-smectite samples, proportional to the concentration of Fe^{2+} added. The uptake of Fe^{2+} as well as U^{VI} was similar on all three smectites, regardless the structural Fe^{3+} content. The comparison of the U-Fe-smectite XANES spectra with U^{IV} (uraninite) and U^{VI} (uranyl nitrate) reference spectra, suggested that after three days of reaction, uranium was present as a mixture of U^{IV}/U^{VI} surface complexes at $\geq$pH 6.5. After 2 weeks' reaction time, adsorbed U was found to be fully reduced to nano-crystalline uraninite (UO_2(s)) at pH 7.6; however, the extent of reduction remained unchanged at pH 6.5. Whether U is directly associated to the clay mineral surface or to Fe could not be evidenced by EXAFS, because no Al/Si nor Fe backscatter were detected. It should be observed; however, that under the experimental conditions applied in this study, Fe surface loadings are far above the adsorption capacity of the edge sites, approaching or even exceeding the cation exchange capacity of the smectites. Indeed, Fe^{2+}–Fe^{3+} oxy(hydr)oxides were detected by XRD and TEM as secondary mineralisation products upon reaction of the nontronites with Fe^{2+}, but apparently had no effect on the partitioning or speciation of uranium. On the iron-poor smectite, no Fe^{2+}–Fe^{3+} oxy(hydr)oxides were detected. The authors also hypothesised that green rust might have formed in the presence of these clay minerals, which was also recently reported by Jones et al. (2017).

These studies have shown that high Fe^{2+} concentrations associated to clay mineral surface can reduce for instance U^{VI}, consequently decreasing the mobility in the natural environment. However, the difference in the identified U surface products depending the experimental system and conditions, also reflects the complexity of such systems and emphasises the need of additional research covering larger ranges of experimental conditions, including other redox-sensitive radionuclides.

References

Alexandrov, V., Rosso, K.M., 2013. Insights into the mechanism of Fe(II) adsorption and oxidation at Fe–clay mineral surfaces from first-principles calculations. J. Phys. Chem. C 117, 22880–22886.

Amaya, T., Chiba, T., Suzuki, K., Oda, C., Yoshikawa, H., Yui, M., 1997. Solubility of Sn(IV) oxide in dilute $NaClO_4$ solution at ambient temperature. MRS Proc 465, 751–758.

Avena, M.J., De Pauli, C.P., 1998. Proton adsorption and electrokinetics of an Argentinean montmorillonite. J. Colloid Interface Sci. 202, 195–204.

Baes, C.F., Mesmer, R.E., 1976. The Hydrolysis of Cations. John Wiley & Sons Inc.

Baeyens, B., Bradbury, M.H., 1997. A mechanistic description of Ni and Zn sorption on Na- montmorillonite. 1. Titration and sorption measurements. J. Contam. Hydrol. 27, 199–222.

Barbier, F., Duc, G., Petit-Ramel, M., 2000. Adsorption of lead and cadmium ions from aqueous solution to the montmorillonite/water interface. Colloids Surf. A Physicochem. Eng. Asp. 166, 153–159.

Beene, G.M., Bryant, R., Williams, D.J.A., 1991. Electrochemical properties of illites. J. Colloid Interface Sci. 147, 358–369.

Behrends, T., Van Cappellen, P., 2005. Competition between enzymatic and abiotic reduction of uranium(VI) under iron reducing conditions. Chem. Geol. 220, 315–327.

Benson, L., 1982. A tabulation and evaluation of ion exchange data on smectites. Environ. Geol. 4, 23–29.

Bergaya, L., Lagaly, G., Vayer, M., 2013. Cation and anion exchange. In: Bergaya, F., Lagaly, G. (Eds.), Chapter 2.11 in Handbook of Clay Science, Techniques and Applications, second ed. Developments in Clay Science, vol. 58. Elsevier, pp. 333–359.

Boland, D.D., Collins, R.N., Payne, T.E., Waite, T.D., 2011. Effect of amorphous Fe(III) oxide transformation on the Fe(II)-mediated reduction of U(VI). Environ. Sci. Technol. 45, 1327–1333.

Bond, D.L., Davis, J.A., Zachara, J.M., 2008. Uranium(VI) release from contaminated vadose zone sediments: estimation of potential contributions from dissolution and desorption. In: Barnett, M.O., Kent, D.B. (Eds.), Adsorption of Metals to Geomedia II. Elsevier, New York, pp. 375–416.

Bourg, I.C., Sposito, G., Bourg, A.C.M., 2007. Modeling the acid-base surface chemistry of montmorillonite. J. Colloid Interface Sci. 312, 297–310.

Bradbury, M.H., Baeyens, B., 1994. Sorption by Cation Exchange. Incorporation of a Cation Exchange Model Into Geochemical Computer Codes. PSI Report Nr. 94–07, Paul Scherrer Institut, Switzerland.

Bradbury, M.H., Baeyens, B., 1997. A mechanistic description of Ni and Zn sorption on Na- montmorillonite, part 2: modelling. J. Contam. Hydrol. 27, 223–248.

Bradbury, M.H., Baeyens, B., 1999. Modelling the sorption of Zn and Ni on Ca-montmorillonite. Geochim. Cosmochim. Acta 63, 325–336.

Bradbury, M.H., Baeyens, B., 2000. A generalised sorption model for the concentration dependent uptake of caesium by argillaceous rocks. J. Contam. Hydrol. 42, 141–163.

Bradbury, M.H., Baeyens, B., 2002. Sorption of Eu on Na- and Ca-montmorillonites: experimental investigations and modelling with cation exchange and surface. Geochim. Cosmochim. Acta 66, 2325–2334.

Bradbury, M.H., Baeyens, B., 2005a. Modelling the sorption of Mn(II), Co(II), Ni(II), Zn(II), Cd(II), Eu(III), Am(III), Sn(IV), Th(IV), Np(V) and U(VI) on montmorillonite: linear free energy relationships and estimates of surface binding constants for some selected heavy metals and actinides. Geochim. Cosmochim. Acta 69, 875–892.

Bradbury, M.H., Baeyens, B., 2005b. Experimental and modelling investigations on Na-illite: acid-base behaviour and the sorption of strontium, nickel, europium and uranyl. PSI Report Nr. 05-02. Paul Scherrer Institut, Villigen, Switzerland.

Bradbury, M.H., Baeyens, B., 2006. Modelling sorption data for the actinides Am(III), Np(V) and Pa(V) on montmorillonite. Radiochim. Acta 94, 619–625.

Bradbury, M.H., Baeyens, B., 2009a. Sorption modelling on illite. Part I: titration measurements and the sorption of Ni, Co, Eu and Sn. Geochim. Cosmochim. Acta 73, 990–1003.

Bradbury, M.H., Baeyens, B., 2009b. Sorption modelling on illite. Part II: actinide sorption and linear free energy relationships. Geochim. Cosmochim. Acta 73, 1004–1013.

Bradbury, M.H., Baeyens, B., Geckeis, H., Rabung, T., 2005. Sorption of Eu(III)/Cm(III) on Ca-montmorillonite and Na-illite. Part 2: surface complexation modelling. Geochim. Cosmochim. Acta 69, 5403–5412.

Brookshaw, D.R., Pattrick, R.A.D., Bots, P., Law, G.T.W., Lloyd, J.R., Mosselmans, J.F.W., Vaughan, D.J., Dardenne, K., Morris, K., 2015. Redox interactions of Tc(VII), U(VI), and Np(V) with microbially reduced biotite and chlorite. Environ. Sci. Technol. 49, 13139–13148.

Brouwer, E., Baeyens, B., Maes, A., Cremers, A., 1983. Cesium and rubidium ion equilibriums in illite clay. J. Phys. Chem. 87, 1213–1219.

Bruggenwert, M., Kamphorst, A., 1982. Survey of experimental information on cation exchange in soil systems. In: Bolt, G.H. (Ed.), Developments in Soil Science. Elsevier, pp. 141–203.

Burns, P.C., Finch, R., 1999. Uranium: Mineralogy, Geochemistry and the Environment. Reviews in Mineralogy, Washington, DC.

Chakraborty, S., Favre, F., Banerjee, D., Scheinost, A.C., Mullet, M., Ehrhardt, J.-J., Brendlé, J., Vidal, L., Charlet, L., 2010. U(VI) sorption and reduction by Fe(II) sorbed on montmorillonite. Environ. Sci. Technol. 44, 3779–3785.

Charlet, L., Tournassat, C., 2005. Fe(II)-Na(I)-Ca(II) cation exchange on montmorillonite in chloride medium: evidence for preferential clay adsorption of chloride–metal ion pairs in seawater. Aquat. Geochem. 11, 115–137.
Choppin, G.R., 2007. Actinide speciation in the environment. J. Radioanal. Nucl. Chem. 273, 695–703.
Choppin, G.R., Rydberg, J., 1980. Nuclear Chemistry (Theory and Applications). Pergamon Press, Oxford.
Churakov, S.V., Dähn, R., 2012. Zinc adsorption on clays inferred from atomistic simulations and EXAFS spectroscopy. Environ. Sci. Technol. 46, 5713–5719.
Dähn, R., Baeyens, B., Bradbury, M.H., 2011. Investigation of the different binding edge sites for Zn on montmorillonite using P-EXAFS—the strong/weak site concept in the 2SPNE SC/CE sorption model. Geochim. Cosmochim. Acta 75, 5154–5168.
de Haan, F.A.M., Zwerman, P.J., 1976. Chapter 10 pollution of soil. In: Bolt, G.H., Bruggenwert, M.G.M. (Eds.), Developments in Soil Science. Elsevier, pp. 192–271.
Duc, M., Gaboriaud, F., Thomas, F., 2005. Sensitivity of the acid-base properties of clays to the methods of preparation and measurement: 1. Literature review. J. Colloid Interface Sci. 289, 139–147.
Dzombak, D., Morel, F.M.M., 1990. Surface Complexation Modeling: Hydrous Ferric Oxide. Wiley Interscience, New York.
Finck, N., Schlegel, M.L., Bauer, A., 2015. Structural iron in dioctahedral and trioctahedral smectites: a polarized XAS study. Phys. Chem. Miner. 42, 847–859.
Fletcher, P., Sposito, G., 1989. The chemical modelling of clay/electrolyte interactions for montmorillonite. Clay Miner. 24, 375–391.
Fox, P.M., Davis, J.A., Kukkadapu, R., Singer, D.M., Bargar, J., Williams, K.H., 2013. Abiotic U(VI) reduction by sorbed Fe(II) on natural sediments. Geochim. Cosmochim. Acta 117, 266–282.
Gaines, G.L., Thomas, H.C., 1953. Adsorption studies on clay minerals. II. A formulation of the thermodynamics of exchange adsorption. J. Chem. Phys. 21, 714–718.
Gates, W.P., Slade, P.G., Manceau, A., Lanson, B., 2002. Site occupancies by iron in nontronites. Clay Clay Miner. 50, 223–239.
Géhin, A., Grenèche, J.M., Tournassat, C., Brendlé, J., Rancourt, D.G., Charlet, L., 2007. Reversible surface-sorption-induced electron-transfer oxidation of Fe(II) at reactive sites on a synthetic clay mineral. Geochim. Cosmochim. Acta 71, 863–876.
Gorski, C.A., Aeschbacher, M., Soltermann, D., Voegelin, A., Baeyens, B., Marques Fernandes, M., Hofstetter, T.B., Sander, M., 2012a. Redox properties of structural Fe in clay minerals. 1. Electrochemical quantification of electron-donating and -accepting capacities of smectites. Environ. Sci. Technol. 46, 9360–9368.
Gorski, C.A., Klüpfel, L., Voegelin, A., Sander, M., Hofstetter, T.B., 2012b. Redox properties of structural Fe in clay minerals. 2. Electrochemical and spectroscopic characterization of electron transfer irreversibility in ferruginous smectite, SWa-1. Environ. Sci. Technol. 46, 9369–9377.
Gorski, C.A., Klüpfel, L.E., Voegelin, A., Sander, M., Hofstetter, T.B., 2013. Redox properties of structural Fe in clay minerals: 3. Relationships between smectite redox and structural properties. Environ. Sci. Technol. 47, 13477–13485.
Grim, R.E., 1953. Clay Mineralogy. McGraw-Hill Book Company, Inc., New York.
Guillaumont, R., Fanghänel, T., Fuger, J., Grenthe, I., Neck, V., Palmer, D.A., Rand, M.H., 2003. Update on the Chemical Thermodynamics of Uranium, Neptunium, Plutonium, Americium and Technetium. Elsevier, Amsterdam.

Hayes, K.F., Redden, G., Ela, W., Leckie, J.O., 1991. Surface complexation models: an evaluation of model parameter estimation using FITEQL and oxide mineral titration data. J. Colloid Interface Sci. 142, 448–469.
Hiemstra, T., Van Riemsdijk, W.H., 1996. A surface structural approach to ion adsorption: the charge distribution (CD) model. J. Colloid Interface Sci. 179, 488–508.
Huang, C.-P., Stumm, W., 1973. Specific adsorption of cations on hydrous [gamma]-Al_2O_3. J. Colloid Interface Sci. 43, 409–420.
Ilton, E.S., Haiduc, A., Moses, C.O., Heald, S.M., Elbert, D.C., Veblen, D.R., 2004. Heterogeneous reduction of uranyl by micas: crystal chemical and solution controls. Geochim. Cosmochim. Acta 68, 2417–2435.
Ilton, E.S., Boily, J.-F., Buck, E.C., Skomurski, F.N., Rosso, K.M., Cahill, C.L., Bargar, J.R., Felmy, A.R., 2010. Influence of dynamical conditions on the reduction of UVI at the magnetite–solution interface. Environ. Sci. Technol. 44, 170–176.
Jaisi, D.P., Liu, C., Dong, H., Blake, R.E., Fein, J.B., 2008. Fe^{2+} sorption onto nontronite (NAu-2). Geochim. Cosmochim. Acta 72, 5361–5371.
Jaisi, D.P., Dong, H., Plymale, A.E., Fredrickson, J.K., Zachara, J.M., Heald, S., Liu, C., 2009. Reduction and long-term immobilization of technetium by Fe(II) associated with clay mineral nontronite. Chem. Geol. 264, 127–138.
Jang, J.-H., Dempsey, B.A., Burgos, W.D., 2008. Reduction of U(VI) by Fe(II) in the presence of hydrous ferric oxide and hematite: effects of solid transformation, surface coverage, and humic acid. Water Res. 42, 2269–2277.
Jeon, B.-H., Dempsey, B.A., Burgos, W.D., Barnett, M.O., Roden, E.E., 2005. Chemical reduction of U(VI) by Fe(II) at the solid–water interface using natural and synthetic Fe(III) oxides. Environ. Sci. Technol. 39, 5642–5649.
Jones, A.M., Murphy, C.A., Waite, T.D., Collins, R.N., 2017. Fe(II) interactions with smectites: temporal changes in redox reactivity and the formation of green rust. Environ. Sci. Technol. 51, 12573–12582.
Kersten, M., Xiao, T., Kreissig, K., Brett, A., Coles, B.J., Rehkamper, M., 2014. Tracing anthropogenic thallium in soil using stable isotope compositions. Environ. Sci. Technol. 48, 9030–9036.
Kirsch, R., Fellhauer, D., Altmaier, M., Neck, V., Rossberg, A., Fanghaenel, T., Charlet, L., Scheinost, A.C., 2011. Oxidation state and local structure of plutonium reacted with magnetite, mackinawite, and chukanovite. Environ. Sci. Technol. 45, 7267–7274.
Kowal-Fouchard, A., Drot, R., Simoni, E., Ehrhardt, J.J., 2004. Use of spectroscopic techniques for uranium(VI)/montmorillonite interaction modeling. Environ. Sci. Technol. 38, 1399–1407.
Kraepiel, A.M.L., Keller, K., Morel, F.M.M., 1999. A model for metal adsorption on montmorillonite. J. Colloid Interface Sci. 210, 43–54.
Kulik, D.A., Aja, S.U., Sinitsyn, V.A., Wood, S.A., 2000. Acid–base surface chemistry and sorption of some lanthanides on K+-saturated Marblehead illite: II. A multisite–surface complexation modeling. Geochim. Cosmochim. Acta 64, 195–213.
Latta, D.E., Boyanov, M.I., Kemner, K.M., O'Loughlin, E.J., Scherer, M.M., 2012. Abiotic reduction of uranium by Fe(II) in soil. Appl. Geochem. 27, 1512–1524.
Latta, D.E., Neumann, A., Premaratne, W.A.P.J., Scherer, M.M., 2017. Fe(II)–Fe(III) electron transfer in a clay mineral with low Fe content. ACS Earth Space Chem. 1, 197–208.
Liger, E., Charlet, L., Van Cappellen, P., 1999. Surface catalysis of uranium(VI) reduction by iron(II). Geochim. Cosmochim. Acta 63, 2939–2955.
Maes, A., Peigneur, P., Cremers, A., 1975. Thermodynamics of transition metal ion exchange in montmorillonite. In: Proc. 5th Int. Clay Conf. Mexico City. 66, 319–333.

Manceau, A., Drits, V.A., Lanson, B., Chateigner, D., Wu, J., Huo, D., Gates, W.P., Stucki, J.W., 2000a. Oxidation-reduction mechanism of iron in dioctahedral smectites: II. Crystal chemistry of reduced Garfield nontronite. Am. Mineral. 85, 153–172.

Manceau, A., Lanson, B., Drits, V.A., Chateigner, D., Gates, W.P., Wu, J., Huo, D., Stucki, J.W., 2000b. Oxidation-reduction mechanism of iron in dioctahedral smectites: I. Crystal chemistry of oxidized reference nontronites. Am. Mineral. 85, 133–152.

Marques Fernandes, M., Baeyens, B., 2018. Cation exchange and surface complexation studies of lead on montmorillonite and illite including competitive adsorption effects. Appl. Geochem. (submitted for publication).

Marques Fernandes, M., Baeyens, B., Bradbury, M.H., 2008a. The influence of carbonate complexation on lanthanide/actinide sorption on montmorillonite. Radiochim. Acta 96, 691–698.

Marques Fernandes, M., Stumpf, T., Rabung, T., Bosbach, D., Fanghänel, T., 2008b. Incorporation of trivalent actinides into calcite: a time resolved laser fluorescence spectroscopy (TRLFS) study. Geochim. Cosmochim. Acta 72, 464–474.

Marques Fernandes, M., Baeyens, B., Daehn, R., Scheinost, A.C., Bradbury, M.H., 2012. U(VI) sorption on montmorillonite in the absence and presence of carbonate: a macroscopic and microscopic study. Geochim. Cosmochim. Acta 93, 262–277.

Marques Fernandes, M., Vér, N., Baeyens, B., 2015. Predicting the uptake of Cs, Co, Ni, Eu, Th and U on argillaceous rocks using sorption models for illite. Appl. Geochem. 59, 189–199.

Marques Fernandes, M., Scheinost, A.C., Baeyens, B., 2016. Sorption of trivalent lanthanides and actinides onto montmorillonite: macroscopic, thermodynamic and structural evidence for ternary hydroxo and carbonato surface complexes on multiple sorption sites. Water Res. 99, 74–82.

McKinley, J.P., Zachara, J.M., Smith, S.C., Turner, G.D., 1995. The influence of uranyl hydrolysis and multiple site-binding reactions on adsorption of U(VI) to montmorillonite. Clay Clay Miner. 43, 586–598.

McKinley, J.P., Zeissler, C.J., Zachara, J.M., Serne, R.J., Lindstrom, R.M., Schaef, H.T., Orr, R.D., 2001. Distribution and retention of 137Cs in sediments at the Hanford Site, Washington. Environ. Sci. Technol. 35, 3433–3441.

Merkel, B.J., Hasche-Berger, A., 2006. Uranium in the Environment Mining Impact and Consequences. Springer, Germany.

Montoya, V., Baeyens, B., Glaus, M.A., Kupcik, T., Marques Fernandes, M., Van Laer, L., Bruggeman, C., Maes, N., Schäfer, T., 2018. Sorption of Sr, Co and Zn on illite: batch experiments and modelling including Co in-diffusion measurements on compacted samples. Geochim. Cosmochim. Acta 223, 1–20.

Nagra, 2002. Technical Report NTB 02-05. Project Opalinus Clay: Safety Report. Demonstration of disposal feasibility (Entsorgungsnachweis) for spent fuel, vitrified high-level waste and long-lived intermediate-level waste. Nagra, Wettingen, Switzerland.

Neck, V., Kim, J.I., 2001. Solubility and hydrolysis of tetravalent actinides. Radiochim. Acta 89, 1–16.

Neumann, A., Olson, T.L., Scherer, M.M., 2013. Spectroscopic evidence for Fe(II)–Fe(III) electron transfer at clay mineral edge and basal sites. Environ. Sci. Technol. 47, 6969–6977.

Neumann, A., Wu, L., Li, W., Beard, B.L., Johnson, C.M., Rosso, K.M., Frierdich, A.J., Scherer, M.M., 2015. Atom exchange between aqueous Fe(II) and structural Fe in clay minerals. Environ. Sci. Technol. 49, 2786–2795.

Nilsson, N., Persson, P., Lövgren, L., Sjöberg, S., 1996. Competitive surface complexation of o-phthalate and phosphate on goethite (α-FeOOH) particles. Geochim. Cosmochim. Acta 60, 4385–4395.

O'Loughlin, E.J., Kelly, S.D., Cook, R.E., Csencsits, R., Kemner, K.M., 2003. Reduction of uranium(VI) by mixed iron(II)/iron(III) hydroxide (green rust): formation of UO2 nanoparticles. Environ. Sci. Technol. 37, 721–727.

Paschoa, A.S., 1998. Potential environmental and regulatory implications of naturally occurring radioactive materials (NORM). Appl. Radiat. Isot. 49, 189–196.

Peretyazhko, T., Zachara, J.M., Heald, S.M., Jeon, B.H., Kukkadapu, R.K., Liu, C., Moore, D., Resch, C.T., 2008. Heterogeneous reduction of Tc(VII) by Fe(II) at the solid–water interface. Geochim. Cosmochim. Acta 72, 1521–1539.

Peretyazhko, T.S., Zachara, J.M., Kukkadapu, R.K., Heald, S.M., Kutnyakov, I.V., Resch, C.T., Arey, B.W., Wang, C.M., Kovarik, L., Phillips, J.L., Moore, D.A., 2012. Pertechnetate (TcO4 –) reduction by reactive ferrous iron forms in naturally anoxic, redox transition zone sediments from the Hanford Site, USA. Geochim. Cosmochim. Acta 92, 48–66.

Peter, A.L.J., Viraraghavan, T., 2005. Thallium: a review of public health and environmental concerns. Environ. Int. 31, 493–501.

Rabung, T., Pierret, M.C., Bauer, A., Geckeis, H., Bradbury, M.H., Baeyens, B., 2005. Sorption of Eu(III)/Cm(III) on a-montmorillonite and Na-illite. Part 1: batch sorption and time-resolved laser fluorescence spectroscopy experiments. Geochim. Cosmochim. Acta 69, 5393–5402.

Savanonda, N., Prongpunyasakul, E., Ratanakorn, S., 1985. Quantitative Analysis of Uranium, Thorium and Potassium From Lignite Ash by Neutron Activation Method. Gordon and Breach, United States.

Schaefer, M.V., Gorski, C.A., Scherer, M.M., 2011. Spectroscopic evidence for interfacial Fe (II)-Fe(III) electron transfer in a clay mineral. Environ. Sci. Technol. 45, 540–545.

Scheinost, A.C., Steudtner, R., Hübner, R., Weiss, S., Bok, F., 2016. Neptunium^{V} retention by siderite under anoxic conditions: precipitation of NpO2-like nanoparticles and of Np^{IV} pentacarbonate. Environ. Sci. Technol. 50, 10413–10420.

Schnurr, A., Marsac, R., Rabung, T., Lutzenkirchen, J., Geckeis, H., 2015. Sorption of Cm(III) and Eu(III) onto clay minerals under saline conditions: batch adsorption, laser-fluorescence spectroscopy and modeling. Geochim. Cosmochim. Acta 151, 192–202.

Schultz, C., Grundl, T., 2004. pH dependence of ferrous sorption onto two smectite clays. Chemosphere 57, 1301–1306.

Shannon, R.D., 1976. Revised effective ionic radii and systematic studies of interatomic distances in halides and chalcogenides. Acta Crystallogr. A 32, 751–767.

Singer, D.M., Chatman, S.M., Ilton, E.S., Rosso, K.M., Banfield, J.F., Waychunas, G.A., 2012. U(VI) sorption and reduction kinetics on the magnetite (111) surface. Environ. Sci. Technol. 46, 3821–3830.

Singhal, J.P., Singh, R.P., Kumar, D., 1975. Thermodynamics of ion exchange equilibria involving Fe^{2+} ion on Na^{+}-montmorillonite. J. Indian Chem. Soc 52, 380–384.

Soltermann, D., 2014. Ferrous Iron Uptake Mechanisms at the Montmorillonite-Water Interface Under Anoxic and Electrochemically Reduced Conditions. PhD Thesis, ETH-Zürich.

Soltermann, D., Fernandes, M.M., Baeyens, B., Dähn, R., Miehé-Brendlé, J., Wehrli, B., Bradbury, M.H., 2013a. Fe(II) sorption on a synthetic montmorillonite. A combined macroscopic and spectroscopic study. Environ. Sci. Technol. 47, 6978–6986.

Soltermann, D., Marques Fernandes, M., Baeyens, B., Daehn, R., Miehé-Brendlé, J., Wehrli, B., Bradbury, M.H., 2013b. Fe(II) Sorption on a synthetic montmorillonite. A combined macroscopic and spectroscopic study. Environ. Sci. Technol. 47, 6978–6986.

Soltermann, D., Baeyens, B., Bradbury, M.H., Fernandes, M.M., 2014a. Fe(II) uptake on natural montmorillonites. II. Surface complexation modeling. Environ. Sci. Technol. 48, 8698–8705.

Soltermann, D., Marques Fernandes, M., Baeyens, B., Dähn, R., Joshi, P.A., Scheinost, A.C., Gorski, C.A., 2014b. Fe(II) uptake on natural montmorillonites. I. Macroscopic and spectroscopic characterization. Environ. Sci. Technol. 48, 8688–8697.
Sposito, G., 1984. The Surface Chemistry of Soils. Oxford University Press, New York, NY.
Środoń, J., MaCarty, D.K., 2008. Surface area and layer charge of smectite from CEC and EGME/H_2O-retention measurements. Clay Clay Miner. 56, 155–174.
Stadler, M., Schindler, P.W., 1993. Modeling of H^+ and Cu^{2+} adsorption on calcium-montmorillonite. Clay Clay Miner. 41, 288–296.
Stumm, W., Sulzberger, B., 1992. The cycling of iron in natural environments: considerations based on laboratory studies of heterogeneous redox processes. Geochim. Cosmochim. Acta 56, 3233–3257.
Stumm, W., Huang, C.P., Jenkins, S.R., 1970. Specific chemical interaction affecting the stability of dispersed systems. Croat. Chem. Acta 42, 223–244.
Stumpf, T., Bauer, A., Coppin, F., Il Kim, J., 2001. Time-resolved laser fluorescence spectroscopy study of the sorption of Cm(III) onto smectite and kaolinite. Environ. Sci. Technol. 35, 3691–3694.
Stumpf, T., Hennig, C., Bauer, A., Denecke, M.A., Fanghanel, T., 2004. An EXAFS and TRLFS study of the sorption of trivalent actinides onto smectite and kaolinite. Radiochim. Acta 92, 133–138.
Tamura, T., Jacobs, D., 1960. Structural implications in cesium sorption. Health Phys. 2, 391–398.
Tombácz, E., Szekeres, M., 2004. Colloidal behavior of aqueous montmorillonite suspensions: the specific role of pH in the presence of indifferent electrolytes. Appl. Clay Sci. 27, 75–94.
Tournassat, C., 2003. Cations–Clays Interactions: The Fe(II) Case. Application to the Problematic of the French Deep Nuclear Repository Field Concept. PhD thesis, Université Joseph-Fourier, Grenoble.
Tournassat, C., Ferrage, E., Poinsignon, C., Charlet, L., 2004. The titration of clay minerals: II. Structure-based model and implications for clay reactivity. J. Colloid Interface Sci. 273, 234–246.
Tournassat, C., Bourg, I.C., Steefel, C.I., Bergaya, F., 2015. Chapter 1—surface properties of clay minerals. In: Tournassat, C., Steefel, C.I., Bourg, I.C., Bergaya, F. (Eds.), Developments in Clay Science. Elsevier, pp. 5–31.
Tournassat, C., Davis, J.A., Chiaberge, C., Grangeon, S., Bourg, I.C., 2016. Modeling the acid–base properties of montmorillonite edge surfaces. Environ. Sci. Technol. 50, 13436–13445.
Tournassat, C., Tinnacher, R.M., Grangeon, S., Davis, J.A., 2018. Modeling uranium(VI) adsorption onto montmorillonite under varying carbonate concentrations: a surface complexation model accounting for the spillover effect on surface potential. Geochim. Cosmochim. Acta 220, 291–308.
Trubert, D., Le Naour, C., Jaussaud, C., Mrad, O., 2003. Hydrolysis of protactinium(V). III. Determination of standard thermodynamic data. J. Solut. Chem. 32, 505–517.
Tsarev, S., Waite, T.D., Collins, R.N., 2016. Uranium reduction by Fe(II) in the presence of montmorillonite and nontronite. Environ. Sci. Technol. 50, 8223–8230.
Turner, G.D., Zachara, J.M., McKinley, J.P., Smith, S.C., 1996. Surface-charge properties and UO22+ adsorption of a subsurface smectite. Geochim. Cosmochim. Acta 60, 3399–3414.
Turner, D.R., Pabalan, R.T., Bertetti, F.P., 1998. Neptunium(V) sorption on montmorillonite: an experimental and surface complexation modeling study. Clay Clay Miner. 46, 256–269.
Vaněk, A., Komárek, M., Vokurková, P., Mihaljevič, M., Šebek, O., Panušková, G., Chrastný, V., Drábek, O., 2011. Effect of illite and birnessite on thallium retention and bioavailability in contaminated soils. J. Hazard. Mater. 191, 170–176.

Vanselow, A.P., 1932. Equilibria of the base-exchange reactions of bentonites, permutites, soil colloids, and zeolites. Soil Sci. 33, 95–114.

Vantelon, D., Montarges-Pelletier, E., Michot, L.J., Pelletier, M., Thomas, F., Briois, V., 2003. Iron distribution in the octahedral sheet of dioctahedral smectites. An Fe K-edge X-ray absorption spectroscopy study. Phys. Chem. Miner. 30, 44–53.

Verschoor, M., Molot, L., 2013. A comparison of three colorimetric methods of ferrous and total reactive iron measurement in freshwaters. Limnol. Oceanogr. Methods 11, 113–125.

Voegelin, A., Pfenninger, N., Petrikis, J., Majzlan, J., Plötze, M., Senn, A.-C., Mangold, S., Steininger, R., Göttlicher, J., 2015. Thallium speciation and extractability in a thallium- and arsenic-rich soil developed from mineralized carbonate rock. Environ. Sci. Technol. 49, 5390–5398.

Wanner, H., Albinsson, Y., Karnland, O., Wieland, E., Charlet, L., Wersin, P., 1994. The acid/base chemistry of montmorillonite. Radiochim. Acta 66–67, 157.

Weast, R.C., Astle, M.J., Beyer, W.H., 1983. CRC Handbook of Chemistry and Physics. CRC Press, Boca Raton, FL.

Wehrli, B., 1990. Redox reactions of metal ions at mineral surfaces. In: Stumm, W. (Ed.), Aquatic Chemical Kinetics: Reaction Rates of Processes in Natural Waters. Wiley, New York, pp. 311–366.

Wendt, S., 2008. Sorption and Direct Speciation of Neptunium(V) on Aluminium Oxide and Montmorillonite. PhD thesis, Johannes Gutenberg University, Mainz, Germany.

Westall, J., 1982. FITEQL: A Computer Program for Determination of Equilibrium Constants From Experimental Data. Department of Chemistry, Oregon State University, Corvallis, OR.

Wick, S., Baeyens, B., Marques Fernandes, M., Voegelin, A., 2018. Thallium adsorption onto illite. Environ. Sci. Technol. 52, 571–580.

Xiao, T., Yang, F., Li, S., Zheng, B., Ning, Z., 2012. Thallium pollution in China: a geo-environmental perspective. Sci. Total Environ. 421–422, 51–58.

Yalçıntaş, E., Scheinost, A.C., Gaona, X., Altmaier, M., 2016. Systematic XAS study on the reduction and uptake of Tc by magnetite and mackinawite. Dalton Trans. 45, 17874–17885.

Zachara, J.M., Smith, S.C., 1994. Edge complexation reactions of cadmium on specimen and soil-derived smectite. Soil Sci. Soc. Am. J. 58, 762–769.

Zachara, J.M., Smith, S.C., Liu, C., McKinley, J.P., Serne, R.J., Gassman, P.L., 2002. Sorption of Cs+ to micaceous subsurface sediments from the Hanford Site. USA, Geochim. Cosmochim. Acta 66, 193–211.

Zhu, Y., Elzinga, E.J., 2014. Formation of layered Fe(II)-hydroxides during Fe(II) sorption onto clay and metal-oxide substrates. Environ. Sci. Technol. 48, 4937–4945.

From transition metal ion complexes to chiral clay minerals

6

Hisako Sato, Akihiko Yamagishi*†
*Graduate School of Science and Engineering, Ehime University, Matsuyama, Japan,
†Faculty of Medicine, Toho University, Tokyo, Japan

6.1 Introduction

"Chirality recognition" on a solid surface is one of the most extensively studied subjects in chemistry (Mallouk and Gavin, 1998; Berthod, 2006). A number of biomimetic systems, such as crown ethers, cryptands, and cyclodextrins are designed as artificial sites to achieve chiral selectivity. In comparison with the molecular systems in homogeneous phases, the surface of a solid provides another possibility for molecular recognition. The clean surface of a metal single crystal, for example, provides information on an atomic scale on how a molecule interacts with a surface, or with its neighbours. A scanning tunnelling microscope (STM), for example, makes it possible to visualise chiral recognition by a single molecule on the surface of a single crystal (Kuhnle et al., 2002). In contrast to these most successful examples, there remain solid materials, such as microcrystalline insulating metal oxides, that are still resistant against microscopic approaches.

In this chapter, the chiral aspects of smectites were considered. In comparison with metals or bulk metal oxides, the unique features of smectites for molecular recognition are: (i) two-dimensional network structures of clay mineral layers, (ii) high cation-exchange capacity (CEC), and (iii) exfoliation into single layers or nanolayers.

In the following, the use of smectites for optical resolution, asymmetric syntheses, and photochemical enantioselective sensing are reviewed. Solid-state vibrational circular dichroism (VCD) is described toward molecular recognition on a clay mineral surface. Attention is focused on how the preceding structural features contributed to the recognition of molecular chirality on a clay mineral surface.

6.2 Stereochemistry of a clay mineral surface

There have been extensive efforts to apply a solid surface as a site for separating chiral substances on a preparative scale (Yamagishi and Sato, 2012). It was reported, for example, that some biological substances, such as amino acids and glucoses, were adsorbed stereoselectively by natural montmorillonite or vermiculite (Bondy and Harrington, 1979; Fraser et al., 2011a,b). The ability of clay minerals to show chiral selectivity for biological molecules has profound implications for chemical evolution.

Developments in Clay Science, Vol. 9. https://doi.org/10.1016/B978-0-08-102432-4.00006-8

Yamagishi and Soma (1981) studied the adsorption of chiral molecules by sodium montmorillonite (Na^+-Mt). As an adsorbate, tris(1,10-phenanthroline)metal(II) ($[M(phen)_3]^{2+}$) (phen = 1,10-phenanthroline) was chosen (Yamagishi and Soma, 1981). The complex was known to be adsorbed strongly by clay minerals. Its rigid, highly symmetric structure provided a sensitive molecular probe to distinguish enantiomers. The enantiomers of $[M(phen)_3]^{2+}$ are denoted as Δ- and Λ-isomers, depending on the direction of helical twisting (Fig. 6.1). The adsorption of enantiomeric $[Ru(phen)_3]^{2+}$ by a clay mineral was studied by measuring the electronic absorption spectrum of a colloidal dispersion. When Na^+-Mt was added to an aqueous solution of Δ-$[Ru(phen)_3]Cl_2$, the metal-to-ligand charge transfer (MLCT) transition band around 450 nm increased its intensity and shifted toward longer wavelengths, with a simultaneous decrease of the π–π* transition band localised in the ligands around 270 nm. These spectroscopic changes were ascribed to the structural deformation of the chelate caused by adsorption on a clay mineral particle. No effect of molecular chirality was detected, however, when the spectral changes of the Δ- and Λ-enantiomers of $[Ru(phen)_3]^{2+}$ were compared.

When the same experiments were performed with a racemic mixture of $[Ru(phen)_3]^{2+}$, the change of the electronic spectrum was distinctly different from the enantiomeric case. The bathochromic shift of the MLCT transition band was more enhanced, and the π–π* transition band split into two components. In addition to these spectroscopic differences, a remarkable difference was observed in the adsorption amounts. For the chelates, the adsorbed amount increased linearly, with the increase of the amount of clay mineral. In the case of the enantiomer, the adsorption was saturated when the ratio of clay mineral-to-metal chelate (denoted as [clay mineral]/[chelate]) attained the value of 2, where [clay mineral] and [chelate] were measured in terms of the CEC equivalent/L and mole/L, respectively. The results were reasonable because $[Ru(phen)_3]^{2+}$ carried two positive charges, and neutralised two negative charges on adsorption by a clay mineral. Contrary to this, in the case of the racemic mixture,

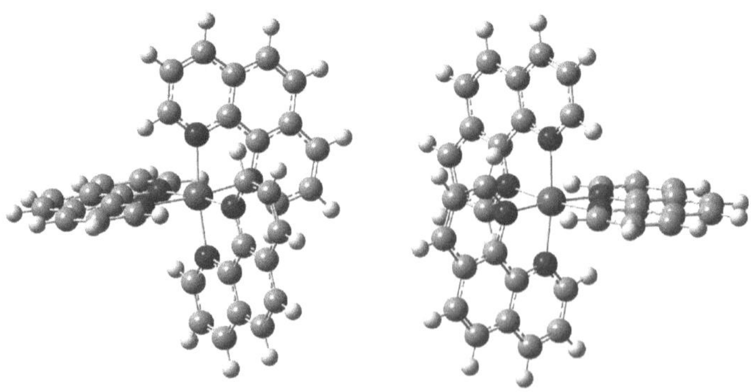

Fig. 6.1 Chiral structures of a tris-chelated metal complex, $[Ru(phen)_3]^{2+}$: (*left*) Δ-isomer and (*right*) Λ-isomer.

the adsorption was saturated already at [clay mineral]/[chelate] = 1. In other words, the adsorbed amount of the racemic chelate was twice the CEC, and anions were included for electric neutrality.

The preceding facts indicated that the chelates interacted with each other on a clay mineral surface in a *stereoselective* way. Here, "*stereoselective* way" means that the situations are different whether the interacting two molecules are a Δ–Δ (or Λ–Λ) enantiomeric pair or a Δ–Λ racemic pair. If the interaction was indifferent to such chirality effects, no difference would have appeared, whether the chelates were added as a pure enantiomer or as a racemic mixture. The fact that the racemic mixture was adsorbed to two times the CEC, and the enantiomers up to the CEC suggested that the chelates preferred to form a racemic pair on a clay mineral surface. Based on the finding, this type of adsorption was denoted as "racemic adsorption." The structural origin of racemic adsorption was investigated by applying a number of spectroscopic techniques, such as XRD (X-ray diffraction) (Taniguchi et al., 1991), electric dichroism (Yamagishi et al., 1994), ESR (Yano et al., 1992), photoelectron spectroscopy (Yamagishi et al., 1982), AFM (Takahashi et al., 2003; Wakabayashi et al., 2004) and theoretical calculation (Sato et al., 1992). The basal distance (d_{001}) was determined by XRD measurements for the adducts of Na^+-Mt and $[M(phen)_3]^{2+}$. Electro-optical measurements were applied to determine the orientation of $[Ru(phen)_3]^{2+}$ bound on the surface of a colloidal clay mineral particle. An electric field pulse was applied to a dispersion containing a colloidal clay mineral and the chelate. The anisotropy in optical absorption (dichroism) was induced by the alignment of clay mineral layers in the direction of the electric field.

The adsorption by Na^+-Mt was investigated over a wide range of chiral chelates (Yamagishi, 1982, 1985a; Yamagishi and Fujita, 1984). When the thin film of an ion-exchange adduct of racemic $[Os(phen)_3]^{2+}$/Mt was analysed by in-plane XRD measurements, the two-dimensional regularity was detected with a periodicity of 0.95 nm (Umemura and Shinohara, 2004). STM was successful only for highly crystalline clay minerals such as hydrotalcite (Yao et al., 1998).

Breu et al. (2001) performed experiments on the intercalation of metal complexes by a smectite. In the case of the intercalation compounds of natural smectites, no 2D long-range order of the interlayer space was observed. On the contrary, in-plane order was detected using a homogeneously charged synthetic fluorohectorite as a host material. With these intercalates, the distinct 2D structures for enantiopure and racemic $[Ru(bpy)_3]^{2+}$(bpy = 2,2′-bipyridine) interlayers, as previously predicted by atomistic simulations, could be proven experimentally (Breu et al., 1999, 2001; Hirsemann et al., 2012; Herling et al., 2017). Recently Kunz et al. (2015) reported the quasiepitaxial growth of $[Ru(bpy)_3]^{2+}$ by confinement in clay mineral layers, yielding polarised emission under the irradiation of visible light. This work is a proof of the principle that the orientation of a transition metal luminophore intercalated within the interlayer spaces of well-organised clay mineral layers can be controlled.

Villemure (1990) reported the adsorption of enantiomeric $[Ru(bpy)_3]^{2+}$ by different smectites. The adsorption of the racemic mixture was lower than that of pure enantiomers. The enantiomer was adsorbed in excess of the CEC.

6.3 Chirality recognition by a clay mineral surface modified with metal complexes

6.3.1 Clay mineral column chromatography for optical resolution

The stereochemical effects on the adsorption of chiral molecules by a smectite led to the development of chromatographic resolution by use of an ion-exchange adduct of a clay mineral and metal complexes (Yamagishi, 1985b; Yamagishi et al., 1992). A spherical aggregate was prepared by spray-drying an aqueous dispersion of synthetic hectorite. Particles of ca. 5 μm were soaked in a methanol solution containing Λ-$[Ru(phen)_3]Cl_2$. The ion-exchange of Λ-$[Ru(phen)_3]^{2+}$ into the clay mineral took place in seconds to produce a yellow solid. The material was packed into a stainless steel tube and used as a column for high-performance liquid chromatography (Nakamura et al., 1988; Kakegawa and Yamagishi, 2005).

The chromatogram of Fig. 6.2 shows the elution of a racemic mixture of 1,1′-bi-2-naphthol (denoted as BINOL) on the column (4 mm (i.d.) × 250 mm) with methanol. Two peaks with equal area were obtained on the base-line separation. From the electronic circular dichroism (ECD) spectra, the less and more retaining peaks contained the pure *R*- and *S*-enantiomers of BINOL, respectively. The separation factor was calculated to be 1.78, which was quite large in comparison with the known chiral columns. The enthalpy change ($-\Delta H$) of adsorption was determined from the temperature dependence of the retention times: $-\Delta H = 3.1$ and 5.7 kJ mol^{-1} for the *R*- and *S*-BINOL, respectively. Thus, the high separation capability of the present column was ascribed to the large difference of adsorption enthalpy (~2.6 kJ mol^{-1}) (Yamagishi et al., 1996).

The clay mineral column exhibited the following characteristics from the practical point of view: (i) performance is stable and reproducible for a long time against the change of temperature or eluting solvents (except for water), (ii) resolution capacity is

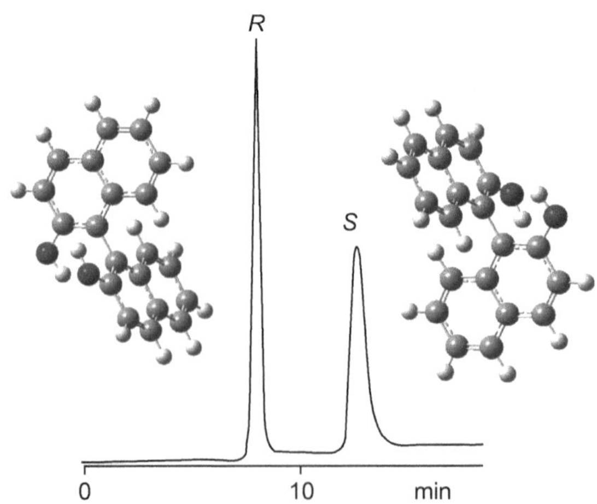

Fig. 6.2 The chromatogram when a racemic mixture of 1,1′-bi-2-naphthol (BINOL) was eluted on a column packed with Λ-[Ru$(phen)_3]^{2+}$/synthetic hectorite with methanol. The first and second peak are *R*- and *S*-enantiomers, respectively.

remarkably high (e.g. 400 mg of [Ru(acac)$_3$] was resolved in a single run on a 20 mm (i.d.) × 250 mm column), (iii) antipodal columns, Δ-[Ru(phen)$_3$]$^{2+}$/clay mineral and Λ-[Ru(phen)$_3$]$^{2+}$/clay mineral, are both available. This is particularly useful in separating a mixture of diastereomeric species. As an example, a complete series of mixed ligand Ru(III) complexes, which are expressed by the formula of Δ- or Λ-[Ru((−)- or (+)-tfac)$_n$(acac)$_{3-n}$] ($n = 1$, 2, and 3, (−)- or (+)-tfac = (−)- or (+)-3-trifluoroacetylcamphorato and acac = acetylacetonato), were prepared in a pure diastereomeric form (Sato et al., 2009) and (iv) the systematic modification of a packing material is possible to tune resolution capability by changing the structure of a preadsorbed chelate (e.g. [Ru(bpy)$_3$]$^{2+}$/clay mineral) (Yamagishi et al., 1996).

A column packed with Δ-[Ru(phen)$_3$]$^{2+}$/synthetic hectorite is commercially available and is used for separating a wide range of organic and inorganic compounds (Ceramosphere Ru-1, Shiseido Ind. Co., Japan). In one recent application, all of the four diastereomers of a star-burst type tetranuclear ruthenium(III) complex were resolved. The diastereomers were differentiated by the parts connecting the core and periphery regions. Thus, the column discriminated local chirality within a large molecule (Fujimoto et al., 2010).

Recently, the clay mineral chiral column was improved by using a silica gel particle coated with a thin film of hectorite. The coating of silica gel particles was carried out based on the recent report by Okada et al. (2016). Hectorite was grown on the surface of silica particles to produce core–shell type silica particles. A column was prepared by adsorbing Δ-[Ru(phen)$_3$](ClO$_4$)$_2$ by ion-exchange. The resulting slurry was poured into a 4 mm (i.d.) × 25 cm stainless column. When racemic [Ru(acac)$_3$] was eluted with methanol at a flow rate of 1.0 mL min^{-1}, two slightly overlapping sharp peaks were obtained at an elution volume of 2.6 and 3.0 mL, respectively. From the ECD spectrum of the fraction collected at each peak, the first and second peaks contained the excess Λ- and Δ-enantiomers, respectively. For comparison, the same molecule was mounted on the column packed with the spray-dried hectorite particles ion-exchanged with Δ-[Ru(phen)$_3$]$^{2+}$ (Ceramosphere RU-1, Shiseido Ind. Co., Japan) and eluted under the same conditions. Two separated peaks were obtained with the elution volumes of 19.1 and 49.2 mL, respectively. From the ECD spectra of the collected fractions, the same enantiomeric order was confirmed. Comparing the chromatographic features between these two columns, it was concluded that the same mechanism was responsible for chiral recognition. The difference of elution was ascribed to the situation that the layer of a clay mineral was less than 5 μm in the former column, while the whole particle consisted of a clay mineral in the latter column. Accordingly, equilibration was attained rapidly within thc thin clay mineral film with no penetration of [Ru(acac)$_3$] into the inner region of the packing particle (Fig. 6.3).

6.3.2 Chiral phosphorescent probes on a clay mineral surface

The surfaces of clay minerals are thought to be unique as a medium for photochemical reactions. Research on this topic is reviewed by Shichi and Takagi (2000). The photophysical processes of tris(2,2′bipyridine) ruthenium(II) ([Ru(bpy)$_3$]$^{2+}$) bound to colloidal Mt particles were studied (Ghosh and Bard, 1983; Thomas, 1988;

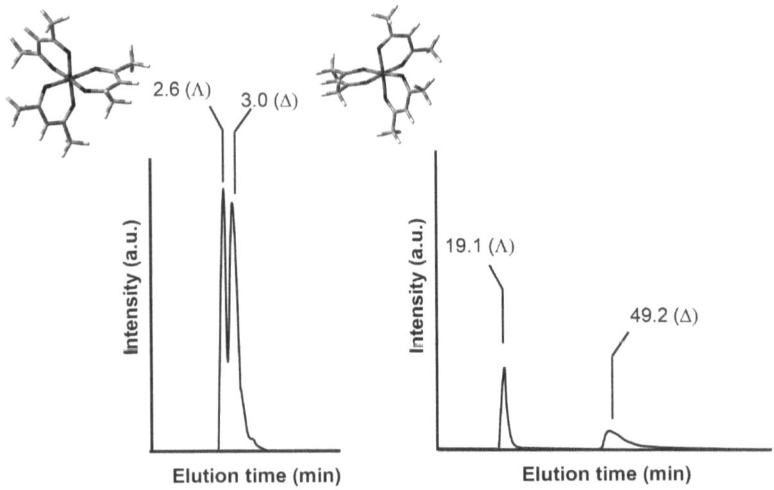

Fig. 6.3 Chromatograms when a racemic mixture of [Ru(acac)$_3$] was eluted with methanol on columns packed with (*left*) Δ-[Ru(phen)$_3$]$^{2+}$-hectorite-coated silica gel, (*right*) and RU-1 (chiral column packed with an ion-exchange adduct of Δ-[Ru(phen)$_3$]$^{2+}$ and synthetic hectorite (Ceramosphere RU-1, Shiseido Ind. Co., Japan)). The flow rate was 1 mL min^{-1}, and the elution was monitored by the absorbance at 346 nm at room temperature.
Modified from Okada, T., Kumasaki, A., Shimizu, K., Yamagishi, A., Sato, H., 2016. Application of hectorite-coated silica gel particles as a packing material for chromatographic resolution. J. Chromatogr. Sci. 54, 1238–1243.

Kamat et al., 1991; Sato et al., 2005). The effects of chirality on the intermolecular interactions of poly(pyridyl) metal chelates adsorbed on a smectite were studied by photochemical measurements (Sato et al., 2005; Sato and Yamagishi, 2007). In most of this research, cationic polypyridyl ruthenium(II) complexes were used as emitting probes.

Recently, attention has been paid to cyclometalated iridium(III) complexes as efficient emitting species in the visible region. In addition to their strongly emitting properties, some iridium(III) complexes are chiral and stable against racemisation at a higher temperature. These characteristic features were utilised for chirality-dependent emission on a clay mineral surface. Photoreactive clay mineral hybrid materials were prepared and used for enantioselective photosensing (Sato et al., 2011a, 2017a; Tamura et al., 2015; Takimoto et al., 2018). The intermolecular transfer of light energy was investigated on a clay mineral surface, using two types of cationic iridium(III) complexes. One was [Ir(dfppy)$_2$(C$_n$-bpy)]$^+$ (**Ir-1**) (dfppyH = 2-(2′,4′-difluorophenyl)pyridine; C$_n$-bpy = 4,4′-dialkyl-2,2′-bipyridine; dialkyl = dimethyl with n = the number of carbon atoms in an alkyl group) and the other [Ir(piq)$_2$(C$_n$-bpy)]$^+$ (piqH = 1-phenyisoquinoline) (**Ir-2**). They were used as a donor and an acceptor, respectively (Fig. 6.4). These two complexes were coadsorbed on colloidally dispersed synthetic saponite (Sap). The efficiency of energy transfer from the donor to the acceptor (denoted as η_{ET}) was obtained from the emission spectra at various donor-to-acceptor ratios (D/A).

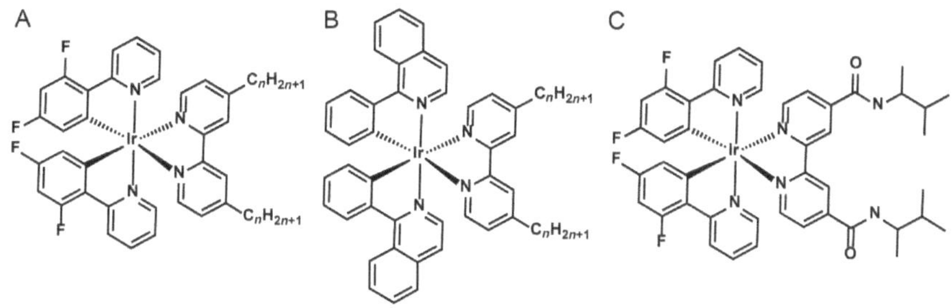

Fig. 6.4 Structures of the Ir(III) complexes, (A) Λ-[Ir(dfppy)$_2$(Cn-bpy)]$^+$ (dfppyH = 2-(4′,6′-difluorophenyl)pyridine; C_n-bpy = 4,4′-dialkyl-2,2′-bipyridine) (**Ir-1**), (B) Λ-[Ir(piq)$_2$(C_n-bpy)]$^+$ (piqH = 1-phenyisoquinoline) (**Ir-2**) and (C) Λ-[Ir(dfppy)$_2$(*R*-pep-bpy)] (*R*-pep-bpy = 4,4′-bis((*R*-1,2-dimethylpropyl)aminocarbonyl-2,2′-bipyridine)) (**Ir-3**).

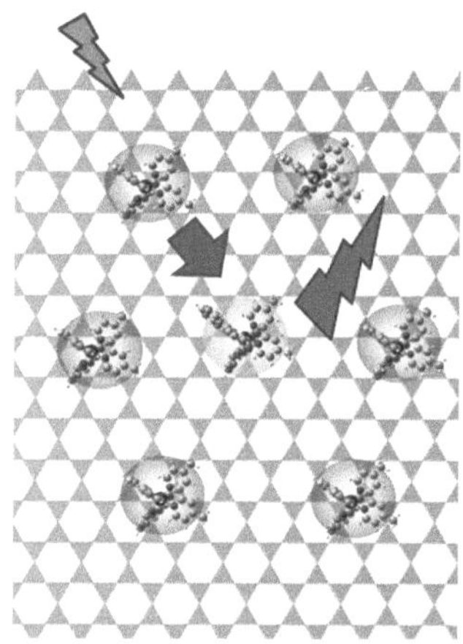

Fig. 6.5 Model of harvesting of light energy on a clay mineral surface.

The results were analysed on the basis of the Förster-type energy transfer mechanism (Lakowicz, 2006). For C_1-bpy, η_{ET} was as high as 0.5 with D/A value higher than 20. The results implied that the photon energy captured by several donor molecules was collected by a single acceptor molecule (i.e., the harvesting of light energy). The drawing in Fig. 6.5 shows how the harvesting of light energy occurs on a clay mineral surface. Enantioselectivity was observed in the presence of iridium(III) complex with C_1-bpy. The results indicated that contact pairs of donor and acceptor molecules were involved during the energy transfer.

A cyclometalated iridium(III) complex with two amido groups bound to ligands was synthesised in order to enhance affinity towards amino acids (**Ir-3**, Fig. 6.4) (Sato et al., 2017a). The complexes were adsorbed on synthetic Sap in colloidal dispersion. Such hybridisation resulted in maintaining high emission levels, even in a medium containing water. Among the investigated amino acid methyl esters, tryptophan methyl ester exhibited remarkable quenching effects on the emission from **Ir-3**.

The possibility of enantioselectivity was examined by using the pure enantiomer of Ir-3 as an emitter. Dinitrobenzoyl alanine methyl ester (DNB-al-Me) was chosen as a chiral quencher. The quenching of emission from Ir-3 was investigated for the following four pairs: Λ-Ir-3/*S*-DNB-al-Me, Λ-Ir-3/*R*-DNB-al-Me, Δ-Ir-3/*S*-DNB-al-Me and Δ-Ir-3/*R*-DNB-al-Me. They gave curved Stern–Volmer plots so that the results were analysed in terms of two quenching constants as shown in Fig. 6.6. The results implied that chiral discrimination was achieved under the coupled contribution of ΔΛ and *RS* chiralities. Enantioselectivity for the preceding systems was examined by dynamic measurements. The decay rate from Λ-Ir-3/Sap increased upon adding DNB-al-Me. The effect was higher for *S*-DNB-al-Me than for *R*-DNB-al-Me. Thus the same enantioselectivity was confirmed by both the stationary and dynamic measurements. The enantioselectivity observed for DNB-al-Me suggested that the phenyl group in DNB-al-Me was attracted by the phenylpyridyl groups in **Ir-3** and that the stereoselectivity appeared through the interaction at the chiral amido groups (*R*-form) in **Ir-3**. The stereoselectivity towards *N*-3, 5-dinitrobenzoyl alanine methyl ester appeared only through the interplay of two chiral centres, *RS* in the ligand and ΔΛ in the metal centre, but not through each chirality alone for **Ir-3**. The results are promising for developing a photosensor of amino acids in biological systems.

6.3.3 Asymmetric catalysis on a clay mineral surface

Asymmetric catalysis driven by visible light is regarded as one of the most promising environmentally friendly synthetic methods (Yamagishi, 1987; Heitbasum et al., 2006; Thomas and Paja, 2008; Huo et al., 2014; Zhong et al., 2017). For clay minerals, one approach is the immobilisation of a chiral catalyst onto a clay mineral surface (Hikita et al., 1989; Fraile et al., 2004, 2008, 2009a,b, 2013; Fujita et al., 2006; Kawasaki et al., 2009). For practical uses, the immobilisation of a chiral species in a solid leads to the advantage of versatile separation of the catalyst from solution as well as avoidance of effluxes of poisonous metals. Only a few heterogeneous

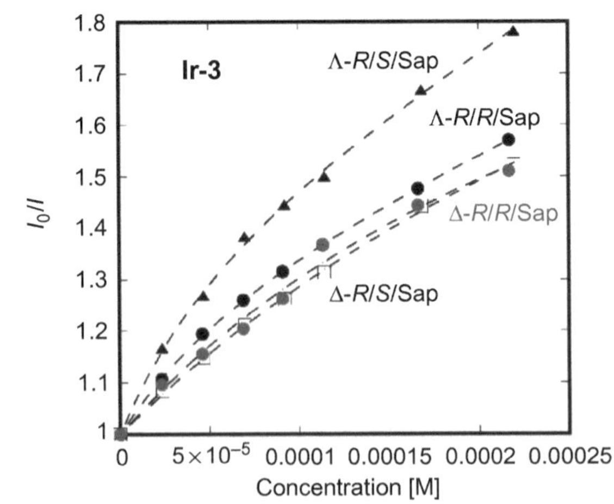

Fig. 6.6 Stern–Volmer plots for the quenching effect of DMB-al-Me on the emission from **Ir-3**; Λ-*R*/*S*-DNB-al-Me/Sap, Λ-*R*/*R*-DNB-al-Me/Sap, Δ-*R*/*S*-DNB-al-Me/Sap and Δ-*R*/*R*-DNB-al-Me/Sap: [Sap] = 2.5×10^{-4} eq. L^{-1} and [**Ir-3**] = 2.0×10^{-6} M. The solvent was 3:1 (v/v) water/methanol. Modified from Sato, H., Tamura, K., Yajima, T., Sato, F., Yamagishi, A., 2017. Chiral phosphorescent probes for amino acids: hybrids of iridium(III) complexes with synthetic saponite. New J. Chem. 41, 2780–2785.

asymmetric catalysts, however, have been reported in photochemistry, probably because of the difficulties in the preparations of suitable solid supports (Yamagishi, 1987).

The photooxidation of α-ethylbenzyl phenyl sulphide was studied on the surface of Mt (Fujita et al., 2006). As a chiral director, chiral metal complexes, $[Ru(bpy)_3]^{2+}$ or $[Ru(phen)_3]^{2+}$, were immobilised on Mt by ion-exchange. It was expected that the photooxidation of a sulphide catalysed by $[Ru(bpy)_3]^{2+}$ took place close to the clay mineral surface, leading to effective stereochemical control by the preadsorbed complex. As summarised in Table 6.1, a chiral sulphoxide was produced with the enantioselectivity (e.e.) of about 43% (Fujita et al., 2006). The chirality of a sulphoxide molecule was determined by the absolute configuration of preadsorbed Ru(II) complexes. It was noted that no chiral induction was attained for the homogenous system (e.g., water/methanol), although the same chiral chelates were used as a photosensitiser. Thus, the steric control on a clay mineral surface was essential for the appearance of chiral induction. The observed stereoselectivity was ascribed to the selection of reaction paths on a clay mineral surface.

The asymmetric synthesis on a clay mineral surface was applied to auto-catalytic systems. An achiral aldehyde was transformed to chiral pyrimidyl alkanol by the catalytic reaction of *i*-dipropylzinc(II) (Pr_2Zn). The reaction was performed in the presence of Δ-$[Ru(phen)_3]^{2+}$/Ht or Λ-$[Ru(phen)_3]^{2+}$/Ht. As shown in Table 6.2, the enantiomeric excess of the product was as high as 60%–70%. Again, chiral selectivity was determined by the absolute configuration of the adsorbed chelates (Kawasaki et al., 2009).

Another approach of using clay minerals in the field of asymmetric syntheses is to immobilise a chiral catalyst on the surface of a clay mineral (Fraile et al., 2009a,b).

Table 6.1 Stereoselectivity of ethylbenzyl phenyl sulphide in various heterogeneous oxidations

		Enantiomer excess (%)		Diastereomer ratio
Catalyst	**Condition**	***SR* – *RS***	***RR* – *SS***	**(*SR* + *RS*):(*RR* + *SS*)**
Δ-$[Ru(phen)_3]^{2+}$/Mt	At 25°C	−30	−29	53:47
Δ-$[Ru(phen)_3]^{2+}$/Mt	At 4°C	−33	−36	51:49

Table 6.2 Heterogeneous enantioselective syntheses of chiral pyrimidyl alkanol by the reaction of *i*-Pr_2Zn with aldehyde in the presence of chiral $[Ru(phen)_3]^{2+}$/synthetic hectorite (Ht)

	Pyrimidyl alkanol		
Intercalator	**Yield (%)**	**Enantiomer excess (%)**	**Configuration**
Δ-$[Ru(phen)_3]^{2+}$/Ht	82	63	*S*
Λ-$[Ru(phen)_3]^{2+}$/Ht	80	69	*R*

As one of the earliest attempts, a well-known hydrogenation catalyst, [(PNNP)Rh (cod)]$^+$ (cod=cyclooctadiene), was ion-exchanged in various clay minerals (hectorite, nontronite, and halloysite) and in bentonite. The catalysts were shown to be active for hydrogenation of several amino acid precursors. The method was extended to other rhodium catalysts such as [(BINAP)Rh(cod)]$^+$ (BINAP=2,2′-bis (diphenylphosphino)-1,1′-binaphthyl) for the hydrogenation of itaconate and its derivatives to succinate. The obtained heterogeneous catalysts attained the selectivity and enantiomer excess comparable to those of the homogeneous reactions using the same catalysts. The recent accomplishment by the noncovalent immobilisation method is reviewed by Fraile et al. (2009a) summarising the immobilisation of chiral catalysts on clay mineral surfaces.

Recently, the asymmetric oxidation of methyl phenyl sulphide was reported using new bis(oxazoline)-vanadyl complexes supported by synthetic Ht (Zid et al., 2017). As an oxidant, *tert*-butyl hydroperoxide or cumylhydroperoxide were used. Activities and enantioselectivities obtained with the heterogeneous catalysts were always lower than in solution. A Cu(II) complex with the same bis(oxaline) ligand was applied for the Diels–Alder reaction. A solid catalyst was prepared by adding synthetic Ht to a methanol solution containing bis(oxaline) and copper triflate. As a Diels–Alder reaction, cyclopentadiene and (*E*)-3-(but-2-enoyl)-oxazolidin-2-one were added in dichloromethane containing the clay mineral catalyst. The reaction yield and enantiomer excess were less than in homogeneous reaction conditions.

6.4 Solid-state VCD towards molecular recognition on a clay mineral surface

VCD spectroscopy is an extension of circular dichroism into the infrared region (Nafie, 2011). The spectroscopy was developed as a means to determine the absolute configuration of a chiral molecule in solution. A large number of VCD peaks provide detailed information concerning the molecular conformation. Even a molecule with no electronic absorption in the UV–visible region can be a target, such as amino acids, sugars, and perfluorinated compounds. From the theoretical viewpoint, density functional theory (DFT) calculation of VCD spectra, which involves the ground electronic states alone, is more reliable than for ECD involving the excited states. The programme analysing VCD spectra is commercially available. The special VCD spectrometer, which was made by Sato et al. (2015) in cooperation with JASCO Co., is shown in Fig. 6.7. The machine was combined with LD (linear dichroism) and named PRESTO-S-2016 (JASCO Co., Japan).

One of the problems in VCD spectroscopy is that the signals are so weak that it takes a long time to obtain a reliable spectrum. Normally, 10^4–10^5 scans were required for one sample. In order to overcome this problem, a strategy is required to enhance the signal. As an example, signal enhancement has been reported for the metal complexes of paramagnetic ions, some gels, and the aggregates of small organic molecules (Domingos et al., 2015; Kurosaki et al., 2015; Sato et al., 2015).

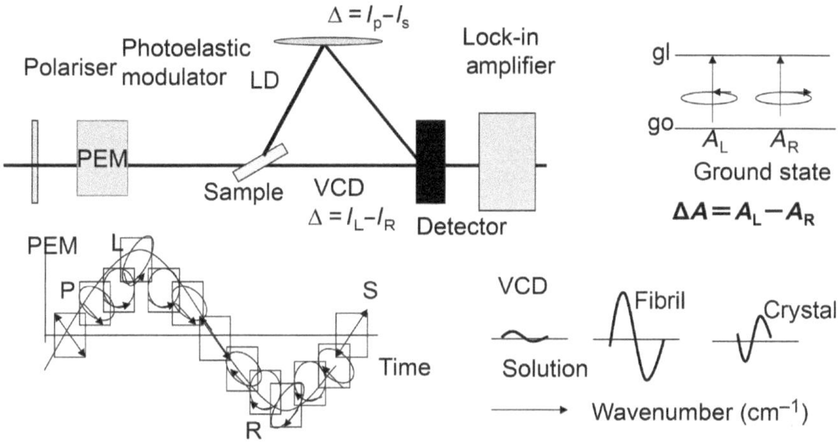

Fig. 6.7 An example of special VCD machine (PRESTO-S-2016: VCD-LD) (JASCO Co., Japan).

VCD spectroscopy was applied to study the vibrational properties of chiral transition metal complexes (Sato et al., 2006, 2013; Mori et al., 2011; Sato and Yamagishi, 2013; Takimoto et al., 2017). The method unravelled various structural aspects that were difficult to attack by means of other spectroscopies, such as the interligand vibronic coupling, the effects of *d*-electron configurations on vibrational energy levels, the contribution of anharmonic vibrational terms, the local chirality within a multinuclear complex and the low-lying electronic states in the infrared region.

Recently the VCD method has been applied for studying chiral molecules adsorbed on a solid surface (Bieri et al., 2007; Sasaki et al., 2013; Zhang et al., 2015; Sato et al., 2017b). In the case of clay minerals as a solid support, Sato et al. (2018a) measured the VCD spectra of chiral metal complexes intercalated in Na^+-Mt. The work was intended to clarify the mechanism of chiral recognition on a clay mineral surface. As for the intermolecular interactions in intercalated states, it seems to be a difficult task to make clear on a molecular scale how molecules recognise each other within the interlayer space of a clay mineral. This is because expandable clay minerals such as smectites are microcrystalline (< 2 μm in size) and electrical insulators. Accordingly, conventional methods such as single crystal X-ray analyses or probe microscopies are hardly applicable. To overcome this difficulty, the VCD method was expected to detect any subtle change of conformation induced by intermolecular interactions. This was the first application of the solid state VCD method to inorganic layered materials.

The solid state VCD spectra were measured on an ion-exchanged adduct of Δ- or Λ-$[Ru(phen)_3]^{2+}$ and Na^+-Mt. The sample was prepared as a KBr pellet, including Mt intercalating $[Ru(phen)_3]^{2+}$. The solid state IR and VCD spectra of the intercalation compounds of $[Ru(phen)_3]^{2+}$/Mt were obtained. The observed peaks showed the opposite signs between the optical antipodes, validating the correctness of the present measurements (Fig. 6.8). When the spectra were compared with those in solution,

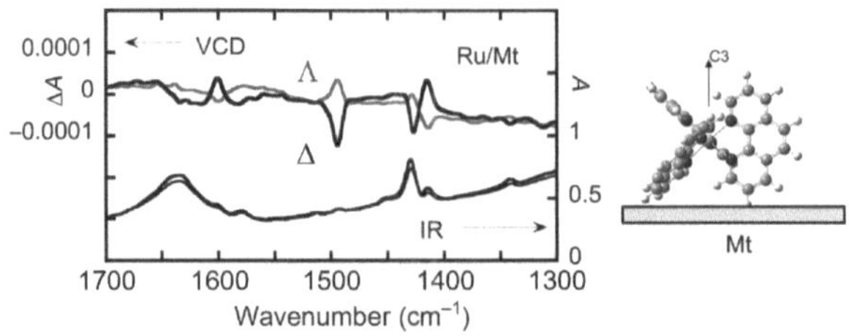

Fig. 6.8 Solid state VCD and IR spectra of $[Ru(phen)_3]^{2+}$ intercalated in Mt. *Bold* and *thin curves* denote the Δ- or Λ-enantiomers, respectively.
Modified from Sato, H. Tamura, K., Takimoto, K., Yamagishi, A., 2018. Solid state vibrational circular dichroism towards molecular recognition: chiral metal complexes intercalated in a clay mineral. Phys. Chem. Chem. Phys. 20, 3141–3147; Sato, H., Yajima, T., Yamagishi, A., 2018. Stereochemical effects on dynamics in two-component systems of gelators with perfluoroalkyl and alkyl chains as revealed by vibrational circular dichroism. Phys. Chem. Chem. Phys. 20, 3210–3215.

it was noted that several peaks showed a sign opposite to those of the solution samples. Previous spectroscopic investigations on an intercalated $[M(phen)_3]^{2+}$ (M = Ru, Cu and Ni) showed that a molecule of this type underwent a structural deformation in the intercalated state, such as the shortening of the C_3 axis (Yano et al., 1992). The distortion often caused the change of spectroscopic properties. Thus, the observed sign inversion of VCD peaks might be rationalised in terms of the deformation of the coordination structure in the intercalated states.

For these samples, a chiral organic molecule, *R*- or *S*-1, 1′-bi-2-naphthol (denoted as *R*- or *S*-BINOL), was coadsorbed to form a three-component intercalation compound. Attention was concentrated on how VCD spectra were affected by the chirality relation between BINOL and $[Ru(phen)_3]^{2+}$. The following four samples were prepared *R*- or *S*-BINOL/Λ-$[Ru(phen)_3]^{2+}$ or Δ-$[Ru(phen)_3]^{2+}$/Mt. In these samples, BINOL and $[Ru(phen)_3]^{2+}$ were included at a molar ratio of ca. 1:2. The powder XRD patterns of (a) Na^+-Mt, (b) Λ-$[Ru(phen)_3]^{2+}$/Mt, (c) *R*-BINOL/Λ-$[Ru(phen)_3]^{2+}$/Mt and (d) *S*-BINOL/Λ-$[Ru(phen)_3]^{2+}$/Mt are shown in Fig. 6.9. The reflections at the lowest diffraction angle gave the d_{001} values of Mt layers. The height of the interlayer space was calculated by subtracting the thickness of one layer (or 0.95 nm) from d_{001}. In case of Λ-$[Ru(phen)_3]^{2+}$/Mt, the replacement of sodium ions with $[Ru(phen)_3]^{2+}$ ions resulted in the expansion of the interlayer space from 0.29 to 0.78 nm. In case of *R*-BINOL/Λ-$[Ru(phen)_3]^{2+}$/Mt, no new reflection appeared, except for the increase of the diffraction background in the range of 2–4 nm. In the case of *S*-BINOL/Λ-[Ru $(phen)_3]^{2+}$/Mt, a new reflection appeared at 2.68 nm, indicating that the coadsorption of *S*-BINOL caused the further expansion of the interlayer space to 1.73 nm. From the chromatographic elution order, Λ-$[Ru(phen)_3]$ interacted with *S*-BINOL more strongly than with *R*-BINOL on a clay mineral surface (Fig. 6.2). Based on this, *S*/Λ or *R*/Δ are a strongly interacting pair, while *R*/Λ or *S*/Δ are a weakly interacting pair (Table 6.3).

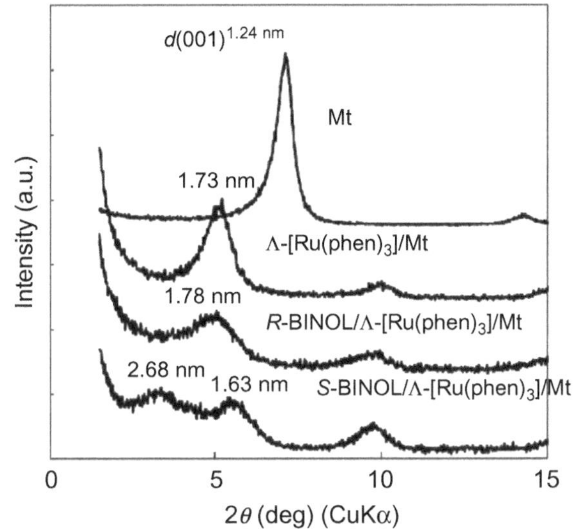

Fig. 6.9 Powder X-ray diffraction patterns of Na^+-Mt, Λ-$[Ru(phen)_3]^{2+}$/Mt, *R*-BINOL/Λ-$[Ru(phen)_3]^{2+}$/Mt (weakly interacting pair) and *S*-BINOL/Λ-$[Ru(phen)_3]^{2+}$/Mt (strongly interacting pair).
Modified from Sato, H. Tamura, K., Takimoto, K., Yamagishi, A., 2018. Solid state vibrational circular dichroism towards molecular recognition: chiral metal complexes intercalated in a clay mineral. Phys. Chem. Chem. Phys. 20, 3141–3147; Sato, H., Yajima, T., Yamagishi, A., 2018. Stereochemical effects on dynamics in two-component systems of gelators with perfluoroalkyl and alkyl chains as revealed by vibrational circular dichroism. Phys. Chem. Chem. Phys. 20, 3210–3215.

Table 6.3 Three-component samples for VCD measurements

Intercalated metal complex	Strongly interacting pair	Weakly interacting pair
Λ-$[Ru(phen)_3]^{2+}$/Mt	*S*-BINOL	*R*-BINOL
Δ-$[Ru(phen)_3]^{2+}$/Mt	*R*-BINOL	*S*-BINOL

Solid state VCD and IR spectra were measured on the intercalation compounds of {*R*-BINOL/Δ-$[Ru(phen)_3]^{2+}$/Mt and *S*-BINOL/Λ-$[Ru(phen)_3]^{2+}$/Mt} and {*R*-BINOL/Λ-$[Ru(phen)_3]^{2+}$/Mt and *S*-BINOL/Δ-$[Ru(phen)_3]^{2+}$/Mt}. They are abbreviated as {*R*/Δ/Mt and *S*/Λ/Mt} (strongly interacting pairs) and {*R*/Λ/Mt and *S*/Δ/Mt} (weakly interacting pairs), respectively. The VCD spectra for the strongly interacting pair of *R*/Δ/Mt (upper) and weakly interacting pair of *R*/Λ/Mt (lower) are shown in Fig. 6.10. The figure also includes the VCD spectra of Δ/Mt or Λ/Mt together with *R*-BINOL in CD_3CN. In the weak interaction pair, the *R*/Λ/Mt was simply the overlap of the peaks assigned to the individual components. The bands in the wavenumber range 1300–1400 cm^{-1} were the VCD peaks assigned to the bending vibrations of OH groups in BINOL. Notably, there was an appreciable difference observed in the

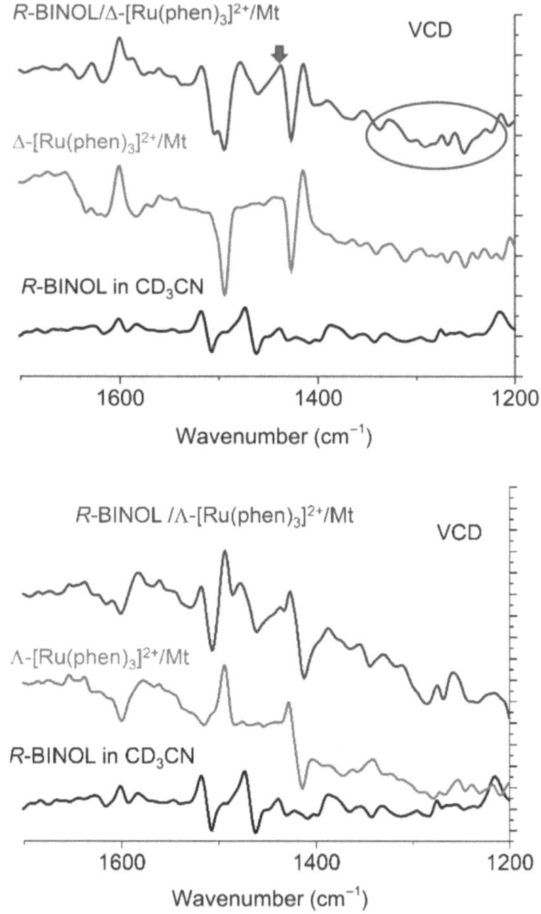

Fig. 6.10 Vibrational circular dichroism on the chirally modified clay mineral surface; (*upper*) strong interaction pair: (*blue*; *dark grey* in the print version) *R*/Δ/Mt, (*red*; *grey* in the print version) Δ/Mt, (*black*) *R*-BINOL; (*lower*) weakly interaction pair: (*blue*; *dark grey* in the print version) *R*/Λ/Mt, (*red*; *grey* in the print version) Λ/Mt, (*black*) *R*-BINOL.

intensity of VCD peaks between the strongly and weakly interacting pairs. It was concluded that BINOL in *R*/Δ/Mt (or *S*/Λ/Mt) as a strongly interacting pair took the *cis–cis* conformation in the intercalated state. To the contrary, no conclusion was derived as to the dominant conformation taken by BINOL in *R*/Λ/Mt (or *S*/Δ/Mt) as a weakly interacting pair. Their VCD peaks were too weak in this region. One possibility was that BINOL took various conformations (*cis–cis*, *cis–trans* and *trans–trans*) at comparable ratios in the intercalated state. Moreover, the behaviour of VCD peak at 1440 cm^{-1} was also noted. The peak was clearly observed in the case of the strongly interacting pair, while it was hardly seen for the weakly interacting pair. In other words, the intensity of the VCD peak (assigned to BINOL) was affected drastically by the chirality relation between the $[Ru(phen)_3]^{2+}$ and BINOL.

The formation of an associate between *R*-BINOL and Δ- or Λ-$[Ru(phen)_3]^{2+}$ (denoted as *R*/Δ or *R*/Λ, respectively) was investigated theoretically. As for the conformation of BINOL, three cases (*cis–cis*, *cis–trans* and *trans–trans*) were compared. The most stable was a *cis–cis* conformation for both *R*/Δ and *R*/Λ. The optimised structures were calculated for these pairs under this conformation of BINOL. *R*-BINOL

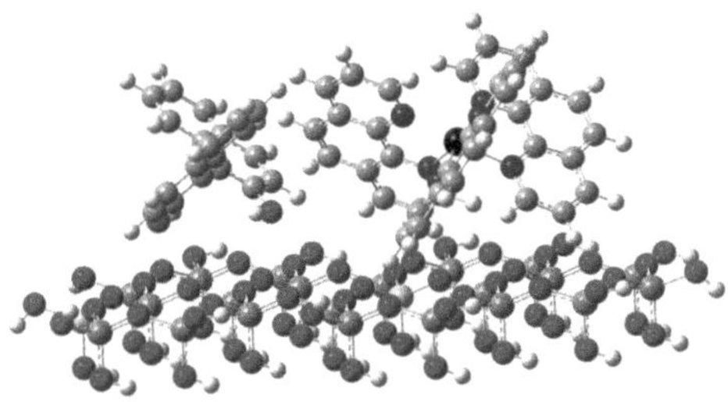

Fig. 6.11 A molecular recognition model on a clay mineral surface.

formed a more compact associate with Δ-$[Ru(phen)_3]^{2+}$ than with Λ-$[Ru(phen)_3]^{2+}$. The results were in agreement with the enantioselectivity of BINOL as observed in chromatographic resolution, although the present calculation included no effect of the clay mineral surface.

From the animation of vibrational motion at $1440\,cm^{-1}$ (the arrow in Fig. 6.10) in the calculated spectra, both *R*-BINOL and Δ-$[Ru(phen)_3]^{2+}$ oscillated simultaneously for *R*/Δ. Because the peak was partially overlapped with the peak of $[Ru(phen)_3]^{2+}$, both of them could be excited vibrationally at this wavelength. To the contrary, no such collaborative vibration was seen for *R*/Λ, in which *R*-BINOL alone oscillated. It should be emphasised that *R*/Δ represented a stereochemically preferred pair to form a compact associate (a strongly interacting pair). Therefore, the tight association between two components may be the main factor for realising collaborative oscillation. In agreement with these theoretical results, the stereoselective enhancement of VCD peaks was observed experimentally as described herein ($1440\,cm^{-1}$). It was suggested that the clay mineral played the role of generating such collaborative effects as a host. The model is shown in Fig. 6.11.

6.5 Summary and future development

Smectites were used as an adsorbent for chirality recognition. The surface of a clay mineral was modified with chiral metal complexes. In the case of polypyridyl tris(chelate) complexes such as $[Ru(phen)_3]^{2+}$ and $[Ru(bpy)_3]^{2+}$, a microscopic vacant space was generated on the surface. This space acted as a binding site with stereoselectivity. As an extension of the approach, chiral sensing was achieved by use of a hybrid of emitting iridium(III) complexes and synthetic Sap. The VCD method was employed to shed light on the mechanism of molecular recognition on a clay mineral surface.

As for the future development of the present approach, the following possibilities are proposed:

i. To pursue the possibility of obtaining a single crystal of chiral inorganic materials. This attempt is significant in both fundamental (as a possible role in generating chiral molecules during chemical evolution) and practical aspects (as a material for chiral support for heterogeneous asymmetric syntheses). The VCD method is promising for determining the absolute configuration of a synthesised chiral inorganic material.

ii. To improve the clay mineral column chromatography by developing newly synthesised complexes as chiral modifiers. One target is to resolve directly nonderivative aliphatic compounds such as amino acids and sugars.
iii. To manipulate several photoresponsive molecules onto a clay mineral surface, in which a donor, a mediator, and an acceptor are arranged in an organised order. In arranging these molecules, molecular chirality can be utilised. Such a system might present a model for light-harvesting in biological photosyntheses.
iv. To apply the solid state VCD to clay minerals and other inorganic materials in order to reveal the mechanism of chiral recognition. The instrument for VCD measurements should be improved to cover a wider range of wavenumbers. Besides, the theoretical simulation should be improved by including the surface of the clay minerals.

Appendix. Basic strategy of applying vibrational circular dichroism (VCD) spectroscopy to solid or film samples

The vibrational circular dichroism (VCD) method covers the infrared and near-infrared regions of circular dichroism spectra. A large number of peaks arising from the normal-mode vibrations are utilised to analyse the molecular vibrational properties of a given molecule in order to determine its detailed conformation. Another advantage of the method is that even a chiral molecule with no electronic absorption in the UV–visible region can be a target, such as amino acids, sugars, and fluorinated compounds.

The simulation based on the DFT calculation is helpful in analysing the observed VCD spectra. Because the theoretical calculation includes the ground electronic state alone, except for the low-lying electronic transitions, the theory is thought to be more reliable and straightforward than in the case of calculating ECD.

In spite of these favourable features, the VCD method has one drawback; the signal is very small. The asymmetric factor of circular dichroism (g-value) is given by $g = \Delta A/A$, in which A and ΔA are absorbance and the difference of absorbance between left- and right-handed circularly polarised lights (or $A = A_{\mathrm{left}} - A_{\mathrm{right}}$), respectively (Fig. 6.7). The ΔA value for VCD is in the region of 10^{-5}–10^{-4} for isolated chiral molecules in solution, which is two or three orders lower than those for ECD (10^{-3}–10^{-2}). In order to overcome the difficulty arising from such small signals, the measurements of VCD spectra are usually carried out by accumulation of $\sim 10^4$ scans. Accordingly, a single measurement requires more than 1 h. Due to this limitation, it is difficult to apply the method to systems involving unstable species such as reaction intermediates.

Recently, the enhancement of VCD signals has been reported (Domingos et al., 2015; Sato et al., 2015, 2018a,b). As for the origin of enhancement, one possible reason is the role of metal ions. In fact, theoretical analyses established that the presence of electronic states of d-orbitals in the near-infrared region contributed to the increase in the intensity of VCD signals (Mori et al., 2011; Sato et al., 2006). Another possible reason is the collaborative interaction in vibrational motion between the molecules in the neighbouring molecular fibres. This mechanism was proposed to explain the enhanced signals during the aggregation of amyloids (Pazderková et al., 2017).

When VCD spectra are recorded on gels or KBr-pellets, attention should be paid to establish that the signals arise from the intrinsic chiral vibrational properties of a molecule. First, the mirror-imaged relation should be maintained between the samples of antipodal molecules. Next, no VCD signal is observed for the samples containing a racemic mixture. It is also necessary to confirm that no change in the spectrum occurs when the cell rotates along the direction of monitoring to eliminate the dichroic effects in samples with anisotropic domains.

In case of a solid sample, a KBr pellet was mounted into a cell with windows of 10, 5, or 3 mm in diameter (Sato et al., 2017a). In the case of films, a Langmuir–Blodgett film of at least 50 layers is transferred onto an infrared-transparent substrate such as calcium fluoride at a constant surface pressure (Sato et al., 2011b).

References

Berthod, A., 2006. Chiral recognition mechanisms. Anal. Chem. 78, 2093–2099.

Bieri, M., Gauter, C., Bürgi, T., 2007. Probing chiral interfaces by infrared spectroscopic methods. Phys. Chem. Phys. Chem. 9, 671–685.

Bondy, S.C., Harrington, M., 1979. L-Amino acids and D-glucose bind stereospecifically to a colloidal clay. Science 203, 1243–1244.

Breu, J., Nilesh, R.N., Catlow, C.R.A., 1999. Atomistic computer modeling of $[Ru(bpy)_3]^{2+}$ and $[Ru(phen)_3]^{2+}$ intercalated into low charge smectites. J. Chem. Soc. Dalton Trans. 835–845.

Breu, J., Stoll, A., Lange, K.G., Probst, T., 2001. Two-dimensional diffraction from enantiopure and racemic monolayers of $[Ru(bpy)_3]^{2+}$ intercalated into synthetic fluorohectorite. Phys. Chem. Phys. Chem. 3, 1232–1235.

Domingos, S.R., Hartl, F., Brma, W.J., Woutersen, S., 2015. Elucidating the structure of chiral molecules by using amplified vibrational circular dichroism: from theory to experimental realization. Chemphyschem 16, 3363–3373.

Fraile, J.M., García, J.I., Herrerías, C.I., Mayoral, J.A., Reiser, O., Socuellamos, A., Werner, H., 2004. The role of binding constants in the efficiency of chiral catalysts immobilized by electrostatic interactions: the case of azabis(oxazoline)-copper complexes. Chemistry 10, 2997–3005.

Fraile, J.M., García, J.I., Jiménez-Osés, G., Mayoral, J.A., Roldán, M., 2008. Surface confinement effects on enantioselective cyclopropanation. Reactions with supported chiral 8-oxazolinylquinoline-copper complexes. Oraganometallics 27, 2246–2251.

Fraile, J.M., García, J.I., Myoral, J.A., 2009a. Noncovalent immobilization of enantioselective catalysts. Chem. Rev. 109, 360–417.

Fraile, J.M., García, J.I., Herrerías, C.I., Mayoral, J.A., Pires, E., 2009b. Enantioselective catalysis with chiral complexes immobilized on nanostructured supports. Chem. Soc. Rev. 38, 695–706.

Fraile, J.M., García, N., Herrerías, C.I., 2013. Support effect on stereoselectivities of vinylogous Mukaiyama – Michael reactions catalyzed by immobilized chiral copper complexes. ACS Catal. 3, 2710–2718.

Fraser, D.G., Greenwell, H.C., Skipper, N.T., Smalley, M.V., Wilkinson, M.V., Demé, B., Heenan, R.K., 2011a. Chiral interactions of histidine in a hydrated vermiculite clay. Phys. Chem. Phys. Chem. 13, 825–830.

Fraser, D.F., Fitz, D., Jakschitz, T., Rode, B.M., 2011b. Selective adsorption and chiral amplification of amino acids in vermiculite clay-implications for the origin of biochirality. Phys. Chem. Phys. Chem. 13, 831–838.

Fujimoto, N., Mori, Y., Yamagishi, A., Sato, H., 2010. Molecular recognition of star-burst tetranuclear Ru(III) complexes on a chirally modified clay surface. Chem. Commun. 46, 5473–5474.

Fujita, S., Sato, H., Kakegawa, N., Yamagishi, A., 2006. Enantioselective photooxidation of a sulfide by a chiral ruthenium(II) complex immobilized on a montmorillonite clay surface: the role of weak interactions in asymmetric induction. J. Phys. Chem. B 110, 2533–2540.

Ghosh, P.K., Bard, A.J., 1983. Clay-modified electrodes. J. Am. Chem. Soc. 105, 5691–5693.

Heitbasum, M., Glorius, F., Escher, I., 2006. Asymmetric heterogeneous catalysis. Angew. Chem. Int. Ed. 45, 4732–4762.

Herling, M.M., Lacher, U., Reiß, M., Seibt, S., Schwedes, M., Kaio, H., Schobert, R., Breu, J., 2017. Sub-micron pore size tailoring for efficient chiral discrimination. Chem. Commun. 53, 1072–1075.

Hikita, T., Tamaru, K., Yamagishi, A., Iwamoto, T., 1989. Production of an optically active sulfoxide by use of colloidally dispersed Δ-$[Ru(bpy)_3]^{2+}$ montmorillonite as a photosensitizer. Inorg. Chem. 28, 2221–2223.

Hirsemann, D., Shylesh, S., De Souza, R.A., Diar-Bakerly, B., Biersack, B., Mueller, D.N., Martin, M., Schobert, R., Breu, J., 2012. Large-scale low-cost fabrication of Janus-type emulsifiers by selective decoration of natural kaolinite platelets. Angew. Chem. Int. Ed. 51, 1348–1352.

Huo, H., Shen, X., Wang, C., Zhang, L., Röse, P., Chen, L.A., Harms, K., Marsch, M., Hilt, G., Meggers, E., 2014. Asymmetric photoredox transition-metal catalysis activated by visible light. Nature 515, 100–103.

Kakegawa, N., Yamagishi, A., 2005. Coadsorption studies of Tris(1,10-phenanthroline)-ruthenium(II) and *N*-methylated alkaloid cation by laponite with an application for a chiral column packing material. Chem. Mater. 17, 2997–3003.

Kamat, P.V., Gopidas, K.R., Mukherjee, T., Joshi, V., Kotkar, D., Pathak, V.S., Ghosh, P.K., 1991. Spectral differences between enantiomeric and racemic tris(bipyridine)ruthenium (II) on layered clays: probable causes. J. Phys. Chem. 95, 10009–10018.

Kawasaki, T., Omine, T., Suzuki, K., Sato, H., Yamagishi, A., Soai, K., 2009. Highly enantioselective asymmetric autocatalysis using chiral ruthenium complex-ion-exchanged synthetic hectorite as a chiral initiator. Org. Biomol. Chem. 7, 1073–1075.

Kuhnle, A., Linderoth, T., Hammer, B., Besenbacher, F., 2002. Chiral recognition in dimerization of adsorbed cysteine observed by scanning tunnelling microscopy. Nature 415, 891–893.

Kunz, D.A., Leitl, M.J., Schade, L., Schmid, J., Bojer, B., Schwarz, U.T., Ozin, G.A., Yersin, H., Breu, J., 2015. Quasi-epitaxial growth of $[Ru(bpy)_3]^{2+}$ by confinement in clay nanoplatelets yields polarized emission. Small 11, 792–796.

Kurosaki, D., Handen, J.D., Dukor, R.K., Nafie, L.A., Lednev, I., 2015. Supramolecular chirality in peptide microcrystals. Chem. Commun. 51, 89–92.

Lakowicz, J.R., 2006. Principles of Fluorescence Spectroscopy, third ed. Springer, New York.

Mallouk, T.E., Gavin, J.A., 1998. Molecular recognition in lamellar solids and thin films. Acc. Chem. Res. 31, 209–217.

Mori, H., Yamagishi, A., Sato, H., 2011. Theoretical study on vibrational circular dichroism spectra of Tris(acetylacetonato)metal(III) complexes: anharmonic effects and low-lying excited states. J. Chem. Phys. 135, 084506.

Nafie, L.A., 2011. Vibrational Optical Activity: Principles and Applications. Wiley, Chichester.

Nakamura, Y., Yamagishi, A., Matsumoto, A., Tohkubo, K., Ohtsu, Y., 1988. High-performance liquid chromatography for optical resolution on a column of an ion-exchange adduct of spherically shaped synthetic hectorite and optically active metal complexes. J. Chromatogr. A 482, 165–167.

Okada, T., Kumasaki, A., Shimizu, K., Yamagishi, A., Sato, H., 2016. Application of hectorite-coated silica gel particles as a packing material for chromatographic resolution. J. Chromatogr. Sci. 54, 1238–1243.

Pazderková, M., Pazderka, T., Shanmugasundaram, M., Dukor, R.K., Lednev, I.K., Nafie, L.A., 2017. Origin of enhanced VCD in amyloid fibril spectra: effect of deuteriation and pH. Chirality 29, 469–475.

Sasaki, T., Hisaki, I., Miyano, T., Tohnai, N., Morimoto, K., Sato, H., Tsuzuki, S., Miyata, M., 2013. Linkage control between molecular and supramolecular chirality in 21-helical hydrogen-bonding networks by using achiral components. Nat. Commun. 4, 1787.

Sato, H., Yamagishi, A., 2007. Application of the ΔΛ isomerism of octahedral metal complexes as a chiral source in photochemistry. J. Photochem. Photobiol. C Photochem. Rev. 8, 67–84.

Sato, H., Yamagishi, A., 2013. VCD studies on chiral characters of metal complex oligomers. Int. J. Mol. Sci. 14, 964–978.

Sato, H., Yamagishi, A., Kato, S., 1992. Monte Carlo simulations for the interactions of metal complexes with the silicate sheets of a clay: comparison of binding states between tris (1,10-phenanthroline)metal(II) and tris(2,2'-bipyridyl)metal(II) chelates. J. Am. Chem. Soc. 114, 10933–10940.

Sato, H., Hiroe, Y., Tamura, K., Yamagishi, A., 2005. Orientation tuning of a polypyridyl Ru(II) complex immobilized on a clay surface toward chiral discrimination. J. Phys. Chem. B 109, 18935–18941.

Sato, H., Taniguchi, T., Monde, K., Nishimura, S.-I., Yamagishi, A., 2006. Dramatic effects of d-electron configurations on vibrational circular dichroism spectra of Tris(acetylacetonato)metal(III). Chem. Lett. 35, 364–365.

Sato, H., Mori, Y., Fukuda, Y., Yamagishi, A., 2009. Syntheses and vibrational circular dichroism spectra of the complete series of [Ru((−)- or (+)-tfac)n(acac)3 − n] (n = 0–3, tfac = 3-trifluoroacetylcamphorato and acac = acetylacetonato). Inorg. Chem. 48, 4354–4361.

Sato, H., Tamura, K., Aoki, R., Kato, M., Yamagishi, A., 2011a. Enantioselective sensing by luminescence from cyclometalated Iridium(III) complexes adsorbed on a colloidal saponite clay. Chem. Lett. 40, 63–65.

Sato, H., Sakurai, T., Yamagishi, A., 2011b. Comparison of vibrational circular dichroism between the Langmuir-Blodgett Films and gels of 12-hydroxyoctadecanoic acid. Chem. Lett. 40, 25–27.

Sato, H., Mori, Y., Yamagishi, A., 2013. Conformational change of a chiral Schiff base Ni(II) complex with a binaphthyl moiety: application of vibrational circular dichroism spectroscopy. Dalton Trans. 42, 6873–6878.

Sato, H., Yajima, T., Yamagishi, A., 2015. Chiroptical studies on supramolecular chirality of molecular aggregates. Chirality 27, 659–666.

Sato, H., Tamura, K., Yajima, T., Sato, F., Yamagishi, A., 2017a. Chiral phosphorescent probes for amino acids: hybrids of iridium(III) complexes with synthetic saponite. New J. Chem. 41, 2780–2785.

Sato, H., Kawamura, I., Yamagishi, A., Sato, F., 2017b. Solid state VCD spectra of isoleucine and its related compounds: effects of interplay between two chiral centers. Chem. Lett. 46, 449–452.

Sato, H., Tamura, K., Takimoto, K., Yamagishi, A., 2018a. Solid state vibrational circular dichroism towards molecular recognition: chiral metal complexes intercalated in a clay mineral. Phys. Chem. Chem. Phys. 20, 3141–3147.

Sato, H., Yajima, T., Yamagishi, A., 2018b. Stereochemical effects on dynamics in two-component systems of gelators with perfluoroalkyl and alkyl chains as revealed by vibrational circular dichroism. Phys. Chem. Chem. Phys. 20, 3210–3215.

Shichi, T., Takagi, K., 2000. Clay minerals as photochemical reaction fields. J. Photochem. Photobiol. C Photochem. Rev. 1, 113–130.

Takahashi, S., Tanaka, R., Wakabayashi, N., Taniguchi, T., Yamagishi, A., 2003. Design of a chiral surface by modifying an anionically charged single-layered Inorganic compound with metal complexes. Langmuir 19, 6122–6125.

Takimoto, K., Watanabe, Y., Mori, S., Sato, H., 2017. Vibrational circular dichroism and single crystal X-ray diffraction analyses of $[Ir(bzq)_2(phen)]^+$ (bzq = benzo[*h*]quinoline; phen = 1, 10-phenanthroline): absolute configuration and role of CH–π interaction in molecular packing. Dalton Trans. 46, 4397–4402.

Takimoto, K., Tamura, K., Watanabe, Y., Yamagishi, A., Sato, H., 2018. Microscopic chiral pockets in a tris(chelated)iridium(III) complex as sites for dynamic enantioselective quenching. New J. Chem. 42, 4818–4823.

Tamura, K., Yamagishi, A., Kitazawa, T., Sato, H., 2015. Harvesting light energy by iridium(III) complexes on a clay surface. Phys. Chem. Phys. Chem. 17, 18288–18293.

Taniguchi, M., Yamagishi, A., Iwamoto, T., 1991. X-ray diffraction and electric dichroism studies on the adsorption of metal complexes by a clay. Inorg. Chem. 30, 2462–2467.

Thomas, J.K., 1988. Photophysical and photochemical processes on clay surfaces. Acc. Chem. Res. 21, 275–280.

Thomas, J.M., Paja, R., 2008. Exploiting nanospace for asymmetric catalysis: confinement of immobilized, single-site chiral catalysts enhances enantioselectivity. Acc. Chem. Res. 41, 708–720.

Umemura, Y., Shinohara, E., 2004. Two-dimensional molecular ordering of Os(II) complexes in organo-clay hybrid ultrathin films. Chem. Commun. 42, 1110–1111.

Villemure, G., 1990. Effect of negative surface-charge densities of smectite clays on the adsorption isotherms of racemic and enantiomeric Tris(2,2′-bipyridyl)ruthenium(II) chloride. Clay Clay Miner. 38, 622–630.

Wakabayashi, N., Nishimura, S.-I., Kakegawa, M., Sato, H., Yamagishi, A., 2004. Racemic adsorption of Tris(1,10-phenanthroline) ruthenium(II) onto a mica surface. Clay Sci. 12, 259–266.

Yamagishi, A., 1982. Racemic adsorption of iron(II) tris(1,10-phenanthroline) chelate on a colloidal clay. J. Phys. Chem. 86, 2472–2479.

Yamagishi, A., 1985a. Racemic adsorption, antiracemization, and induction of optical activity of metal chelates in the presence of a colloidal clay. Inorg. Chem. 24, 1689–1695.

Yamagishi, A., 1985b. Chromatographic resolution of enantiomers having aromatic groups by an optically active clay-chelate adduct. J. Am. Chem. Soc. 107, 732–733.

Yamagishi, A., 1987. Optical resolution and asymmetric syntheses by use of adsorption on clay minerals. J. Coord. Chem. 16, 131–211.

Yamagishi, A., Fujita, N., 1984. Racemic adsorption of a bis(chelated)cobalt(III) complex by colloidally dispersed sodium montmorillonite. J. Colloid Interface Sci. 100, 136–142.

Yamagishi, A., Sato, H., 2012. Stereochemistry and molecular recognition on the surface of a smectite clay. Clay Clay Miner. 60, 411–419.

Yamagishi, A., Soma, M., 1981. Optical resolution of metal chelates by use of adsorption on a colloidal clay. J. Am. Chem. Soc. 103, 4640–4642.

Yamagishi, A., Tanaka, K., Toyoshima, I., 1982. Auger electron spectroscopic evidence for the racemic adsorption of tris(1,10-phenanthroline)iron(II) on sodium montmorillonite. J. Chem. Soc. Chem. Commun. 343–345.
Yamagishi, A., Makino, H., Nakamura, Y., Sato, H., 1992. Separation of isomers of cobalt(III) complexes by liquid chromatography on a column packed with a clay-ruthenium(II) complex adduct. Clay Clay Miner. 40, 359–361.
Yamagishi, A., Taniguchi, M., Takahashi, M., Asada, C., Matsushita, N., Sato, H., 1994. Electric dichroism studies on adsorption of Tris(5,5'-dimethyl-2,2'-bipyridyl)ruthenium(II) by a colloidally dispersed clay. J. Phys. Chem. 98, 7555–7561.
Yamagishi, A., Taniguchi, M., Imamura, Y., Sato, H., 1996. Clay column chromatography for optical resolution: selectivities of Λ-$[Ru(phen)_3]^{2+}$ and Λ-$[Ru(bpy)_3]^{2+}$ laponite columns towards 1,1′-binaphthol. Appl. Clay Sci. 11, 1–10.
Yano, Y., Watanabe, T., Taniguchi, M., Yamagishi, A., 1992. Electron spin resonance studies on Tris (1,10-phenthroline) Co(II) adsorbed by hectorite and saponite. Clay Sci. 8, 381–391.
Yao, Y., Taniguchi, M., Nakata, M., Takahashi, M., Yamagishi, A., 1998. Electrochemical scanning tunneling microscopy observation of ordered surface layers on an anionic clay-modified electrode. Langmuir 14, 2890–2895.
Zhang, S.-Y., Li, D., Guo, D., Zhang, H., Shi, W., Cheg, P., Wojtas, L., Zaworotko, M.J., 2015. Synthesis of a chiral crystal form of MOF-5, CMOF-5, by chiral induction. J. Am. Chem. Soc. 137, 15406–15409.
Zhong, R., Lindfhorst, A.C., Groche, F.J., Kuhn, F.E., 2017. Immobilization of N-heterocyclic carbine compounds: a synthetic perspective. Chem. Rev. 117, 1970–2058.
Zid, T.B., Fadhli, M., Khedher, I., Fraile, J.M., 2017. New bis(oxazoline)-vanadyl complexes, supported by electrostatic interaction in laponite clay, as heterogeneous catalysts for asymmetric oxidation of methyl phenyl sulfide. Microporous Mesoporous Mater. 239, 167–172.

Organic pollutant adsorption on clay minerals

7

Jean-François Lambert
Laboratoire de Réactivité de Surface, Sorbonne Université- CNRS (UMR 7197), Paris, France

7.1 Pollutants? Definitions and scope of the chapter

In a chapter on pollutants adsorption on clay minerals, it seems a logical first step to define what is meant by 'pollutant.' A more precisely defined term is '**xenobiotic**': this word refers to any anthropogenic substance that is not normally present in a given ecosystem[a] (or not in significant quantities) and is introduced into it as a consequence of human activities. 'Pollutant' is a more emotionally charged name, as is indeed the whole issue. It is impossible to separate the scientific data on environmental pollution from economic, political, and sociocultural aspects, as shown by two examples among many (see Scheme 7.1):

- Daminozide is a growth regulator sprayed on apples. It is also used to improve their visual appearance. In 1989, legitimate concerns about its carcinogenic effects were hyped by the media, resulting in a panic known as the 'Alar scare.'
- At the time of writing this chapter, a protracted political and legal battle has concerned the banning of glyphosate, an organophosphorous derivative of glycine bearing a phosphonate group (brand name Roundup), as a pesticide in Europe.

Organic pollutants may be more or less susceptible to biodegradation, that is, to the transformation by living organisms into harmless compounds, and the most resistant ones are called persistent or recalcitrant pollutants. Partial transformation of pollutants may give rise to long-lasting 'bound residues' (Gevao et al., 2000).

The most harmful effects of pollutants arise when they are ingested by humans, and by plants and animals in the food chain. Therefore, the noxiousness of a particular molecule will depend on its bioavailability, that is, the fraction that is available for incorporation into living matter. Bioavailability depends on the distribution between the various compartments of the ecosystem. Fig. 7.1 is a highly schematised view of compartments in soils.

Estimating this parameter correctly is crucial for establishing a rational basis for environmental pollution norms. It is sometimes argued that existing norms based on overall pollutant contents in the soil are unnecessarily stringent because a large fraction of the pollutant stock is so strongly adsorbed as to be unavailable. However, detailed

[a] Originally, 'xenobiotic' referred to substances foreign to an individual organism, but the generalisation to ecosystems is now generally accepted.

Developments in Clay Science, Vol. 9. https://doi.org/10.1016/B978-0-08-102432-4.00007-X

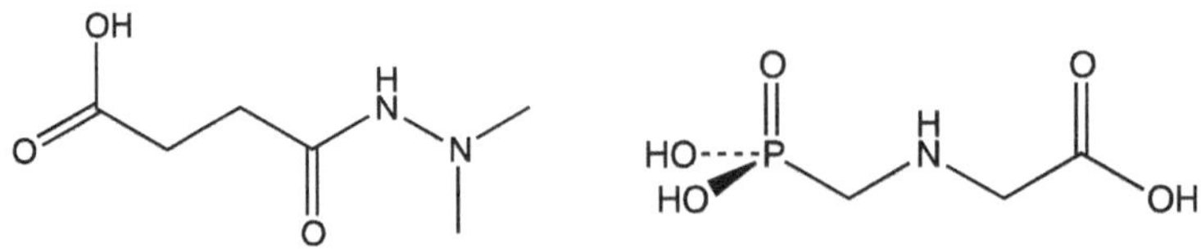

Scheme 7.1 Two xenobiotic molecules, diaminozide (Alar) and glyphosate (Roundup), that have given rise to much controversy.

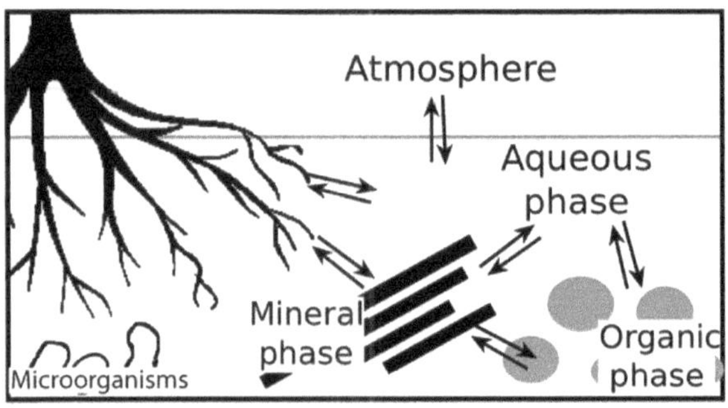

Fig. 7.1 Exchange phenomena between the different compartments of soil.

research indicates that such claims are oversimplified because the bioavailability is highly dependent on the nature of the adsorbents and their interaction with the pollutant molecules (Boyd et al., 2011b).

Clay minerals are one component of the mineral phase, probably the most expansive one from the point of view of surface area. The exchanges of xenobiotic molecules between them and the aqueous phase (groundwater) are adsorption–desorption phenomena. Thus, both the quality of drinking water and the effective toxicity of pollutants depend on their adsorption on clay minerals, interacting with the other exchange processes in the biotope. Therefore, it is important to study the adsorption of pollutants on clay minerals for at least two reasons:

(i) to understand their mobility and bioavailability in soils and other real-life systems and
(ii) to control them through the use of clay minerals as adsorbents, in the frame of water purification, as amendments applied in the field or as controlled release formulations (Yusoff et al., 2016) used to 'smooth out' the concentration profiles of pesticides as a function of time, avoiding a sudden concentration peak followed by washing out.

Clay minerals are already used in water purification cartridges or as lining materials in landfills (Koutsopoulou et al., 2010), to avoid leakage of dangerous substances. In both cases, their use has emerged empirically, and the retention of organic pollutants is only one of the desirable properties of clay minerals: another one is the reduced hydraulic conductivity they afford. The fact that these technologies are already developed constitutes a strong incentive for the fundamental understanding of clay mineral–organics interactions. Other domains of clay science may overlap with the adsorption of organic pollutants, such as inorganic–organic nanocomposites. For example, nanocomposites composed of clay minerals and epoxy resin are certainly relevant for the study of bisphenol A adsorption because those resins are formed by polymerisation of the bisphenol monomer, sometimes directly in the clay mineral interlayer space (Lan et al., 1996).

This chapter is centred on adsorption into natural or ion-exchanged clay minerals, with only limited coverage of organoclays (clay minerals exchanged with surfactant molecules, making their interlayer space hydrophobic). Of course, research on adsorbing materials is not limited to naturally occurring materials, and more or less complicated formulations have been tested for pollutants fixation, in which the clay mineral is combined with one or several other components. This chapter does not attempt to cover this field, nor the adsorption of organic molecules by soils of which clay minerals are only one component; the task would certainly need more than one book chapter. A recent review of pollutants adsorption by organoclays has been published by Park et al. (2011). A more general review of pesticides in soils is published by Arias-Estévez et al. (2008). The reference *Handbook of Clay Science* contains two chapters closely related to the topic of the present work (Nir et al., 2013; Yuan et al., 2013). There are also articles dealing with specific classes of compounds, for example, antibiotics in soils (Thiele-Bruhn, 2003). As regards the type of pollutants, this chapter is limited to simple organic molecules; some natural toxins that may be regarded as pollutants are complex proteins. Proteins adsorption on clay minerals is covered in Chapter 8 of this book. Heavy metals, including radionuclides, are also important nonorganic pollutants, and a special chapter is dedicated to them (see Chapter 5 of this book).

The role of clay minerals with respect to organic molecules is not limited to their immobilisation, reversible or not. They can and do catalyse different transformations through heterogeneous catalysis. These may be beneficial, as when adsorbed pollutants are mineralised to CO_2 in photocatalytic processes. But sometimes, reactions initiated on clay minerals cause transformations to even more dangerous compounds, such as dioxins. Conversely, adsorption may stabilise or protect pollutants against transformations that would occur if they were dissolved in the aqueous phase. Again, this may be undesirable, or desirable, in the case of controlled release formulations in which the pesticide should not be inactivated before being made bioavailable. The question is briefly covered in Section 7.5, but it has clear overtones in Chapter 9 of this book on *Clay Mineral Catalysts*.

Throughout the text, many 'semitrivial' names (see Section 7.2.2. for a definition) or trade names are used for polluting substances. The reader is invited to refer to Appendix, which contains a lexicon of pollutants whose adsorption on clay minerals has been studied (not exhaustive).

7.2 Classification of pollutants

The number of compounds released into the environment by humans is enormous; for example, it is often claimed that more than 10,000 different substances are used as dyes alone. There are several ways to classify pollutants.

7.2.1 Classification by origin

Fuels, detergents, industrial surfactants, dyes, pharmaceuticals (pharmaceutically active compounds), including veterinary pharmaceuticals, antibiotics (which are not only pharmaceuticals but are included in animal food), personal care products (PCP), and explosives are all released as waste, either at industrial sites, or for some of them, after use by the consumers, in a more diffuse manner (wastewaters).

Even micropollutants, released in small quantities, may cause significant health problems, either because of locally high concentrations, or because of their high potency. Thus, endocrine-disrupting compounds have been the object of considerable concern because they can interfere with hormonal pathways, even at a very low concentration. They include contraceptives, but also substances manufactured for quite different purposes, such as bisphenol A, used in epoxy resins. Such substances whose noxiousness has only been recently recognised are sometimes called emerging pollutants (e.g. perfluorinated compounds). The seriousness of the problem is likely to increase as effluents from wastewater are more and more often recycled, due to the shortage of drinking water in many locations.

In contrast, pesticides are spread mostly voluntarily in the field. They include herbicides, insecticides, fungicides, bactericides, nematicides and general disinfectants. Some substances are also applied to crops to regulate plant growth and development.

In addition, some molecules may be considered as xenobiotic because they represent a disruption from the normal state of the ecosystem, yet have a natural origin, such as wildfires, volcanoes or volatile organic compounds produced by vegetation.

Some industrial organic molecules may not be very harmful, but their transformation can result in extremely toxic byproducts, for example, when pyrolysis of chlorinated molecules forms polychlorinated dibenzodioxins and dibenzofurans.

7.2.2 Classification by chemical nature

A cursory look at the structures represented in Appendix will convince the reader that extremely diverse molecules are included within the group of organic pollutants, and have been studied in the context of clay mineral adsorption.

Due to their structural complexity, most pesticides and pharmaceuticals are not designated by their official IUPAC name. Most often, a 'semitrivial' denomination is used to identify the active chemical, such as glyphosate (reported in Scheme 7.1) for *N*-(phosphonomethyl)glycine. The term 'semitrivial' means that the name 'glyphosate,' while being simpler than the official, IUPAC-sanctioned nomenclature, is still chemically motivated, as it alludes to two functional groups of the molecule, the <u>gly</u>cine moiety at one end, and the <u>phosphate</u> group at the other end. Thus, 'glyphosate' is the name under which the substance is found in most databases, but for the general public, it is better known by the trade name *Roundup* (actually Roundup is a mixture that contains surfactants, in addition to glyphosate as an active ingredient). Often the denomination is an abbreviation of the IUPAC nomenclature name, for example, DDT for dichlorodiphenyltrichloroethane (this compound may also be called *clofenotane*) or DCMU for (3-(3,4-dichlorophenyl)-1,1-dimethyl urea)—its tradename is *Diuron*. There is no one-to-one correspondence; the same substance can have many different trade names, and a trade name may not correspond to a single chemical, but to a commercial formulation of several active species, for example, the herbicide *Resolva* is actually a combination of diquat (its IUPAC name is 6,7-dihydrodipyrido[1,2-*a*:2′,1′-*c*]pyrazinediium dibromide!) and glyphosate.

Often, several pesticides or drugs are modifications of the same basic structure. This is quite logical because it is more likely that a new, active compound will be discovered by modification of an existing one than by random, de novo synthesis. Therefore, they may be loosely grouped in classes or families based on the recognition of common structural features. This classification is quite arbitrary, depending on what features are considered as relevant. For example, the basic units of phenylureas and acetanilides have a lot in common. A given molecule may well belong to two classes. Picloram, belonging to the picolinic acid class, is also an organochlorine—a rather broad class potentially including all molecules with at least one C—Cl bond. Metsulfuron-methyl is both a sulfonylurea and a triazine.

Broader distinctions are made according to basic chemical properties, the most general one being between ionic and nonionic molecules (Gevao et al., 2000); the abbreviation NOC is often used for nonionic organic compounds. Within ionic compounds, one further distinguishes between cationic, acidic, and basic ones, which may seem puzzling. In fact, 'cationic' designates molecules that have a permanent positive charge, independent of the protonation state, such as diquat and paraquat (see Scheme 7.2), or the dyes methylene blue (MB), acriflavine, crystal violet, and malachite green. Therefore, they are supplied as salts containing compensating ions, often chlorides.

On the other hand, many molecules have moieties that take part in acido-basic equilibria, such as carboxylic acids (–COOH) or amines (NR_3). Their speciation will be pH dependent. In the first case, if the pH is inferior to the pK_a of the carboxylic acid/carboxylate couple (2–4 most often), the carboxylic acid form will predominate, and the molecule will be neutral, and if the pH is superior to the pK_a, the deprotonated form will predominate, and the molecule will be anionic. In the same way, when the pH is inferior to the pK_a of the ammonium/amine couple, the molecule will be protonated, and thus cationic. Of course, many molecules have more than one acido-basic moiety, and their pH-dependent speciation will span more than two different charge states. For example, the antibiotic ciprofloxacin has two pK_a, and its speciation is reminiscent of that of amino acids: the predominant form is a cation at pH $\leq$ 6, a zwitterion (bearing

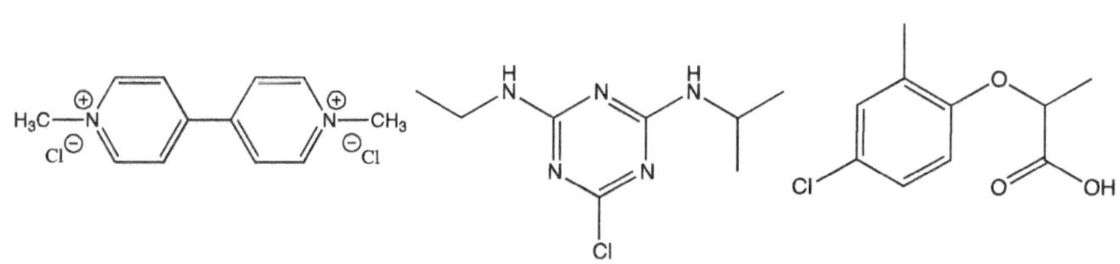

Scheme 7.2 Paraquat, a molecule with pH-independent cationic charge, here as the chloride salt; atrazine, a basic molecule, cationic at low pH; mecoprop, an acidic molecule, anionic at high pH.

both one positive and one negative charge) between pH 6 and 9, and an anion at pH $\geq$ 9 (Wang et al., 2011). Tetracyclines have four pK_a, so that as the pH increases, the net charge of the molecule goes from +1 to −3 (Figueroa et al., 2004). Molecules of all these classes (cationic, acidic, basic) are fairly soluble in water (e.g. glyphosate, $12\,g\,L^{-1}$ at 25°C).

Other pollutants, such as the PAH (polyaromatic hydrocarbons) or DDT, do not have either a permanent charge or acido-basic groups. They may still differ in hydrophilicity, and therefore in water solubility, due to the presence of polar groups, for example, parathion, with NO_2, P—O— and P=S groups, is about 100 times more soluble than pyrene (a PAH). Attempts have been made to correlate the adsorption of pesticides on clay minerals with their hydrophilicity (Polati et al., 2006), but using a single parameter has only little predictive valuc.

Many functional groups present in pollutants have good ligand properties, allowing them to form coordinative bonds with metal cations (not only transition metal ions, but also alkaline and alkaline earth ions). This is particularly true for amine and carboxylate groups, but also, to a lesser degree, for carbonyl, ether, sulphonate, and even chlorine groups. If several such groups are present in close vicinity and correctly oriented, the molecule may act as a polydentate ligand, stabilising the metal–ligand complex by a chelate effect.

Ligands are, by definition, Lewis bases. Lewis bases are not necessarily strong Brönsted bases, that is, they do not have to be able to undergo complete proton transfer from a Brönsted acid. However, hydrogen bonding may be viewed as partial proton transfer, and therefore those groups that can act as ligands are also susceptible to act as H-bond receptors. Conversely, Brönsted acidic groups can act as H-bond donors, and so can X—H groups that have no measurable Brönsted acidity.

One last type of functional group often present in organic pollutants is aromatic rings. They are a defining feature of PAH, and are often present in dyes too, whose spectroscopic properties are then the consequence of highly conjugated π electron systems. At first sight, they would seem to be typical hydrophobic groups, unable to establish anything but van der Waals interactions, and thus susceptible to segregation out of water-rich regions (see Section 7.3). But in fact, the electrons in aromatic orbitals can give rise to a specific coordination to metal cations, where the 'naked' cation is nested above the aromatic ring to maximise orbital overlap (Dougherty, 1996). In such a complex, all carbons participate in the bonding to the same extent, so that the 'hapticity', that is, the number of C atoms bound to the cation, is equal to the number of atoms in the ring, for example, six for a benzene ring (η^6).

7.3 Pollutant adsorption mechanisms on clay minerals: An overview

General discussions of the adsorption mechanisms of organic molecules have sometimes been offered, be it in soil constituents (Gevao et al., 2000), or more specifically, in clay minerals (Mortland, 1970; Johnston, 1996; Lagaly, 2001; Cornejo et al., 2008; Boyd et al., 2011a; Thiebault et al., 2016). Unfortunately, there is no agreement on the terminology, so the same mechanisms are sometimes designated by different words.

Furthermore, authors may emphasise different distinctions, which they perceive as being more fundamental (Kulik, 2009). However, adsorption phenomena belong to the realm of chemistry, and may be treated in familiar chemical terms, with close parallels in homogeneous chemistry (both solution and solid state). Even the classical distinction between physisorption and chemisorption has more to do with the existence of separate scientific communities than with a clear-cut separation of adsorption mechanisms; and while in limiting cases, chemisorption may appear quite distinct from physisorption, in reality, there is a continuum of situations where one smoothly transforms into the other.

7.3.1 Electrostatic interaction and ion exchange

If a molecule in solution has a net electric charge, whether intrinsic or due to acid–base equilibration, it will undergo an electrostatic interaction with a charged surface, such as that of clay minerals. One speaks of electrostatic adsorption when the energetics of the phenomenon can be described by modelling the molecule as a point charge. Electrochemists have long known how to treat this situation in the frame of the *double layer* model. The surface is modelled as a plane with uniform charge density, and the effect on the neighbouring electrolyte solution is to concentrate ions of the opposite charge while repelling those of the same charge. This effect is, of course, most pronounced in the immediate vicinity of the surface, and quickly decays with distance—over a few nanometres. This treatment may approximately describe the interaction of charged organic molecules with the external basal plane of a clay mineral particle, the surface charge being negative in this case. It is probably most appropriate for clay minerals with octahedral substitution such as montmorillonite (Mt), while in tetrahedrally substituted clay minerals (e.g. saponite or beidellite), the effect of substitutional charge is more localised, and it is an oversimplification to view it as spread out on the surface. Furthermore, the basal surface of clay mineral particles is not infinitely extended, and the energetics will be perturbed near the clay mineral edges; even when considering only electrostatic effects, the latter have a pH-dependent charge that will be positive at low pH.

At any rate, in this situation the organic molecules would be excluded if negatively charged (this phenomenon is sometimes called negative adsorption), and if positively charged, they would be spatially distributed in exactly the same way as other compensating cations of the same charge (Na^+, K^+, ...). Doubly charged ions, such as paraquat or diquat, would, however, be favoured over singly charged alkali metal ions (Valisko and Boda, 2007), and thus, are expected to be adsorbed preferentially in the case of pure electrostatic interaction. The situation is more complicated with molecules that bear several charges. In particular, if they are of opposite signs (zwitterion), an electric dipole results, and although the net charge is zero, the preferential orientation of the dipole may result in a favourable interaction with the charged surface (Chavez et al., 1996). Thus, electrostatic interaction is hardly ever a trivial matter. However, in a first approximation, if a molecule only interacts with the surface through electrostatic interactions, its spectroscopic properties will be unaffected with respect to those it has in solution.

Ion exchange cannot be regarded as equivalent to electrostatic adsorption. It refers to replacement of compensating cations (in the case of clay minerals) inside a solid phase. It corresponds to a process that can be written as a chemical equilibrium, for example (the reader can refer to Chapter 4 of this book for more details on ion exchange):

$$P^{+}_{aq} + Na^{+} - \text{clay mineral} = Na^{+}_{aq} + P^{+} - \text{clay mineral} \quad (7.1)$$

Here, P^+ is the cationic pollutant, and the subscript 'aq' (for 'aqueous phase') subsumes all the interactions between the ion and the solvent molecules in solution. In the same way, 'Na^+-clay mineral' and 'P^+-clay mineral' refer to all the interactions between the ion under consideration, and the clay mineral matrix, most of the ions being in the interlayer space. These interactions, of course, include electrostatic attraction, and in some cases, this is sufficient to explain differences in the adsorption behaviours of related molecules. For example, in a recent study of dyes adsorption by Mt, it was found that the cationic MB was adsorbed in quantities three times higher than the neutral Rhodamine B, while the anionic Orange II was not adsorbed—but the equally anionic sulforhodamine G was adsorbed (Fang et al., 2018).

Other interactions than electrostatic are certainly present, even for the original Na^+—which is generally coordinated by water molecules in the interlayer space. Thus, ion exchange must be viewed as a complex process involving several different kinds of interactions. Note that each molecule fixed in the clay mineral by ion exchange releases one compensating cation into the solution, which provides an easy way to quantify the phenomenon (e.g. Grauer et al., 1987b; Li et al., 2011a, 2011b; Wang et al., 2011; Fang et al., 2018). The possibility of ion exchange for a given molecule depends on its acid–base speciation, with acidic pH favouring the protonated forms, and thus the presence of a cationic charge (Özcan and Özcan, 2004). For example, diphenhydramine is only ion exchanged at $pH < pK_a$, when its tertiary amine is protonated to the ammonium (Li et al., 2011a). Nalidixic acid is much more adsorbed on Mt at $pH < pK_a$ of its carboxylic acid group, when it is neutral, than at $pH > pK_a$ where the negative charge of the carboxylate causes a repulsion with the Mt layers. On neutral kaolinite, on the other hand, the difference is not pronounced (Wu et al., 2013a).

7.3.2 Covalent bonds and coordinative bonding

From a physical point of view, of course covalent bonds are a consequence of the electrostatic interaction too, but electrons in molecules must be treated with quantum mechanics (Schrödinger's equation or an approximation thereof). This results in the archetypical chemical bond, namely two electrons (a bonding pair) between two atoms, that is schematised with a hyphen in organic structures—but it may also result in less familiar forms of bonding.

It is perfectly conceivable that adsorption of a pollutant organic molecule involves the formation of localised chemical bonds, for example, with silanol or aluminol groups at the edges of clay mineral layers. Indeed, in the case of glycine on smectite (the situation is relevant even though glycine is not a pollutant), it has been hypothesised that adsorption could involve the condensation of the carboxylic acid moiety with such groups on the clay mineral edge, to produce R–CO–O–Si and R–CO–O–Al 'mixed anhydride' bonds (Collins et al., 1988). However, further studies revealed that this reaction is only favourable if coupled with the opening of strained siloxane rings that exist on fumed silicas, but not on clay mineral edges (Rimola et al., 2013). More generally, the author is not aware of any instance where such a mechanism has been clearly demonstrated for organic pollutants on clay minerals.

There is, however, one type of chemical bond that is highly relevant to pollutants/clay mineral systems, namely, the coordinative bond with metal cations. In pristine clay minerals, compensating cations are most often alkali or alkaline earth ones, although $Fe^{2+/3+}$ can also be present. The association of Lewis basic groups of ligand molecules with transition metal cations is the subject of coordination chemistry, and the splitting in energy of *d* orbitals under the effect of the ligand field is at the origin of interesting properties, including a specific energetic stabilisation called the crystal field stabilisation energy (CFSE). In the case of alkali and alkaline earth cations, the *d* orbitals are vacant, so that the CFSE is zero. However, coordination of various ligands does occur, although it is more difficult to study. For example, the calcium ion in the aqueous phase, denoted as $Ca^{2+}{}_{aq}$, is actually a complex with 6 or 8 water molecules, $[Ca(H_2O)_n]^{2+}$ (Bruzzi and Stace, 2017). In the clay mineral interlayer too, the original compensating cations are usually present as coordination complexes, as revealed, for example, by solid-state NMR—the coordination can vary according to the activity of water. If an organic molecule containing some ligand groups is introduced in the system, it will compete with water for coordination to the ion, and possibly form a coordination complex, or several different ones depending on the degree of hydration. The importance of this type of bonding was recognised early on by Tahoun and Mortland (1966) and by Jacobs et al. (1972). For example, 2,4-D is expected to form a monodentate complex with Ca^{2+} in a Ca^{+}-Mt through its carboxylate group in the presence of water, but a bidentate one (thus occupying two coordination positions) in thoroughly dry conditions (Tunega et al., 2007). The process of displacing some of the original water ligands to allow coordination of the organic molecule is called ligand exchange.

Here mention must be made of 'ion–dipole interactions' that are sometimes listed as a different type of interaction. In fact, they fall within the realm of coordinative bonding. In fact, the successful model of the 'crystal field' precisely rests on the idealisation of the coordinative bond as an ion–dipole interaction, because the ligand directs an electron-rich part of the molecule toward the central ion. The real coordinative bond is ionocovalent, and the partial covalence can be accounted for in more precise, molecular orbitals models. Is there, then, any need to maintain the term of ion–dipole interaction? The answer may be 'no' if one is talking to chemists, for whom 'coordinative bonding' will do perfectly; but 'yes' when talking to physicists, who may be more familiar with this view of the interaction.

In some cases, it is assumed that organic molecules are coordinated to the compensating cation, and that the latter is also coordinated to the basal surface of the clay mineral, for example, to the siloxanes in the 6-ring of oxygen atoms. This is occasionally called an 'ion bridge'. This has been proposed, for example, in the adsorption of methyl orange on Ca^{2+}-palygorskite (Yang et al., 2018), or of alprazolam and diazepam in Mg^{2+}-vermiculites (Palace Carvalho et al., 2014).

7.3.3 Hydrogen bonds, van der Waals interactions, and hydrophobic effects

Here, these different effects are treated together because they correspond to weaker interactions than the preceding ones. It might then be thought that adsorption through these mechanisms counts as physisorption. That would be a mistake, because even if individual interactions are weak, several of them may be present between different groups of the adsorbed pollutant and the surface, or they may coexist with stronger interactions (see Section 7.3.4).

Hydrogen bonding is a possibility to take into account in most cases (see Chapter 1 of this book for details). As previously underlined, many functional groups of common organic pollutants can act as H-bond donors and/or acceptors. They would certainly find partners on the clay mineral surface. Although the siloxanes of the basal surfaces are only weak H-bond acceptors, more diverse groups are found on the clay mineral edges. Silanols and aluminols can act as donors and as acceptors, and in other materials such as amorphous silicas, it is well established that 'associated silanols' are engaged at the same time in both types of interactions, each one reinforcing the other in a cooperative play (Rimola et al., 2013). Water molecules bound to the exposed octahedral cations on the edges are expected to be strong H-bond donors due to the electron-withdrawing consequences of their coordinative bonding to the cation. The same holds for hydration water attached to the compensating cations in the interlayer—it has long been known that the acidity of interlayer water is strongly enhanced (Mortland and Raman, 1968; Poinsignon et al., 1978). Finally, the structural OH groups of the octahedral sheets of TOT and TO clay minerals may be accessible through the ditrigonal cavities in the tetrahedral sheet, and thus play a role in H-bonding too.

H-bond networks of more or less complexity may exist, a simple example being the formation of water bridges to metal ions or to silanols. Fusi et al. (1988) observed that fluazifop-*p*-butyl forms such a water-bridged link to the compensating cations in conditions of high relative humidity that can be reversibly transformed to direct coordinative bonding upon heating at moderate temperatures.

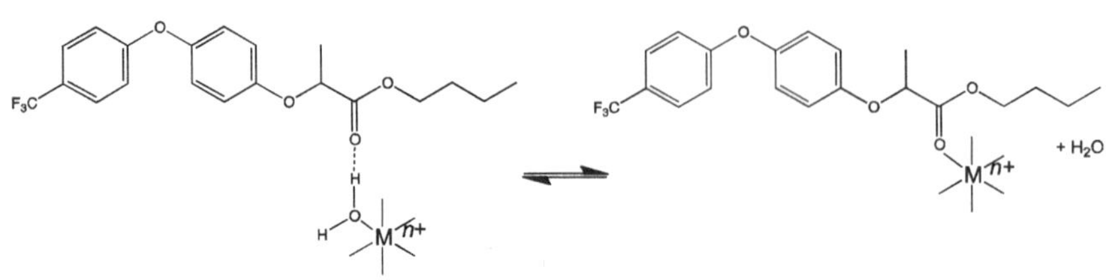

These two competing adsorption modes are sometimes called 'outer sphere complexes' and 'inner sphere complexes', respectively.

Water bridges were also invoked to explain the interaction of atrazine with smectites (Laird, 1996; Xu et al., 2001), or of alachlor (Bosetto et al., 1993) and metolachlor (Bosetto et al., 1994) with Mt.

'Hydrophobic interactions' refer to the fact that molecules and other assemblies with little affinity to water prefer to stay together, and segregate from the water solution. This effect is well known in micelles formation and may play an important role too in the adsorption of molecules containing aromatic rings (which is a frequent feature of organic pollutants, as already mentioned). It is probably predominant in molecules such as the PAH that have no polar functional groups. On the side of the adsorbent, the siloxanes of the basal planes by themselves are considered to be hydrophobic, providing suitable partners for aromatic rings; but of course, in the case of smectites or vermiculites, there will also be compensating cations in the vicinity of these surfaces, and it may be unfavourable to displace them. A potential misunderstanding may arise here. There is nothing in physics that corresponds to 'hydrophobic interactions' as distinct from other interactions between molecules and chemical species. To be sure, hydrocarbon chains will be subject to the various types of weak interactions lumped together under the name of 'van der Waals forces' (see Chapter 1 of this book). They account for a slight attraction, of the order of only 1 or 2 $kJ\,mol^{-1}$ for a pair of atoms. These figures rise, of course, as the van der Waals interactions are summed up over all atoms in the structure, but this does not seem enough to hold an assembly together. The solution to this difficulty is in taking into account the whole system involved in the adsorption process, as explained in the next paragraph.

7.3.4 Putting it all together: The importance of water

A simple way of writing an adsorption reaction for a molecule A is

$$A_{\text{aq}} + {}^{*} = A - {}^{*} \tag{7.2}$$

where * stands for a surface site, the special configuration of atoms that interacts with molecule A upon adsorption. Whether adsorption is favoured or not will depend on the sign of the standard free enthalpy of the adsorption reaction, $\Delta_{\text{ads}}G^{\circ}$. In order to evaluate this parameter, all the contributions to the energy of the system have to be taken into account. Now contrary to what happens for adsorption from the gas phase, in the case of adsorption from aqueous solutions, all partners have significant interactions with water molecules—these interactions may be diverse, but at any rate, a more realistic depiction of 'A_{aq}' would be $A(H_2O)_m$. In the same way, the surfaces sites, whether occupied or not, are hydrated, so that instead of Eq. (7.2), one should write

$$A(H_2O)_m + {}^{*}(H_2O)_n = A - {}^{*}(H_2O)_p + (p - m - n)\,(H_2O) \tag{7.3}$$

in which a number of water molecules are liberated into the aqueous phase. If both molecule A and the surface are hydrophobic, these water molecules will be more stabilised when they are in the bulk solvent than they were initially when interacting with

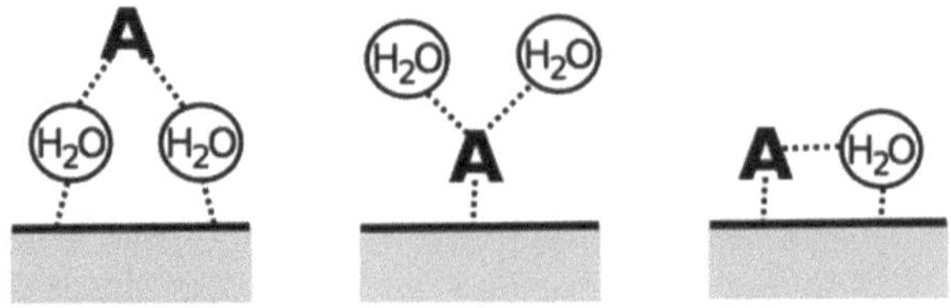

Fig. 7.2 Three modes of coadsorption of an organic molecule A with water.

A and *. This will suffice to make $\Delta_{ads}G°$ negative, and the adsorption favourable. This is the main origin of the hydrophobic effect.

Eq. (7.3) also has the merit of indicating that adsorbing molecules are in competition with pre-adsorbed water. They may, in fact, adsorb without establishing direct bonds to the surface, as in the case of water bridges mentioned previously. Schematically, the three situations could be considered for the coadsorption of an organic molecule with water (Fig. 7.2).

Another general remark is that because of the complexity of some organic pollutants that often exhibit on the same molecule groups of different reactivity, and of the rich chemistry of clay mineral surfaces that also bear different types of sites, an organic molecule may establish several interactions with the surface at the same time. This is difficult to evidence experimentally, except in an indirect manner when the hypothesis of a single interaction fails to explain all the features of the adsorption data. However, the increasing use of molecular modelling suggests that multiple interactions indeed often occur. This raises the interesting possibility of a kind of interactional complementarity between an organic molecule and an inorganic surface, similar to the one that may exist with an enzyme: some molecules could be selectively adsorbed from the solution if they have a pattern of functional groups that matches the distribution of surface adsorption sites.

7.4 Tools of the trade: Experimental studies of organic pollutants adsorption

Several methods are available for the study of pollutants adsorption on clay minerals. Most of them are non-specific and can be used for other adsorbates too. One can distinguish between those providing macroscopic and molecular-level information.

7.4.1 Macroscopic characterisation

7.4.1.1 Isotherms, thermodynamics, and kinetics

In the case of soils, the adsorption of pollutants is often studied through the use of breakthrough curves obtained from columns or lysimeters (dynamic adsorption tests) that are relevant for practical application in the field. The relatively high amounts of adsorbing materials that are needed for these studies generally prevent their use for pure clay minerals, although there are some exceptions (Cabrera-Lafaurie et al., 2015).

The adsorption of pollutants can be measured in batch experiments, even for small amounts of clay minerals. For low concentrations, radioactive labelling may be used (^{14}C labelling, Nowara et al., 1997); otherwise, any method that can quantify the concentration in the aqueous phase before and after adsorption will allow determination of the adsorbed amounts from the solution depletion. This can often be done by UV–visible spectroscopy. In this way, 'adsorption capacities' expressed, for example, in mg/g, can be obtained, and they are listed, for example, for dyes on various clay minerals (Ngulube et al., 2017). They can be fairly high, up to 500 mg/g clay mineral, but in general they vary with concentration in the aqueous phase, pH, and so forth. At times, 'negative adsorption' is observed, that is, the pollutant concentration in the solution apparently increases upon contact with the clay mineral. This is the case, for example, for aldicarb on Mt, while positive adsorption occurred in illite and kaolinite (Supak et al., 1977), or to a smaller extent, atrazine on smectites (Barriuso et al., 1994). It only makes sense if one considers that the clay mineral retains an amount of 'bound water' that will not be probed by ordinary analytical techniques, and where concentration conditions may differ from the bulk. If a molecule is excluded from the bound water layers due to unfavourable interactions, its concentration will necessarily increase in the bulk.

An empirical parameter that is often tabulated for various pesticides in soils (Clausen et al., 2001; Delle Site, 2001; Drillia et al., 2005) is the distribution coefficient K_d. It simply consists in the ratio between the amount of pesticide fixed by the soil (which may be given in mole per gram, $\mathrm{mol\,g^{-1}}$) and the concentration remaining in solution at equilibrium (expressed in $\mathrm{mol\,L^{-1}}$): in this case the units of K_d will be $\mathrm{L\,g^{-1}}$. Alternatively the adsorbed amount may be expressed as a surface concentration ($\mathrm{mol\,m^{-2}}$), taking into account the surface area of the soil (S in $\mathrm{m^2\,g^{-1}}$), the units of K_d will then be $\mathrm{L\,m^{-2}}$. At any rate, this corresponds to a simple picture of adsorption as an exchange between two phases, the aqueous solution (aq) and the solid soil:

$$A_{aq} \rightleftharpoons A_{soil} \tag{7.4}$$

where K_d is then the selectivity constant of this exchange. As such, this model does not include any limit on the adsorption capacity of the soil, which is not realistic, as shown in the following paragraph. Thus, these values should be considered more as practical guidelines for soil amendment than as fundamental properties. They will depend on the intensive thermodynamic parameters of the aqueous solution, such as the pH (Weber et al., 2004). They will be concentration-independent only in the range where the adsorption isotherm is linear (e.g. Langmuir at low concentrations, see the following).

Adsorption isotherms can easily be established by plotting the adsorbed amounts (Γ_{eq}, if expressed as surface concentrations) as a function of the concentration in the aqueous phase C_{eq} (after the adsorption equilibrium is established). They can be modelled using one of a few classical models of physical chemistry. For a basic presentation of these models, see Chapter 8 of this book. The most commonly used one for organic pollutants is the Freundlich isotherm:

$$\Gamma_{eq} = K_f C_{eq}^{\frac{1}{n}} \tag{7.5}$$

which depends on the two parameters K_f and n. One can check if the data fit this model by representing them on a logarithmic scale.

The Langmuir model is also commonly tested:

$$\Gamma_{eq} = \Gamma_m \frac{bC_{eq}}{1 + bC_{eq}} \tag{7.6}$$

Like the Freundlich model, it also depends on two parameters, but their interpretation is more intuitive: b is the equilibrium constant of the adsorption reaction and Γ_m is a saturation coverage (or a saturation quantity if the adsorbed amounts are expressed per gram instead of per unit surface).

The same criticisms can be made as for the application of these models to proteins. The Freundlich isotherm has a heuristic value in that it allows one to summarise the data obtained with a single, simple equation. K_f is related to the adsorbed amounts, and therefore comparing the K_f values of different clay minerals for the same pollutant, or of different pollutants on the same clay mineral, may give a general impression of the relative adsorption capacities. On the other hand, the physical bases for the Freundlich model are not clear, apart from the fact that the adsorption energy is not constant with the amount adsorbed, in opposition to the Langmuir model. This could be due to the existence of different adsorption sites, to interactions between neighbouring adsorbed organic molecules, or to multilayer adsorption (in this case though, the BET model would probably be more appropriate). The very fact that so many adsorption data for various pollutants may be fitted by a Freundlich isotherm suggests that this does not tell us anything fundamental about the adsorption process.

In some instances, it has been observed that the Langmuir isotherm is a better fit than the Freundlich one, for example, for ciprofloxacin on Mt, rectorite, and illite (Wang et al., 2011), where the adsorption mechanism seemed to be ion exchange, and the saturation quantity indeed corresponded well to the CEC. Similar cases are pyrene on sepiolite (Sabah and Ouki, 2017), diphenhydramine on Mt (Li et al., 2011a), MB on Mt (Gürses et al., 2004), levofloxacine on Fe-PILC (Liu et al., 2015b), glyphosate on Mt (Khoury et al., 2010), and to a lesser extent, tramadol and doxepin on Mt (Thiebault et al., 2015). In their comparative study of dyes adsorption on Mt, Fang et al. (2018) found that the Langmuir model was more appropriate for the cationic MB, which was ion exchanged; but for the zwitterion Rhodamine B, the Freundlich model gave a better fit. On palygorskite, Yang et al. (2018) found a better fit with Langmuir for cationic dyes and, unexpectedly, also for anionic ones. Other studies on related systems could not discriminate between the two models, for example, for methyl violet adsorption on halloysite (Liu et al., 2011a), bisphenol-A on palygorskite-Mt formulations (Berhane et al., 2016), prometryn on halloysite (Grabka et al., 2017), or crystal violet on Mt (Sarma et al., 2016). The Freundlich model was found to be better for gemfibrozil, mefenamic acid, and naproxen on vermiculite (Dordio et al., 2017), naphthalene on bentonite (Kaya et al., 2013), and for Congo Red on bentonite (Vimonses et al., 2009). Initial Langmuir adsorption followed by further adsorption (possibly due to the inception of a multilayer) was invoked for nonionic pesticide on kaolinite and Mt (Torrents and Jayasundera, 1997). The general impression from these studies is that the Langmuir model fits well in situations dominated by ion exchange, with a saturation adsorption corresponding to the CEC; a better fit of the Freundlich model, and especially of the BET one (appropriate for multilayers), would indicate significant adsorption on the edges (Fang et al., 2018). A more refined, two-step Langmuir model was applied to understand glyphosate adsorption on Mt (Khoury et al., 2010), which seemed to imply both adsorption on the edges and intercalation. Of course, deconvoluting two adsorption mechanisms that occur simultaneously requires high precision data. The reader must be cautioned that fit does not prove a mechanism, and that all studies mentioned here must be considered as empirical.

When it applies, the Langmuir model gives access to the adsorption equilibrium constant b, and from there to $\Delta_{\text{ads}}G°$ ($\Delta_{ads}G° = -RT\ln b$).

Because $\Delta_{\text{ads}}G° = \Delta_{\text{ads}}H° - T\Delta_{\text{ads}}S°$, a study of adsorption at different temperatures should allow one to separately estimate the enthalpic ($\Delta_{\text{ads}}H°$) and entropic ($\Delta_{\text{ads}}S°$) contributions to the adsorption reaction. This is not done systematically. For MB adsorbed on a Turkish clay mineral, $\Delta_{\text{ads}}H°$ and $\Delta_{\text{ads}}S°$ were -8.0 and $+25.5\,\text{kJ}^{-1\circ}\text{mol}^{-1}$, respectively, meaning that enthalpic and entropic effects were both favourable and contributed equally to the adsorption at RT (Gürses et al., 2004). For crystal violet adsorbed on several Mt, $\Delta_{\text{ads}}H°$ ranged from -18.1 to $-21.6\,\text{kJ}\,\text{mol}^{-1}$, and $\Delta_{\text{ads}}S°$ from $+49$ to $+60\,\text{kJ}^{-1}\,\text{mol}^{-1}$ (Sarma et al., 2016).

The very idea of an adsorption isotherm supposes that adsorption is a reversible phenomenon that can be described by an equilibrium. Some systems show partial irreversibility, manifested by a hysteresis between adsorption and desorption branches; a hysteresis index was proposed, and defined as $\text{HI} = \frac{c^{\text{des}} - c^{\text{ads}}}{c^{\text{des}}}$, where c^{ads} and c^{des} refer to the concentration of adsorbed pollutant measured in the adsorption branch and in the desorption branch, respectively, for the same concentration in the solution. The apparent irreversibility may be due to the fact that the desorption process has kinetics that are too slow to be measured in practice. In the case of carbamazepine adsorbed on smectites, hysteresis was attributed to the inception of π–π interactions between the aromatic moieties (Zhang et al., 2010). Such phenomena are, of course, important for controlled release formulations. For atrazine adsorbed on smectites, hysteresis was reported to be positively correlated to the layer charge (Barriuso et al., 1994). It was also observed for alachlor, metolachlor, and linuron adsorbed on Na^+-Mt and kaolinite (Torrents and Jayasundera, 1997), or bisphenol A and ciprofloxacin adsorbed on palygorskite-Mt (Berhane et al., 2016).

The kinetics of adsorption are also often evaluated, and they are fitted to either first-order (or Lagergren, i.e. pseudo-first order) or second-order models. As for the adsorption isotherms, a basic presentation of adsorption kinetics models is found in Chapter 8 of this book. A review of second-order kinetics for pollutants adsorption may be found in Ho and McKay (1999), but it does not deal primarily with pure clay minerals. The Elovich model is not often tested, and when it is, it does not seem particularly appropriate (Kaya et al., 2013; Fil et al., 2014). While the pseudo second-order model often gives a good fit to the data, once again this has a mostly heuristic value. The main conclusion regarding adsorption mechanism is that diffusion is not rate limiting. Adsorption is generally completed in, at most, a few hours; although in high layer charge clay minerals such as vermiculite, very slow adsorption kinetics have been reported, up to 10 days for completion (Rytwo et al., 2009). The knowledge of at least an order of magnitude for the kinetics is important for practical applications, and a prerequisite for equilibrium adsorption measurements such as those needed to obtain isotherms. Adsorption kinetics, like adsorption isotherms, are generally evaluated by solution depletion in batch experiments; however, for an IR-based method, allowing to follow adsorption kinetics in situ.

7.4.1.2 Thermal analysis

Thermal analysis methods include thermogravimetric analysis (TGA) and differential thermoanalysis (DTA). TGA simply records the sample mass as a function of time and allows one to evidence those thermal transformations that are accompanied by mass loss, such as desorption or calcination for clay mineral–organic systems. These events are best revealed by plotting the DTG (derivative of the TG signal). TGA may be coupled with EGA (evolved gas analysis), by mass spectrometry, or sometimes IR. DTA measures thermal effects, whether endo- or exothermic, and may detect phenomena that are not accompanied by mass changes, such as phase transitions.

TGA may allow one to estimate the total amount of pesticide bound to a clay mineral, if carried out under an oxidising atmosphere where the organic matter is burned out (Ibarguren et al., 2014; Wanyika, 2014). Under an inert atmosphere too, it usually shows peaks attributable to the adsorbed organic molecule (Tomic et al., 2016), but in that case, part of the organic matter often pyrolyses transforming to inert carbon that remain adsorbed on the clay mineral particles, so that quantification is more difficult.

By comparison with the TG of the bulk xenobiotic, it allows one to also observe stabilisation, or conversely, catalysed thermal decomposition. Thus, several PAH were observed to desorb from bentonite, but at temperatures significantly above their bulk boiling point (up to 500°C), indicating that they interacted strongly with the clay mineral surface (Biache et al., 2015). At these temperatures, part of the PAH decomposed to lower mass aromatics instead of desorbing intact.

Because the elimination of physisorbed water occurs in a separate temperature range from that of organic molecules decomposition, TG also permits, if conducted in carefully controlled conditions, one to observe the displacement of bound water upon adsorption of the organic molecules (Wanyika, 2014), revealing competition between them.

DTA has been used to discriminate between free and Mt-bound forms of alachlor (Nasser et al., 1997), and evidences the competitive replacement of water with alachlor. The melting peak of promethazine was suppressed when it was deposited on Mt (Ambrogi et al., 2014), confirming the absence of bulk crystals (in contrast, a physical mixture of these and the Mt, with the same global content, did show the melting transition).

7.4.1.3 Electrophoretic mobility

Particles in a dispersion will migrate in a potential difference, and the measurement of this effect gives access to a quantity called the zeta potential. This depends on the charge of the particles—not exactly the surface charge, but that of the particle accompanied by a few layers of bound water. Zeta potential measurements are often used to characterise the surface properties of dispersed clay mineral nanoparticles as a function of pH. Amphoteric nanoparticles, in the absence of specific adsorption, are positively charged at low pH and negatively charged at high pH; at the isoelectric point (IEP), the net charge is zero. Smectites do not have an IEP because their charge is mostly substitutional in origin, and the zeta potential is negative, irrespective of the pH.

In principle, when a charged species is adsorbed on a smectite, the zeta potential will be modified accordingly, and this technique might therefore provide a means to follow adsorption in situ. The data are rather scarce; ofloxacin-loaded activated halloysite had positive zeta potentials, indicating that the surface of halloysite nanotubes is positively charged due to the adsorption of ofloxacin (Wang et al., 2013). Zeta potential measurements are also reported in Pessagno et al. (2008), Maqueda et al. (2013), and Ibarguren et al. (2014), but not much commented upon.

7.4.2 Molecular level characterisation

7.4.2.1 X-ray diffraction (XRD)

XRD can indicate if a fraction of the pollutants has precipitated as bulk crystallites, as opposed to being molecularly adsorbed. This situation is perhaps unlikely to be met in equilibrium adsorption studies, but it can certainly occur in attempts to prepare clay mineral–xenobiotic composites for controlled release. Thus, Fraile et al. (2016) observed that crystallites of dexamethasone form upon attempts to deposit high loadings of the corresponding molecule on Laponite for a controlled release drug therapy application. Crystallites were also observed in controlled release formulations of amitrole with methoxy-modified halloysite and kaolinite (Tan et al., 2015), and their presence was not necessarily detrimental to release efficiency. In such cases, it may be surmised that the clay mineral has a saturation loading for the considered xenobiotic. Forcing the deposition of higher loadings in the solid phase by evaporation of the clay mineral–solvent system results in the excess part precipitating more or less independently from the clay mineral surface.

Apart from this, the main information that can be drawn from the X-ray diffractograms of pollutants–clay minerals systems is the d_{001} value, which is the sum of the clay mineral layer thickness (assumed to be invariable) and the interlayer thickness. In smectites, the interlayer spaces can swell, and this permits intercalation of more or less bulky molecules, which is revealed by an increase of d_{001}. However, the data may be ambiguous because, depending on the water activity, smectites can exist in different hydration states: one can observe an anhydrous form, a 1-layer form and a 2-layer form, with increasing d_{001} values. Because the XRD reflections of clay minerals are rather broad, it may be difficult to decide if a given sample has a d_{001} significantly different from that of the hydrated forms. In this case, the best procedure is to determine if there is a collapse of the layers upon dehydration (Grauer et al., 1987b). If d_{001} stays high at low water activities, it probably means that large organic molecules in the interlayer space oppose collapse. With big molecules, though, the swelling may be large enough that no ambiguity remains, as with enrofloxacin, which swells Mt up to d_{001} values of more than 19 Å (Nowara et al., 1997), which cannot be due to water alone. Diphenhydramine caused the swelling of Mt to 17 Å when adsorbed in high amounts, and because this reflection was not observed for a physical mixture in the same conditions, this was attributed to intercalation (Li et al., 2011a).

There are many examples of xenobiotic intercalation, and unsurprisingly they include a lot of positively charged molecules, such as MB (Hang and Brindley, 1970), which forms the basis of a popular method of surface area measurement for clay minerals in aqueous dispersions. They also include molecules that can be protonated to the cationic form, as already mentioned herein; other examples include benzimidazole in Mt (Lombardi et al., 2006), and chloroanilines in Mt (Angioi et al., 2005).

The same molecule can give different basal spacings due to different stacking modes in the interlayer space, which is the case for MB in Na^{+}-smectites (Hang and Brindley, 1970). It has been proposed that the entrance of picloram into the Mt interlayer space may occur in two different arrangements, perpendicular and planar, depending on the xenobiotic concentration (Marco-Brown et al., 2015). Donepezil intercalated forming double layers in Mt, but only a single layer in Laponite and saponite (Park et al., 2008).

Alachlor expanded the interlayer spaces of Mt, but not those of pillared Mt (Nasser et al., 1997) where the pillars are inorganic species that may bind strongly enough to the layers to prevent them from moving apart. In the same way, ciprofloxacin swelled Mt and rectorite, but not illite (Wang et al., 2011). There are many studies that compare the adsorption of substances on Mt and on kaolinites, and in a lot of cases Mt swells while kaolinite does not (see Table 7.1 for examples).

Table 7.1 Studies reporting the intercalation of pollutants in montmorillonite, with comparison to kaolinite

References	Pollutants studied
Torrents and Jayasundera (1997)	Alachlor, metolachlor, linuron
Gianotti et al. (2008)	2,4,6-TCA, chlorophenol
Angioi et al. (2005) and Gianotti et al. (2008)	Chloroanilines
Oudou and Hansen (2002)	Pyrethroids
Shareef et al. (2006)	Endocrinedisruptors
Styszko et al. (2015)	Bisphenol A, triclosan
Gimsing and Borggaard (2002)	Glyphosate

Even for smectites, swelling will depend on the original compensating cation, being more limited and more difficult with Ca^{2+}-clay minerals than with Na^{+}-clay minerals (Hang and Brindley, 1970; Boyd et al., 2011a).

Swelling was also reported in conditions when the speciation of the xenobiotic in solution is supposed to favour a neutral form. First, it must be remarked that the local speciation inside the clay mineral layers—and also in the vicinity of the external surfaces—may be altered with respect to that in solution. Consider a molecule involved in an acid–base equilibrium:

$$A_{aq} + H^{+}{}_{aq} = AH^{+}{}_{aq} \tag{7.7}$$

Suppose further that the conditions in solution favour the unprotonated form (which means that $pH > pK_a$ of the couple). If the ion exchange of species AH^{+} into the clay mineral (along reaction (7.8) below) is favoured, that is, if its interactions with the clay mineral layers are more favourable than those of the cation in the pristine clay mineral, then reaction (7.7) will be drawn to the right according to Le Châtelier's principle, and the global process (7.9) may be favourable.

$$AH^{+}{}_{aq} + Na^{+} - \text{clay mineral} = Na^{+}{}_{aq} + AH^{+} - \text{clay mineral} \tag{7.8}$$

$$A_{aq} + H^{+}{}_{aq} + Na^{+} - \text{clay mineral} = Na^{+}{}_{aq} + AH^{+} - \text{clay mineral} \tag{7.9}$$

In that case, the pH will increase upon ion exchange, because H^{+} will be removed from the solution (or alternatively, OH^{-} will be formed because the H^{+} come from the dissociation of water). Thus, a complete study of adsorption should include measuring the release of the pristine compensating cations (to check if cation exchange occurs), and measuring the pH change (to obtain information on the comparative interactions of both cations with the clay mineral matrix).

However, intercalation is not synonymous with ion exchange. The driving force for the migration of molecules into the interlayer space may be another one, such as a coordinative interaction with the compensating cations (Damonte et al., 2007), or the establishment of hydrophobic interactions with the basal planes in the interlayer space. For example, DNC molecules were intercalated in Mt interlayer spaces with their aromatic groups lying flat between the two layers (Sheng et al., 2002). In this case, the d_{001} of the intercalate remained at the low value of 12 Å, even under conditions of high humidity, meaning that the intercalated species opposed not only collapse, but swelling too—in order to stay able to interact with both layers simultaneously. In the same study, it was also observed that exchange of the Mt with weakly hydrated cations (K^{+}, Cs^{+}) resulted in higher DNC adsorption than when starting with more hydrophilic cations (Na^{+}, Ca^{2+}), meaning that coordination of the $-NO_2$ groups to the cations, by ligand exchange with the originally present H_2O ligands, contributed to the driving force for intercalation.

Not only neutral molecules such as DNC or 2,4,6-TCA and 4-chlorophenol (Gianotti et al., 2008) but also negatively charged species can be intercalated by a ligand exchange mechanism, and this has been observed with glyphosate at least at high concentrations

(Damonte et al., 2007). Indeed glyphosate, a polyacid that may develop up to three negative charges, intercalated in Mt in conditions where its global charge was −1 or −2 (Khoury et al., 2010). However, locally, it also carried one positive charge on the amine group, so that this may again be a case of cooperative interactions—the negative phosphate group interacting with the cation through a coordinative bond, while the $-NH_2{}^+-$ group would interact with the negatively charged clay mineral layers.

7.4.2.2 Transmission electron microscopy (TEM)

TEM may be used as a complementary technique to XRD to observe swelling—directly on micrographs when stacked layers in smectite-based nanocomposites may be very clearly visible. A more original application is to use the lateral resolution of microanalysis (SEM-EDS, scanning electron microscopy with energy-dispersive X-ray spectroscopy) in order to differentiate adsorption on the edges from adsorption on the planar surfaces (Joseph-Ezra et al., 2014). For PAH adsorbed on Cu^{2+}-Mt, measurement of the organic carbon contents indicated that adsorbed molecules concentrated at the outermost part of the clay mineral particles, which is compatible with adsorption on edge sites.

7.4.2.3 UV–visible absorption and fluorescence

UV–visible spectroscopy may, of course, be used to quantify pollutants in solution, but in this chapter the concern is its use for characterisation of the adsorbed pollutants. Dyes are, of course, good candidates for detection in the UV–visible, and visually perceptible colour changes may occur upon contact with clay mineral dispersions. Dyes may also be fluorescent, allowing a study in the adsorption, as well as the emission mode. Sometimes, the aggregation of dye molecules can give rise to new absorption bands, a phenomenon called metachromasy. In fact, for MB, several bands are observed corresponding to different association states of the dye molecules—monomers, dimers, and agglomerates. On a series of Mt whose charge had been modified by the Hofmann–Klemen effect, Bujdák and Komadel (1997) found that low layer charges favoured the monomers, while high layer charges favoured agglomerates—even though they worked in conditions of low MB loading where the loadings for all smectites were the same. They could also evidence a slow reorganisation of the adsorbed MB (over a few days) that could not have been observed otherwise because the total adsorbed amount did not change.

On the fundamental level, metachromasy is due to the interaction of the π electron systems of two or more dye molecules. It has been observed that the π electrons of a single adsorbed dye molecule can also establish similar interaction, probably involving the oxygen atoms of the basal plane, and undergo a blue shift; this was shown by for several dyes, including MB (Yariv and Lurie, 1971), acridine orange (Cohen and Yariv, 1984), and pyronin (Grauer et al., 1987a). Rhodamine B, on the other hand, showed a red shift instead on adsorption to Mt, perhaps because steric hindrance prevented the interaction of π electrons with basal planes. In these studies, the effect of molecule–molecule interactions is difficult to disentangle from that of molecule–clay mineral interactions.

A new fluorescence polarisation method was applied to evaluate the preferential orientation of rhodamine 6G adsorbed into Laponite (López Arbeloa and Martínez, 2006). The fluorescence dichroic ratio was used to evaluate the orientation of the fluorescent species adsorbed in the clay mineral films. It was found that the plane of the polyaromatic ring was tilted 30 degrees from the basal plane, and a minor amount of coplanar rhodamine dimers was also observed.

7.4.2.4 Vibrational spectroscopy

Vibrational spectroscopy is one of the main techniques of identification and characterisation of organic molecules. It has been available for a long time, and as a consequence, it was applied to organic molecules on clay minerals since the first studies of these systems.

Clay minerals strongly absorb IR in the 1200–900 cm^{-1} region due to lattice vibrations, and to a smaller extent, in the 3800–3200 cm^{-1} region due to OH stretching. They have a window of transparency between 3000 and 1300 cm^{-1} (except for the δ_{HOH} of adsorbed H_2O at about 1630 cm^{-1}). This is particularly fortunate because this region includes several diagnostic bands of organic pollutants, such as the amide I and II, the stretching ($\nu_{C=O}$) of carboxylic acid groups, the asymmetric and symmetric stretchings (ν_{as} and ν_s) of carboxylates, and the bending vibrations (δ_{as} and δ_s) of amines and ammoniums. Other, less common functional groups may also be observed, such as the P=S of some organophosphorous compounds (Sánchez Camazano and Sánchez Martín, 1980) or the N=O in nitro compounds (Johnston et al., 2001). The near IR portion of the spectrum potentially contains useful information too, in the form of overtone and combination bands, but it is less often explored, an exception being (Tomic et al., 2012).

IR in the transmission mode can be used to verify the presence of adsorbed organic molecules in the solid, and to some extent, to quantify them. Absolute quantification is almost impossible because the intensities of the IR bands depend on their extinction coefficients, which are difficult to estimate a priori. But relative quantification is possible, for example, in DNB adsorption on smectites; Johnston et al. (2001) found that the intensity of the N=O stretching was proportional to the amount of adsorbed DNB, independently derived by HPLC. In the same way, in a study of dibenzo-*p*-dioxin on smectites Rana et al. (2009) found that the intensity of dioxin bands correlated well with the adsorbed amount from HPLC—in addition, the relaxation of selection rules for the dioxin molecule allowed one to garner information on its interaction with Cs^+ ions. Relative quantification is also possible at the solid–solution interface by using a different experimental setting (attenuated total reflectance or ATR mode). In this way, the intensities of the aromatic and COO bending bands were used to follow in situ the kinetics of picloram adsorption on Mt (Marco-Brown et al., 2014).

Most often, the bands of the adsorbing substance can be easily recognised, meaning that adsorption occurs without degradation. There may still be significant band shifts or other modifications; however, that provide useful information on the clay mineral/organic molecule interaction. For example, IR easily distinguishes carboxylic acid from carboxylates (Nowara et al., 1997) and this allows us to know the predominant species of the acid–base couple in the adsorbed state. The protonation state of amines and other nitrogen containing groups is also accessible. For example, N-heterocycles such as benzimidazole and thiabendazole in Mt showed a band at $1624\,cm^{-1}$, indicating that the nitrogen was protonated to the cationic form (other evidence suggested ion exchange with intercalation into the interlayer spaces).

The position of IR bands is sensitive to more subtle effects than protonation/deprotonation. Coordination to cations has a strong influence on the vibrations of the coordinated moieties: coordinative bonds of C=O groups usually give rise to shifts of their stretching vibrations higher than $40–50\,cm^{-1}$, while smaller shifts are assigned to H-bonding effects (Fusi et al., 1988). In the heterocycle nalidixic acid, ring vibrations involving the $C{=}N_1$ are observed at 1475 and $1521\,cm^{-1}$ upon adsorption, they shifted by +13 and $+8\,cm^{-1}$, respectively, which was attributed to coordination to the cation through the N_1 atom (Wu et al., 2013a).

Nasser et al. (1997) interpreted IR band shifts as indicative of H-bonding of both the amine and carbonyl moieties of alachlor, supposedly with water molecules in the interlayer.

The vibrations of the clay mineral lattice itself may be affected by adsorption. Rytwo et al. (2002) observed the δ_{XOH} bending mode and the ν_{OH} stretching in sepiolite and followed them upon the addition of cationic xenobiotics. These authors observed band shifts and broadenings with MB, but not with Diquator Paraquat, and deduced that only the first compound interacted with the silanol groups (through H-bonding).

More precise techniques can give information on the orientation of the adsorbed molecule with respect to the clay mineral matrix. Sheng et al. (2001) prepared self-supported oriented films of DNP containing smectite (SWy-2), and measured the dependence of the in-plane and out-of-plane NO_2 vibrations (ν_{sym} and ω_{oop}, respectively) for incident beam angles of 0 and 45 degrees. The intensity of the first one decreased at 45 degrees, while that of the second one increased, indicating that the molecules were oriented parallel to the clay mineral layers. This technique may be compared with the fluorescence polarisation method previously mentioned. In another study by Rana et al. (2009), polarised ATR revealed that dioxin molecules were tilted, this time by 65 degrees, with respect to the basal planes.

Of course, IR spectroscopy can detect the appearance of new molecules in conditions that lead to adsorbed pesticide degradation. Thus, the slow adsorption of crystal violet molecules on vermiculite led to the appearance of methylene groups that are not present in the original dye (Rytwo et al., 2009), pointing at degradation of the molecule in soft conditions.

7.4.2.5 X-ray photoelectron spectroscopy (XPS)

XPS is a spectroscopic technique able to provide information on the atomic composition of the topmost surface layers of a sample. While its capabilities for getting chemical information remain limited, it can still be useful in conjunction with other techniques, beyond confirming that the stoichiometric ratios in the surface organic matter (e.g. S/C or N/C) are compatible with those of the adsorbed pollutant molecule. The position of the N 1s peak is sensitive to the degree of protonation of the nitrogen atom, and therefore XPS can discriminate between the acidic and basic forms of a couple involving these atoms. It has been applied, for this purpose, to thiabendazole and benzimidazole on Mt (heterocyclic N, Lombardi et al., 2006) and glyphosate on Mt (secondary amine, Khoury et al. 2010). A more complex situation was observed with picloram that has a primary amine as well as a heterocyclic nitrogen (pyridine ring), both susceptible of protonation. Marco-Brown et al. (2015) argued that through a quantitative comparison of the spectra of picloram adsorbed on Mt and a physical mixture, they could discriminate the contributions of the two species, and evidence coordinative bonding of the heterocyclic N, as well as protonation of the amine. When interpreting XPS spectra, however, one must keep in mind that they are obtained in conditions of ultrahigh vacuum, where the speciation may be different from that in solution (for example, glycine is a zwitterion in solution, but a neutral molecule under vacuum).

7.4.2.6 Solid-state nuclear magnetic resonance (NMR)

Solid-state NMR has a huge potential to characterise the interactions of organic molecules with inorganic matrices (Bonhomme et al., 2014). Yet, surprisingly, it is used much less often in the field of organic pollutants adsorption than, for example, in that of clay mineral based nanocomposites. There are, of course, some exceptions. Fraile et al. (2016) recorded the NMR spectra of dexamethasone in solution, in bulk crystals, and adsorbed on Laponite. They observed significant shifts in the signals of some carbons upon adsorption and interpreted them as due to the interaction of the C3 carbonyl and C11 alcohol groups (see numbering in scheme of Appendix) with electron-withdrawing centres on the clay mineral surface, either through coordination to the compensating cations or through H-bonding.

Advanced solid-state NMR techniques were applied to pollutants on clay mineral systems by Maciel's group. A variable temperature study of the deuterium NMR line shapes of D-enriched benzene gave access to rich information on the local motions and mobility of this pollutant in Ca^{2+}-Mt (Xiong and Maciel, 1999). It was concluded that at low temperature (−75°C), benzene is able to coordinate to the interlayer Ca^{2+} ions through its π electrons (π–cation complex). As the temperature increased, more and more benzene molecules gained freedom from this coordinative bonding, and at room temperature, about one half of the benzene molecules are desorbed from the cation and reorient freely.

A previous study of benzene adsorption in Cu^{2+}-Mt had involved solid-state ^{13}C NMR (Hinedi et al., 1993). Only poor quality spectra were obtained due to paramagnetic interaction with Cu^{2+} centres, but signals with chemical shifts in the aliphatic region were obviously present, showing that benzene was chemically unstable in this environment. Indeed, NMR has been applied most often to situations where the adsorbed pollutants underwent some type of degradation (see Section 7.5). Thus, ^{31}P NMR was used to explore the decomposition of chlorpyrifos to inorganic phosphate (Seger and Maciel, 2006); ^{13}C solid-state NMR played a role in identifying the reaction products between two pollutants in Mt, namely CCl_4 and benzene (Tao and Maciel, 1998), or the photodegradation pathways of trichloroethylene in Ca^{2+}-Mt (Tao et al., 1999).

7.4.3 Molecular modelling

Applications of molecular modelling to adsorption phenomena are treated in detail in Chapter 3 of this book. With the huge increase of computing power, molecular modelling studies of pollutant/clay mineral systems are becoming increasingly common. They include density functional theory (DFT), molecular mechanics, and hybrid methods that sometimes allow one to model systems containing many atoms. Occasionally, molecular dynamics (MD) simulations are also included. Because rather weak interactions such van der Waals and H-bonds often play a determining role in adsorption, it is important to use dispersion-corrected computational methods (Belzunces et al., 2017). Molecular modelling should be particularly appropriate to test cooperative adsorption scenarios where the organic molecule establishes several different interactions with the surface, especially if already available experimental data constrain the possibilities—otherwise, the adsorption scenarios might simply be too numerous to test all of them.

Chatterjee et al. (2002) studied the bonding of dioxins to Mt and beidellite in the frame of the hard–soft acid–base (HSAB) theory. They defined a local softness as a reactivity index and used it to predict the interaction of organic molecules with the structural OH in the octahedral layer as a function of layer substitution (the molecules acted as H-bond acceptors through their chlorine atoms). They predicted beidellite to be the best adsorbent, through formation of a H-bond between a chlorine and the octahedral OH. The weakness of this approach is that the clay mineral was represented by a small cluster that did not include edge sites. Dioxin adsorption on smectites with varying compensating cations was studied by MD (Liu et al., 2012), and the experimentally observed adsorption trends were reproduced by considering a complexation of the cations by the oxygen in the heterocycle.

In a study of triazine herbicides in Mt and saponite, Aggarwal et al. (2006) found a favourable interaction with the aromatic cycles interacting with both clay mineral layers, and the N and Cl atoms of the organic molecule simultaneously coordinated to K^+. At the same time, the alkyl groups were well positioned to interact with 'hydrophobic microsites' on smectite surfaces (the siloxane planes). Bonding of the pollutant molecule to the clay mineral was thus a highly cooperative phenomenon. This study combined MD with XRD data indicating that dioxin molecules were intercalated in the interlayer space, and adsorption isotherms that showed inter alia a much higher loading on Cs^+-exchanged clay minerals. The latter observation is consistent with a ligand-exchange mechanism because water is easier to displace from bulky, chaotropic cations.

Tunega et al. (2007) studied the adsorption of 2,4-D on Ca^2-Mt using MD. These authors compared adsorption on a dry external surface, on a hydrated one, and upon intercalation. Coordinative binding of the COO^- moiety to the Ca^{2+} compensating cation is favourable in the first two cases: it is bidentate in the absence of water, monodentate in its presence. This study paid attention to the H-bonding of water, with the adsorbed organic molecule, the clay mineral surface groups, and among H_2O molecules; multiple H-bonding (H-bonding lattice) made an important energetic contribution to the adsorption. However, Ca^{2+} could not be considered as forming 'cation bridges' with the surface, as previously proposed. Meleshyn and Tunega (2011), using a periodic model of the basal plane, found that phenanthrene as a model PAH could favourably interact with a compensating Na^+ cation through cation–π complexation; however, when present, water outcompeted phenanthrene, suggesting that PAH can only be significantly retained by clay minerals in dry soils. Rather different conclusions were drawn by Hou et al. (2015) in a study with periodic DFT including explicit water molecules: for aromatics (toluene and anisole) adsorption on Mt, cation–π interactions played an important role, in subtle cooperation with H-bonds to the siloxanes, of the organic molecule itself, and also of the competing water molecules. Overall, and somewhat counter intuitively, the presence of water reinforced the adsorption of the aromatic molecules.

Shapley et al. (2013) using DFT and force fields, tried to distinguish between 'inner sphere' and 'outer sphere' complexation of PCDD molecules on pyrophyllite and Mt, that is, direct coordination or coordination through a water bridge. The free energy of adsorption did not depend much on the choice of the compensating cations. Diffusion coefficients were also evaluated: on pyrophyllite, PCDD was predicted to be more favourably adsorbed, but also more mobile on the basal planes than on Mt, corresponding to a picture where it efficiently displaces water from the first adsorption layer, while on Mt it is presumably 'pinned down' by localised charges. It must be underlined that this study included a modelling of the edge sites and not only of the siloxane planes.

Palace Carvalho et al. (2014) studied the interaction of benzodiazepines on Mg^{2+}-vermiculite (periodic DFT, without edge sites, limited number of explicit water molecules). These authors found that the energetically most favourable adsorption modes were through direct coordination to a nitrogen atom from the heterocyclic ring, and through the formation of a H-bond with a water molecule that was itself bound to the surface (this is different from what was called 'water bridges' previously).

Liu et al. (2015a) noticed that traditional MD methods were too expensive computationally to accurately model transport processes in clay minerals. These authors applied enhanced MD methods that sample the conformational space faster for the adsorption of PCB (polychlorinated biphenyls) and PAH on Mt. They found that PCB were most strongly adsorbed in the interlayer spaces, if their aromatic rings could interact with siloxane planes on both sides at once. This was more difficult with cations such as Na^+ that induced swelling and easier with those, such as K^+, that were not strongly hydrated. Thus, the experimental finding was similar to that of Aggarwal et al. (2006), but received a different explanation. The authors also noted that pyrophyllite, without any interlayer cations, probably provided the best conditions to guarantee an interlayer hydrophobic environment.

In a recent study, Belzunces et al. (2017) explored different interaction modes of Atrazine on Ca^{2+}-Mt: coordination of Cl to the Ca^{2+} cation provided the strongest adsorption, and on a pyrophyllite surface lacking these cations atrazine molecules were only weakly adsorbed.

7.5 Adsorption: A gateway to reactivity

As already mentioned, molecules adsorbed on clay minerals may be subject to chemical reactions that do not occur, or occur only slowly, in water solution.

Clay minerals are indeed good catalysts (Kloprogge, 1998; Vaccari, 1998), and they have been used in the industry for a long time, especially for acid-catalysed reactions; some clay minerals may also act as redox catalysts, especially if they contain Fe. The first step in heterogeneously catalysed mechanisms is usually the adsorption of one or several reagents on the surface, and this is why adsorption cannot be separated from reactivity.

Under ambient conditions, Fe^{3+}-Mt catalysed the in situ formation of OCDD (octachloro-*p*-dibenzodioxin) from pentachlorophenol (Gu et al., 2008):

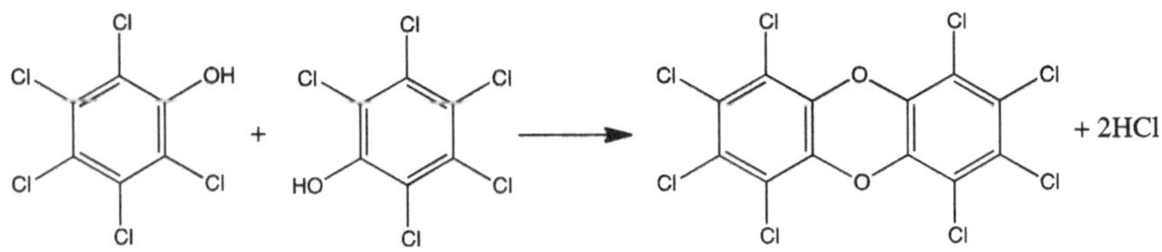

The joint IR and DFT study suggested that the mechanism involved reversible, water activity-dependent coordination of the phenol to Fe^{3+}, followed by one-electron transfer forming a phenol radical. While this transformation is detrimental because it forms a very toxic compound, similar chemistry can lead to beneficial effects. DDT is a persistent pollutant whose degradation is very difficult. It has long been known that the interaction with bentonite or vermiculite favours its activation by dehydrohalogenation (loss of HCl) (López-González and Valenzuela-Calahorro, 1970):

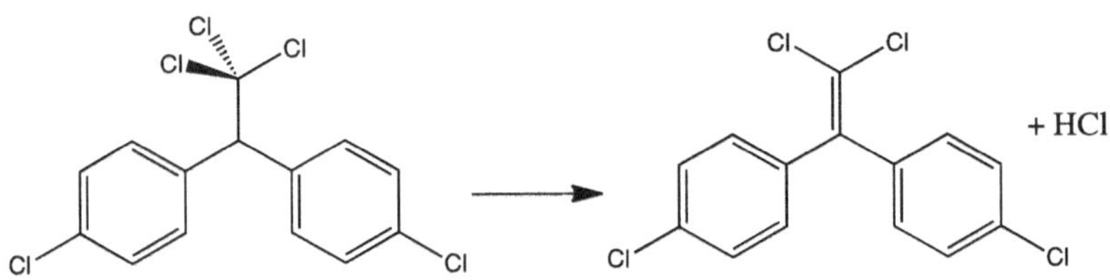

Benzene in Cu^{2+}-Mt is very reactive, giving rise to polymerisation products upon drying, but also phenols and other oxidation products, and dearomatisation. The NMR study by Hinedi et al. (1993) has already been mentioned; the Cu^{2+} ions were essential for benzene degradation, and the presence of water inhibited their activity, no doubt by preventing access to the coordination sphere (Joseph-Ezra et al., 2014). Electron transfer to Cu^{2+} was hypothesised to produce a radical cation susceptible to polymerisation and oxidation; phenanthrene was more reactive than pyrene.

Electron transfer steps such as the ones postulated herein may initiate other catalytic reactions. The endocrine disruptor 17β-estradiol underwent oxidative polymerisation following a one-electron transfer to Fe^{3+} in an ion-exchanged Mt (Qin et al., 2015); the Fe^{2+} ions formed in the initial step were then reoxidised by atmospheric O_2, providing a closed reaction sequence.

The final oxidant could also be an intentionally introduced chemical such as hydrogen peroxide (H_2O_2). Thus, Fe-pillared clays (Fe-PILC) were active for methyl orange discolouration by catalytic wet peroxide oxidation (CWPO) (Muñoz et al., 2017).

This can be considered as a Fenton reaction, where Fe^{x+} catalyses oxidation of organic matter by H_2O_2. Ye and Lemley (2008) compared the Fenton degradation of three pesticides (anionic, neutral, and cationic) on SWy-2 Mt. Paraquat was protected by the smectite against oxidation by the hydroxyl radicals because it was more strongly adsorbed than the other molecules. There are many reports of the use of clay minerals or modified clay minerals as Fenton catalysts (Garrido-Ramirez et al., 2010) and they cannot all be mentioned here.

Other transformations more likely involve acid catalysis. In a typical example (Sánchez Camazano and Sánchez Martín, 1991), on Ca^{2+}-hectorite, the organophosphate azinphosmethyl was hydrolysed along

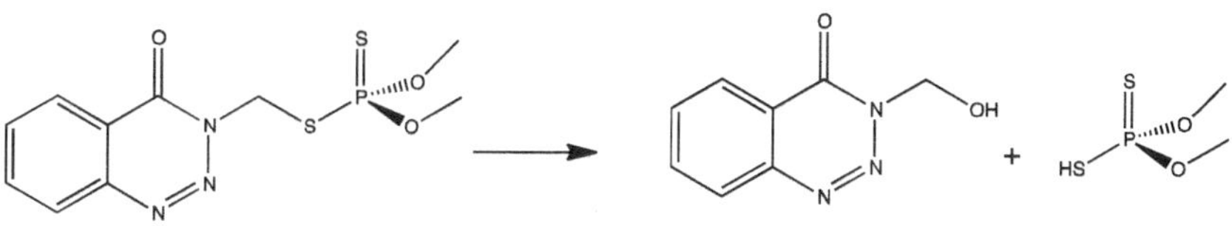

The molecule was activated for nucleophilic attack by coordination to the interlayer cations (Lewis acid catalysis); related smectites with higher layer charges were not efficient for hydrolysis, presumably due to interlayer crowding. Wei et al. (2001) observed the hydrolysis of several carbamates such as Carbosulfan, where the addition of Mt as a catalyst increased the reaction rate more than 200-fold; illite, vermiculite, and beidellite were less efficient. The catalysis was assigned either to Brönsted or to Lewis acidity (resp. strong H-bonding or metal ion coordination of the molecule). It is noteworthy that an analogue of carbamates with a somewhat different structure was actually protected against hydrolysis by Mt rather than activated, underlining the possible complexity of these effects.

In Al-pillared Mt, the tertiary amide group of Alachlor was slowly hydrolysed (Nasser et al., 1997), giving a secondary amide (and not an amine). This degradation pathway is also confirmed by the work of Bosetto et al. (1993).

Sometimes two different pollutants can react together in the presence of clay minerals. Tao and Maciel (1998) studied a mixture of benzene and CCl_4 on Ca^{2+}- and Zn^{2+}-Mt and kaolinite. These authors found that the metal cations could activate CCl_4 (and probably other organochlorines) by abstraction of a Cl^- to give a carbocation, susceptible of nucleophilic attack on the aromatic; this process could be repeated with other chlorines, yielding products such as dibenzophenone and triphenylmethanol. Here, the crucial factor in the catalytic activity is the Lewis acidity of the metal cations: thus, dechlorination can be realised both by redox (cf. supra) and acidic catalysis.

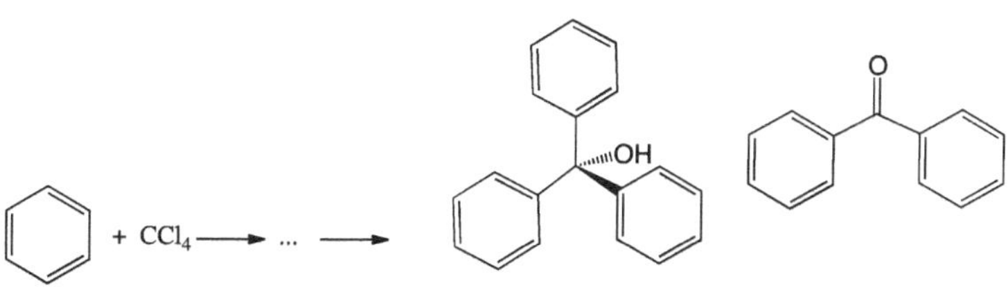

Some pollutants are slowly degraded by photochemical reactions, and clay minerals may also be active as photocatalysts. The photodegradation of trichloroethylene was observed on Ca^{2+}-Mt by Tao et al. (1999), giving 40% mineralisation in 16 days—but forming toxic phosgene ($COCl_2$, used as a chemical weapon) in addition to CO_2! This involved a radical mechanism of Cl abstraction, with dichloroacetic acid as an intermediate. Photodegradation of bisphenol A was observed in the presence of Mt (Liu et al., 2008), as well as of paracetamol in the presence of Mt (Liu et al., 2010), and chloramphenicol and sulfamethoxazole on Mt, kaolinite and rectorite (Liu et al., 2011b). In this case the clay mineral containing systems underwent 40%–70% mineralisation in 3 h, as opposed to hardly any degradation in the absence of clay minerals. Upon photolysis of carbofuran on Moroccan clay minerals, Mountacer et al. (2014) observed both a hydroxylation of the aromatic ring, and a hydrolysis of the carbamate C—O bond.

On the other hand, photoprotection was observed in some systems: Ambrogi et al. (2014) recorded a slowing down of promethazine degradation by a factor of five as the drug was deposited on Mt. When it exists, the photodegradation effect could be used to enhance the stability of a drug in controlled delivery systems. Several such examples are listed in Yuan et al. (2013).

7.6 Conclusion

From this brief summary, and even more from the list of pollutant molecules provided in the next section, it can be seen that a lot of information is available on individual clay mineral–pollutant systems that have been studied for more than 50 years. This information is somewhat disparate, however. Many studies give quantitative data on adsorbed amounts as a function of environmental conditions, and they constitute a valuable database for practical applications, from environmental remediation to controlled release. But in order to understand the adsorption phenomenon at a fundamental level—and thus hopefully to be able one day to control it, and to predict it for new molecules—one has to combine macroscopic and molecular-level information.

At the macroscopic level, one may propose a few guidelines for adsorption studies—the measurement of adsorbed amount of an organic molecule should be combined with estimates of the amount of cations released, and of the pH drift, to determine, respectively, if ion exchange occurs, and if the acid–base speciation is modified.

At the molecular level, vibrational spectroscopy (IR) has long been available and allows us to observe changes of the functional groups brought about by their interaction with the surface through coordinative or H-bonding. XPS can also provide such information to a more limited degree; so can UV–visible absorption and fluorescence, but they are not applicable to all molecules. Solid-state NMR has a great potential, but has been little applied so far, and it is hoped that more NMR studies of pollutants adsorption will be undertaken in the future.

Spectroscopic techniques are often 'short-sighted' in that they shed light only on localised interactions. It is getting rather clear, though, that pollutant adsorption on clay minerals is often a cooperative phenomenon, because many molecules bear several functional groups with different reactivities. The clay mineral surface on its side is also a heterogeneous system able to induce different types of bonding. This is why it is difficult to obtain a complete picture of the adsorption, and especially of its energetics. In this respect, the rapid development of molecular modelling techniques is likely to advance our understanding of adsorption phenomena, as many possible conformations of the adsorbed molecule can be explored. For best results, though, theoretical modelling has to be combined with experimental investigations, and such integrated studies are still rare.

As a final note, one should be wary of overgeneralising the results obtained in pollutant–clay mineral interaction studies. Due to the complexity of the adsorption process, each system is a particular case and should be studied in ists own respect, as witnessed by the opposing effects of clay minerals observed in degradation/protection studies.

Appendix: A list of pollutant structures

The following pages list the main organic pollutants whose adsorption has been studied on raw clay minerals, inorganic cation exchanged clay minerals, and some pillared clays. Substances are listed in alphabetical order; when several different substances belong to a generally recognised class or chemical family, this is indicated in bold, while the structures are compared under the heading corresponding to the class.

2,4-D: 2,4-Dichlorophenoxyacetic acid, selective herbicide, **phenoxyalkanoic** family (Sannino et al., 1997; Clausen et al., 2001; Tunega et al., 2007; Li et al., 2009; Bakhtiary et al., 2013; Werner et al., 2013)

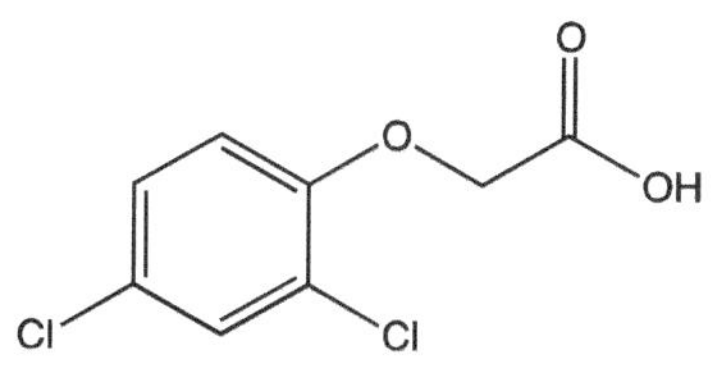

Other phenoxyalkanoic herbicides include 2,4-DB, 2,4-DP, 2,4,5-T, and MCBA (Akçay and Yurdakoç, 2000).

2,4,6-TCA: Trichloroaniline, used as a fungicide (Gianotti et al., 2008) or sometimes trichloroanisole

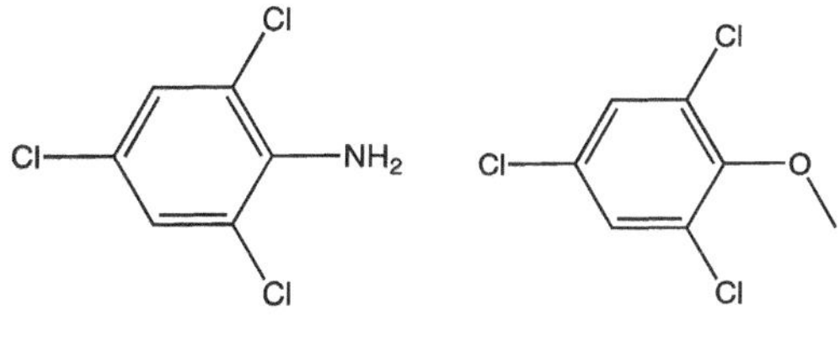

Acetaminophen: an **acetanilide**, trade name paracetamol, an analgesic pharmaceutical (Liu et al., 2010; Thiebault et al., 2016)

Acetanilides

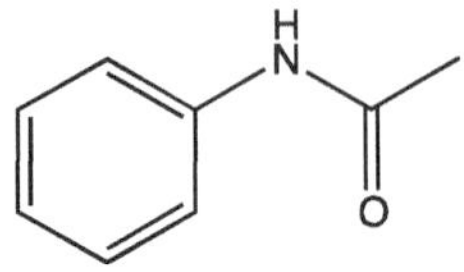

Examples

Acetaminophen

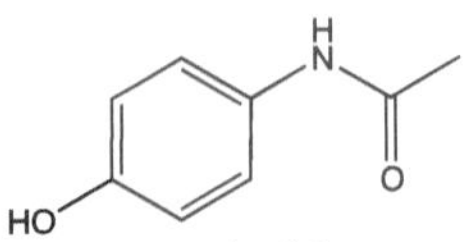

Acetochlor

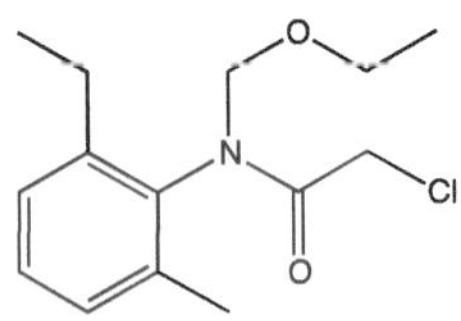

Alachlor

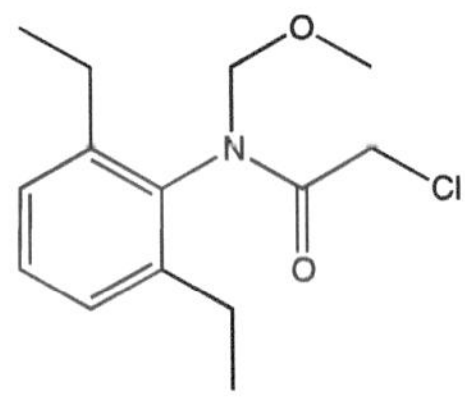

Metolachlor

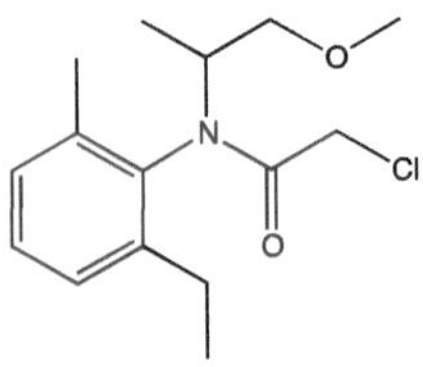

Propachlor

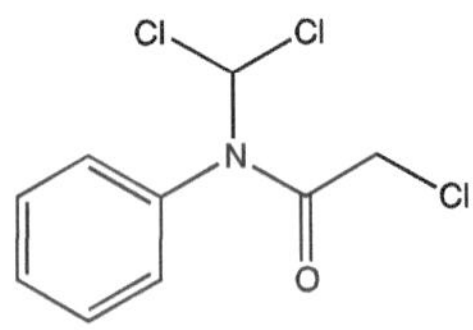

Propanil

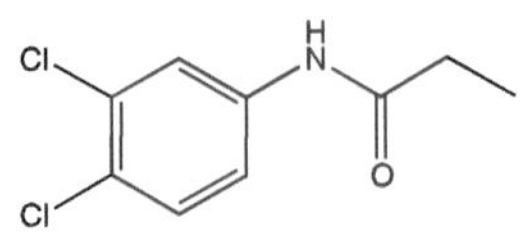

Acetochlor: an **acetanilide**, herbicide, third most frequent herbicide in natural waters (Li et al., 2009; Tomic et al., 2016)

Acid blue 25: a dye (Özcan and Özcan, 2004)

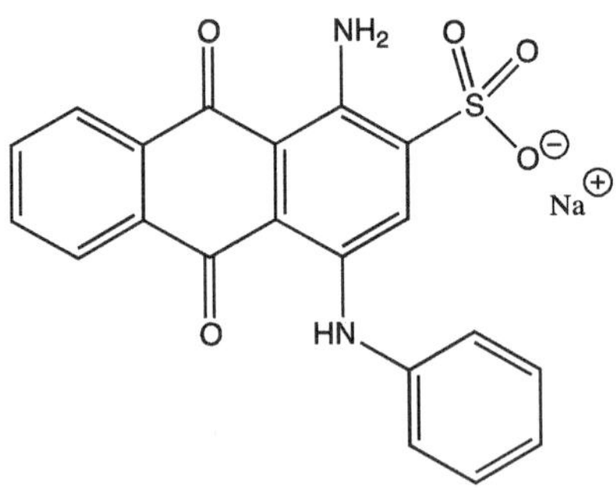

Acriflavine: an orange dye and antiseptic (Mishael et al., 1999)

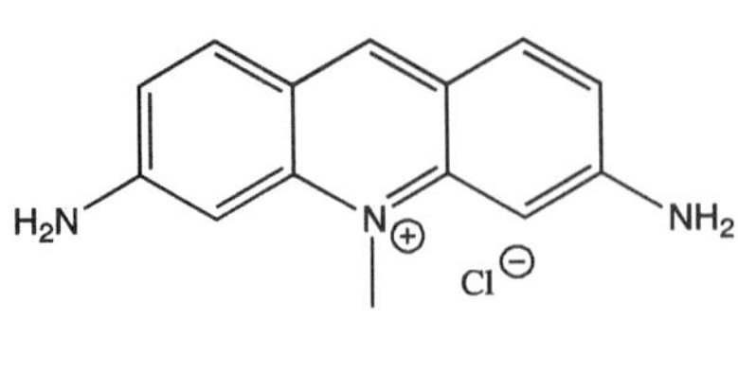

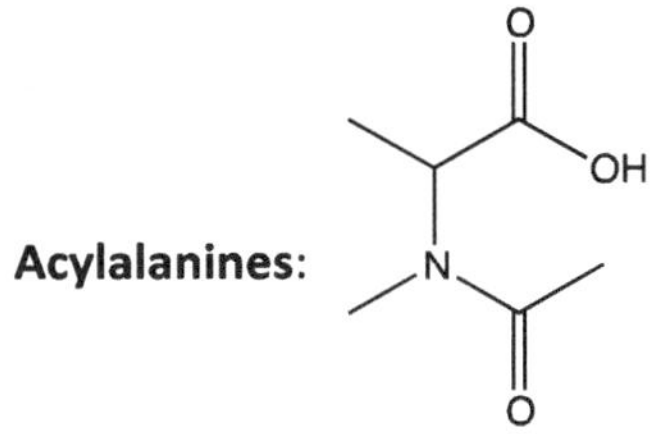

<u>Example</u>:

Metalaxyl

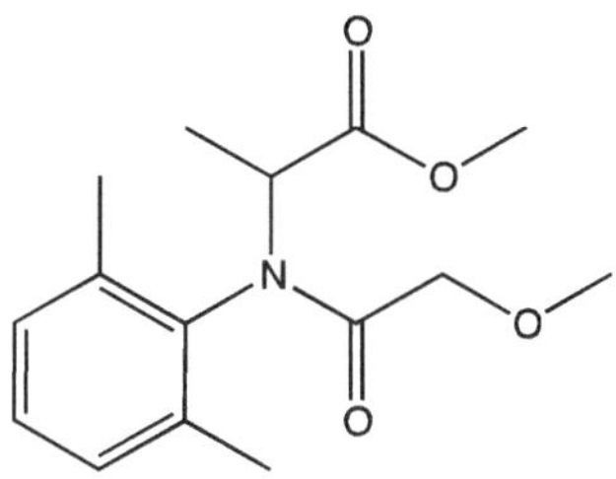

(compare with acetochlor, where a different functional group is underlined)

Alachlor: a chloro**acetanilide**, herbicide (banned in Europe, second most used in the United States) (Bosetto et al., 1993; Nasser et al., 1997; Torrents and Jayasundera, 1997; Gerstl et al., 1998; Xu et al., 2001; Li et al., 2006; Tejada et al., 2017)

Aldicarb: a **carbamate** insecticide, active substance in Temik (Supak et al., 1977; Wei et al., 2001)

Alprazolam: a **benzodiazepine**, psychoactive pharmaceutical, tradename Xanax (Palace Carvalho et al., 2014)

Ametryn, a **triazine**, herbicide (Pissinati de Rezende et al., 2013)

Aminopyralid, **picolinic acid** class, herbicide, tradenames Milestone, Banish, etc. (Fast et al., 2010)

Amitrole, a **triazole**, nonselective cropherbicide (Gu et al., 2015; Tan et al., 2015)

AOPP, see fluazifop

Astrazonred, a dye (Fil et al., 2014; Stawinski et al., 2017)

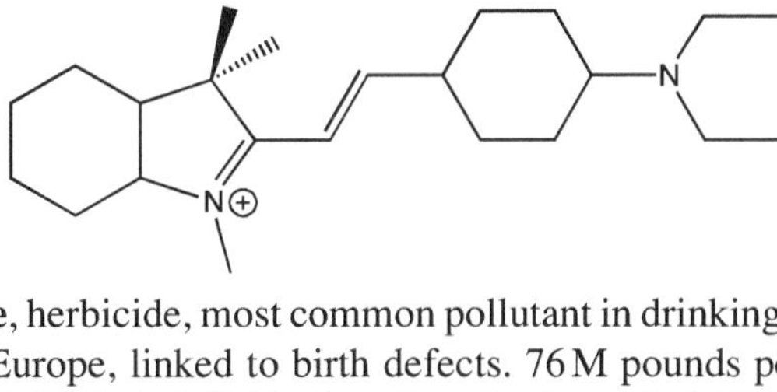

Atrazine, a **triazine**, herbicide, most common pollutant in drinking water in the United States, banned in Europe, linked to birth defects. 76 M pounds per year

(Barriuso et al., 1994; Laird, 1996; González-Pradas et al., 1999; Konstantinou et al., 2000; Clausen et al., 2001; Sheng et al., 2001; Xu et al., 2001; Aggarwal et al., 2006; Li et al., 2006; Pissinati de Rezende et al., 2013; Park et al., 2014; Belzunces et al., 2017)

Azinphos-methyl, an organophosphorous insecticide (Sánchez Camazano and Sánchez Martín, 1991)

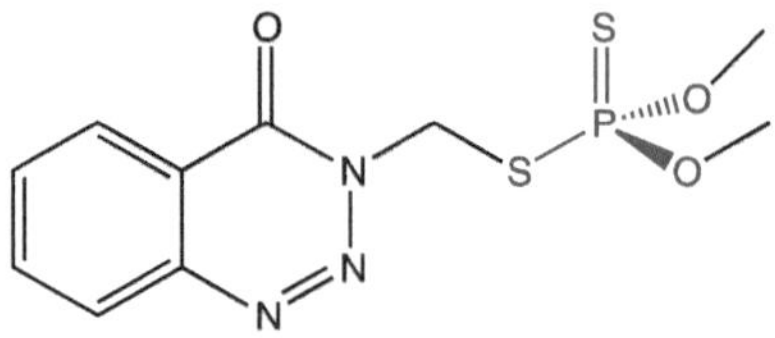

Bentazone, a **triazine** (Clausen et al., 2001)

Benzimidazole (Lombardi et al., 2006)

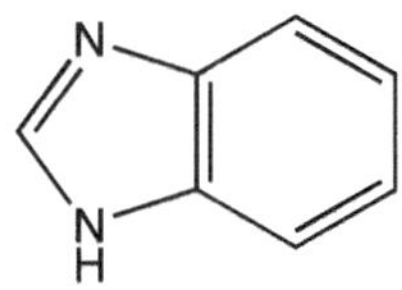

Benzodiazepins: a class of psychoactive pharmaceuticals including Valium

Core structure

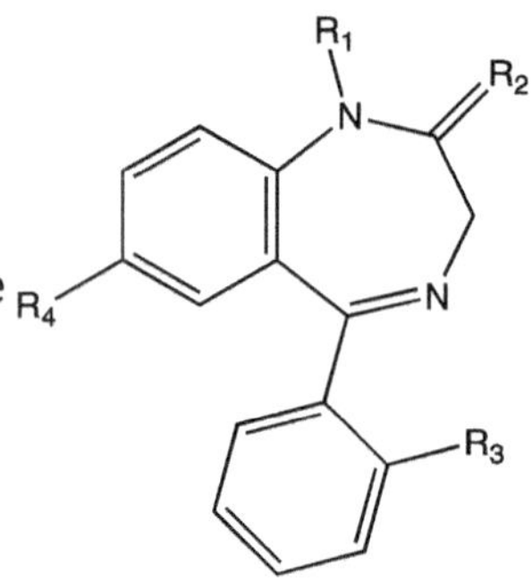

Examples:

Diazepam

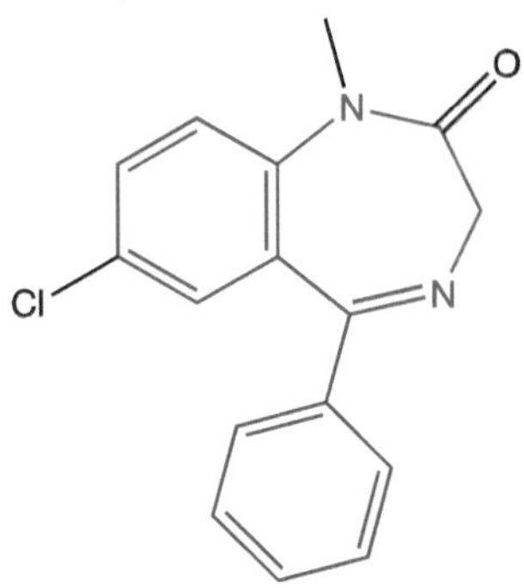

Alprazolam

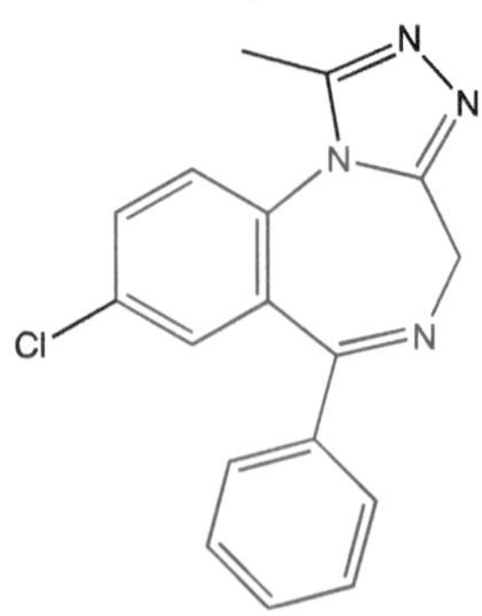

Bisphenol A: used to manufacture epoxide resin. Endocrine-disrupting pollutant (Atun and Sismanoglu, 1996; Shareef et al., 2006; Liu et al., 2008; Styszko et al., 2015; Berhane et al., 2016)

Carbamates, a class of insecticides inhibiting acetylcholinesterase,

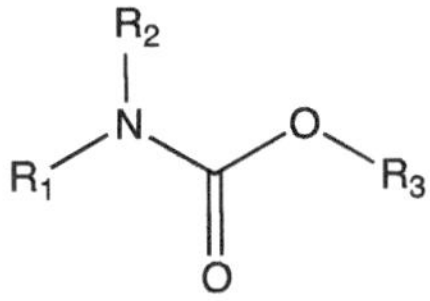

Examples:

Aldicarb

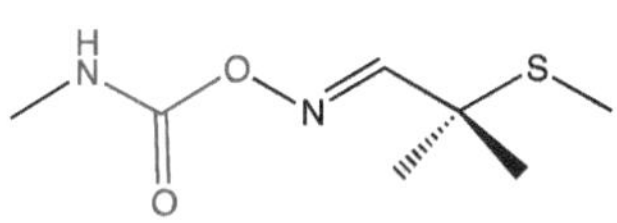

Carbaryl

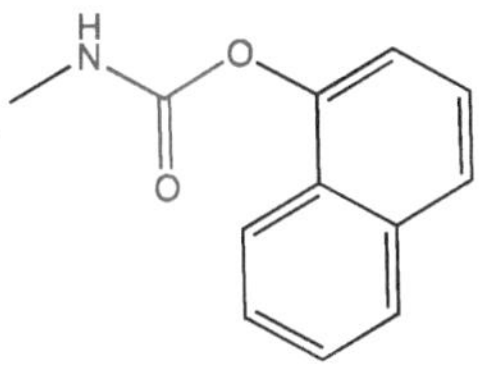

Carbofuran

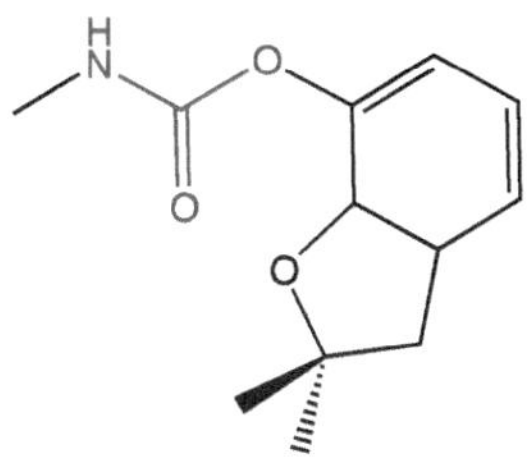

Carbosulfan

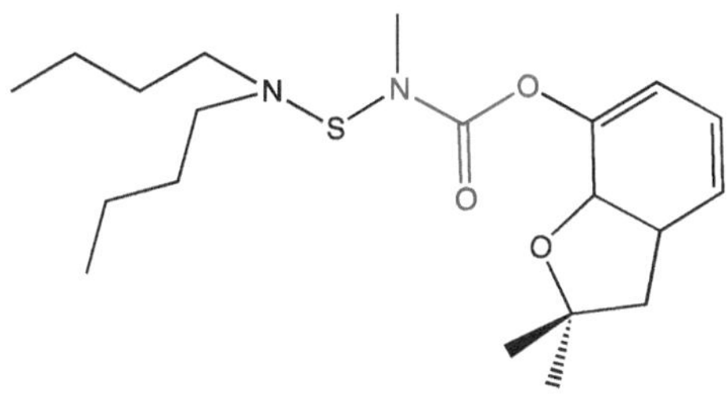

Carbamazepin, trade name Tegretol, generic pharmaceutical (psychoactive) (Drillia et al., 2005; Cabrera-Lafaurie et al., 2015; Thiebault et al., 2016)

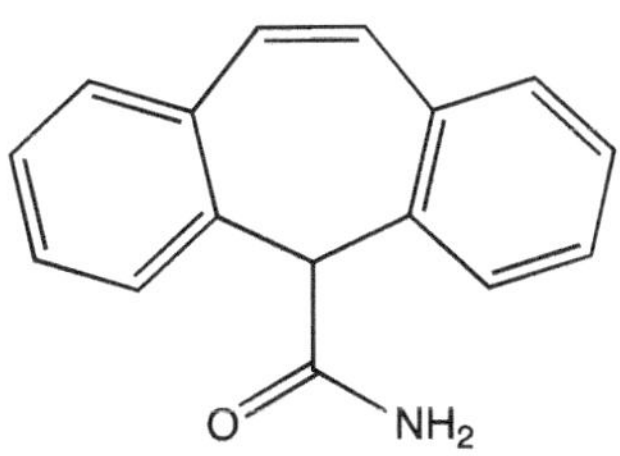

Carbaryl, a **carbamate**, trade name Sevin (Sheng et al., 2001; Ye and Lemley, 2008)

Carbendazim, a benzimidazole fungicide (Dios Cancela et al., 1992)

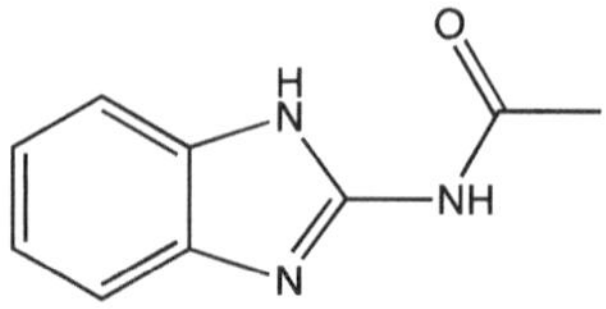

Carbofuran: a **carbamate**, systemic insecticide, banned in Europe and Canada (Wei et al., 2001; Mountacer et al., 2014)

Carbosulfan: a **carbamate**, systemic insecticide, banned in Europe (Wei et al., 2001)

Chloramphenicol: an antibiotic (Liu et al., 2011b)

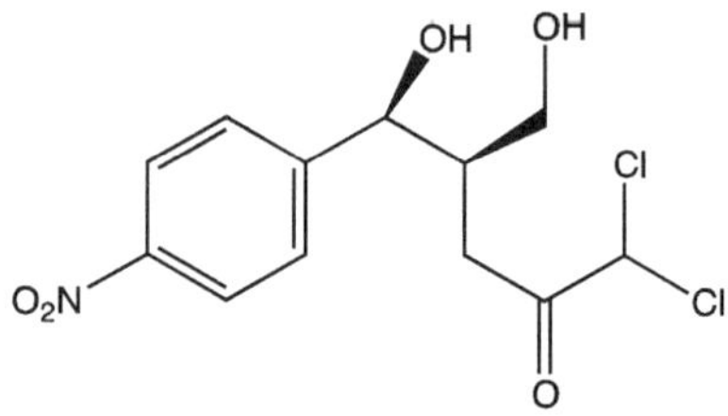

Chloroanilin: important chemical intermediate, building block of several pesticides, including chlorhexidine

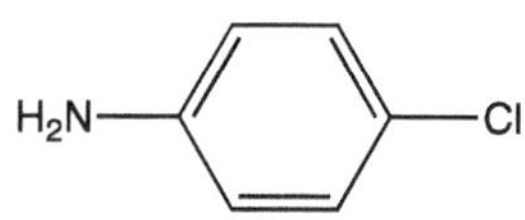

4-chloro (Angioi et al., 2005) isomer; 3-chloro, 4-chloro, 3,4-dichloro (Szczepanik et al., 2014); also see TCA)

Chlorpyrifos, an organophosphate pesticide (Seger and Maciel, 2006)

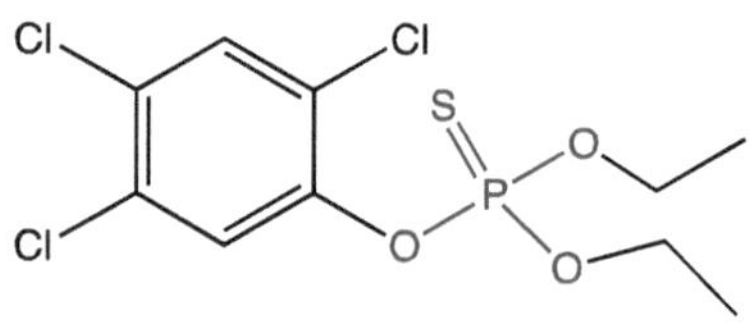

Ciprofloxacin: a fluoro**quinolone** (fluoroquinolone carboxylic acid) (Nowara et al., 1997; Wu et al., 2010; Wang et al., 2011; Li et al., 2011b; Wu et al., 2013b; Berhane et al., 2016; dos Santos et al., 2017)

Clofibric acid, a herbicide, auxine antagonist (and metabolite of the cholesterol-lowering drug clofibrate) (Cabrera-Lafaurie et al., 2015)

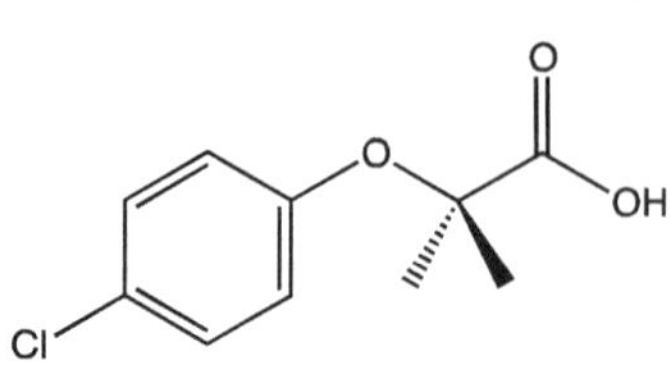

Congo Red, a dye (and pH indicator), carcinogenic (Vimonses et al., 2009)

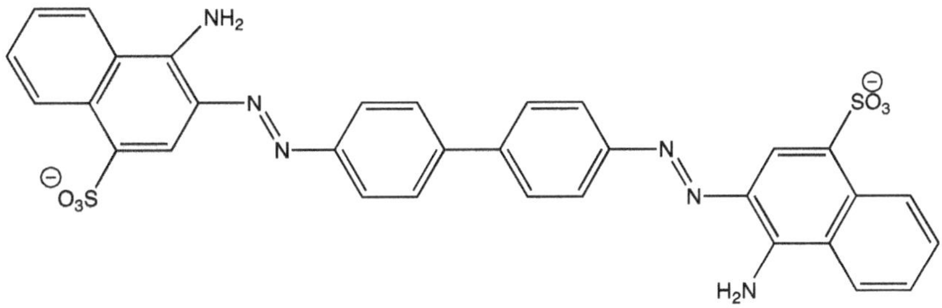

Crystal violet or *Methyl violet*, a dye, also general antiseptic (Mishael et al., 1999; Rytwo et al., 2009; Zhu et al., 2014; Yang et al., 2018)

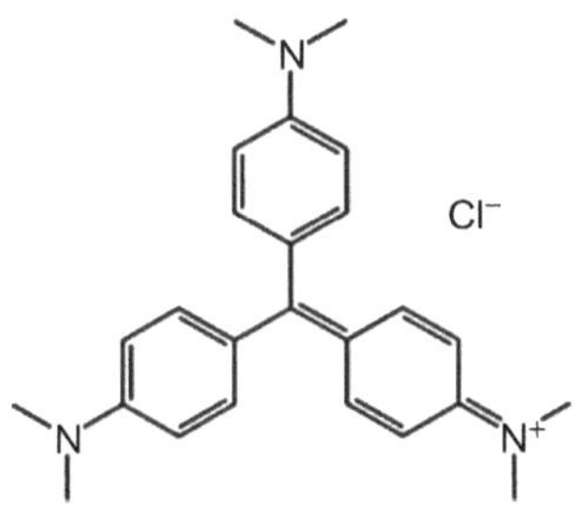

Cypermethrine: an insecticide of the pyrethroid class (synthetic molecules related to the naturally occurring pyrethrines); (Oudou and Hansen, 2002)

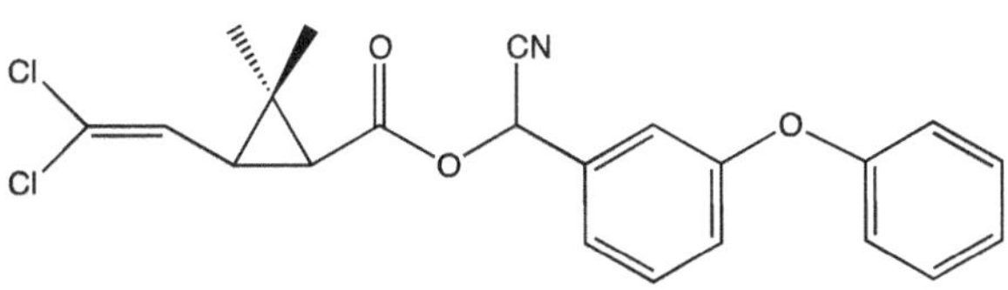

DCBN or *Dichlobenil* (2,6-dichlorobenzonitrile) a herbicide interfering with cellulose synthesis (Sheng et al., 2001; Li et al., 2006)

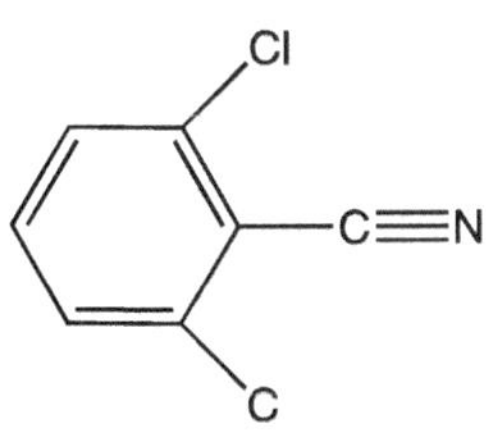

DCMU, tradename Diuron, a phenyl**urea**, herbicide (photosynthesis inhibitor) (Sheng et al., 2001; Li et al., 2006; Maqueda et al., 2013)

DDT: dichlorodiphenyltrichloroethane, an organochlorine pesticide (López-González and Valenzuela-Calahorro, 1970; Dai et al., 2008)

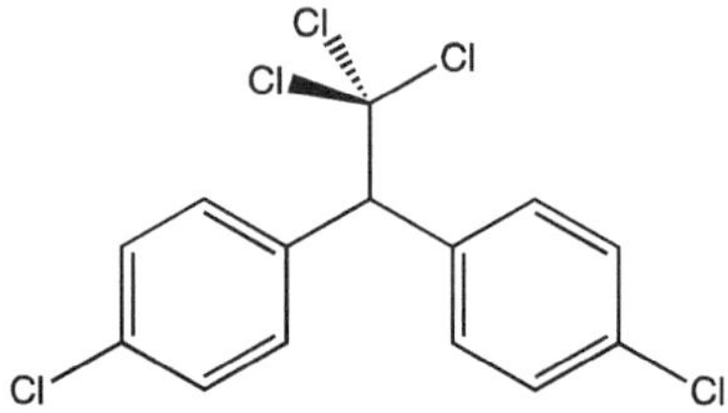

Dexamethasone, a synthetic hormone, cortico**steroid** family (Fraile et al., 2016)

Diazepam, a **benzodiazepin**e (Palace Carvalho et al., 2014)

Dibenzofurans: The basis compound is by itself harmless, but polychlorinated dibenzofurans (waste products of chlorinated compounds pyrolysis) are toxic (Chatterjee et al., 2002)

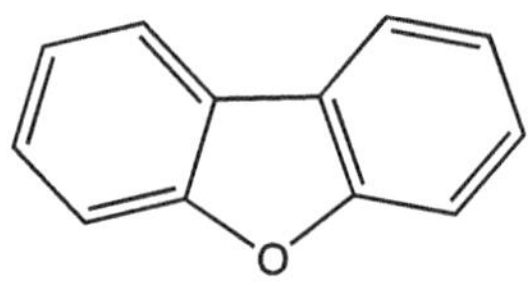

Diclofenac, pharmaceutical, anti-inflammatory, anionic pollutant, trade name Voltaren (Drillia et al., 2005; Styszko et al., 2015; Thiebault et al., 2016)

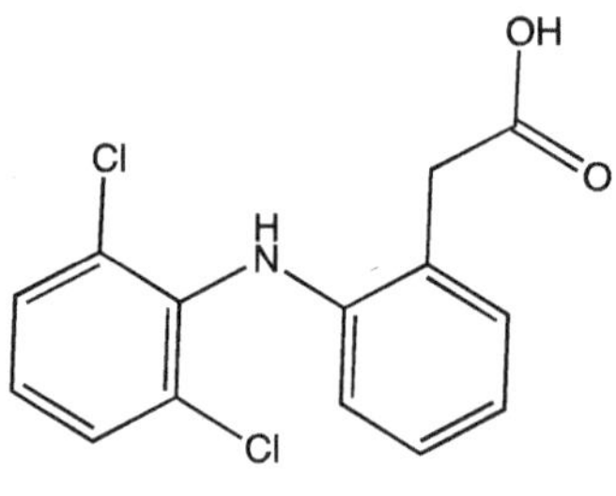

Dioxins

Dioxins 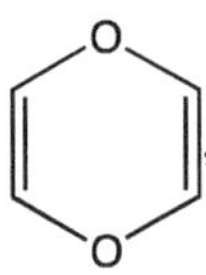

, dibenzodioxins 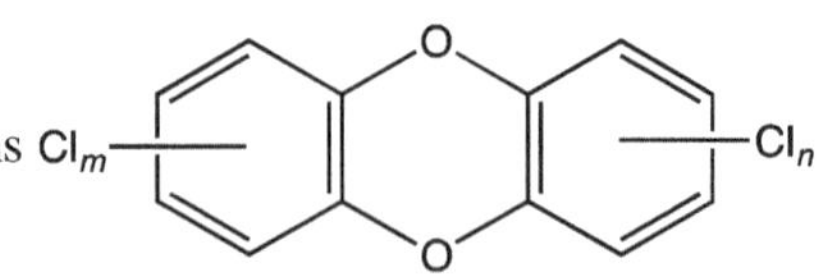

are industrial byproducts, from the combustion of plastics include TCDD, OCDD (tetra- and octachloro), and corresponding Bromo compounds (TBDD) (Chatterjee et al., 2002).

Diphenhydramine: an antihistaminic drug (Li et al., 2011a)

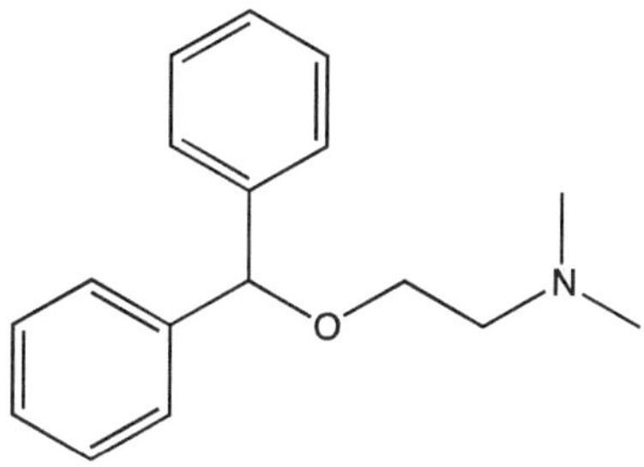

Diquat, bipyridilium halide, nonselective herbicide, trade names Aquacid, Dectrone, etc. (Rytwo et al., 2002)

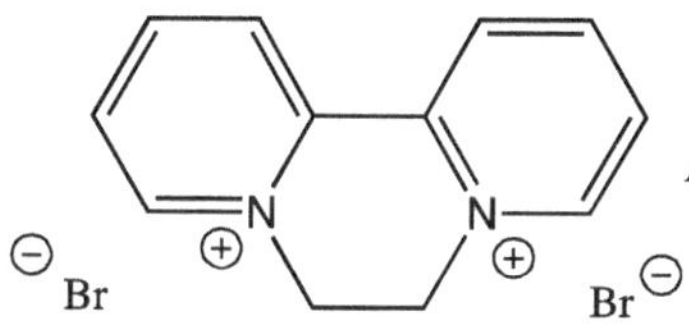

Also see paraquat

DNB, dinitrobenzene (Johnston et al., 2001)

DNOC, dinitroorthocresol, a herbicide (Johnston et al., 2001; Sheng et al., 2001, 2002)

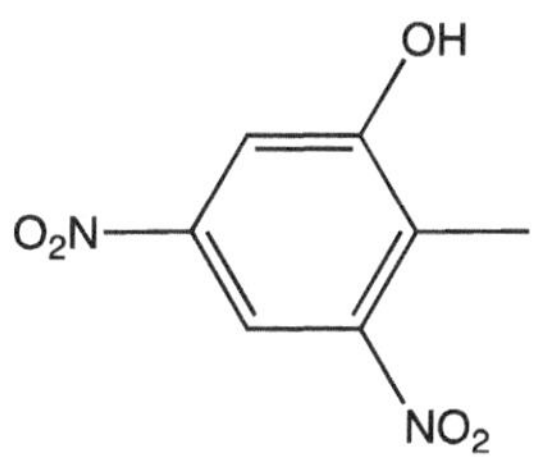

DNP, dinitrophenol (Sheng et al., 2002)

The sec-butyl derivative of DNP has the trade name Dinoseb (Johnston et al., 2001)

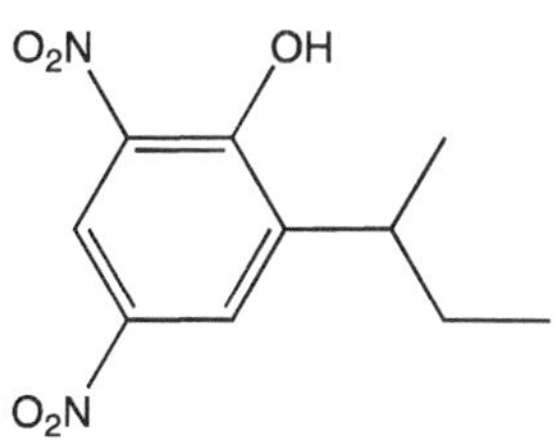

Donepezil, a drug used in the treatment of Alzheimer disease (Park et al., 2008)

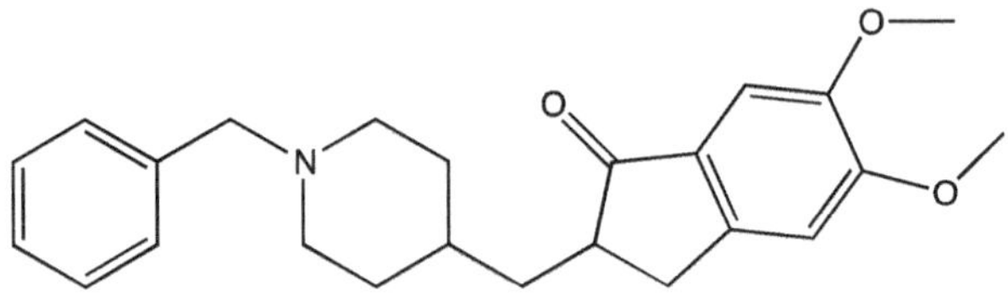

Doxepin: a **tricyclic**, antidepressant (Thiebault et al., 2015, 2016)

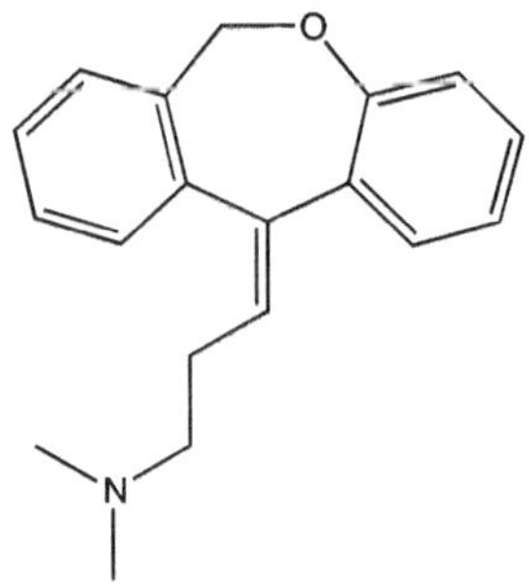

Enrofloxacin, a fluoro**quinolone** (Nowara et al., 1997; Sturini et al., 2015)

Fluazifop-p-butyl: a propionate, selective herbicide (Fusi et al., 1988)

NB: more specifically an AOPP, aryloxyphenoxy propionate

Flumioxazin: N-phenylphthalimide herbicide (Ferrell et al., 2005)

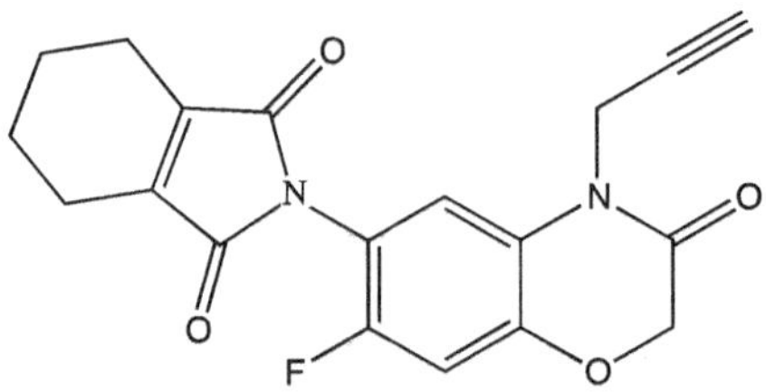

Gemfibrozil, pharmaceutical brand name Lopid: a fibrate (amphiphilic carboxylic acid), used to lower lipid levels

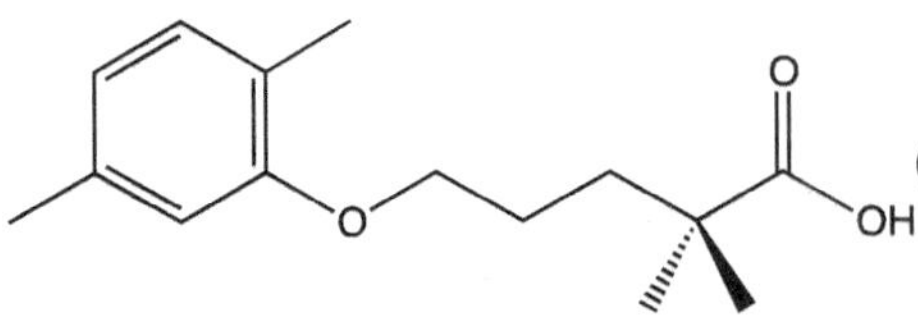

(Thiebault et al., 2016; Dordio et al., 2017)

Glyphosate: an organophosphorous herbicide, trade name Roundup (Gimsing and Borggaard, 2002; Damonte et al., 2007; Pessagno et al., 2008; Khoury et al., 2010)

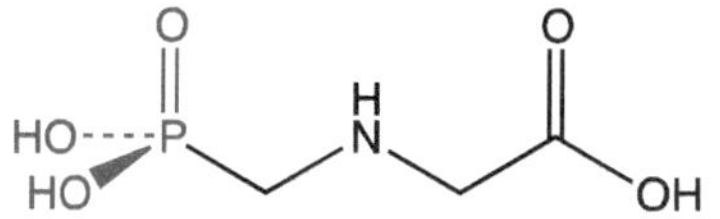

Ibuprofen, a **propionic acid** class pharmaceutical, antiinflammatory drug, trade names Brufen, Alvin, Motril (Behera et al., 2012; Styszko et al., 2015; Thiebault et al., 2016)

Imazaquin: an **imidazolinone** herbicide (Undabeytia et al., 2013; Park et al., 2014)

Imidacloprid: a neonicotinoid insecticide. Little toxicity for mammals (Cox et al., 1998; González-Pradas et al., 1999; Cox et al., 2001; Liu et al., 2002)

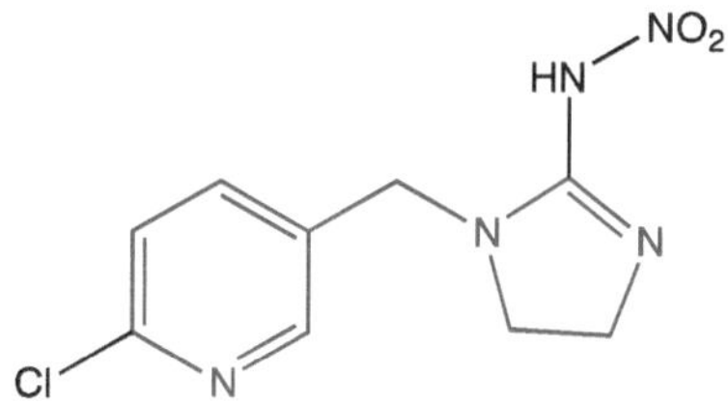

Imidazolinone: a heterocyclic structure

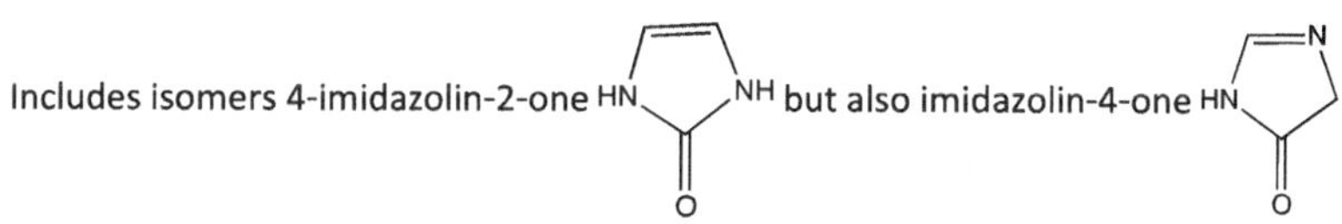

<u>Example</u>

Imazaquin

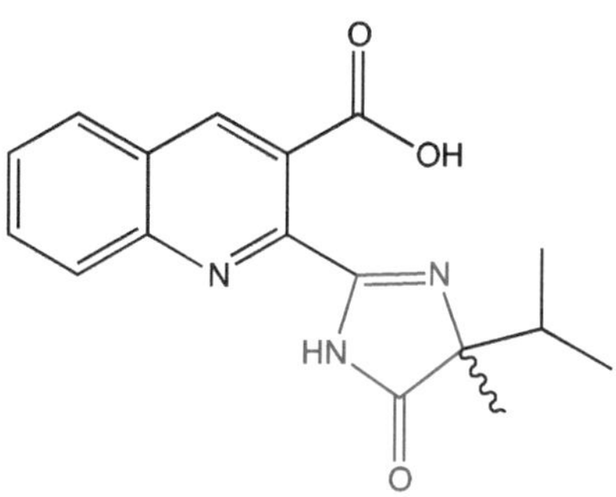

Isoproturon: a dimethyl**urea** (González-Pradas et al., 1999; Clausen et al., 2001)

Ketoprofen: **propionic acid** class antiinflammatory drug (Styszko et al., 2015; Thiebault et al., 2016)

Levofloxacin, a fluro**quinolone** (Nowara et al., 1997; Liu et al., 2015b)

Linuron, a **phenylurea** class herbicide, inhibitor of photosystem II (Torrents and Jayasundera, 1997)

Malachite green: a cationic dye used for food colouring (and pH indicator), also biocide in aquaculture, antihelminthic, medical disinfectant. Reported to be a carcinogenic and mutagenic compound (Tahir and Rau, 2006; Kiani et al., 2011; Elmoubarki et al., 2015)

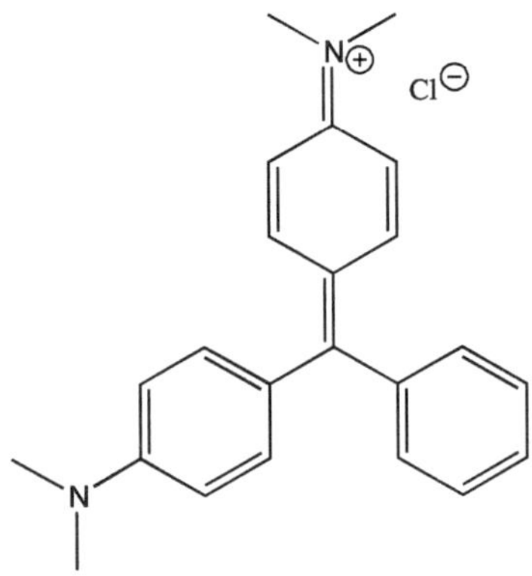

compare with methyl green and crystal violet

Mecoprop, a general purpose herbicide considered as anionic (Clausen et al., 2001; Ye and Lemley, 2008)

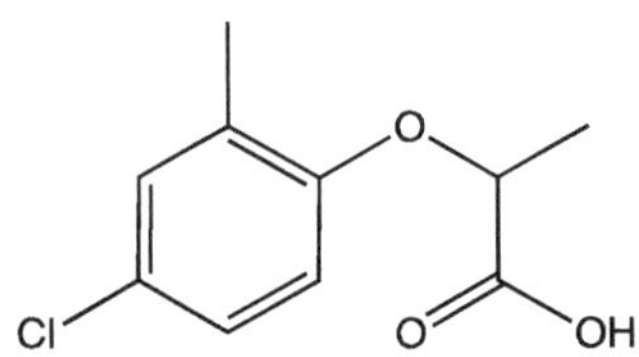

Mefenamic acid: a pharmaceutical, painkiller (Dordio et al., 2017)

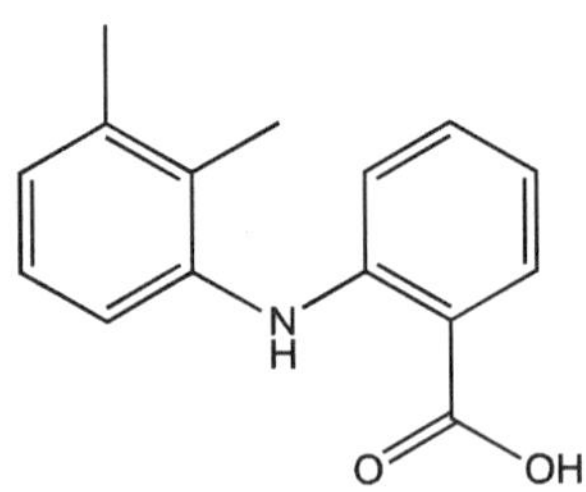

Metalaxyl: a fungicide, trade name Ridomil Gold or Mefenoxam (Wanyika, 2014)

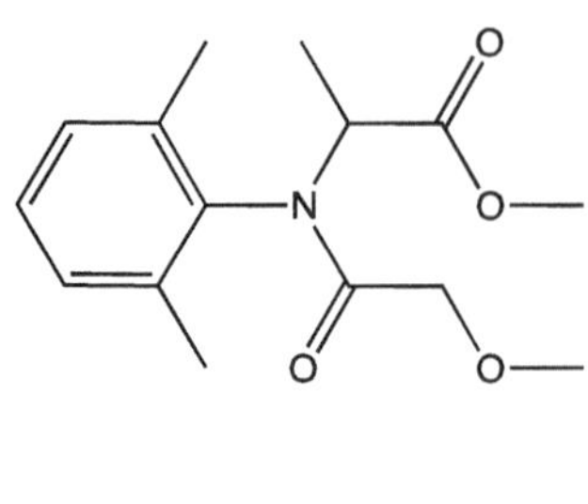

Methyl green: a dye, used for staining DNA (Rytwo et al., 2002, 2009)

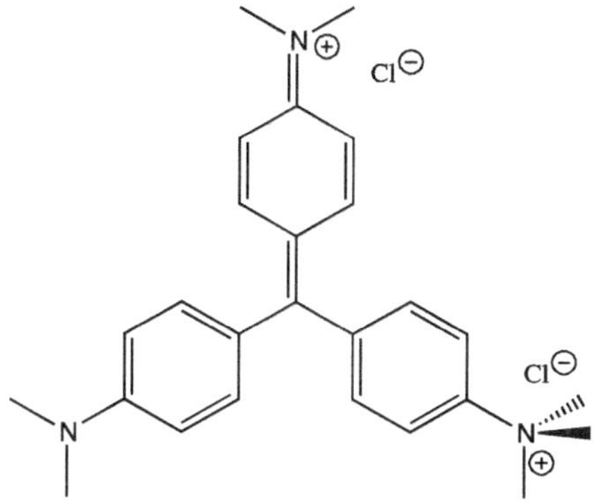

compare with malachite green and crystal violet

Methyl violet, see crystal violet

Methyl orange, a diazodye (Elmoubarki et al., 2015; Muñoz et al., 2017; Yang et al., 2018)

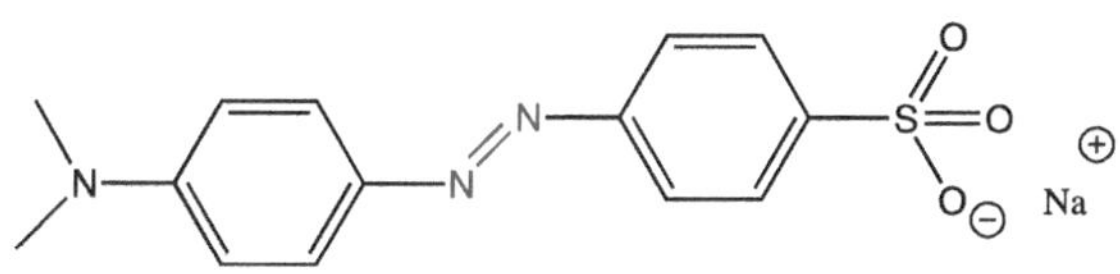

Methyleneblue, or *MB*, a dye (Bujdák and Komadel, 1997; Mishael et al., 1999; Gürses et al., 2004; Elmoubarki et al., 2015; Stawinski et al., 2017; Fang et al., 2018; Yang et al., 2018)

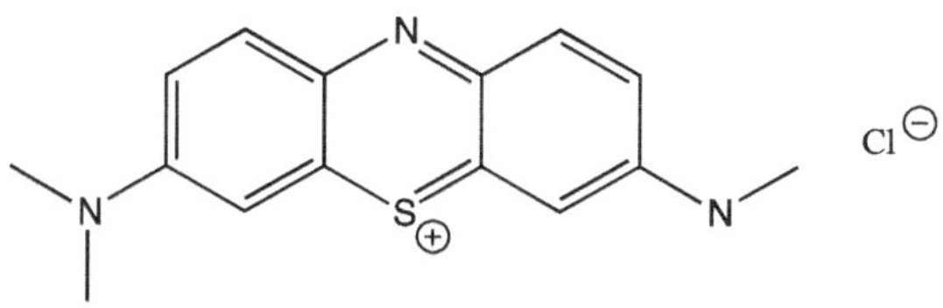

Metolachlor: an **acetanilide**, enantioselective (S isomer is active), herbicide. Also exhibits atropoisomerism (Torrents and Jayasundera, 1997)

Metsulfuron-methyl: a sulfonyl**urea**, inhibits cell division in shoots

Metribuzin: a **triazine** class herbicide, photosynthesis inhibitor (Aggarwal et al., 2006)

Molinate: anazepam, commercial pesticide

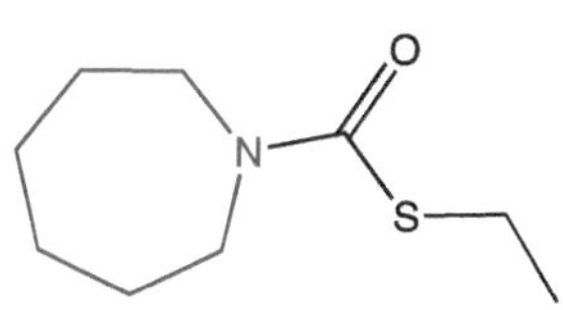

Nalidixic acid: a **quinolone** antibiotic (Wu et al., 2013a)

Naproxen: a **propionic acid** class nonsteroidal antiinflammatory drug (NSAID; Thiebault et al., 2016; Dordio et al., 2017)

Nitrophenols: pollutants, from explosive and fabrics factories (Boyd et al., 2001)

Example: *p*-nitrophenol (Park et al., 2013) 

Orange II: a dye (Fang et al., 2018)

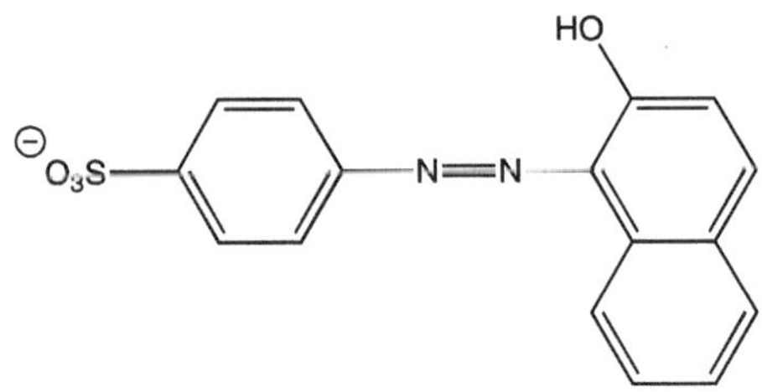

PAH: **polyaromatic** hydrocarbons

Examples:

Naphthalene

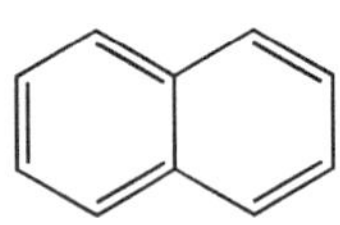

Phenanthrene

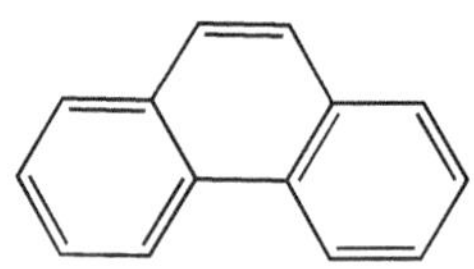

Pyrene

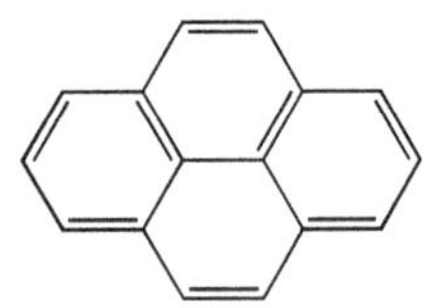

Paraquat: a viologen, cationic pesticide, nonselective herbicide, involved in Parkinson's disease (Rytwo et al., 2002; Ye and Lemley, 2008; Gu et al., 2015)

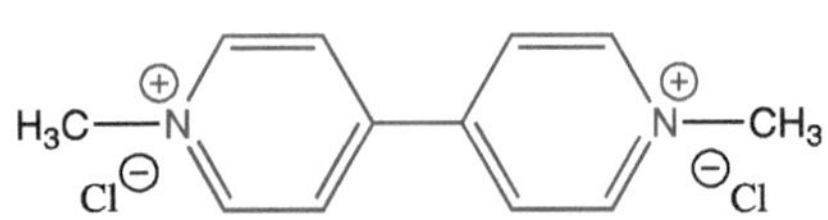

Parathion, an organophosphate, insecticide, acaricide (Sheng et al., 2001)

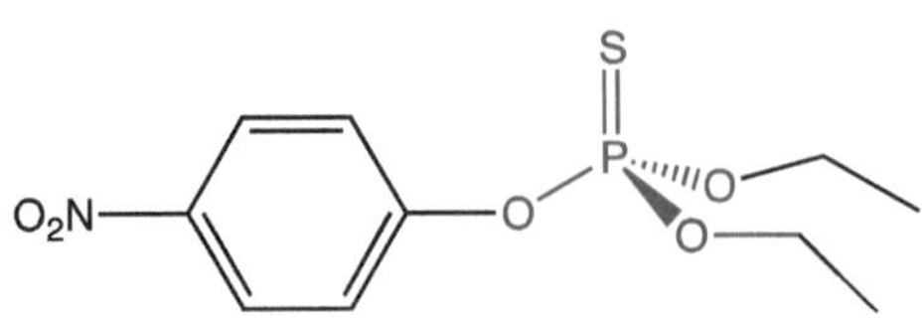

Phenanthrene, a **PAH** pollutant (Joseph-Ezra et al., 2014)

Phosmet, a phthalimide organophosphate (Sánchez Camazano and Sánchez Martín, 1980)

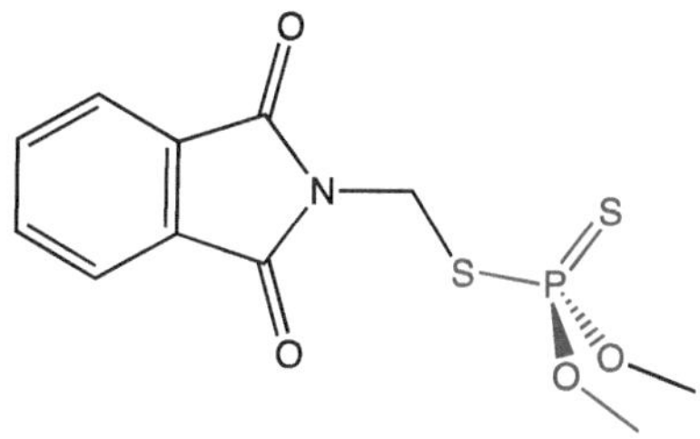

Picloram: a herbicide of the **picolinic acid** class, brand name Tordon, present in Agent White and Agent Orange (Fast et al., 2010; Marco-Brown et al., 2012, 2014, 2015, 2017)

Picolinic acids, a class of herbicides, based on the core structure

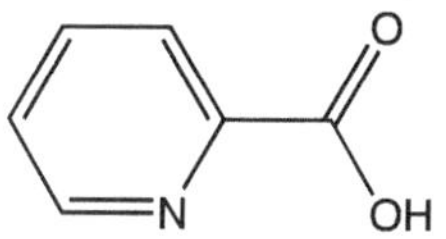

Examples:

Aminopyralid

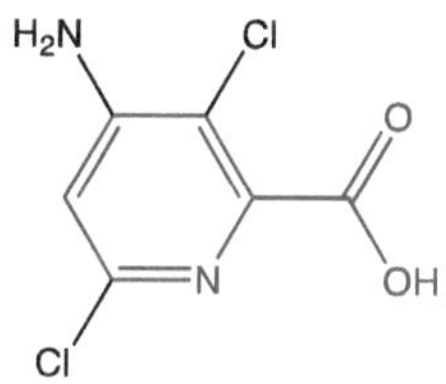

Picloram

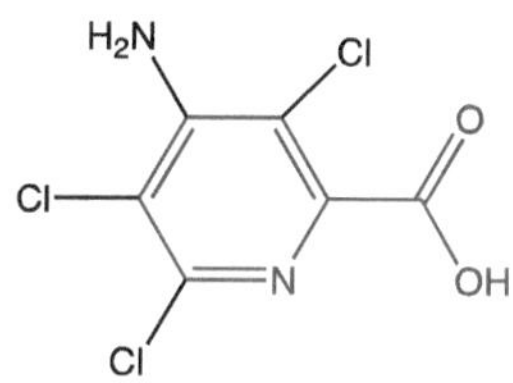

Piroxicam: a NSAID (nonsteroid antiinflammatory drug), sensitive to light (Ambrogi et al., 2012)

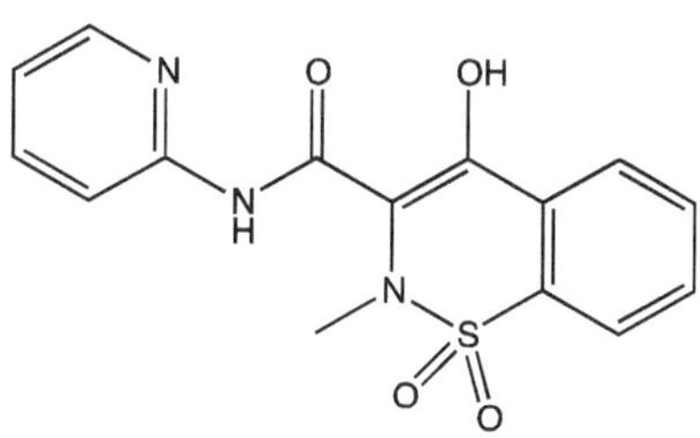

Promethazine (Ambrogi et al., 2014)

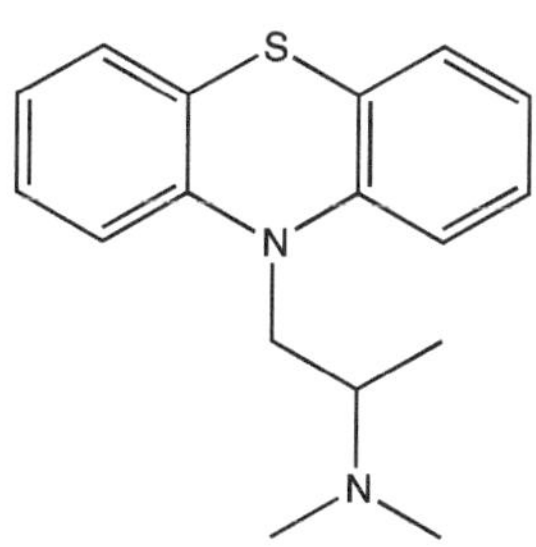

Prometryn, a **triazine**, herbicide (Konstantinou et al., 2000; Grabka et al., 2017)

Propachlor, an **acetanilide**, herbicide, forbidden in Europe (Konstantinou et al., 2000)

Propanil, an **acetanilide**, herbicide, several thousand tonnes per year (Konstantinou et al., 2000)

Propionates, or **propionic acid** class, a family of drugs containing the group

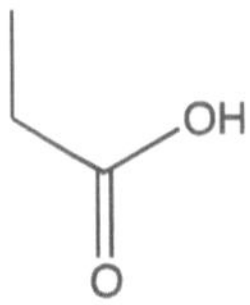

Examples

Fluazifop-*p*-butyl

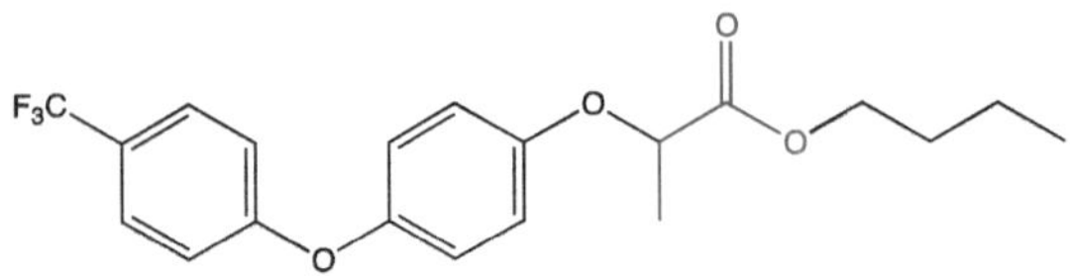

Ibuprofen

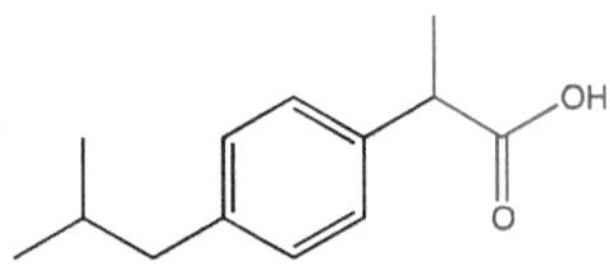

Ketoprofen

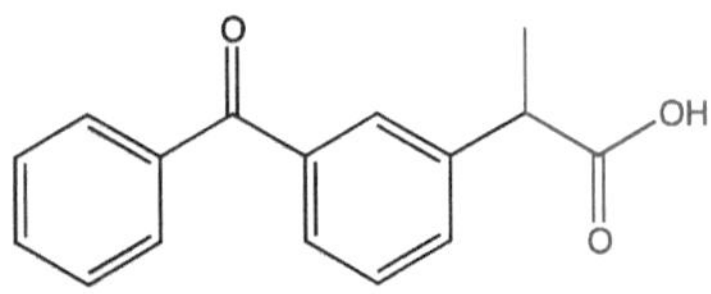

Naproxen

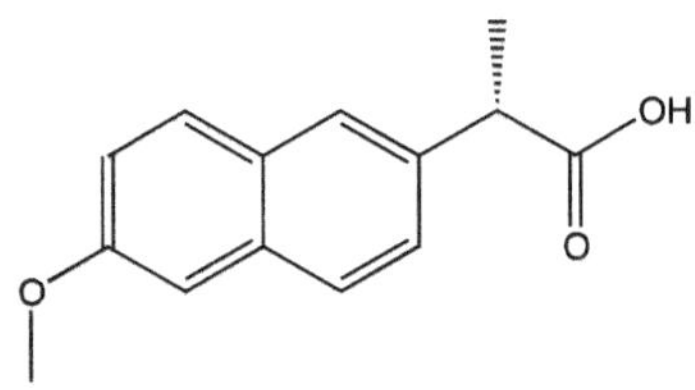

Pyrene: a **PAH** pollutant (Joseph-Ezra et al., 2014)

Pyronin B, a dye (Grauer et al., 1987a)

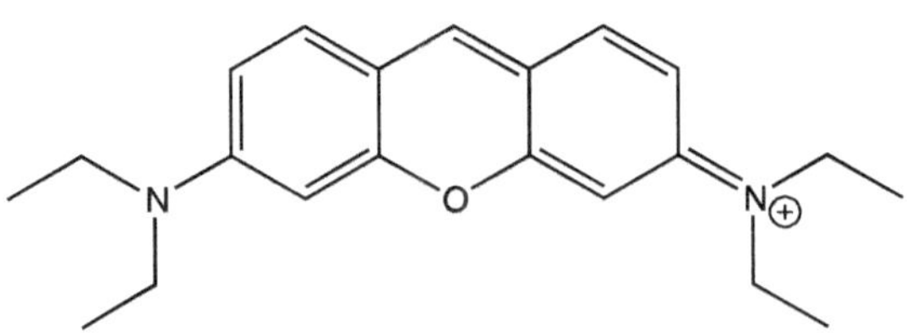

Quinolones: a class of antibiotics, core structure

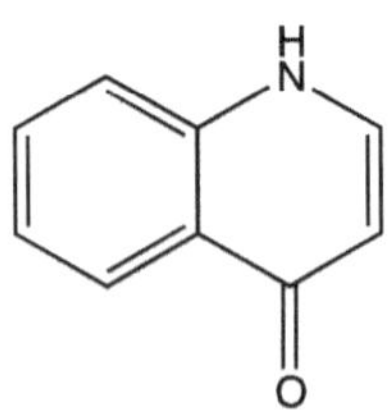

<u>Examples</u>:

Ciprofloxacin

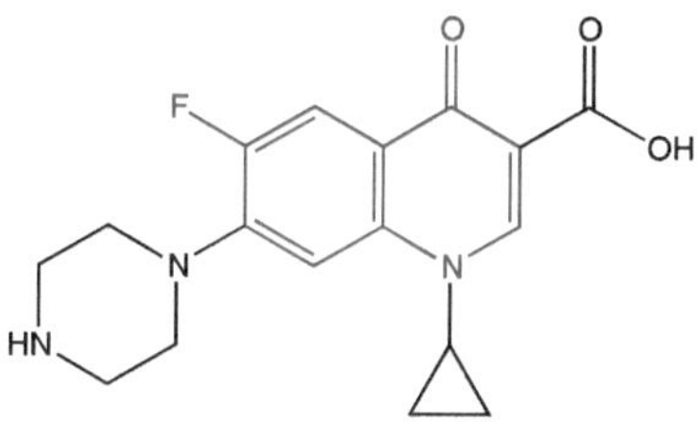

Enrofloxacine

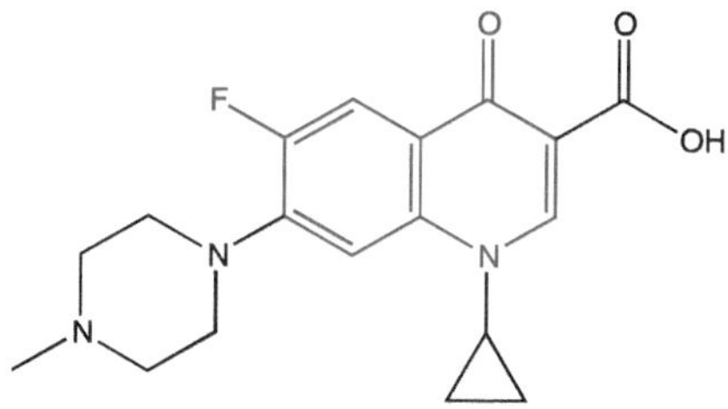

Levofloxacine

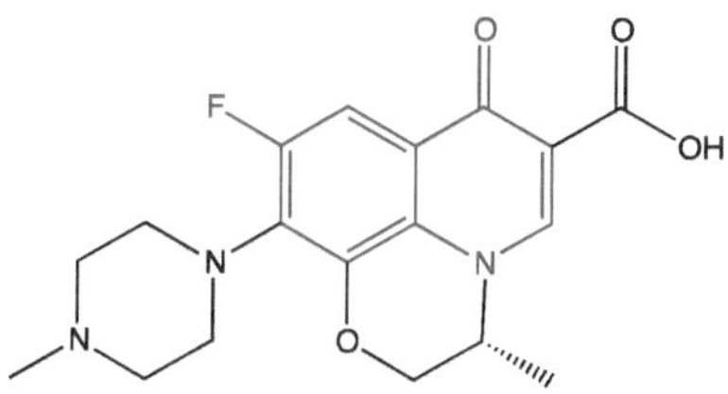

Nalidixic acid

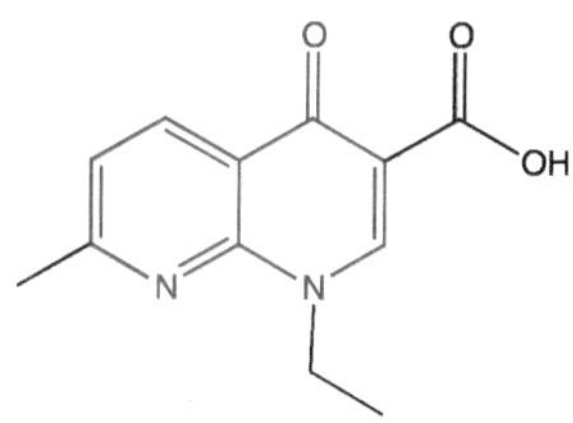

Norfloxacine

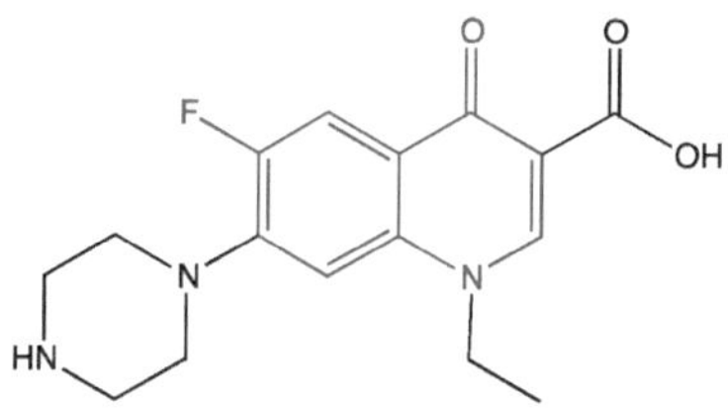

ciprofloxacine, enrofloxacine, levofloxacine, and norfloxacineare **fluoro**quinolones
Rhodamine B, a dye, zwitterionic (Grauer et al., 1987b; Fang et al., 2018)

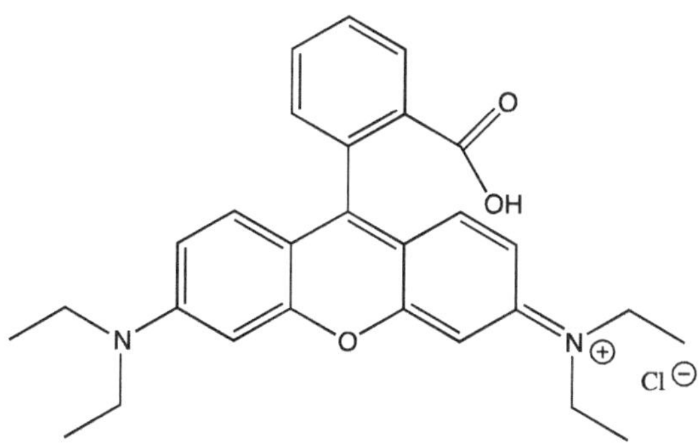

Salicylic acid, a pharmaceutical, trade name Aspirin (Cabrera-Lafaurie et al., 2015)

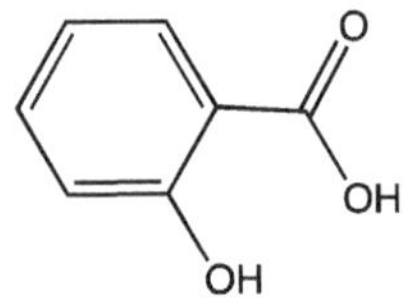

Simazine: a **triazine** class herbicide, trade names Gesatop, Princep, (Aggarwal et al., 2006; Pissinati de Rezende et al., 2013)

Steroid hormones: endocrine disruptors, for example, from birth control pills (Shareef et al., 2006)

Examples:

Dexamethasone

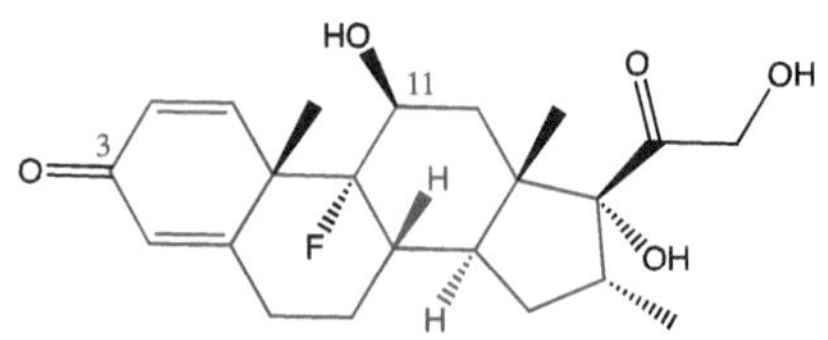

Estrone

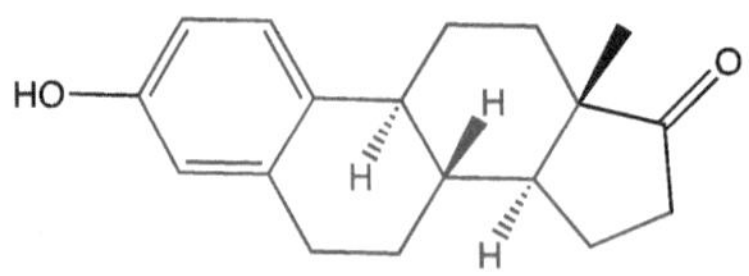

17α ethynylestradione

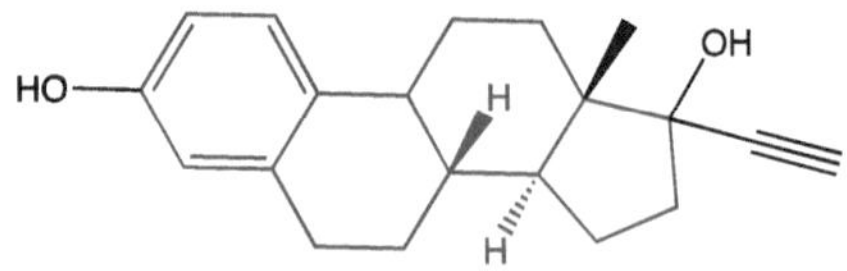

Sulfamethazine, a **sulfonamide** antibiotic (Gao and Pedersen, 2005)

Sulfamethoxazole, a **sulfonamide** antibiotic (Liu et al., 2011b)

Sulfathiazole, a **sulfonamide**, antimicrobial (aquariums) (Kahle and Stamm, 2007)

Sulfonamides or sulfa drugs: a class of antibacterial substances that also includes other types of drugs, based on the $-SO_2-NR_2$ moiety (see also **urea**s, sulfonylurea)

Examples:

Sulfamethazine

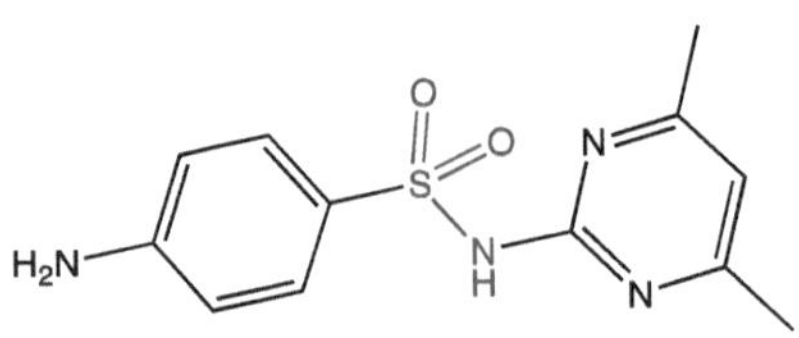

Sulfamethoxazole

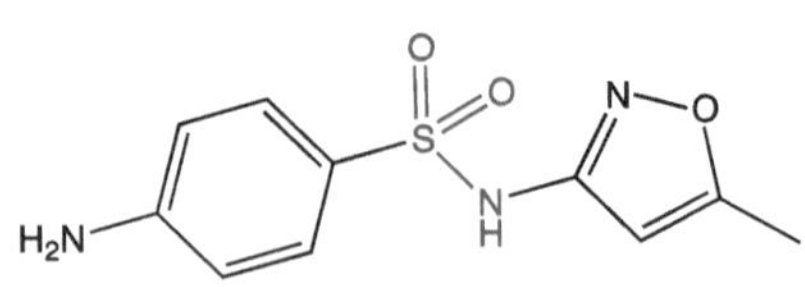

Sulfathiazole

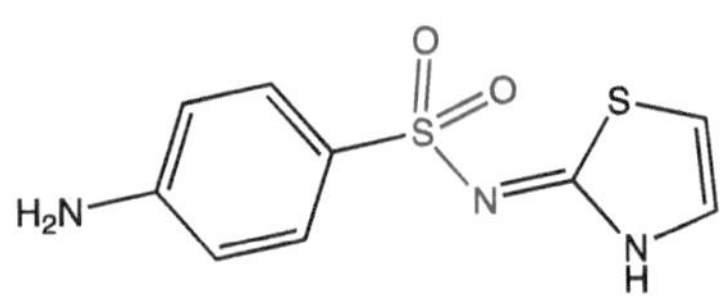

(the imino form is shown; it is in equilibrium with an amino form)

Sulforhodamine G: a dye, anionic (Fang et al., 2018)

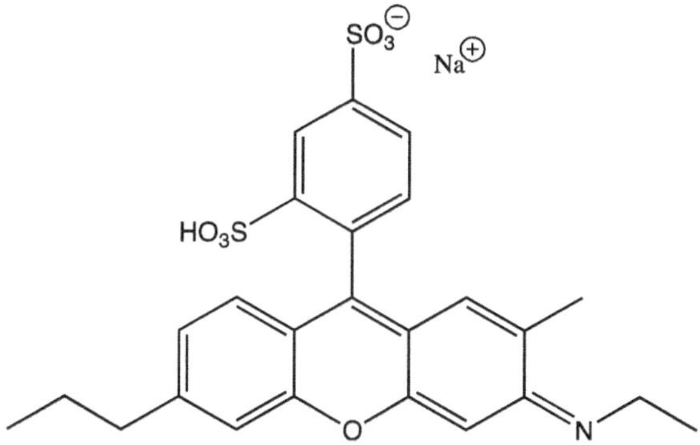

Tetracyclins: octahydrotetracene-2-carboxamide derivatives (Figueroa et al., 2004): includes basic compound + oxytetracycline and chlortetracycline

Examples:

Chlortetracycline

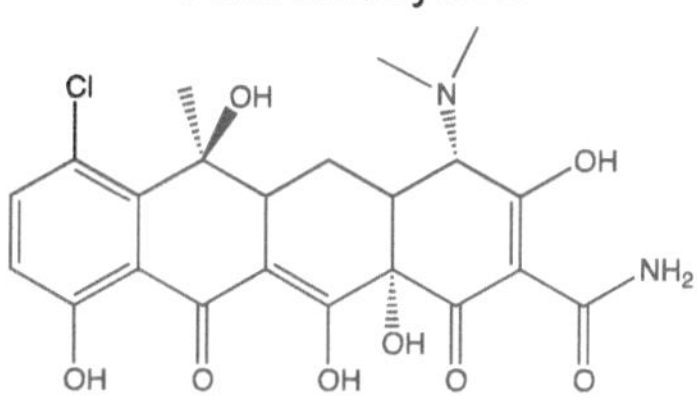

Oxytetracycline

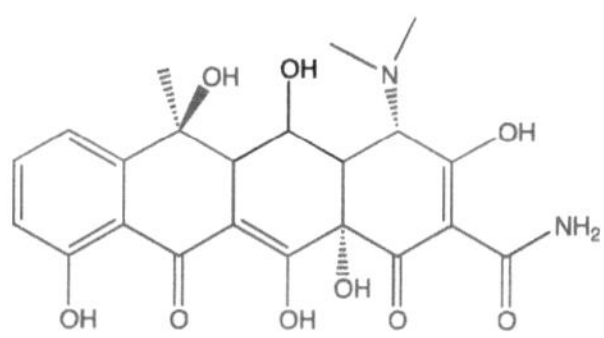

Tetracycline (Liu et al., 2011b)

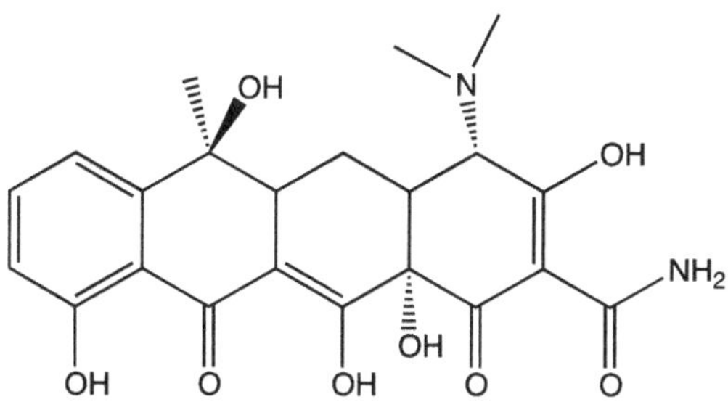

Thiabendazole: a fungicide (Lombardi et al., 2006)

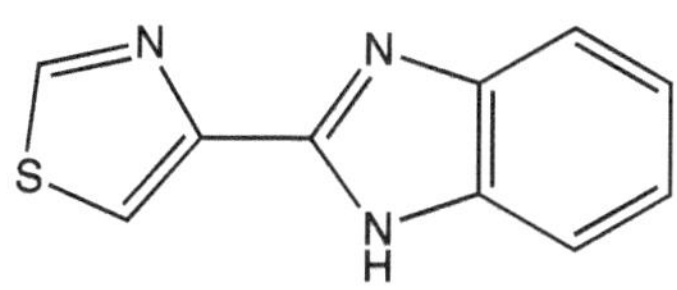

Tramadol: opioid of the 'benzenoid class', pharmaceutical, brand name Ultram (Thiebault et al., 2015, 2016)

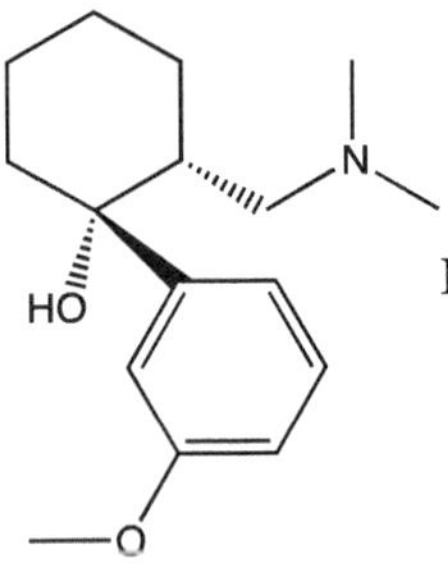

NB: the (1R, 2R) isomer is shown

Triazines: a class of pesticides based on a heterocyclic nucleus; comparison of ≠ triazines (Weber, 1966; Zarpon et al., 2006)

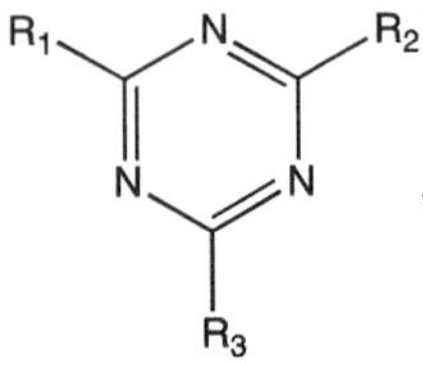

, sometimes other isomers

Examples:

Ametryn

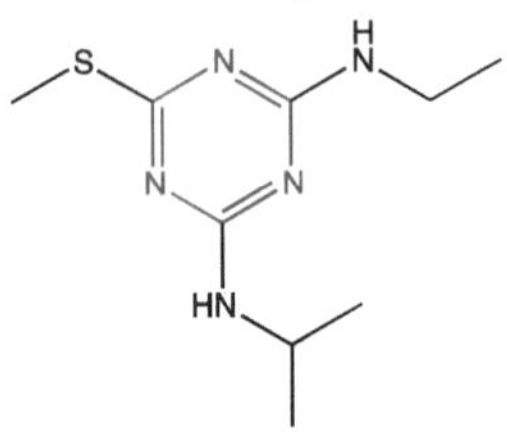

Bentazone

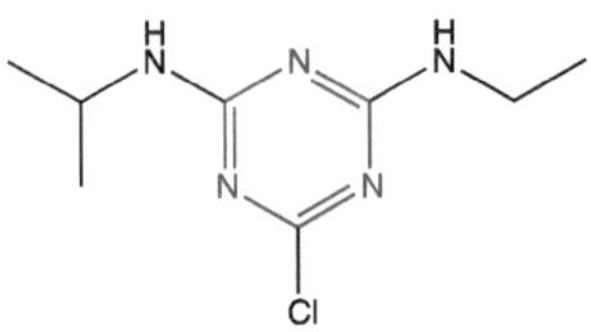

Atrazine

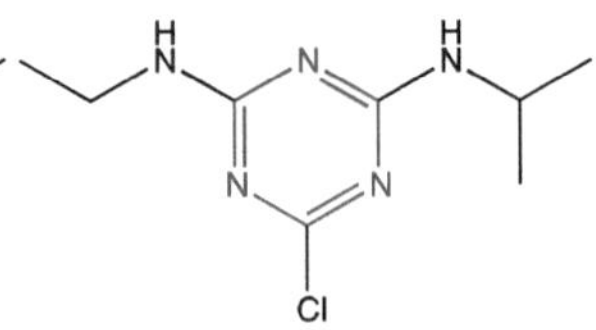

Metribuzine

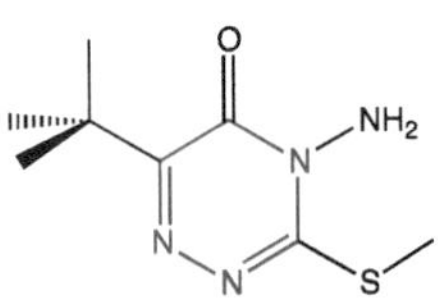

Prometryne

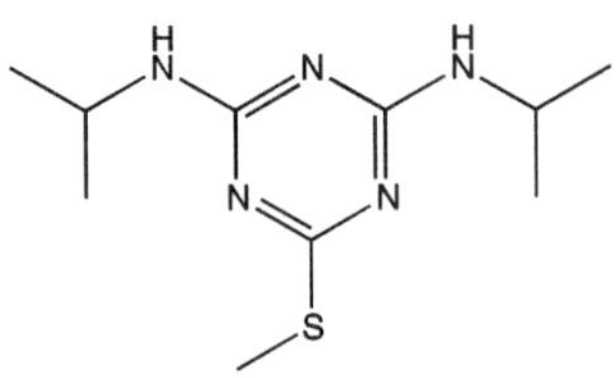

Simazine

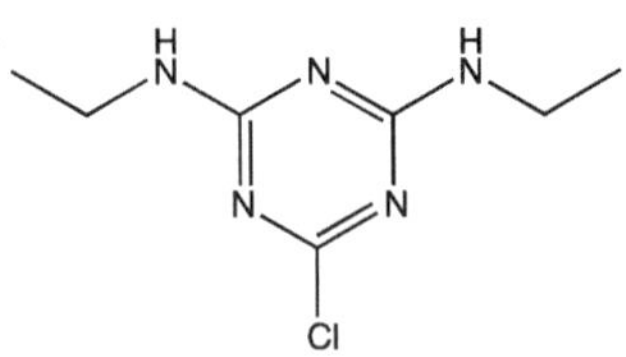

Triazoles: a class of pesticides based on a five-membered heterocyclic nucleus

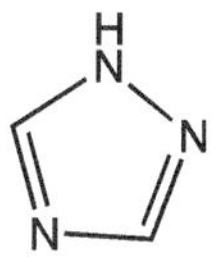

Example:

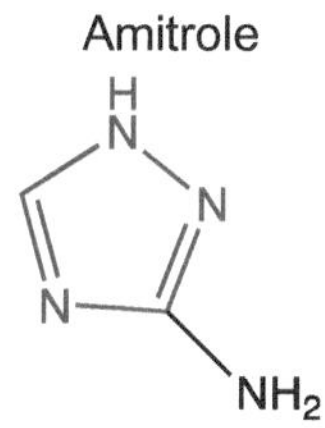

Trichloroethylene, an industrial solvent (Tao et al., 1999) 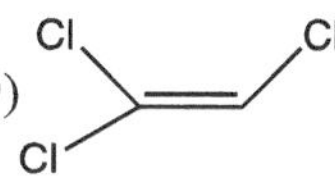

Triclopyr: a pyridine, insecticide, and fungicide (Pusino et al., 1994)

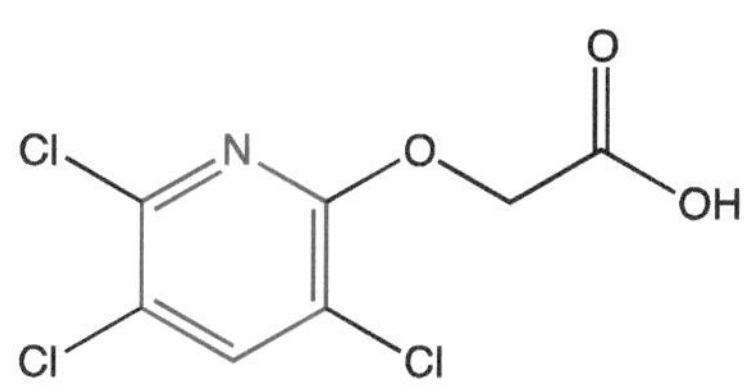

Triclosan: antibacterial, antifungal, also found in PCP and toys (Styszko et al., 2015)

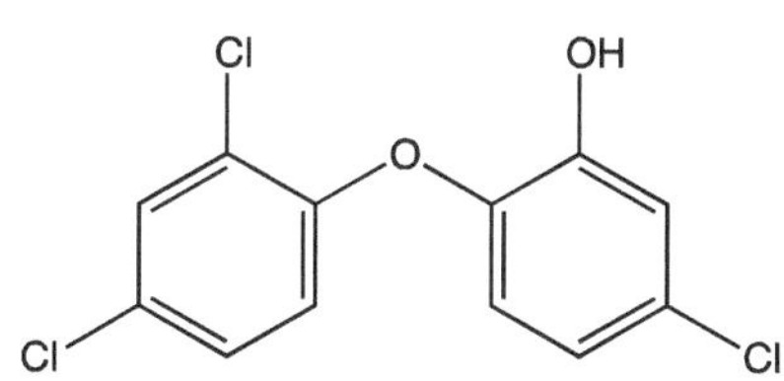

Urea: core structure 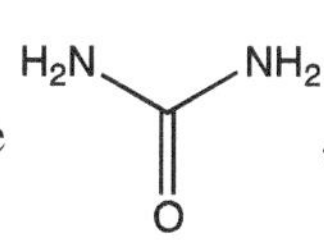

, present in several herbicides

Examples:

DCMU, a phenylurea

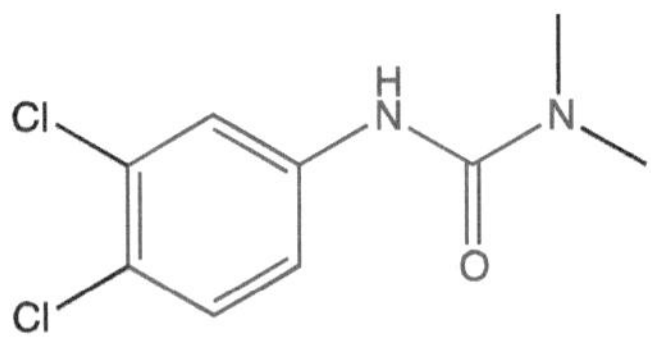

Isoproturon, a phenylurea (both also dimethylureas)

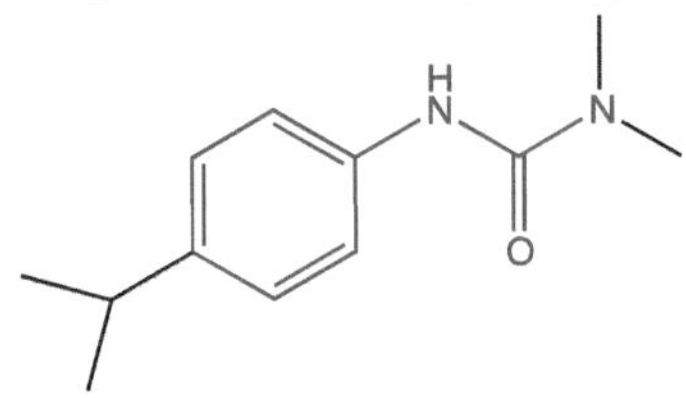

Linuron, a phenylurea

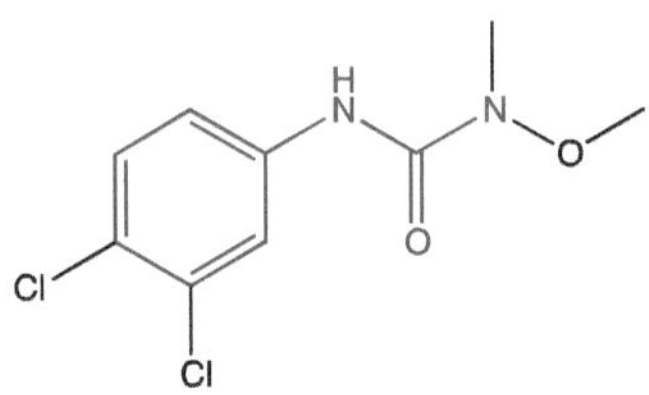

Metsulfuron-methyl, a sulfonylurea (also a triazine)

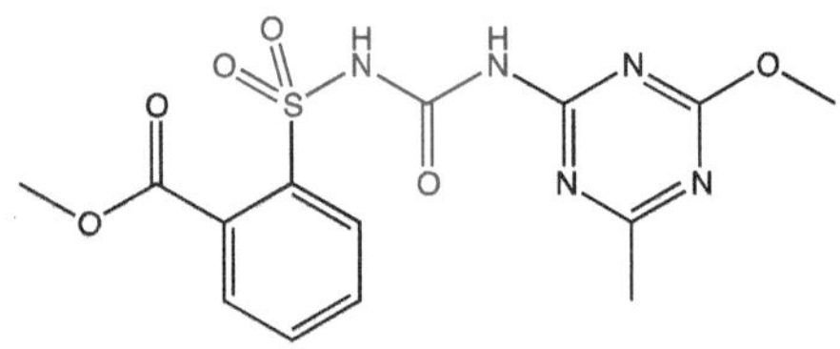

References

Aggarwal, V., Li, H., Teppen, B.J., 2006. Triazine adsorption by saponite and beidellite clay minerals. Environ. Sci. Technol. 25 (2), 392–399.

Akçay, G., Yurdakoç, K., 2000. Removal of various phenoxyalkanoic acid herbicides from water by organo-clays. Acta Hydrochim. Hydrobiol. 28 (6), 300–304.

Ambrogi, V., Latterini, L., Nocchetti, M., Pagano, C., Ricci, M., 2012. Montmorillonite as an agent for drug photostability. J. Mater. Chem. 22, 22743.

Ambrogi, V., Nocchetti, M., Latterini, L., 2014. Promethazine – montmorillonite inclusion complex to enhance drug photostability. Langmuir 30, 14612–14620.

Angioi, S., Polati, S., Roz, M., Rinaudo, C., Gianotti, V., Gennaro, M.C., 2005. Sorption studies of chloroanilines on kaolinite and montmorillonite. Environ. Pollut. 134, 35–43.

Arias-Estévez, M., López-Periago, E., Martínez-Carballo, E., Simal-Gandara, J., García-Río, J.-C.M.L., 2008. The mobility and degradation of pesticides in soils and the pollution of groundwater resources. Agric. Ecosyst. Environ. 123, 247–260.

Atun, G., Sismanoglu, T., 1996. Adsorption of 4,4′-iso propylidene diphenol and diphenylolpropane 4,4′ dioxyacetic acid from aqueous solution on kaolinite. J. Environ. Sci. Health A 31 (8), 2055–2069.

Bakhtiary, S., Shirvani, M., Shariatmadari, H., 2013. Adsorption–desorption behavior of 2,4-D on NCP-modified bentonite and zeolite: implications for slow-release herbicide formulations. Chemosphere 90, 699–705.

Barriuso, E., Laird, D.A., Koskinen, W.C., Dowdy, R.H., 1994. Atrazine desorption from smectites. Soil Sci. Soc. Am. J. 58, 1632–1638.

Behera, S.K., Oh, S.Y., Park, H.S., 2012. Sorptive removal of ibuprofen from water using selected soil minerals and activated carbon. Int. J. Environ. Sci. Technol. 9, 85–94.

Belzunces, B., Hoyau, S., Benoit, M., Tarrat, N., Bessac, F., 2017. Theoretical study of the atrazine pesticide interaction with pyrophyllite and Ca^{2+} montmorillonite clay surfaces. J. Comput. Chem. 38, 133–143.

Berhane, T.M., Levy, J., Krekeler, M.P.S., Danielson, N.D., 2016. Adsorption of bisphenol A and ciprofloxacin by palygorskite-montmorillonite: effect of granule size, solution chemistry and temperature. Appl. Clay Sci. 132–133, 518–527.

Biache, C., Lorgeoux, C., Saad, A., Faure, P., 2015. Behavior of PAH/mineral associations during thermodesorption: impact for the determination of mineral retention properties towards PAHs. Anal. Bioanal. Chem. 407, 3509–3516.

Bonhomme, C., Gervais, C., Laurencin, D., 2014. Recent NMR developments applied to organic–inorganic materials. Progr. Nucl. Magn. Res. Spectr. 77, 1–48.

Bosetto, M., Arfaioli, P., Fusi, P., 1993. Interactions of alachlor with homoionic montmorillonites. Soil Sci. 155 (2), 105–113.

Bosetto, M., Arfaioli, P., Fusi, P., 1994. Adsorption of metolachlor on homoionic montmorillonites. Agrochimica 38 (1–2), 14–24.

Boyd, S.A., Sheng, G.Y., Teppen, B.J., Johnston, C.T., 2001. Mechanisms for the adsorption of substituted nitrobenzenes by smectite clays. Environ. Sci. Technol. 35, 4227–4234.

Boyd, S.A., Johnston, C.T., Laird, D.A., Teppen, B.J., Li, H., 2011a. Comprehensive study of organic contaminant adsorption by clays: methodologies, mechanisms, and environmental implications. In: Biophysico-Chemical Processes of Anthropogenic Organic Compounds in Environmental Systems. John Wiley & Sons, Inc.

Boyd, S.A., Johnston, C.T., Pinnavaia, T.J., Kaminski, N.E., Teppen, B.J., Li, H., Khan, B., Crawford, R.B., Kovalova, N., Kim, S.-S., Shao, H., Gu, C., Kaplan, B.L.F., 2011b. Suppression of humoral immune responses by 2,3,7,8-tetrachlorodibenzo-p-dioxin intercalated in smectite clay. Environ. Toxicol. Chem. 30 (12), 2748–2755.

Bruzzi, E., Stace, A.J., 2017. Experimental measurements of water molecule binding energies for the second and third solvation shells of $[Ca(H_2O)_n]^{2+}$ complexes. R. Soc. Open Sci. 4(1), 160671.

Bujdák, J., Komadel, P., 1997. Interaction of methylene blue with reduced charge montmorillonite. J. Phys. Chem. B 101, 9065–9068.

Cabrera-Lafaurie, W.A., Román, F.R., Hernández-Maldonado, A.J., 2015. Single and multicomponent adsorption of salicylic acid, clofibric acid, carbamazepine and caffeine from water onto transition metal modified and partially calcined inorganic–organic pillared clay fixed beds. J. Hazard. Mater. 282, 174–182.

Chatterjee, A., Iwasaki, T., Ebina, T., 2002. 2:1 dioctahedral smectites as a selective sorbent for ioxins and furans: reactivity index study. J. Phys. Chem. A 106, 641–648.

Chavez, P., Ducker, W., Israelachvili, J., Maxwell, K., 1996. Adsorption of dipolar (zwitterionic) surfactants to dipolar surfaces. Langmuir 12, 4111–4115.

Clausen, L., Fabricius, I., Madsen, L., 2001. Adsorption of pesticides onto quartz, calcite, kaolinite, and α-alumina. J. Environ. Qual. 30, 846–857.

Cohen, R., Yariv, S., 1984. Metachromasy in clay minerals—sorption of acridine orange by montmorillonite. J. Chem. Soc. Faraday Trans. 1 (80), 1705–1715.

Collins, J.R., Loew, G.H., Luke, B.T., White, D.H., 1988. Theoretical investigation of the role of clay edges in prebiotic peptide bond formation. Orig. Life Evol. Biosph. 18, 107–119.

Cornejo, J., Celis, R., Pavlovic, I., Ulibarri, M.A., 2008. Interactions of pesticides with clays and layered double hydroxides: a review. Clay Miner. 43, 155–175.

Cox, L., Koskinen, W.C., Celis, R., Yen, P.Y., Hermosin, M.C., Cornejo, J., 1998. Sorption of imidacloprid on soil clay mineral and organic components. Soil Sci. Soc. Am. J. 62 (4), 911–915.

Cox, L., Hermosin, M.C., Koskinen, W.C., Cornejo, J., 2001. Interactions of imidacloprid with organic and inorganic-exchanged smectites. Clay Miner. 36, 267–274.

Dai, R.-L., Zhang, G.-Y., Gu, X.-Z., Wang, M.K., 2008. Sorption of 1,1,1-trichloro-2,2-bis (p-chlorophenyl) ethane (DDT) by clays and organoclays. Environ. Geochem. Health 30, 479–488.

Damonte, M., Sánchez, R.M.T., Afonso, M.d.S., 2007. Some aspects of the glyphosate adsorption on montmorillonite and its calcined form. Appl. Clay Sci. 36, 86–94.

Delle Site, A., 2001. Factors affecting sorption of organic compounds in natural sorbent/water systems and sorption coefficients for selected pollutants. A review. J. Phys. Chem. Ref. Data 30 (1), 187–439.

Dios Cancela, G., Romero Taboada, E., Sánchez-Rasero, F., 1992. Carbendazim adsorption on montmorillonite, peat and soils. J. Soil Sci. 99–111.

Dordio, A.V., Miranda, S., Prates Ramalhoa, J.P., Palace Carvalhoa, A.J., 2017. Mechanisms of removal of three widespread pharmaceuticals by two clay materials. J. Hazard. Mater. 323, 575–583.

dos Santos, E.C., Rozynek, Z., Hansen, E.L., Hartmann-Petersen, R., Klitgaard, R.N., Løbner-Olesen, A., Michels, L., Mikkelsen, A., Plivelic, T.S., Bordallo, H.N., Fossum, J.O., 2017. Ciprofloxacin intercalated in fluorohectorite clay: identical pure drug activity and toxicity with higher adsorption and controlled release rate. RSC Adv. 7, 26537–26545.

Dougherty, D.A., 1996. Cation–π interactions in chemistry and biology: a new view of benzene, Phe, Tyr, and Trp. Science 271 (5246), 163–168.

Drillia, P., Stamatelatou, K., Lyberatos, G., 2005. Fate and mobility of pharmaceuticals in solid matrices. Chemosphere 60, 1034–1044.

Elmoubarki, R., Mahjoubi, F.Z., Tounsadi, H., Moustadraf, J., Abdennouri, M., Zouhri, A., El Albani, A., Barka, N., 2015. Adsorption of textile dyes on raw and decanted Moroccan clays: kinetics, equilibrium and thermodynamics. Water Res. Ind. 9, 16–29.

Fang, Y.F., Zhou, A., Yang, W., Araya, T., Huang, Y.P., Zhao, P., Johnson, D., Wang, J.Z., Ren, Z.Y.J., 2018. Complex formation via hydrogen bonding between rhodamine B and montmorillonite in aqueous solution. Sci. Rep. 8, 229.

Fast, B.J., Ferrell, J.A., MacDonald, G.E., Krutz, L.J., Kline, W.N., 2010. Picloram and aminopyralid sorption to soil and clay minerals. Weed Sci. 58 (4), 484–489.

Ferrell, J. A., Vencill, W. K., Xia, K., Grey, T. L., 2005. Sorption and desorption of flumioxazin to soil, clay minerals and ion-exchange resin. 61, 40–46.

Figueroa, R.A., Leonard, A., MacKay, A.A., 2004. Modeling tetracycline antibiotic sorption to clays. Environ. Sci. Technol. 38, 476–483.

Fil, B.A., Yilmaz, M.T., Bayar, S., Elkoca, M.T., 2014. Investigation of adsorption of the dyestuff astrazon red violet 3RN (basic violet 16) on montmorillonite clay. Braz. J. Chem. Eng. 31 (1), 171–182.

Fraile, J.M., Garcia-Martin, E., Gil, C., Mayoral, J.A., Pablo, L.E., Polo, V., Prieto, E., Vispe, E., 2016. Laponite as carrier for controlled in vitro delivery of dexamethasone in vitreous humor models. Eur. J. Pharm. Biopharm. 108, 83–90.

Fusi, P., Franci, M., Bosetto, M., 1988. Interaction of fluazifop-butyl and fluazifop with smectites. Appl. Clay Sci. 3, 63–73.

Gao, J.A., Pedersen, J.A., 2005. Adsorption of sulfonamide antimicrobial agents to clay minerals. Environ. Sci. Technol. 39, 9509–9516.

Garrido-Ramirez, E.G., Theng, B.K.G., Mora, M.L., 2010. Clays and oxide minerals as catalysts and nanocatalysts in fenton-like reactions—a review. Appl. Clay Sci. 47 (3–4), 182–192.

Gerstl, Z., Nasser, A., Mingelgrin, U., 1998. Controlled release of pesticides into soils from clay–polymer formulations. J. Agric. Food Chem. 46, 3797–3802.

Gevao, B., Semple, K.T., Jones, K.C., 2000. Bound pesticide residues in soils: a review. Environ. Pollut. 108, 3–14.

Gianotti, V., Benzi, M., Croce, G., Frascarolo, P., Gosetti, F., Mazzucco, E., Bottaro, M., Gennaro, M.C., 2008. The use of clays to sequestrate organic pollutants. Leaching experiments. Chemosphere 73, 1731–1736.

Gimsing, A.L., Borggaard, O.K., 2002. Competitive adsorption and desorption of glyphosate and phosphate on clay silicates and oxides. Clay Min. 37, 509–515.

González-Pradas, E., Villafranca-Sánchez, M., Socías-Viciana, M., Fernández-Pérez, M., Urena-Amate, M., 1999. Preliminary studies in removing atrazine, isoproturon and imidacloprid from water by natural sepiolite. J. Chem. Technol. Biotechnol. 74, 417–422.

Grabka, D., Raczyńska-Żak, M., Czech, K., Słomkiewicz, P.M., Jóźwiak, M.A., 2017. Modified halloysite as an adsorbent for prometryn from aqueous solutions. Appl. Clay Sci. 114, 321–329.

Grauer, Z., Grauer, G.L., Avnir, D., Yariv, S., 1987a. Metachromasy in clay minerals—sorption of pyronin-Y by montmorillonite and laponite. J. Chem. Soc. Faraday Trans. 1 (83), 1685–1701.

Grauer, Z., Malter, A.B., Yariv, S., Avnir, D., 1987b. Sorption of rhodamine B by montmorillonite and laponite. Colloids Surf. 25, 41–46.

Gu, C., Li, H., Teppen, B.J., Boyd, S.A., 2008. Octachlorodibenzodioxin formation on Fe(III)-montmorillonite clay. Environ. Sci. Technol. 42, 4758–4763.

Gu, Z., Gao, M.L., Lu, L.F., Liu, Y.N., Yang, S.F., 2015. Montmorillonite functionalized with zwitterionic surfactant as a highly efficient adsorbent for herbicides. Ind. Eng. Chem. Res. 54, 4947–4955.

Gürses, A., Karaca, S., Dögar, Ç., Bayrak, R., Açıkyıldız, M., Yalçın, M., 2004. Determination of adsorptive properties of clay/water system: methylene blue sorption. J. Colloid Interface Sci. 269, 310–314.

Hang, P.T., Brindley, G.W., 1970. Methylene blue adsorption by clay minerals. II (clay-organic studies XVIII). Clays Clay Miner. 18, 203–212.

Hinedi, Z.R., Johnston, C.T., Erickson, C., 1993. Chemisorption of benzene on Cu-montmorillonite as characterized by FTIR and ^{13}C MAS NMR. Clays Clay Miner. 41, 87–94.

Ho, Y.S., McKay, G., 1999. Pseudo-second order model for sorption processes. Process Biochem. 34, 451–465.

Hou, X.-J., Li, H.Q., He, P., Li, S.P., Liu, Q.F., 2015. Molecular-level investigation of the adsorption mechanisms of toluene and aniline on natural and organically modified montmorillonite. J. Phys. Chem. A 119, 11199–11207.

Ibarguren, C., Naranjo, P.M., Stötzel, C., Audisio, M.C., Shama, E.L., Farfán Torres, E.M., Müller, F.A., 2014. Adsorption of nisin on raw montmorillonite. Appl. Clay Sci. 90, 88–95.

Jacobs, P.A., Theng, B.K.G., Uytterhoeven, J.B., 1972. Quantitative infrared spectroscopy of amines in synthetic zeolites X and Y. II. Adsorption of amines on Na-hydrogen zeolites X and Y. J. Catal. 26, 191–201.

Johnston, C.T., 1996. Sorption of organic compound on clay minerals: a surface functional group approach. In: Sawhney, B. (Ed.), Organic Pollutants in Soil Environments. Clay Minerals Society, Boulder, CO.

Johnston, C.T., De Oliveira, M.F., Teppen, B.J., Sheng, G.Y., Boyd, S.A., 2001. Spectroscopic study of nitroaromatic-smectite sorption mechanisms. Environ. Sci. Techol. 35, 4767–4772.

Joseph-Ezra, H., Nasser, A., Mingelgrin, U., 2014. Surface interactions of pyrene and phenanthrene on Cu-montmorillonite. Appl. Clay Sci. 95, 348–356.
Kahle, M., Stamm, C., 2007. Time and pH-dependent sorption of the veterinary antimicrobial sulfathiazole to clay minerals and ferrihydrite. Chemosphere 68, 1224–1231.
Kaya, E.M.O., Ozcan, A.S., Gök, O., Ozcan, A., 2013. Adsorption kinetics and isotherm parameters of naphthalene onto natural- and chemically modified bentonite from aqueous solutions. Adsorption 19, 879–888.
Khoury, G.A., Gehris, T.C., Tribe, L., Sánchez, R.M.T., Afonso, M.d.S., 2010. Glyphosate adsorption on montmorillonite: an experimental and theoretical study of surface complexes. Appl. Clay Sci. 50, 167–175.
Kiani, G., Dostali, M., Rostami, A., Khataee, A.R., 2011. Adsorption studies on the removal of Malachite Green from aqueous solutions onto halloysite nanotubes. Appl. Clay Sci. 54, 34–39.
Kloprogge, J.T., 1998. Synthesis of smectites and porous pillared clay catalysts: a review. J. Porous Mater. 5 (1), 5–41.
Konstantinou, I.K., Albanis, T.A., Dimitrios, M., Petrakis, E., Pomonis, P.J., 2000. Removal of herbicides from aqueous solutions by adsorption on Al-pillared clays, Fe–Al pillared clays and mesoporous alumina aluminum phosphates. Water Res. 34 (12), 3123–3136.
Koutsopoulou, E., Papoulis, D., Tsolis-Katagas, P., Kornaros, M., 2010. Clay minerals used in sanitary landfills for the retention of organic and inorganic pollutants. Appl. Clay Sci. 49, 372–382.
Kulik, D.A., 2009. Thermodynamic concepts in modeling sorption at the mineral–water interface. Rev. Mineral. Geochem. 70 (1), 125–180.
Lagaly, G., 2001. Pesticide–clay interactions and formulations. Appl. Clay Sci. 18, 205–209.
Laird, D.A., 1996. Interactions between atrazine and smectite surfaces. In: Meyer, M.T., Thurman, E.M. (Eds.), Herbicide Metabolites in Surface Water and Groundwater. American Chemical Society, Washington, DC.
Lan, T., Kaviratna, D., Pinnavaia, T. J., 1996. Epoxy self-polymerization in smectite clays. 57, 6–8, 1005-1010.
Li, H., Teppen, B.J., Laird, D.A., Johnston, C.T., Boyd, S.A., 2006. Effects of increasing potassium chloride and calcium chloride ionic strength on pesticide sorption by potassium- and calcium-smectite. Soil Sci. Soc. Am. J. 70 (6), 1889–1895.
Li, J.F., Li, Y.M., Lu, J.H., 2009. Adsorption of herbicides 2,4-D and acetochlor on inorganic–organic bentonites. Appl. Clay Sci. 46, 314–318.
Li, Z.H., Chang, P.-H., Jiang, W.-T., Jean, J.-S., Hong, H.L., Liao, L.B., 2011a. Removal of diphenhydramine from water by swelling clay minerals. J. Colloid Interface Sci. 360, 227–232.
Li, Z.H., Hong, H.L., Liao, L.B., Ackley, C.J., Schulz, L.A., MacDonald, R.A., Mihelich, A.L., Emard, S.M., 2011b. A mechanistic study of ciprofloxacin removal by kaolinite. Colloids Surf. B 88, 339–344.
Liu, W.P., Zheng, W., Gan, J.Y., 2002. Competitive sorption between imidacloprid and imidacloprid-urea on soil clay minerals and humic acids. J. Agric. Food Chem. 50, 6823–6827.
Liu, Y.X., Zhang, X., Guo, L., Wu, F., Deng, N.S., 2008. Photodegradation of bisphenol A in the montmorillonite KSF suspended solutions. Ind. Eng. Chem. Res. 47, 7141–7146.
Liu, Y.X., Wan, K., Deng, N.S., Wu, F., 2010. Photodegradation of paracetamol in montmorillonite KSF suspension. React. Kinet. Catal. Lett. 99, 493–502.
Liu, R.C., Zhang, B., Mei, D.D., Zhang, H.Q., Liu, J.D., 2011a. Adsorption of methyl violet from aqueous solution by halloysite nanotubes. Desalination 268, 111–116.

Liu, Y.X., Lu, X.J., Wu, F., Deng, N.S., 2011b. Adsorption and photooxidation of pharmaceuticals and personal care products on clay minerals. React. Kinet. Catal. Lett. 104, 61–73.
Liu, C., Johnston, C.T., Boyd, S.A., Teppen, B.J., 2012. Relating clay structural factors to dioxin adsorption by smectites: molecular dynamics simulations. Soil Sci. Soc. Am. J. 76 (1), 110–120.
Liu, C., Gu, C., Yu, K., Li, H., Teppen, B.J., Johnston, C.T., Boyd, S.A., Zhou, D.M., 2015a. Integrating structural and thermodynamic mechanisms for sorption of PCBs by montmorillonite. Environ. Sci. Technol. 49, 2796–2805.
Liu, Y.N., Dong, C.X., Yuan, H.W.W.H., Li, K.B., 2015b. Adsorption of levofloxacin onto an iron-pillared montmorillonite (clay mineral): kinetics, equilibrium and mechanism. Appl. Clay Sci. 118, 301–307.
Lombardi, B.M., Sanchez, R.M.T., Eloy, P., Genet, M., 2006. Interaction of thiabendazole and benzimidazole with montmorillonite. Appl. Clay Sci. 33, 59–65.
López Arbeloa, F., Martínez Martínez, V., 2006. Orientation of adsorbed dyes in the interlayer space of clays. 2 fluorescence polarization of rhodamine 6G in laponite films. Chem. Mater. 18, 1407–1416.
López-González, J.d.D., Valenzuela-Calahorro, C., 1970. Associated decomposition of DDT to DDE in the diffusion of DDT on homoionic clays. J. Agric. Food Chem. 19 (3), 520–523.
Maqueda, C., dos Santos Afonso, M., Morillo, E., Torres Sánchez, R.M., Perez-Sayago, M., Undabeytia, T., 2013. Adsorption of diuron on mechanically and thermally treated montmorillonite and sepiolite. Appl. Clay Sci. 72, 175–183.
Marco-Brown, J.L., Barbosa-Lema, C.M., Torres Sánchez, R.M., Mercader, R.C., Afonso, M.d. S., 2012. Adsorption of picloram herbicide on iron oxide pillared montmorillonite. Appl. Clay Sci. 58, 25–33.
Marco-Brown, J.L., Areco, M.M., Torres Sánchez, R.M., Afonso, M.d.S., 2014. Adsorption of picloram herbicide on montmorillonite: kinetic and equilibrium studies. Colloids Surf. A 449, 121–128.
Marco-Brown, J.L., Trinelli, M.A., Gaigneaux, E.M., Torres Sánchez, R.M., Afonso, M.d.S., 2015. New insights on the structure of the picloram–montmorillonite surface complexes. J. Colloid Interface Sci. 444, 115–122.
Marco-Brown, J.L., Undabeytia, T., Torres Sánchez, R.M., Afonso, M.d.S., 2017. Slow-release formulations of the herbicide picloram by using Fe–Al pillared montmorillonite. Envir. Sci. Pollut. Res. 24, 10410–10420.
Meleshyn, A., Tunega, D., 2011. Adsorption of phenanthrene on Na-montmorillonite: a model study. Geoderma 169, 41–46.
Mishael, Y.G., Rytwo, G., Nir, S., Crespin, M., Annabi-Bergaya, F., Damme, H.V., 1999. Interactions of monovalent organic cations with pillared clays. J. Colloid Interface Sci. 209, 123–128.
Mortland, M.M., 1970. Clay–organic complexes and interactions. Adv. Agron. 22, 75–117.
Mortland, M.M., Raman, K.V., 1968. Surface acidity of smectites in relation to hydration, exchangeable-cation and structure. Clays Clay Miner. 16, 393–398.
Mountacer, H., Atifi, A., Wong-Wah-Chung, P., Sarakha, M., 2014. Degradation of the pesticide carbofuran on clay and soil surfaces upon sunlight exposure. Environ. Sci. Pollut. Res. 21, 3443–3451.
Muñoz, H.-J., Blanco, C., Gil, A., Vicente, M.-Á., Galeano, L.-A., 2017. Preparation of Al/Fe-pillared clays: effect of the starting mineral. Materials 10, 1364.
Nasser, A., Gal, M., Gerstl, Z., Mingelgrin, U., Yariv, S., 1997. Adsorption of alachlor by montmorillonites. J. Therm. Anal. Calorim. 50, 257–268.

Ngulube, T., Gumbo, J.R., Masindi, V., Maity, A., 2017. An update on synthetic dyes adsorption onto clay based minerals: a state-of-art review. J. Environ. Manag. 191, 35–57.
Nir, S., El-Nahhal, Y., Undabeytia, T., Rytwo, G., Polubesova, T., Mishael, Y., Rabinovitz, O., Rubin, B., 2013. Clays, clay minerals, and pesticides. In: Lagaly, F.B.a.G. (Ed.), Handbook of Clay Science Part B: Techniques and Applications. Elsevier, Amsterdam.
Nowara, A., Burhenne, J., Spiteller, M., 1997. Binding of fluoroquinolone carboxylic acid derivatives to clay minerals. J. Agric. Food Chem. 45, 1459–1463.
Oudou, H.C., Hansen, H.C.B., 2002. Sorption of lambda-cyhalothrin, cypermethrin, deltamethrin and fenvalerate to quartz, corundum, kaolinite and montmorillonite. Chemosphere 49, 1285–1294.
Özcan, A.S., Özcan, A., 2004. Adsorption of acid dyes from aqueous solutions onto acid-activated bentonite. J. Colloid Interface Sci. 276, 39–46.
Palace Carvalho, A.J., Dordio, A.V., Prates Ramalho, J.P., 2014. A DFT study on the adsorption of benzodiazepines to vermiculite surfaces. J. Mol. Model. 20, 2336.
Park, J.K., Choy, Y.B., Oh, J.-M., Kim, J.Y., Hwang, S.-J., Choy, J.-H., 2008. Controlled release of donepezil intercalated in smectite clays. Int. J. Pharm. 359, 198–204.
Park, Y., Ayoko, G.A., Frost, R.L., 2011. Application of organoclays for the adsorption of recalcitrant organic molecules from aqueous media. J. Colloid Interface Sci. 354, 292–305.
Park, Y., Frost, R.L., Ayoko, G.A., Morgan, D.L., 2013. Adsorption of p-nitrophenol on organoclays. A thermoanalytical study. J. Therm. Anal. Calorim. 111, 41–47.
Park, Y., Sun, Z.M., Ayoko, G.A., Frost, R.L., 2014. Removal of herbicides from aqueous solutions by modified forms of montmorillonite. J. Colloid Interface Sci. 415, 127–132.
Pessagno, R.C., Sánchez, R.M.T., Afonso, M.d.S., 2008. Glyphosate behavior at soil and mineral–water interfaces. Environ. Pollut. 153, 53–59.
Pissinati de Rezende, E.I., Peralta-Zamora, P.G., Jardim, W.d.F., Vidal, C., Abate, G., 2013. Sorption and preconcentration of the herbicides atrazine, simazine, and ametryne on montmorillonite. Anal. Chem. 46 (3), 439–451.
Poinsignon, C., Cases, J.M., Fripiat, J.J., 1978. Electrical-polarization of water molecules adsorbed by smectites. An infrared study. J. Phys. Chem. 82, 1855–1860.
Polati, S., Angioi, S., Gianotti, V., Gosetti, F., Gennaro, M.C., 2006. Sorption of pesticides on kaolinite and montmorillonite as a function of hydrophilicity. J. Environ. Sci. Health B 41 (4), 333–344.
Pusino, A., Liu, W.P., Gessa, C., 1994. Adsorption of triclopyr on soil and some of its components. J. Agric. Food Chem. 42 (4), 1026–1029.
Qin, C., Troya, D., Shang, C., Hildreth, S., Helm, R., Xia, K., 2015. Surface catalyzed oxidative oligomerization of 17β-estradiol by Fe^{3+}-saturated montmorillonite. Environ. Sci. Technol. 49, 956–964.
Rana, K., Boy, S.A., Teppen, B.J., Li, H., Liu, C., Johnston, C.T., 2009. Probing the microscopic hydrophobicity of smectite surfaces. A vibrational spectroscopic study of dibenzo-p-dioxin sorption to smectites. Phys. Chem. Chem. Phys. 11, 2976–2985.
Rimola, A., Costa, D., Sodupe, M., Lambert, J.-F., Ugliengo, P., 2013. Silica surface features and their role in the adsorption of bio-molecules: computational modeling and experiments. Chem. Rev. 113, 4216–4313.
Rytwo, G., Tropp, D., Serban, C., 2002. Adsorption of diquat, paraquat and methyl green on sepiolite: experimental results and model calculations. Appl. Clay Sci. 20, 273–282.
Rytwo, G., Gonen, Y., Huterer-Shveky, R., 2009. Evidence of degradation of triarylmethine dyes on texas vermiculite. Clays Clay Miner. 57 (5), 555–565.
Sabah, E., Ouki, S., 2017. Adsorption of pyrene from aqueous solutions onto sepiolite. Clays Clay Miner. 65 (1), 14–26.

Sánchez Camazano, M., Sánchez Martín, M.J., 1980. Interaction of phosmet with montmorillonite. Soil Sci. 129, 115–118.
Sánchez Camazano, M., Sánchez Martín, M.J., 1991. Hydrolysis of azinphosmethyl induced by the surface of smectites. Clays Clay Miner. (6), 609–613.
Sannino, F., Violante, A., Gianfreda, L., 1997. Adsorption–desorption of 2,4-D by hydroxy aluminium montmorillonite complexes. Pestic. Sci. 51, 429–435.
Sarma, G.K., Gupta, S.S., Bhattacharyya, K.G., 2016. Adsorption of crystal violet on raw and acid-treated montmorillonite, K10, in aqueous suspension. J. Environ. Manag. 171, 1–10.
Seger, M.R., Maciel, G.E., 2006. NMR investigation of the behavior of an organothiophosphate pesticide, chlorpyrifos, sorbed on montmorillonite clays. Environ. Sci. Technol. 40, 797–802.
Shapley, T.V., Molinari, M., Zhu, R.L., Parker, S.C., 2013. Atomistic modeling of the sorption free energy of dioxins at clay–water interfaces. J. Phys. Chem. C 117, 24975–24984.
Shareef, A., Angove, M.J., Wells, J.D., Johnson, B.B., 2006. Sorption of bisphenol A, 17α-ethynylestradiol and estrone to mineral surfaces. J. Colloid Interface Sci. 297, 62–69.
Sheng, G., Johnston, C.T., Teppen, B.J., Boyd, S.A., 2001. Potential contributions of smectite clays and organic matter to pesticide retention in soils. J. Agric. Food Chem. 49, 2899–2907.
Sheng, G.Y., Johnston, C.T., Teppen, B.J., Boyd, S.A., 2002. Adsorption of dinitrophenol herbicides from water by montmorillonites. Clays Clay Miner. 50 (1), 25–34.
Stawinski, W., Wegrzyn, A., Danko, T., Freitas, O., Figueiredo, S., Chmielarz, L., 2017. Acid–base treated vermiculite as high performance adsorbent: insights into the mechanism of cationic dyes adsorption, regeneration, recyclability and stability studies. Chemosphere 173, 107–115.
Sturini, M., Speltini, A., Maraschi, F., Rivagli, E., Pretali, L., Malavasi, L., Profumo, A., Fasani, E., Albini, A., 2015. Sunlight photodegradation of marbofloxacin and enrofloxacin adsorbed on clay minerals. J. Photochem. Photobiol. A 299, 103–109.
Styszko, K., Nosek, K., Motak, M., Bester, K., 2015. Preliminary selection of clay minerals for the removal of pharmaceuticals, bisphenol A and triclosan in acidic and neutral aqueous solutions. C.R. Chim. 18, 1134–1142.
Supak, J.R., Swoboda, A.R., Dixon, J.B., 1977. Adsorption of aldicarb by clays and soil organo–clay complexes. Soil Sci. Soc. Am. J 42 (2), 244–248.
Szczepanik, B., Słomkiewicz, P., Garnuszek, M., Czech, K., 2014. Adsorption of chloroanilines from aqueous solutions on the modified halloysite. Appl. Clay Sci. 101, 260–264.
Tahir, S.S., Rau, N., 2006. Removal of a cationic dye from aqueous solutions by adsorption onto bentonite clay. Chemosphere 63, 1842–1848.
Tahoun, S.A., Mortland, M.M., 1966. Complexes of montmorillonite with primary, secondary and tertiary amides. 2. Coordination of amides on surface of montmorillonite. Soil Sci. 102 (5), 314.
Tan, D.Y., Yuan, P., Annabi-Bergaya, F., Dong, F., Liu, D., He, H.P., 2015. A comparative study of tubular halloysite and platy kaolinite as carriers for the loading and release of the herbicide amitrole. Appl. Clay Sci. 114, 190–196.
Tao, T., Maciel, G.E., 1998. ^{13}C NMR study of co-contamination of clays with carbon tetrachloride and benzene. Environ. Sci. Technol. 32, 350–357.
Tao, T., Yang, J.J., Maciel, G.E., 1999. Photoinduced decomposition of trichloroethylene on soil components. Environ. Sci. Technol. 33, 74–80.
Tejada, M., Morillo, E., Gómez, I., Madrid, F., Undabeyti, T., 2017. Effect of controlled release formulations of diuron and alachlorherbicides on the biochemical activity of agricultural soils. J. Hazard. Mater. 322, 334–347.

Thiebault, T., Guégan, R., Boussafir, M., 2015. Adsorption mechanisms of emerging micro-pollutants with a clay mineral: case of tramadol and doxepine pharmaceutical products. J. Colloid Interface Sci. 453, 1–8.
Thiebault, T., Boussafir, M., Le Forestier, L., Le Milbeau, C., Monnina, L., Guégan, R., 2016. Competitive adsorption of a pool of pharmaceuticals onto a raw clay mineral. RSC Adv. 6, 65257.
Thiele-Bruhn, S., 2003. Pharmaceutical antibiotic compounds in soils—a review. J. Plant Nutr. Soil Sci. 166, 145–167.
Tomic, Z.P., Ašanin, D., Đurovic, R., Đordevic, A., Makreski, P., 2012. Near-infrared spectroscopy study for determination of adsorbed acetochlor in the organic and inorganic bentonites. Spectrochim. Acta A 98, 47–52.
Tomic, Z.P., Kaluderovic, L., Nikolic, N., Markovic, S., Makreski, P., 2016. Thermal investigation of acetochlor adsorption on inorganic and organic-modified montmorillonite. J. Therm. Anal. Calorim. 123, 2313–2319.
Torrents, A., Jayasundera, S., 1997. The sorption of nonionic pesticides onto clays and the influence of natural organic carbon. Chemosphere 35 (7), 1549–1565.
Tunega, D., Gerzabek, M.H., Haberhauer, G., Lischka, H., 2007. Formation of 2,4-complexes on montmorillonites—an ab initio molecular dynamics study. Eur. J. Soil Sci. 58, 680–691.
Undabeytia, T., Galán-Jiménez, M.C., Gómez-Pantoja, E., Vázquez, J., Casal, B., Bergaya, F., Morillo, E., 2013. Fe-pillared clay mineral-based formulations of imazaquin for reduced leaching in soil. Appl. Clay Sci. 80–81, 382–389.
Vaccari, A., 1998. Preparation and catalytic properties of cationic and anionic clays. Catal. Today 41 (1–3), 53–71.
Valisko, M., Boda, D., 2007. Selective adsorption of ions with different diameter and valence at highly charged interfaces. J. Phys. Chem. C 111 (43), 15575–15585.
Vimonses, V., Lei, S.M., Jin, B., Chowd, C.W.K., Saint, C., 2009. Kinetic study and equilibrium isotherm analysis of Congo Red adsorption by clay materials. Chem. Eng. J 148, 354–364.
Wang, C.-J., Li, Z.H., Jiang, W.-T., 2011. Adsorption of ciprofloxacin on 2:1 dioctahedral clay minerals. Appl. Clay Sci. 53, 723–728.
Wang, Q., Zhang, J.P., Wang, A.Q., 2013. Alkali activation of halloysite for adsorption and release of ofloxacin. Appl. Surf. Sci. 287, 54–61.
Wanyika, H., 2014. Controlled release of agrochemicals intercalated into montmorillonite interlayer space. Sci. World J. 2014, 656287.
Weber, J.B., 1966. Molecular structure and pH effects on the adsorption of 13 s-triazine compounds on montmorillonite clay. Am. Mineral. 51, 1657–1670.
Weber, J.B., Wilkerson, G.G., Reinhardt, C.F., 2004. Calculating pesticide sorption coefficients (Kd) using selected soil properties. Chemosphere 55, 157–166.
Wei, J., Furrer, G., Kaufmann, S., Schulin, R., 2001. Influence of clay minerals on the hydrolysis of carbamate pesticides. Environ. Sci. Technol. 35, 2226–2232.
Werner, D., Garratt, J.A., Pigott, G., 2013. Sorption of 2,4-D and other phenoxy herbicides to soil, organic matter, and minerals. J. Soil. Sediment. 13, 129–139.
Wu, Q.F., Li, Z.H., Hong, H.L., Yin, K., Tie, L.Y., 2010. Adsorption and intercalation of ciprofloxacin on montmorillonite. Appl. Clay Sci. 50, 204–211.
Wu, Q.F., Li, Z.H., Hong, H.L., 2013a. Adsorption of the quinolone antibiotic nalidixic acid onto montmorillonite and kaolinite. Appl. Clay Sci. 74, 66–73.
Wu, Q.F., Li, Z.H., Hong, H.L., Li, R.B., Jiang, W.-T., 2013b. Desorption of ciprofloxacin from clay mineral surfaces. Water Res. 47, 259–268.

Xiong, J., Maciel, G.E., 1999. Deuterium NMR studies of local motions of benzene adsorbed on Ca-montmorillonite. J. Phys. Chem. B 103, 5543–5549.

Xu, J.C., Stucki, J.W., Wu, J., Kostka, J.E., Sims, G.K., 2001. Fate of atrazine and alachlor in redox-treated ferruginous smectite. Environ. Toxicol. Chem. 20 (12), 2717–2724.

Yang, R., Li, D.W., Li, A.M., Yang, H., 2018. Adsorption properties and mechanisms of palygorskite for removal of various ionic dyes from water. Appl. Clay Sci. 151, 20–28.

Yariv, S., Lurie, D., 1971. Metachromasy in clay minerals. Part I. Sorption of methylene-blue by montmorillonite. Isr. J. Chem. 9, 537–552.

Ye, P., Lemley, A.T., 2008. Adsorption effect on the degradation of carbaryl, mecoprop, and paraquat by anodic fenton treatment in an SWy-2 montmorillonite clay slurry. J. Agric. Food Chem. 56, 10200–10207.

Yuan, G.D., Theng, B.K.G., Churchman, G.J., Gates, W.P., 2013. Clays and clay minerals for pollution control. In: Lagaly, F.B.a.G. (Ed.), Handbook of Clay Science Part B: Techniques and Applications. Elsevier, Amsterdam.

Yusoff, S.N.M., Kamari, A., Aljafree, N.F.A., 2016. A review of materials used as carrier agents in pesticide formulations. Int. J. Environ. Sci. Technol. 13 (12), 2977–2994.

Zarpon, L., Abate, G., dos Santos, L.B.O., Mas, J.C., 2006. Montmorillonite as an adsorbent for extraction and concentration of atrazine, propazine, deethylatrazine, deisopropylatrazine and hydroxyatrazine. Anal. Chim. Acta 579, 81–87.

Zhang, W.H., Ding, Y.J., Boyd, S.A., Teppen, B.J., Li, H., 2010. Sorption and desorption of carbamazepine from water by smectite clays. Chemosphere 81, 954–960.

Zhu, R.L., Chen, Q.Z., Liu, H.Y., Ge, F., Zhu, L.F., Zhua, J.X., He, H.P., 2014. Montmorillonite as a multifunctional adsorbent can simultaneously remove crystal violet, cetyltrimethylammonium, and 2-naphthol from water. Appl. Clay Sci. 88–89, 33–38.

Protein adsorption on clay minerals

8

Maguy Jaber, Jean-François Lambert†, Sébastien Balme‡*
*Sorbonne Université, Laboratoire d'Archéologie Moléculaire et Structurale, UMR 8220, BP225, Paris Cedex 05, France, †Laboratoire de Réactivité de Surface, Sorbonne Université-CNRS (UMR 7197), Paris, France, ‡Institut Europeen des Membranes, UMR 5635, Universite de Montpellier, ENSCM, CNRS, Montpellier Cedex 5, France

8.1 Introduction

Proteins are major components of living matter, constituting one of the two main classes of biological polymers (the other being nucleic acids). Chemically, they are long chains of amino acids bound together by peptidic, or amide, links (–CO–NH–). Functionally, they are the workhorses of the living cell, where they play a number of indispensable roles. First, as enzymes, they catalyse most of the reactions of metabolism, but they are also active in signalling pathways, as structural elements (collagen in connective tissues, keratin in hair and feathers), and to generate mechanical forces (actin and tubulin in muscles, etc.). Because of this ubiquity, they are liberated in the environment in high amounts, mostly through the activity of microorganisms (and upon cell death and lysis), but also as by-products of human activities (e.g. vaccine protein components, Rytwo et al., 2010).

In soils, proteins interact with organic and mineral constituents, the latter including clay minerals, which are the dominant fraction on a specific surface area basis. Obviously, it is of utmost importance for soil science, and more generally for agriculture and environmental science, to understand how the interaction of proteins with clay minerals will modify their activity and control their bioavailability. There are, however, other incentives to study clay mineral–protein systems, as they may have specific industrial applications, for example, to stabilise emulsions (Ashby and Binks, 2000; Ganley et al., 2017), or in the food industry (Farahnaky et al., 2014), such as in the removal of "protein hazes" from wine to clear it (Sarmento et al., 2000). In particular, it is often desirable to "heterogenise" protein systems; that is, to attach them to a solid support, in order to control their activity. Thus, enzymes are extremely efficient catalysts for a number of chemical reactions, but it is difficult to recover them for a second use because they are dissolved in the reaction medium together with the reagents and products; if they could be immobilised on solid particles without losing their specific activity, their separation from the solution would become a trivial matter. Immobilisation of proteins also allows us to use their molecular recognition abilities in order to build sensors; immunoproteins are particularly good candidates for this type of application.

Developments in Clay Science, Vol. 9. https://doi.org/10.1016/B978-0-08-102432-4.00008-1

The specific subject of protein adsorption on clay minerals has not often been reviewed. An article by Yu et al. (2013) addressed the adsorption on clay minerals of the two main types of biopolymers, proteins, and nucleic acids, mainly from the point of view of the adsorption mechanism.

There are, however, a number of more general studies of proteins on surfaces such as those by Haynes and Norde (1994), Nakanishi et al. (2001), or Rabe et al. (2011) and, especially for biosensors, (Kasemo, 2002), from which the interested reader will obtain useful insights.

This chapter will discuss the general concepts important in understanding protein adsorption on clay minerals (Section 8.2) and the experimental techniques available for this study (Section 8.3). The determining factors for protein adsorption will be discussed in Section 8.4, and finally, an overview of the main results obtained for some classes of proteins will be provided (Section 8.5).

8.2 General considerations on protein adsorption

8.2.1 Binding force/binding site

The conformation of proteins results from three or four levels of structure. The primary structure is the sequence of amino acids in the protein chain. The secondary structure refers to the local folding in regular structures, such as α-helices or β-sheets. The tertiary structure is the global folding of the protein into a three-dimensional shape. It is imposed by a sum of intrachain interactions (electrostatic, hydrogen bonds, hydrophobic interactions, sometimes aided by sulphide bridges) and corresponds to a minimum in energy. The quaternary structure is the self-assembly of several monomer units that is often necessary to build a functional protein; for example, haemoglobin in human adults consists of a supramolecular tetrahedral arrangement of four protein chains.

In fact, different partially folded states usually exist, corresponding to local energy minima (Jahn and Radford, 2005). The conversion between these different conformations requires energy input (temperature or irradiation) or a modification of the energy landscape by a chemical agent (e.g. pH, urea, surfactant, ligand, ionic strength, and solvent). Protein denaturation corresponds to a modification of secondary and/or tertiary structures following such stimuli, which is often followed by a loss of activity. The adsorption of proteins on clay minerals is a complex phenomenon (Nakanishi et al., 2001). It results from various interactions with the surface, but also with the solvent and ions present in solution (Haynes and Norde, 1994). The interactions that drive the adsorption process cover a large scale of energy from covalent bonds (i.e. disulphide bridge 320 kJ mol^{-1}), electrostatic interactions (35–90 kJ mol^{-1}), hydrogen bonds (8–40 kJ mol^{-1}), van der Waals interactions, and hydrophobic interactions (4–12 kJ mol^{-1}).

Many studies consider that the clay mineral is hydrophilic, and that the adsorption process involves the exchange of interlayer cations. Following this description, the protein adsorption is considered only through electrostatic and hydrophilic interactions. This description is far from reality because the interlayer space and the edges have a different chemical structure. In addition, the hydrophobic interactions are totally underestimated, as illustrated by Staunton and Quiquampoix (1994). These authors examined the adsorption of bovine serum albumin (BSA) and methylated BSA on montmorillonite (Mt). They found an increase of adsorption of the more hydrophobic protein (i.e. methylated BSA). This can be understood as follows: the hydrophilicity of Mt is attributed to compensating cations. When these cations are replaced by proteins in the interlayer space through ion-exchange, the remaining interlayer siloxane surface takes on increased hydrophobic character.

Protein adsorption (at constant pressure and temperature) on clay mineral is usually spontaneous, meaning that the Gibbs energy $\Delta_{ads}G$ is negative.

$$\Delta_{ads}G = \Delta_{ads}H - T\Delta_{ads}S < 0 \tag{8.1}$$

where $\Delta_{ads}H$ and $\Delta_{ads}S$ are respectively the variation of enthalpy and entropy during the adsorption process.

Protein adsorption involves different subprocesses that contribute to the negative value of $\Delta_{ads}G$ (Staunton and Quiquampoix, 1994), (i) the change in the state of hydration of adsorbent and protein surface; (ii) the redistribution of charged groups; (iii) the rearrangement in the protein structure; and (iv) the loss of configurational entropy (Garland et al., 2012).

8.2.2 Soft and hard proteins

The concept of "hard" and "soft" proteins was proposed by Arai and Norde (1990). It assumes that the protein behaviour in the adsorption process is driven by its conformational rigidity or "hardness". The "hard" proteins have a high structural stability, contrary to "soft" proteins. The structural stability is directly connected to the unfolding energy that is larger for hard proteins (60 kJ mol^{-1} for lysozyme) than for soft proteins (21 kJ mol^{-1} for α-lactalbumin (Koo et al., 2013)). This unfolding energy depends of the structural domain of protein. Basically, the energy necessary to unfold an α-helix is lower than that for a β-sheet. Thus, soft proteins are characterised by an α-helix/β-sheet ratio higher than that of hard proteins. Table 8.1 lists a number of

Table 8.1 Examples of soft and hard proteins

Hard	Soft
Streptavidin	BSA
Avidin	Human serum albumin
α-Chemotrypsin	α-Lactalbumin
Cytochrome *c*	Myoglobin
Lysozyme	
Pyruvate kinase	
Subtilisin	
RNase (Ribonuclease)	

proteins according to their hard or soft character. However, complex proteins such as fibrinogen are excluded from this classification (Brandes et al., 2006).

The adsorption of "hard" proteins is simpler to describe (Fig. 8.1). It occurs on hydrophilic surfaces when the electrostatic interactions are favourable. Under this condition, the process is driven by the electrostatic attraction $\Delta H < 0$. On the contrary, under electrostatic repulsion, the $T\Delta S$ contribution has to overcome the $\Delta H > 0$. The hard protein adsorbs on hydrophobic surfaces. In this case, ΔH can be positive or negative, but it is small, and thus the driving force will be the change in entropy $T\Delta S > 0$. $\Delta S > 0$ involves a structural modification of the adsorbed protein explaining that hard protein adsorption on surfaces with similar charge and hydrophobic surfaces is usually not favourable. The adsorption of hard proteins weakly modifies the structure. It occurs primarily on the external surface of clay minerals (Rangel-Porras et al., 2015). The hard protein adopts preferential orientation called "side-on" or "end-on", which is driven by its charge distribution and dipole moment (Hristova and Zhivkov, 2015; Fig. 8.2). After adsorption, a reorientation end-on to side-on is possible (Wertz and Santore, 2002; Daly et al., 2003).

The adsorption process of "soft" proteins involves a structural modification, typically a partial or total unfolding (Daly et al., 2003) that increases the entropy ($T\Delta S > 0$) (Fig. 8.3). Thus, both the ΔH and $T\Delta S$ components have to be considered to predict the thermodynamics of the process (Lyklema and Norde, 1996). The low structural energy of soft proteins allows an optimisation of interactions between internal regions and the clay mineral surface by an increase of the number of contact points. Accordingly, soft proteins can adsorb on both hydrophobic and hydrophilic surfaces, even when the electrostatic interaction is unfavourable. Soft proteins can occupy the interlayer space, but also bind to the external surface of clay mineral.

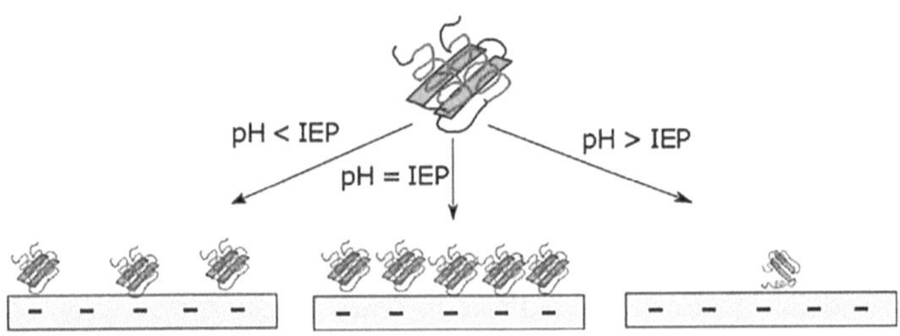

Fig. 8.1 Sketch of hard protein adsorption on clay minerals as a function of pH.

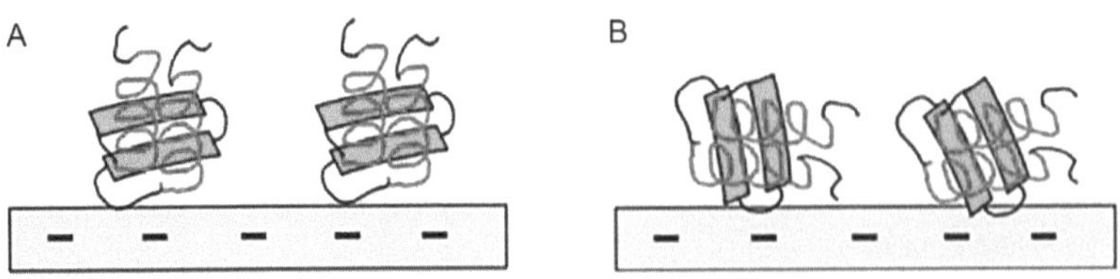

Fig. 8.2 Sketch of hard protein orientation end-on (A) and side-on (B).

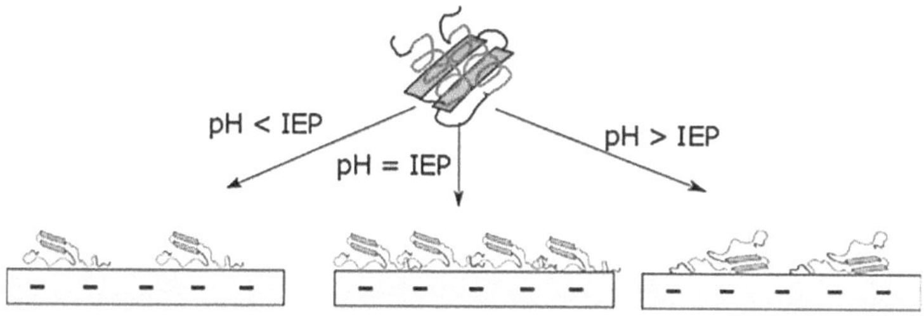

Fig. 8.3 Sketch of soft protein adsorption on clay minerals as a function of pH. The unfavourable electrostatic interaction at pH > IEP causes significant changes in the adsorbed proteins configuration.

8.3 Methodology to study protein adsorption

8.3.1 Interfacial concentration

The study of protein adsorption requires the determination of amounts of proteins adsorbed. While it is possible is to estimate the adsorbed amount by analysis of the solid phase, for example, through thermal analysis (Johnston et al., 2012), this is more often performed by the supernatant depletion method. This method consists of contacting a definite amount of clay mineral with a protein solution. After a time t, the solid phase is separated from the liquid one by centrifugation, and the concentration of protein in solution, $C(t)$, is measured. The interfacial concentration at time t, $\Gamma(t)$, is deduced from the following relation:

$$\Gamma(t) = \frac{V(C_0 - C(t))}{m} \tag{8.2}$$

where V is the volume of the protein solution, C_0 is the initial protein concentration, and m is the mass of the clay mineral.

When t is large enough, one can assume that an equilibrium surface concentration, Γ_{eq}, is reached; an adsorption isotherm is obtained by plotting Γ_{eq} as a function of C_{eq}.

The proteins usually contain aromatic amino acids (tryptophan or Trp, phenylalanine or Phe), which absorb UV light. Their quantification in the solution is commonly carried out by absorbance measurement at 280 nm. This method is direct, nondestructive, and does not require chemical treatment. However, some caveats are in order. The molar extinction coefficient of protein depends on external factors such as pH, solvent, ion strength, temperature, or presence of detergent, which can modify the protein structure and/or the environment of aromatic amino acids. A poor phase separation leaves clay mineral particles in solution, which can falsify the measurements due to diffuse light scattering. There are several alternative methods involving the labelling with dyes that adsorb in visible light (Lozzi et al., 2001). These methods usually improve the detection limit and the interferences of clay mineral layers on protein quantification. The most used one is the dye-binding Bradford assay. It is based on the interaction between dissociated groups of Coomassie blue and basic amino acid residues (i.e. arginine or Arg, lysine or Lys and histidine or His) (Bradford, 1976; Zor and Selinger, 1996). Coomassie blue binding shifts the maximum of absorbance from 465 to 610 nm. The main disadvantage of this method is the incompatibility with the surfactant, such as sodium dodecyl sulphate, Tween 80, or Triton X-100 (Compton and Jones, 1985). The Lowry assay is a redox spectrophotometric method (Lowry et al., 1951). It is based on the biuret reaction using Folin–Ciocalteu phenol reagent. The biuret reaction consists of the reduction of Cu^{2+} to Cu^{+}, which complexes a peptide to give a purplish-violet colour. Despite several improvements on the original method, the Lowry assay is limited by several factors, such as the interference of chelating agent, reducing agent, and free thiols. This method is also not compatible with the presence of surfactant and K^{+} ions due to precipitation. The Bicinchoninic acid assay is similar to the Lowry one (Smith et al., 1985). However, it is more sensitive, and compatible with up to 5 mass% of surfactant. The comparison between these methods shows that the Lowry assay is the least affected by the interference of the clay mineral, and thus affords the best estimate of protein concentration in the supernatant (Lozzi et al., 2008).

8.3.1.1 Adsorption isotherms

The isotherm allows to quantify the impact of different parameters, such as the intercalated ions or molecules, clay mineral treatment, and pH on the equilibrium of protein adsorption. Usually, attempts are made to fit the experimental data to a specific model, such as the Langmuir or Freundlich models, and often little comment is provided on the fits obtained. However, it should be kept in mind that each model is based on specific physicochemical assumptions, even if they are not explicitly recalled.

The Langmuir model assumes that the surface has a well-defined number of adsorption sites (Γ_m) that are energetically equivalent, and that the adsorption process is a dynamic equilibrium described by a constant similar to a thermodynamic equilibrium constant, which may be denoted as b, or sometimes K_L (Langmuir constant). The equilibrium relation is then

$$\Gamma_{eq} = \Gamma_m \frac{bC_{eq}}{1 + bC_{eq}} \tag{8.3}$$

or in terms of a fractional coverage θ defined as $\theta = \Gamma_{eq}/\Gamma_m$,

$$\theta = \frac{bC_{eq}}{1 + bC_{eq}} \tag{8.4}$$

and it depends on two parameters, b, the Langmuir constant and Γ_m, the maximum surface concentration or saturation coverage. It can be linearised, for which two equivalent equations are used:

$$-\text{plot}\frac{1}{\Gamma_{eq}}\text{ as a function of }\frac{1}{C_{eq}}:\frac{1}{\Gamma_{eq}}=\frac{1}{b\Gamma_m}\left(\frac{1}{C_{eq}}\right)+\frac{1}{\Gamma_m} \tag{8.5}$$

or

$$-\text{plot}\frac{C_{eq}}{\Gamma_{eq}}\text{ as a function of }C_{eq}:\frac{C_{eq}}{\Gamma_{eq}}=\frac{1}{b\Gamma_m}+\frac{1}{\Gamma_m}C_{eq} \tag{8.6}$$

To be meaningful, the linearisation requires several points reported in the low concentration range of the isotherm. Eq. (8.6) allows a more precise evaluation of Γ_m, which is extracted from the slope instead of the intercept.

Table 8.2 reports numerous examples from the literature where the Langmuir equation is suitable to describe experimental data. These examples highlight that the Langmuir model can be used for both soft and hard proteins. However, the use of this model can be debated because the hypotheses of the Langmuir model are not satisfied: in fact, the surface of the clay mineral is not homogeneous, and the protein adsorption is often irreversible.

The Freundlich equation is an empirical model that does not assume a homogeneous surface, but a continuous distribution of adsorption energies. As shown in Table 8.3, it can be used to describe the adsorption of hard and soft proteins on different kinds of clay minerals. The expression for the adsorption isotherm is:

$$\Gamma_{eq}=K_f C_{eq}{}^{1/n} \tag{8.7}$$

where n and K_f are Freundlich's constants. The coefficient K_f is related to the amount of protein adsorbed, the equilibrium concentration should appear as a dimensionless number (i.e. as $C_{eq}/C°$, where $C°$ is a reference concentration).

The Freundlich equation can be linearised by using a log–log plot:

$$\log\Gamma_e=\frac{1}{n}\log C_e+\log K_f \tag{8.8}$$

The parameter n is a dimensionless number related to the heterogeneity of the clay mineral. When $1/n$ tends to zero, the surface becomes more heterogeneous. If $n=1$, all surface sites are equivalent, and one obtains a linear isotherm, as in the initial stages of Langmuir adsorption.

Eq. (8.7) shows that Freundlich adsorption never reaches a saturation value, which is of course physically unrealistic. An alternative (Eq. 8.9), including a maximum of surface coverage (Γ_m), has been proposed to describe ion-exchange reactions in the soil (Sposito, 1980).

$$\log\frac{\Gamma_{eq}}{(\Gamma_m-\Gamma_{eq})}=\frac{1}{n}\log C_{eq}+\log\frac{K_s}{\Gamma_m} \tag{8.9}$$

Table 8.2 Examples of adsorption systems that can be described by Langmuir isotherm

Clay and/or clay mineral	Cation	Protein	Isotherm model	K_l	Unit	Γ_m	Unit	References
Mt	Na^+	Bt toxin (Cry1Aa)	Langmuir	3.57	L/g	1.36	g/g	Helassa et al. (2009)
Kaolinite	Na^+	Bt toxin (Cry1Aa)	Langmuir	5.26	L/g	0.027	g/g	Helassa et al. (2009)
Mt	Na^+	Bt toxin (Cry1Aa)	Langmuir	1.68	L/g	4.03	g/g	Zhou et al. (2005)
Kaolinite	Na^+	Bt toxin (Cry1Aa)	Langmuir	88.1	L/g	0.46	g/g	Zhou et al. (2005)
Bentonite	Na^+	Alkaline phosphatase	Langmuir	4.00	mL/U[a]	3.21	U/g[a]	Ghiaci et al. (2009a)
Bentonite	CTMAB	Alkaline phosphatase	Langmuir	2.88	mL/U[a]	3.23	U/g[a]	Ghiaci et al. (2009a)
Kaolinite	Na^+	BSA	Langmuir	4.66	L/g	53	mg/g	Duarte-Silva et al. (2014)
Kaolinite	Na^+	α-Lactalbumin	Langmuir	5.52	g/L	108	mg/g	Duarte-Silva et al. (2014)
Kaolinite	Na^+	β-Lactoglobulin	Langmuir	19.1	g/L	540	mg/g	Duarte-Silva et al. (2014)
Bentonite	Natural	Albumin egg	Langmuir	7.07	mL/mg	1430	mg/g	Sun et al. (2007)
Kaolinite	Natural	BSA	Langmuir	0.21	g/L	21.8	mg/g	Barral et al. (2008)
Kaolinite	Natural	α-Lactalbumin	Langmuir	1.57	g/L	111	mg/g	Barral et al. (2008)
Kaolinite	Natural	β-Lactoglobulin	Langmuir	0.22	g/L	45.24	mg/g	Barral et al. (2008)
Bentonite	Chitosan	α-Amylase	Langmuir	1.387	L/g	52.63	mg/g	Baysal et al. (2014)
Bentonite	Na^+	Lipase	Langmuir	0.16	L/g	80.43	mg/g	Dong et al. (2013a)
Bentonite	CTAB	Lipase	Langmuir	0.172	L/g	65.97	mg/g	Dong et al. (2013a)
Bentonite	DTAB	Lipase	Langmuir	0.68	L/g	69.94	mg/g	Dong et al. (2013a)
Bentonite	OTAB	Lipase	Langmuir	0.184	L/g	59.81	mg/g	Dong et al. (2013a)
Mt	$Ca2^+$	Peroxidase	Langmuir	0.096	L/g	0.68	mg/g	Lozzi et al. (2001)
Mt	Ca^{2+}	Catalase	Langmuir	0.6	μmol/g	3.1	μM/L	Calamai et al. (2000)
Mt	HDTMA	Catalase	Langmuir	1.4	μmol/g	17.8	μM/L	Calamai et al. (2000)
Mt	Pyridinium	Catalase	Langmuir	1.7	μmol/g	25	μM/L	Calamai et al. (2000)

[a] U is the amount of enzyme which produced 1 μL of *p*-nitrophenol, per minute at 37°C and at pH 10, from *p*-nitrophenyl phosphate.

Table 8.3 Examples of adsorption systems that can be described by a Freundlich isotherm and relevant parameters

Clay and/or clay mineral	Cation	Protein	Isotherm model	*K*	Units	*n*	References
Mt	Ca^{2+}	DNA polymerase	Freundlich	12.58	mL/mg	0.4875	Vettori et al. (1999)
Kaolinite	Ca^{2+}	DNA polymerase	Freundlich	6.0548	mL/mg	0.288	Vettori et al. (1999)
Bentonite	Na^{+}	Lipase	Freundlich	1.49	mL/g	1.32	Ghiaci et al. (2009b)
Bentonite	CTMAB	Lipase	Freundlich	2.52	mL/g	1.55	Ghiaci et al. (2009b)
Bentonite	NH_4^{+}	Gelatin	Freundlich	36	mL/g	1.9	Shamsuddin et al. (2014)
Sepiolite	Natural	BSA	Freundlich	1.55	mL/mg	1.87	Alkan et al. (2006)
Kaolinite	Natural	BSA	Freundlich	0.91	mL/mg	1.27	Alkan et al. (2006)
Kaolinite	Natural	Catalase	Freundlich	0.91	mL/mg	1.55	Demirbaş et al. (2013)
Kaolinite	Natural	Casein	Freundlich	1.87	mL/mg	2.52	Demirbaş et al. (2013)
Mt	Na^{+}	Peroxidase	Freundlich	0.12	mL/mg	1.10	Lozzi et al. (2001)

Another alternative is to introduce the empirical factor of heterogeneity n in the Langmuir equation. The "Freundlich–Langmuir" model provides:

$$\Gamma_{\text{eq}} = \Gamma_{\text{m}} \frac{K_{\text{LF}} {C_{\text{eq}}}^{1/n}}{1 + K_{\text{LF}} {C_{\text{eq}}}^{1/n}} \tag{8.10}$$

provided $1/n$ is known, it can be linearised as:

$$\frac{1}{\Gamma_{\text{eq}}} = \frac{1}{{C_{\text{eq}}}^{1/n}} \left(\frac{1}{K_{\text{LF}} \Gamma_{\text{m}}} \right) + \frac{1}{\Gamma_{\text{m}}} \tag{8.11}$$

The hybrid "Freundlich–Langmuir" approach seems particularly efficient to describe the gelatin adsorption on calcium and sodium bentonite (Shamsuddin et al., 2014). It is less often used than the Langmuir and Freundlich models, despite their approximations, because its fitting depends on three parameters instead of two. Anyway, all of these models may be criticised because they do not take into account the fundamental problem of the irreversibility of protein adsorption. There is not a simple model that provides a real description of the complexity of protein adsorption on clay minerals. The isotherms mentioned previously can be used to provide a rough "quantification" yielding empirical parameters such as a (pseudo)equilibrium constant or a maximum interfacial concentration. These parameters can be used to evaluate the influence of different experimental conditions, intercalated molecule, material, etc. However, even when a good fit is obtained, it should not be inferred that the adsorption mechanism has been understood.

8.3.1.2 Adsorption kinetics

Many kinetic studies of clay mineral–protein adsorption are empirical, and attempt to fit the simplest possible rate law on the adsorption data. For example, the rate law of (pseudo) first-order adsorption is:

$$\frac{d\Gamma(t)}{dt} = k_1 \left(\Gamma_{\text{eq}} - \Gamma(t) \right) \tag{8.12}$$

and by integrating:

$$\Gamma(t) = \Gamma_{\text{eq}} \left(1 - e^{-k_1 t} \right), \quad \text{or} \quad \ln \left(\Gamma_{\text{eq}} - \Gamma(t) \right) = \ln \Gamma_{\text{eq}} - k_1 t \tag{8.13}$$

A first-order phenomenon may be characterised by the half-time of adsorption $t_{1/2}$:

$$t_{1/2} = \frac{\ln 2}{k_1} \tag{8.14}$$

A first-order law is expected for irreversible adsorption on a limited number of homogeneous surface sites. A (pseudo) first-order has been reported for BSA and catalase adsorption on sepiolite and kaolinite and casein on kaolinite (Demirbaş et al., 2013).

(Pseudo)second-order adsorption equation answers to:

$$\frac{d\Gamma(t)}{dt} = k_2\left(\Gamma_{\text{eq}} - \Gamma(t)\right)^2 \tag{8.15}$$

with the integrated form:

$$\frac{t}{\Gamma(t)} = \frac{1}{k_2\Gamma_{\text{eq}}^2} + \frac{1}{\Gamma_{\text{eq}}}t \tag{8.16}$$

In principle, a second-order adsorption should correspond to a situation where the protein needs two adjacent free surface sites to adsorb.

The physical basis for applying these two models is flimsy. These models assume that the surface is homogeneous (only one type of adsorption site), and that there are no diffusional limits. Both assumptions are highly questionable in the case of clay minerals. In particular, it is likely that adsorptions on the external surface, on the edges and in the interlayer space, are not energetically equivalent.

The Elovich model also assumes that the adsorption rate is determined by surface reactions, but it allows to take into account the heterogeneity of the surfaces. The rate law is:

$$\frac{d\Gamma(t)}{dt} = \beta e^{-\alpha\Gamma(t)} \tag{8.17}$$

where α and β, respectively, represent an empirical constant connected with the activation energy distribution, and the initial adsorption rate (Aharoni and Suzin, 1982).

Integrating gives:

$$\Gamma(t) = \frac{1}{\beta}\ln(1 + \alpha\beta t) \tag{8.18}$$

The Elovich equation can be simplified (Chien and Clayton, 1980) assuming that $\alpha\beta t \gg 1$:

$$\Gamma(t) = \frac{1}{\beta}\ln(\alpha\beta) + \frac{1}{\beta}\ln(t) \tag{8.19}$$

so that a plot of $\Gamma(t)$ as a function of $\ln(t)$ will provide the adsorption constants.

It must be underlined that the Elovich equation cannot hold for long adsorption times, as the adsorbed amount would tend to infinity. It is best used for the initial steps of the adsorption.

Schmidt and Martínez (2016) have used the Elovich model to describe the adsorption of BSA on Mt.

8.3.2 Protein location

8.3.2.1 Transmission electronic microscopy

In order to ascertain the protein location on clay mineral surfaces, different techniques can be used, including X-ray diffraction (XRD) and transmission electronic microscopy (TEM). The measurement of the distance between dark and clear fringes in TEM micrographs (Fig. 8.4) allow the determination of the interlayer thickness, and therefore the possibility for protein orientation, compression, and large structural modification (Lepoitevin et al., 2014).

8.3.2.2 X-ray diffraction

In general, in the case of protein intercalation in the interlayer space of smectites, the (001) reflection is observed at lower angles than in the starting clay mineral, indicating an increase of the *d* value due to interlayer swelling. The precise value of the interlayer space may provide indications as to the protein orientation, and large structural conformational changes. In fact, the interlayer space expansion upon protein intercalation is often significantly smaller than the thickness of the protein in its native space, suggesting a significant degree of unwinding (Zhou et al., 2002; Szabó et al., 2007). Heterogeneity in the stacking of the layers can occur, inducing a broadening of the (001) reflection. Conversely, higher orders of the (001) reflection are observed when the protein–clay mineral complexes are very well ordered, a situation that has been observed in layer-by-layer construction of hybrid films (Szabó et al., 2007). In some situations where the interlayer spacing after protein intercalation is very high, it may occur that the (001) reflection falls at diffraction angles that are too low to be observed with ordinary instruments, but the (002) reflection may be used instead to determine the spacing. This is the case for hen egg white lysozyme intercalation into saponite (Sap), where the d_{001} value after intercalation reaches the unusually high value of 4.4 nm (Johnston et al., 2012).

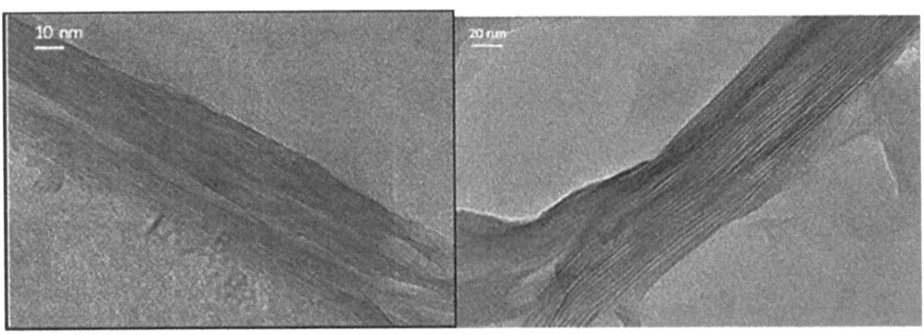

Fig. 8.4 TEM micrograph of a Mt BSA nanocomposite.

8.3.3 Structural modification of proteins

8.3.3.1 Fourier transformed infrared

Fourier transform infrared (FTIR) is a powerful tool to study protein structure and quantify the proportions of α-helices, β-sheets, turns, and disordered regions (Arrondo et al., 1993; Barth, 2007). The amide groups of the protein backbone have several vibrational modes active in the mid-IR region. An absorption band around 3300 cm^{-1} (3310–3270 cm^{-1} range), called the "amide A" band, corresponds to a simple N–H stretching. Because this vibration only involves the N–H group, its wavenumber is insensitive to the conformation of the polypeptide backbone, but it depends on the hydrogen-bonding state of the NH group. It is accompanied by a smaller band (around 3100–3030 cm^{-1}), weaker than the first one, probably due to a Fermi resonance with other vibrational modes of the protein.

The "amide I" vibration absorbs around 1650 cm^{-1}. It mainly involves a C=O stretching vibration, but the out-of-phase CN stretching, the CCN deformation, and the NH in-plane bending all have minor contributions. Thus, the amide I position depends on the precise shape of the backbone, and therefore it is the most commonly used vibration to analyse the secondary structure of proteins. The "amide II" absorbs around 1550 cm^{-1}. It is mainly due to the contribution of NH in-plane bending and CN stretching vibrations; the CO in-plane bend and the CC and NC stretching have minor contributions. The amide II band is affected by the protein structure, but the correlation between structure and frequency is more ambiguous than with the amide I. There is also an "amide III" band around 1200–1400 cm^{-1}, whose position depends not only on the backbone structure, but also on the side chains: its analysis is more complex.

The amide I vibration is almost exclusively used for routine evaluation of protein folding, unfolding, and aggregation, which may all occur upon adsorption on clay minerals. One often observes a broad band in the amide I region, which may be analysed into several components corresponding to at least four structural types (Table 8.4; Barth and Zscherp, 2002). There are two methods to elucidate the secondary structure of protein. The first one consists of fitting the amide I band with different component bands.

Table 8.4 Position of the amide I vibration for the main types of secondary structures encountered in proteins

	Band position in ^1H_2O/cm^{-1}		**Band position in D_2O/cm^{-1}**	
Secondary structure	**Average**	**Range**	**Average**	**Range**
α-Helix	1654	1648–1657	1652	1642–1660
β-Sheet	1633	1623–1641	1630	1615–1638
	1684	1674–1695	1679	1672–1694
Turns	1672	1662–1686	1671	1653–1691

From Barth, A., Zscherp, C., 2002. What vibrations tell about proteins. Q. Rev. Biophys. 35 (4), 369–430.

Then the position of each component is identified by band-narrowing techniques. The second method consists of calculating a basic spectrum from the known protein structure. Then the amide I band is decomposed and compared with the calculated spectrum.

In contrast to circular dichroism, which is the most efficient way to characterise the protein structure (Pelton and McLean, 2000), there are no interferences from light scattering in IR due to the colloidal dispersion. This makes IR suitable to elucidate the structural modifications induced by the protein adsorption on clay minerals. However, FTIR has to be performed on both the protein in solution, and adsorbed on the clay mineral under exactly the same conditions (buffer, pH, ionic strength). Basically, the FTIR spectra recorded on the dry clay mineral are not suitable for a relevant conformation analysis. There are two sampling techniques to record the IR spectrum of proteins, in solution or adsorbed on colloidal clay mineral particles. The first one is in the transmission mode, through a cuvette containing the clay mineral dispersion between two windows of ZnSe, BaF_2 of CaF_2. The major drawback of this method is the strong absorption of water near $1645\,cm^{-1}$, which overlaps with the amide I band (Venyaminov and Prendergast, 1997). In order to allow spectral analysis, the protein concentration has to be high (around 0.1–1 mM) and the optical length small (around 5 μm). Substitution of H_2O for D_2O shifts the water band to around $1210\,cm^{-1}$. This allows to increase the path length to 50 μm and to decrease the protein concentration in proportion. The second method is to work in the attenuated total reflectance (ATR) mode (Glassford et al., 2013). The protein solution, or the clay mineral–protein dispersion, is deposited on the crystal (e.g., ZnSe, diamond) surface and left to dry, and a droplet of buffer is deposited on the film (Oberg and Fink, 1998). This method presents several advantages, such as the possibility to change the buffer. However, the preparation of the film can be difficult, and drying can cause structural modifications. Between these two methods, the transmission measurement in D_2O is the most relevant to study the effects of protein adsorption on clay minerals, even if it is experimentally more complicated (Baron et al., 1999; Servagent-Noinville et al., 2000).

8.3.3.2 Fluorescence

Fluorescence is a powerful method to study the interaction of proteins with clay minerals. The fluorescence of proteins is caused by their Trp residues. The UV absorption of Trp comes from $\pi \rightarrow \pi^*$ transitions in the indole moieties. The band at 220 nm corresponds to the ^1B_b transition; the band at 260–290 nm is due to the ^1L_a and ^1L_b overlapping transitions (Strickland and Billups, 1973; Albinsson et al., 1989). The ^1L_a, which is sensitive to the polarity of light, is the main contributor of fluorescence emission. In aqueous solution, the Trp fluorescent emission lies at about 350 nm with an absolute quantum yield around 0.13. The photophysical properties of the Trp residues are directly affected by its neighbourhood. In the case of protein, they can be more or less quenched depending on their environment. As an example, the lysozyme sequence contains six Trp residues where the contributions of two of them represent 92% of the total fluorescence emission at 345 nm (Nishimoto et al., 1998, 1999). The Trp is thus a local probe that can inform on the protein-folding state, as well as on local structural modifications. When a Trp is located at the active site of an enzyme, its fluorescence can be used to evaluate the site accessibility after adsorption.

Four aspects of the fluorescence phenomenon can be experimentally studied: each one of them can provide information about the clay mineral–protein interactions.

(i) The *position* (wavelength at the emission maximum) reflects changes in the energetics of the excited and ground states. In practice, the position of the emission peak depends on the environment of Trp: it can vary by more than 40 nm, revealing large changes in the protein structure. In particular, protein unfolding results in increased exposure of Trp residues to the aqueous environment, which is revealed by a red shift of the fluorescence. Conversely, when a Trp becomes less exposed to the polar environment, a blue-shift occurs. This simple method is suitable to evidence the large structural modifications of soft proteins induced by adsorption. For instance, a shift towards shorter wavelengths was reported for BSA adsorbed on Mt. On the contrary, for hard proteins such as lysozyme, the emission spectrum is unmodified because the protein is not denaturated by the adsorption (Lepoitevin et al., 2014).

(ii) The *intensity* of the emission is related to the quantum yield q. The latter is defined as the ratio between the number of emitted and absorbed photons. It depends on the probability of various relaxation processes. The fluorescence intensity of Trp is extremely sensitive to its interaction with the environment. In a protein, the quantum yield can vary from close to 0 up to 0.5. The interaction with a solid or a ligand can directly affect the fluorescence of Trp by a collisional quenching or an energy transfer mechanism. The Trp fluorescence is also sensitive to the polarity and the accessibility of the solvent.

(iii) The fluorescence *lifetime* is a kinetic parameter that characterises the average time a molecule spends in the excited state before emitting a photon. The fluorescence decay of Trp is complex, and usually three processes with different lifetimes are required to fit the experimental data. The analysis of fluorescence lifetimes is particularly relevant to characterise weak structural modifications that occur close to the Trp (Fig. 8.5). It allows to interpret the modification of the activity of enzymes such as lysozyme upon adsorption on clay minerals, as well as to assess the protein orientation (Bouaziz et al., 2017).

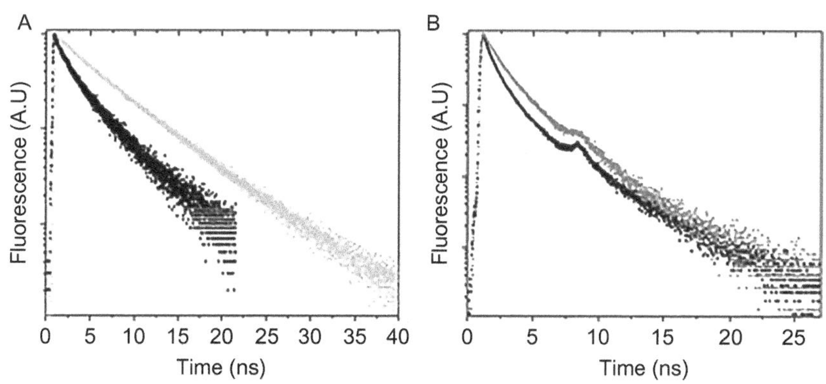

Fig. 8.5 Fluorescence decay of the soft protein BSA (A) and the hard protein lysozyme (B), in solution (*grey*) and adsorbed onto Mt (*black*).

(iv) Fluorescence *anisotropy* is a vectorial quantity reflecting the directional change of the emission transition moment with respect to the excitation state one. The intrinsic anisotropy r_0 varies from 0.4 to -0.2. These limiting values are reached when the absorption and emission transitions dipoles are parallel or perpendicular, respectively. From the fluorescence anisotropy, it is possible to access correlation times that correspond to the rotational diffusion coefficient. The correlation times of proteins are also complex to analyse. Indeed, the depolarisation can stem from the rotation of a Trp, of a chain, or of the global protein. However, protein immobilisation can be directly characterised by a blocking of the depolarisation function (Balme et al., 2017).

8.3.3.3 Nuclear magnetic resonance

Solid-state nuclear magnetic resonance (NMR) of ^{13}C and ^{1}H has the potential to yield important structural information on nanocomposite materials, including clay mineral-based ones (Bonhomme et al., 2014). It has not often been applied to clay mineral–protein systems due to the complexity of the signal of most proteins, but polypeptides (poly-Lys and poly-Glu) on Mt have been characterised by advanced NMR methods, including 2D techniques (^{1}H-^{27}Al HETCOR) (Gougeon et al., 2006). They revealed partial intercalation of the peptidic chains for the positively charged poly-Lysine, but not for the negatively charged poly-glutamate. More recently, a study of β-lactoglobulin on Mt (Assifaoui et al., 2014) reported separate ^{13}C peaks assigned to backbone carbonyls, aromatic side chain carbons (Trp, Tyrosinor Tyr), Arg residues, backbone α-CH carbons, and aliphatic side-chain carbons.

An early, creative application of NMR was to follow the cation exchange of BSA into clay minerals by adsorbing it on a Mn-exchanged Mt, and to observe the release of the paramagnetic Mn^{2+} cations through their broadening effect on the ^{31}P NMR signal of phosphates in solution (Quiquampoix and Ratcliffe, 1992).

8.4 Parameters that influence protein adsorption

8.4.1 Clay mineral structure

Depending on the clay mineral structure, different interactions between the protein and the surface occur. Hydrophilic and hydrophobic clay minerals can be distinguished as follows:

(i) Isomorphic substitutions in the tetrahedral or octahedral sheets, as well as the presence of defects due to the distortion induced by the difference between the bond lengths confer a hydrophilic character to the clay mineral. Moreover, the exchangeable cations, depending on their charge and ionic radius, are often surrounded by water molecules conferring a global hydrophilic effect to clay minerals.

(ii) On the contrary, in the absence of isomorphous substitution (talc, pyrophyllite), the surface is hydrophobic because the surface oxygens are exclusively implicated in the Si–O bonds, which are covalent and weakly polarised.

Thus, in general, clay minerals can be classified according to a charge criterion that will directly influence their interactions with biomolecules such as proteins.

On neutral clay minerals, water molecules interact more strongly with each other than with the siloxane surface (Rotenberg et al., 2011). This kind of surface has a higher affinity for nonpolar molecules and solvents. In the case of charged clay minerals such as smectites, the location of the permanent charge dominates the process of adsorption. Tetrahedral isomorphous substitutions exert their electrostatic influence on a small region of the surface, leaving much of the distant siloxanes unaffected and ready to interact with nonpolar molecules, while in the case of octahedral substitution, the effect of the negative charge is "spread out" laterally, therefore conferring to the whole surface a global hydrophilic character and a low affinity for nonpolar molecules.

Besides the siloxane-containing basal surfaces, clay minerals also contain hydroxyl groups that should be taken into account due to their reactivity with water molecules via hydrogen bonding. In the case of kaolinite (a TO clay mineral), one side of the layer presents siloxanes, while the opposite surface presents hydroxyls. Therefore, the two opposite sides have different reactivities. Taking the example of water, the siloxane surface interacts very weakly with water molecules, while the hydroxylated surface has a high affinity to water.

There are different types of protonated groups that are not limited to the TO, but are also present in the TOT clay minerals: those that are located at the edges of the clay layers. These protonated groups, resulting from the interruption of the clay mineral lattices, are silanol groups (Si–OH) at the edge of the tetrahedral sheets, with ligands completing the coordination sphere of the octahedral cations. These groups have a very different behaviour from surface sites behaviour on the basal surfaces, because they can behave as Brönsted acids and/or bases. Their charge is pH dependent: silanols are deprotonated to silanolates $Si–O^-$. Ligands on octahedral cations (e.g. Mg^{2+}) can be present as $Mg^{2+}(OH_2)$ or $Mg^{2+}(OH^-)$, and so forth. Of course, the charge state of these groups affects the interactions with any water or organic molecules. Moreover, the groups in the tetrahedral sheet are more accessible to organic molecules than those in the octahedral one (Stackhouse et al., 2004). Particular attention should be paid to edge groups, because they are more reactive than the siloxane groups located on the basal surface.

The pH is one of the most important parameters for the interaction between clay minerals and proteins because it affects both the surface charge of the inorganic matrix, and the charge of the protein (Tran and James, 2012). The adsorption of proteins on the clay mineral surface is pH dependent. The positively charged proteins can intercalate via a cationic exchange process in the interlayer space of smectites (Johnston et al., 2012). The adsorption of negatively charged proteins is more complicated, but possible through dispersive interactions between the siloxane surface and the functional groups of the protein. These dispersive interactions should overcome electrostatic repulsions between the clay mineral surface and protein. Another type of interaction can occur at acidic pH between the anionic groups and clay mineral edges via electrostatic interactions, because the aluminium exposed at these sites is positively charged in these conditions. The dependency of protein adsorption on the pH decreases as the ionic force increases. Another possible bonding mode is that of "ligand exchange", by which the anionic group of the protein enters the inner coordination sphere of edge aluminium to form a coordination complex.

The swelling of the clay mineral in an aqueous dispersion is among the parameters that govern the protein adsorption, as well as the size of the protein (molar mass). Swelling clay minerals will interact via their internal and external surfaces with the proteins, while nonswelling clay minerals will only interact via their external surface. Several examples in the literature are reported on Mt, as an example of a swelling clay mineral, and kaolinite and illite for nonswelling ones (Fusi et al., 1989; Zhou et al., 2005; Barral et al., 2008). The structure and size of the protein strongly influence the way the adsorption occurs.

8.4.2 Interlayer cation

The cations present in the system have an impact on both protein and clay mineral structure. "Kosmotropic", that is, order-generating cations (Li^+, Na^+, Mg^{2+}, Ca^{2+}, and Ba^{2+}) stabilise the protein structure in solution. They bind water molecules strongly through coordinative bonds, and thus favour the swelling of clay minerals. In the interlayer space, the cation concentration is high, and this prevents protein unfolding in the adsorbed state, as well (Helassa et al., 2011b). The "chaotropic" cations (K^+, Rb^+ and Cs^+) have the opposite effect: they destabilise the protein structure. They have a low affinity for water molecules, and thus hinder clay mineral swelling.

The adsorption of lysozyme and protamine on Na^+-Sap occur on the edges and external surface of Sap, and in the interlayer space. Lysozyme is not intercalated in Cs^+-Sap (Johnston et al., 2012). For protamine, which is a small protein, the intercalation is partial (Szabó et al., 2008).The reason is that Cs^+ is less hydrated, and thus the smectite is less prone to swelling. The intercalation of protein in the interlayer space increases the maximum of interfacial concentration, and the affinity constant, as shown for the lysozyme (Lepoitevin et al., 2014). The interlayer cations impact more of the adsorption process of hard protein than that of soft ones. Indeed, because soft protein is more prone to unfolding, they can penetrate the interlayer space more easily. This can be illustrated by the low difference of interfacial concentration observed for BSA when chaotropic and kosmotropic cations are intercalated inside Mt.

For *kosmotropic cations*, the valence also plays a role on the protein adsorption. Usually, divalent cations such as Ca^{2+} promote the protein adsorption, as reported for alkaline phosphatase and lysozyme on Sap (Carrasco et al., 1995) and Mt, respectively. The amount of horseradish peroxidase adsorbed is three times higher on Na^+-Mt than on Ca^{2+}-Mt, due to a partial interaction in the interlayer space. This intercalation dramatically inhibits the enzymatic activity due to inaccessibility of the substrate (Calamai et al., 2000; Lozzi et al., 2001). On the contrary, for lysozyme, the adsorbed amount is larger on Ca^{2+}-Mt than on Na^+-Mt (Lepoitevin et al., 2014). The intercalated cation also influences the loading of negatively charged proteins. The adsorption of alkaline phosphatase (IEP 5.7) at pH 7 on homoionically exchanged Sap reveals a larger interfacial concentration for Ca^{2+} than Na^+ after washing, while the enzymatic activity is better for Na^+-Sap (Carrasco et al., 1995). This behaviour was interpreted by a structural effect of the cation.

At low ionic strength, the α-helix content of *Bacillus thuringiensis* (Bt) in pesticidal protein (Cry1Aa) decreases, inducing an oligomerisation. It was surprisingly reported that monomeric Cry1Aa keeps its structure when it is adsorbed on Mt at low ionic strength (Helassa et al., 2011a). Actually, the local concentration of countercations at the interface is high. This can prevent the unfolding of the α-helix that occurs at low salt concentration.

8.4.3 pH of the adsorption solution

The pH directly influences both the global charge and structure of proteins. Its effect depends on the type of interaction between the protein and the clay mineral. If the electrostatic interactions dominate the adsorption, the dependence on pH will be high (Causserand et al., 1997). Even if a precise law cannot be formulated, this is a general tendency for all proteins.

The adsorption of a protein is optimal at pH close to its isoelectric point (IEP). Examples are DNA polymerase (Vettori et al., 1999), monomeric Bt Cry1A (Carrasco et al., 1995) or laccase (Hu et al., 2007). This behaviour is assigned to the overall zero charge of the protein, which minimises the electrostatic repulsion (Helassa et al., 2011a). Below the IEP, the adsorption increases with pH. Above the IEP, it decreases with increasing pH. In this case, the dependence is most pronounced when the electrostatic interaction dominates (Causserand et al., 1997). Indeed, to counterbalance the electrostatic repulsion, the proteins have to optimise the number of weak interactions. This general law also depends on the protein "hardness". For BSA, a soft protein, the adsorption induces a decrease in the α-helix/β-sheet ratio, whatever the pH. At pH close to the IEP, external and internal α-helix domains are unfolded in order to provide a maximum of hydrophobic interactions with the clay mineral. This explains that the maximum of interfacial concentration of BSA is observed at pH close to IEP (Staunton and Quiquampoix, 1994). At pH lower than the IEP (2.9), the behaviour of adsorbed BSA follows the same trend, and the protein is mainly unfolded. At pH higher than the IEP (6.7), the negative charge of BSA makes the adsorption less favourable. Despite this electrostatic repulsion, adsorption occurs due the unfolding of the peripheral domains of α-helix (Servagent-Noinville et al., 2000). In this case, the protein unfolding is partial.

Because the hard proteins are less prone to unfolding, the compensation of electrostatic repulsive effect is not counterbalanced by the optimisation of hydrophobic interaction with the clay mineral, as occurs for soft proteins (Tran and James, 2012). As an example, the adsorption of α-chemotrypsin does not induce a strong modification of secondary structure at pH below or close to the IEP. At pH above the IEP, the electrostatic repulsions would make one think that adsorption is not possible. However, it can still occur and is accompanied by a weak peripheral structural modification of protein to optimise the interaction with the clay mineral. This partial unfolding, combined with the internal hydration in the vicinity of the catalytic centre, are at the origin of the loss of enzymatic activity reported at pH 10 (Tran and James, 2012). The global behaviour of hard proteins is that the decrease of adsorption at pH above IEP is sharper than for soft proteins (Tran and James, 2012).

This general description does not claim to predict the behaviour of all proteins. In the case of Bt CryA1, the optimal pH of adsorption is close to the IEP (between 4.4 and 5.5) for Mt and kaolinite (Helassa et al., 2009). However, opposing results were reported by Zhou et al. (2005) where the optimal pH is 9 for Mt. In addition, this description does not take into account other effects of pH on the protein, such as oligomerisation.

8.5 An overview of the adsorption of different proteins

8.5.1 Bovine serum albumin

BSA may be considered as a reference for soft proteins and is often used to understand the behaviour of proteins at solid/liquid interfaces. The behaviour of BSA in solution and in interaction with Mt was studied as a function of pH (Servagent-Noinville et al., 2000). In general, the protein was denatured in the presence of clay minerals, but more so with Mt than with talc. BSA was methylated to elucidate the balance between hydrophobic and electrostatic interactions in the adsorption processes. Further investigations have allowed researchers to demonstrate that the interfacial concentration of the protein plays a role on its conformation (Staunton and Quiquampoix, 1994; Schmidt and Martínez, 2016). Besides the interest it generates as a tool to investigate fundamental issues, BSA serves as a model to evaluate the suitability of clay minerals to remove the protein in wine (Salazar et al., 2006; Sarmento et al., 2000), as well as the performance of membranes for protein separation (Tran et al., 2012). From the biological point of view, albumin is the major protein in blood, with a concentration between 35 and 50 g/L. Thus, it is systematically used to evaluate the interference of plasma. As an example, the functionalisation of a clay mineral with His allows improvement of its specific interaction with IgG (immunoglobulin G) in order to carry out specific separation (Öztürk et al., 2007).

8.5.2 Enzymes

Enzymes are the proteins for which adsorption is most often studied on clay minerals. Enzyme immobilisation on clay minerals can either occur spontaneously, or be induced by specific manipulations in the laboratory when the immobilisation is desirable to build specific devices, such as heterogenised enzymatic catalysts or sensors. Especially, attempts have often been made to transpose immobilisation procedures that have proved efficient on other surfaces, such as glass or silica. The most popular one involves acid activation of the clay mineral to produce silanols, functionalisation of the latter by a silane bearing an amino group (such as aminopropyltrimethoxysilane or APTMS), and covalent grafting of the protein on the latter using a glutaraldehyde cross-linker (see Fig. 8.6). The efficiency of glutaraldehyde as a linker depends on its condensation with amine groups, on both the solid support and the protein. One cannot help being sceptical about the use of glutaraldehyde on supports that have not been previously functionalised by APTMS (Šekuljica et al., 2016). At any rate, these directed chemical procedures completely change the nature of the clay mineral–protein interaction, invalidating many of the remarks that were made in Section 8.4, for spontaneous adsorption.

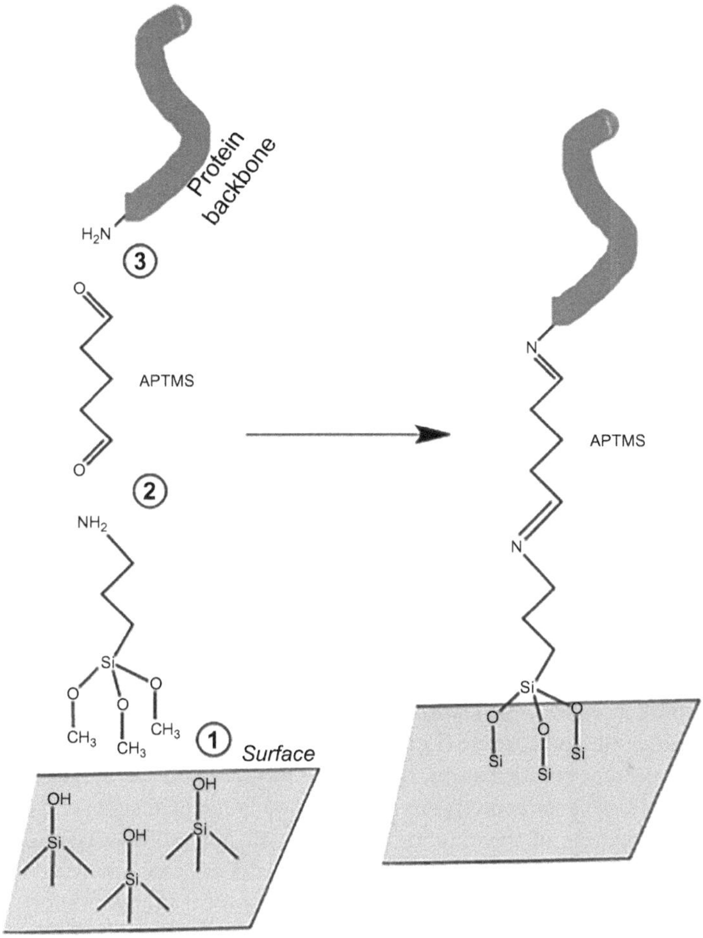

Fig. 8.6 The strategy for immobilisation of proteins on silicic surfaces by grafting, using a glutaraldehyde (APTMS) spacer. Step (1) is the functionalisation of surface silanols by an amine-bearing molecule such as APTMS. Step (2) is the glutaraldehyde addition. Step (3) is the protein grafting by amine group.

This elaborate procedure is meant to avoid close contact between the enzyme and the surface; the two being separated by a linker, in the hope that the enzyme will function in the same way as in an aqueous solution. In some articles, it is called a "covalent procedure", as opposed to an "adsorption procedure", but this is misleading. The term "grafting" vs spontaneous adsorption would be preferable. In the case of spontaneous adsorption, enzymes belonging to the "hard" category are more likely to keep their structural integrity, while "soft" ones may lose the structural features responsible for their catalytic activity. Another condition must be satisfied, however, for the immobilised enzymes to work satisfactorily: the active site of the enzyme has to be accessible to the substrate, meaning that the protein has to be correctly oriented.

Quantitatively, enzymatic activity in solution usually obeys the Michaelis–Menten kinetics:

$$v = \frac{V_{\max}[S]}{K_M + [S]} \tag{8.20}$$

where $[S]$ is the substrate concentration, and the reaction rate v depends on two parameters, the Michaelis constant K_M (inversely proportional to the enzyme–substrate affinity) and a maximum rate $V_{\max}$.

Immobilisation preserves the Michaelis behaviour of the enzyme; however, the kinetic constants are often affected. Typically, $V_{\max}$ decreases and K_M increases, meaning that the efficiency of the enzyme is lower after adsorption (Vettori et al., 1999; Rao et al., 2000). This partial loss of enzymatic activity is attributable to weak structural modification, as well as an unfavourable orientation. However, the complexity of clay mineral–protein interactions does not allow one to predict the enzymatic activity a priori. On the positive side, the adsorption of enzymes on clay minerals usually improves their thermal stability (Rao et al., 2000; Yeşiloğlu, 2005; Cengiz et al., 2012): a possible explanation is that thermal unfolding of the secondary/tertiary structure of the enzyme is made more difficult by its multipoint adsorption on the clay mineral surface (reduction of conformational flexibility). The same point has been made for enzyme adsorption on other surfaces (Cetinus and Oztop, 2000).

Another interesting feature is that the optimal pH of activity can shift by more than one unit upon adsorption. This was observed for lipase (Yeşiloğlu, 2005), catalase (Calamai et al., 2000), and urease (Leprince and Quiquampoix, 1996). This is ascribed to changes in protein conformation, but also to the fact that the local pH may differ from the bulk pH, because protons are concentrated in the diffuse layer to compensate for the negative charge of the clay mineral. For values significantly higher or lower than the optimal pH, the activity falls down to zero. This phenomenon is not always reversible, as shown in the case of β-D-glucosidase on Mt (Quiquampoix, 1987).

One last general feature, which is actually the justification for immobilising enzymes, is that adsorption allows reuse of the catalytic system several times because of the ease of catalyst/products separation.

Several clay mineral modifications have been explored in order to improve the efficiency of enzyme immobilisation. The most common ones are intercalation of amino acids, such as cysteine (Cys) or Trp, and intercalation of surfactants such as cetyltrimethylammonium bromide (CTAB) or nonionic surfactants. Both procedures open up the interlayer space and also change its character from hydrophilic to hydrophobic. Sometimes quite complicated multicomponent nanocomposites are used, which are beyond the scope of this chapter.

The following paragraphs list recent studies according to the type of enzymes immobilised.

Lipases catalyse the hydrolysis of numerous fatty acyl esters at the oil/water interfaces. As with many enzymes, the adsorption of lipase allows reuse of the enzyme in processes of depollution or food transformation, and to improve its thermal stability. For example, while the half-life of free lipase from *Candida rugosa* at 60°C is only 5 min, it is 17 min after adsorption on bentonite (Yeşiloğlu, 2005). The optimal pH to hydrolyse olive oil is also shifted, from about 6 for free lipase to 7.5 for the immobilised one. The main drawback is the important loss of activity; around 70% of absorbed lipase, explained by a low transfer rate of the substrate (Yeşiloğlu, 2005). Several strategies were explored in order to facilitate substrate diffusion. The total exfoliation of bentonite layers permitted the stabilisation of a "house-of-cards" nanocomposite structure for bentonite lipase nanocomposites, where adsorbed enzymes were easily accessible (Cao et al., 2016).

Another strategy is the intercalation of surfactants, which creates a hydrophobic environment stabilising the enzymatic activity. Indeed, CTAB intercalation allows retaining 75% of activity of absorbed *C. rugosa* lipase, compared with the free enzyme (Ghiaci et al., 2009b). An improvement of the catalytic activity was also reported for dodecyltrimethylammonium bromide (DTAB), CTAB, and octadecyltrimethylammonium bromide (OTAB). CTAB was also used to enhance bovine pancreatic lipase activity. Indeed, on Na^+-bentonite, the enzyme loses 80% of its activity, while it keeps a 50% activity when Na^+ is exchanged with CTAB (Dong et al., 2013b). In organic media, the catalytic activity of *C. rugosa* lipase is improved when it is bound covalently on the Mt surface. On the contrary, in aqueous media, immobilisation by adsorption is more efficient (Reshmi and Sugunan, 2013). Surfactants such as DTAB, CTAB, and OTAB limit the loss of activity of lipase from *C. rugosa* after reuse and incubation at 50°C (Dong et al., 2013a). The intercalation of benzyltriethyl ammonium in Mt improves the thermal stability of enzymes in the range between 35°C and 45°C (Díaz Ramos et al., 2014). After four uses, the activity of free lipase B from *Candida antarctica* is lower than 10%, while it is about 60% and 40% for OTAB-exchanged Laponite and Mt, respectively (Tzialla et al., 2010).

Phosphatase plays a key role in the phosphorous cycle in the soil. It is secreted by bacteria and plants and cleaves the phosphate moiety of phosphorylated organic molecules, forming inorganic phosphate that can be assimilated. In the soil, these secreted enzymes can establish interactions with the clay mineral fraction. The studies of both acid and alkaline phosphatase are of interest for the understanding of the phosphate cycle in the soil, as well as for the degradation of organophosphorus pollutants such as pesticides. The acid phosphatases have been less frequently studied than the alkaline phosphatases. Their specific activity is enhanced between 4% and 48%, and the V_{max} is improved to about 30% after immobilisation on clay minerals. The most efficient clay mineral is allophane, compared with Mt (Calabi-Floody et al., 2012). After adsorption on Mt, the activity of acid phosphatase from a potato decreases to 20%, compared with that of the free enzyme in solution (Vettori et al., 1999). On Na^+-sepiolite, the adsorption of alkaline phosphatase reduces its activity to around 50%, while on Na^+-bentonite, it is close to that of the free enzyme in solution. For both clay minerals, the intercalation of a double layer of CTMAB allows recovery of the enzymatic activity, while a monolayer decreases or inhibits it (Ghiaci et al., 2009b; Sedaghat et al., 2009b). The adsorption of alkaline phosphatase (IEP 5.7) at pH 7 on homoionically exchanged Sap reveals a larger interfacial concentration for Ca^{2+} than Na^+ after washing, while the enzymatic activity is better for Na^+-Sap than for Ca^{2+}-Sap. As previously mentioned, the interest of immobilising enzymes on clay minerals lies in the improvement of thermal stability and enzymatic activity. Alkaline phosphatase is a perfect counter example. Indeed, after adsorption on Na^+-Mt or Na^+-Sap, a faster decrease of enzyme stability was observed, as compared with its stability in solution. After 2 days at 37°C, the activity of alkaline phosphatase loaded on Na^+-Sap is 32.3%, compared with 94.2% for the same enzyme in solution (Carrasco et al., 1995). On Na^+-Mt, a quick loss of activity of about 50% was reported after about 1 h. The adsorption of alkaline phosphatase on Na^+-Mt occurs on the external surface. However, the adsorption of protamine allows the intercalation of the enzyme and stabilises the fast protein unfolding at 37°C (Tang et al., 1993).

Glycoside hydrolases catalyse the hydrolysis of glycosidic bonds, such as those between hexose monomers in complex carbohydrates. *Lysozyme* catalyses the cleavage of the glycosidic linkage β (1-4) between *N*-acetylglucosamine and *N*-acetylmuramic acid in peptidoglycans, resulting in antibiotic properties against Gram-positive bacteria whose cell wall contains these molecules and is thus destroyed by lysozymes. In addition, lysozyme is a hard protein, and thus has served as a model for numerous studies of adsorption. On Mt, the amount of lysozyme absorbed is larger on Ca^{2+}-Mt than on Na^+-Mt (Lepoitevin et al., 2014). The intercalation of Trp in bentonite allows an increase in the enzyme loading from a few $mg\,g^{-1}$ to $200\,mg\,g^{-1}$ at pH 10 (Kalburcu et al., 2015). Modification of Mt by a nonionic surfactant allowed tuning of the adsorption mechanism: lysozyme was intercalated in both cases, but in the end-on and side-on orientations, respectively (Balme et al., 2017).

It may be difficult to successfully immobilise lysozyme, because its substrate is located on the bacterial membrane, and therefore only the enzymes adsorbed on the outer clay mineral surfaces can access their target. On layered double hydroxide, most lysozyme molecules were adsorbed on the external surfaces; but even so, a slight decrease of antibacterial activity was observed (by 6%, and around 22% for the fibrillated form), which is likely due to a weak modification of the active site (Bouaziz et al., 2017).

α-Amylase allows the hydrolysis of long-chain polysaccharides. The immobilisation of *Bacillus subtilis* α-amylase on Na^+-Mt decreases the activity to 49%, as compared with amylase in solution (Gopinath and Sugunan, 2007). This number was interpreted as an effectiveness factor η, giving information on diffusional limitations, that is, other factors, such as protein denaturation, were not considered. The efficiency factor could be somewhat improved (to 0.59) when the enzyme is covalently grafted. On Na^+-bentonite, V_{max} is divided by a factor of 5 compared with V_{max} of the enzyme in solution; the optimal temperature shifts from 37°C (enzyme in solution) to 45°C and 60°C for Na^+ and CTAB exchanged clay minerals, respectively. The immobilisation on clay minerals also allows maintenance of the activity with time (Sedaghat et al., 2009a).

Catalase and peroxidase. Catalase is an abundant enzyme that decomposes H_2O_2. Its immobilisation on bentonite and sepiolite induces a shift of optimal pH from 7 to about 9. Interestingly, V_{max} decreases by a factor of about 4; but on the other hand, the affinity with the substrate is improved (Cengiz et al., 2012). The adsorption on clay minerals also prevents thermal degradation. The intercalation of Cys in bentonite layers increases the interfacial concentration of catalase from 2.7 to 175 mg/g.

However, the activity is not improved, because K_M is close to that of the enzyme in solution (Öztürk et al., 2007).

All enzymes are not suitable for application after adsorption on clay mineral. *α-Chemotrypsin* is a hard protein; its function is to perform proteolysis by hydrolysing peptide bonds in proteins. After adsorption on Mt, a partial loss of enzymatic activity and a shift of the optimal pH from 8 to 9 (Baron et al., 1999) are observed. The study of protein structure reveals that a partial unfolding, combined with the internal hydration in the vicinity of catalytic centre is at the origin of this result. The intercalation of the enzyme in the interlayer space is not always desirable. Indeed, although horseradish peroxidase adsorbed on Na-Mt only undergoes a partial intercalation, this is enough to make the enzymatic site inaccessible to the substrate (Lozzi et al., 2001), dramatically inhibiting the enzymatic activity. Some enzymes such as DNA polymerase from *Thermus aquaticus* are not suitable, because it loses its activity after adsorption on both kaolinite and Mt, even if it is not intercalated.

Oxidases: laccase. Laccase catalyses the one-electron oxidation of phenols by O_2 and intervenes in the synthesis of lignin and its further modifications, including the formation of lacquer (hence the name). It is used in the food industry because of its ability to cross-link polysaccharides, and in environmental remediation to activate polyaromatic hydrocarbons and start their degradation. Laccase has been immobilised on kaolin, both by simple adsorption, and by grafting (acid activation, functionalisation by a silane bearing an amino group, and use of glutaraldehyde as cross-linker, cf. Hu et al., 2007). The same method was also applied for immobilisation on Mt (Chang et al., 2016). Adsorption allows reusing the enzyme, which makes application possible as a phenol sensor (Songurtekin et al., 2013).

8.5.3 Structural proteins: collagen/gelatin/fibrinogen

Collagen is the most abundant protein in many animals; it plays a structural role as an extracellular protein in connective tissues. The molar mass of collagen components is about 100 and 200 kDa for α and β chains, respectively. Collagen and its composites are the object of significant interest due to their role in wound care, for example, as burns dressing, bone scaffolding and also as a binder in paintings.

Gelatin is a denaturated (partly hydrolysed) form of collagen used in the industry. It consists of a heterogeneous mixture of single and multistranded polypeptides, containing sequences of repeated triplets (glycine–X–Y), where *X* and *Y* are often proline and hydroxyproline. Historically, gelatin is one of the first proteins with which interaction with clay minerals has been studied (Talibudeen, 1950). While gelatin extraction may be desirable for purification, it would be interesting to kill two birds with one stone and form bioplastic nanocomposites in the same process. Bentonites rapidly extracted gelatin from food by-products (stickwater); the loading depended on pH and compensating cation, Ca^{2+}-bentonite being most efficient at pH 7–9 and Na^{+}-bentonite at pH 3–5.2 (Shamsuddin et al., 2014).

Gelatin nanocomposites were prepared with Mt (Zheng et al., 2002) or laponite (Xavier et al., 2015) alone, or in more complex formulations with chitosan (a functionalised polysaccharide obtained from chitine). The swelling properties of gelatin were decreased by combining with Mt (Li et al., 2003). If chitosan was added, highly porous scaffolds were obtained with good mechanical properties (Zheng et al., 2007) and good biocompatibility (Zhuang et al., 2007). In some of these studies, clay minerals are a minority component, for example, 5–10 mass% in a gelatin hydrogel (Rao, 2007), especially for applications to the food and packaging industry (Bae et al., 2009), where Cu^{2+}-exchanged Mt can be used to add antibiotic activity (Martucci and Ruseckaite, 2017). Thus, it is to be expected that a significant part of the polypeptide chains is not interacting with the clay mineral component, in contrast to more classical adsorption studies. Biohybrid (i.e. nanocomposite) materials of the same type were also obtained from Mt mixed with zein, the storage protein of corn (Alcântara et al., 2016). The optimisation of large-scale physical parameters of the nanocomposites, such as the Young modulus, does not necessarily correlate with the nanoscale interactions between its components: for example, in a series of comparable gelatin-based bionanocomposites, it was observed that gelatin only intercalated in Mt while this was not the case with the high-layer charge vermiculite, nor with the anionic hydrotalcite (Fernandes et al., 2009).

Fibronectin is another extracellular protein that is involved in cell spreading (McGrath, 2007). Attempts were made to use it in order to cultivate cells in microfluidic reactors. It turned out that satisfactory adhesion and cell development could be insured by using multicomponent architectures where fibronectin was deposited on top of clay mineral–PDDA mixtures (Mehta et al., 2007; PDDA is a polymeric ammonium-based surfactant). No physical insight on the interaction of the different components was offered.

8.5.4 Toxins

Bt produces proteins that are toxic to some insects, such as the Cry1A toxin. The genes responsible for the production of these proteins may be introduced into transgenic crops, and its persistence in soils is crucial to assess its effect on the environment (Pagel-Wieder et al., 2004; Fu et al., 2008, 2009; Helassa et al., 2009, 2011b). The insecticidal activity of this protein is related to its ability to oligomerise. Adsorption on Mt prevented oligomerisation (Helassa et al., 2011b) by protecting a specific α-helix domain from unfolding; a step necessary for the formation of oligomers. Thus, clay minerals retain the toxic protein in the soil, but in a form that is less potent than in solution. Attempts have been made to use these properties to prepare adsorption formulations of Bt toxin, and it was shown that clay minerals protected the active site from UV degradation (Zhou et al., 2008; Zhou, 2011). Beyond Mt, kaolinite and rectorite have been tested for this purpose.

Ricin is another toxin obtained from castor beans that has been militarised for chemical war. Highly charged Mt effectively retains ricin, apparently by intercalation in the interlayer spaces, up to 0.7 g ricin per g Mt (Jaynes et al., 2005). It could be useful for remediation measures following, for example, a terrorist attack.

Prions, the agents responsible for bovine spongiform encephalopathy (mad cow disease) are also proteins, and Mt has been shown to adsorb up to its own weight of prion proteins PrP, making it likely that soils can play a role in the transmission of the disease. The prions adsorption mechanism is a combination of electrostatic interaction through positively charged N-terminal domains, and hydrophobic interactions with the basal siloxane surface, and is accompanied by a loss in α-helices concomitant to an increase in β-structures (Rigou et al., 2006).

8.6 Conclusion

Protein adsorption on clay minerals is a thriving field of science. As compared with other classes of adsorbates studied in this book, proteins are characterised by a high level of structural complexity. Thus, studies of their adsorption are concerned not only with the mechanism of their interaction with clay minerals and the effects of adsorption on the clay mineral matrix (such as swelling), but also with its effects on the secondary, tertiary, and quaternary structures of the proteins. This is why the concept of hard and soft proteins plays an important role in understanding proteins adsorption. In a nutshell, it allows us to establish a distinction between proteins that essentially keep their structure intact on the one hand, and those that unfold (or change their folding) upon adsorption on the other hand. Experimental techniques to evaluate the folding state of the protein chains and the presence of secondary structure motives in situ (i.e. while interacting with the clay mineral) have a special importance, and among them, IR spectroscopy and fluorescence methods have been the most informative so far.

Another consequence of the structural complexity is that a protein chain can establish many interactions with the adsorbing surface. Electrostatic interactions certainly play a role, but if they were the only factor at play, protein adsorption would always be favoured at low pH because of the positive charges on the protein when $pH < IEP$. This is not true, and in many cases adsorption is most favoured at a pH close to the IEP. This is because many "weak" interactions, such as H-bonding and hydrophobic bonding, can cooperate to overcome electrostatic effects. The sheer size of protein chains makes it more likely that cooperative weak interactions will dominate the energetic picture.

From a practical point of view, clay mineral protein adsorption studies may be motivated by two main fields of application. They may be aimed at scavenging or water purification, in which case technical parameters such as the maximum loading or the speed of adsorption will be the most important concerns. This is why empirical treatments of the adsorption data are valuable, even though their theoretical justification is unclear, as discussed in this chapter—for example, fitting the adsorbed amounts to a Freundlich isotherm, or the adsorption vs time data to first-order or second-order kinetics.

In other studies, the aim is to load a protein, most often an enzyme, by attaching it to a solid surface, enabling easy separation from the operating solution and reuse. In that case, minimal interaction with the surface seems desirable because the activity of the protein is highly dependent on its conformation, and any unfolding/refolding will be detrimental. This leads to adsorption procedures based on surface functionalisation, aiming at a limited number of controlled covalent bonds with the surface, through the use of an intercalated linker.

For the material scientist, clay mineral protein systems represent a fascinating type of nanocomposite in which the two components interact in ways that sometimes have no parallel in other types of hybrid materials. The existing body of information on various clay mineral protein systems will no doubt lead to new insights for the preparation of nanocomposites based on clay minerals.

References

Aharoni, C., Suzin, Y., 1982. Application of the Elovich equation to the kinetics of occlusion: analysis of experimental data from the literature. J. Chem. Soc., Faraday Trans. I 78, 2320–2327.

Albinsson, B., Kubijsta, M., Norden, B., Thulstrup, E.W., 1989. Near-ultraviolet electronic transitions of the tryptophan chromophore: linear dichroism, fluorescence anisotropy, and magnetic circular dichroism spectra of some indole derivatives. J. Phys. Chem. 93 (18), 6646–6654.

Alcântara, A.C.S., Darder, M., Aranda, P., Ruiz-Hitzky, E., 2016. Effective intercalation of zein into Na-montmorillonite: role of the protein components and use of the developed biointerfaces. Beilstein J. Nanotechnol. 7, 1772–1782.

Alkan, M., Demirbaş, Ö., Doğan, M., Arslan, O., 2006. Surface properties of bovine serum albumin—adsorbed oxides: adsorption, adsorption kinetics and electrokinetic properties. Microporous Mesoporous Mater. 96 (1–3), 331–340.

Arai, T., Norde, W., 1990. The behavior of some model proteins at solid-liquid interfaces 1. Adsorption from single protein solutions. Colloids Surf. 51 (C), 1–15.

Arrondo, J.L.R., Muga, A., Castresana, J., Goni, F.M., 1993. Quantitative studies of the structure of proteins in solution by Fourier-transform infrared spectroscopy. Prog. Biophys. Mol. Biol. 59, 23–56.

Ashby, N.P., Binks, B.P., 2000. Pickering emulsions stabilised by laponite clay particles. Phys. Chem. Chem. Phys. 2 (24), 5640–5646.

Assifaoui, A., Huault, L., Maissiat, C., Roullier-Gall, C., Jeandet, P., Hirschinger, J., Raya, J., Jaber, M., Lambert, J.-F., Cayot, P., Gougeon, R.D., Loupiac, C., 2014. Structural studies of adsorbed protein (betalactoglobulin) on natural clay (montmorillonite). RSC Adv. 4 (105), 61096–61103.

Bae, H.J., Park, H.J., Hong, S.I., Byun, Y.J., Darby, D.O., Kimmel, R.M., Whiteside, W.S., 2009. Effect of clay content, homogenization RPM, pH, and ultrasonication on mechanical and barrier properties of fish gelatin/montmorillonite nanocomposite films. LWT—Food Sci. Technol. 42 (6), 1179–1186.

Balme, S., Guégan, R., Janot, J.-M., Jaber, M., Lepoitevin, M., Dejardin, P., Bourrat, X., Motelica-Heino, M., 2017. Structure, orientation and stability of lysozyme confined in layered materials. Soft Matter 7, 3188–3196.

Baron, M., Revault, M., Servagent-Noinville, S., Abadie, J., Quiquampoix, H., 1999. Chymotrypsin adsorption on montmorillonite: enzymatic activity and kinetic FTIR structural analysis. J. Colloid Interface Sci. 214 (2), 319–332.

Barral, S., Villa-García, M.A., Rendueles, M., Díaz, M., 2008. Interactions between whey proteins and kaolinite surfaces. Acta Mater. 56 (12), 2784–2790.

Barth, A., 2007. Infrared spectroscopy of proteins. Biochim. Biophys. Acta 1767, 1073–1101.

Barth, A., Zscherp, C., 2002. What vibrations tell us about proteins. Q. Rev. Biophys. 35 (4), 369–430.

Baysal, Z., Bulut, Y., Yavuz, M., Aytekin, Ç., 2014. Immobilization of α-amylase via adsorption onto bentonite/chitosan composite: determination of equilibrium, kinetics, and thermodynamic parameters. Starch/Stärke 66 (5–6), 484–490.

Bonhomme, C., Gervais, C., Laurencin, D., 2014. Recent NMR developments applied to organic–inorganic materials. Prog. Nucl. Magn. Reson. Spectrosc. 77, 1–48.

Bouaziz, Z., Soussan, L., Janot, J.-M., Lepoitevin, M., Bechelany, M., Djebbi, M.A., Amara, A.B.H., Balme, S., 2017. Structure and antibacterial activity relationships of native and amyloid fibril lysozyme loaded on layered double hydroxide. Colloids Surf. B Biointerfaces 157, 10–17.

Bradford, M.M., 1976. A rapid and sensitive method for the quantitation of microgram quantities of protein utilizing the principle of protein-dye binding. Anal. Biochem. 72, 248–254.

Brandes, N., Welzel, P.B., Werner, C., Kroh, L.W., 2006. Adsorption-induced conformational changes of proteins onto ceramic particles: differential scanning calorimetry and FTIR analysis. J. Colloid Interface Sci. 299 (1), 56–69.

Calabi-Floody, M., Velásquez, G., Gianfreda, L., Saggar, S., Bolan, N., Rumpel, C., Mora, M.L., 2012. Improving bioavailability of phosphorous from cattle dung by using phosphatase immobilized on natural clay and nanoclay. Chemosphere 89 (6), 648–655.

Calamai, L., Lozzi, I., Stotzky, G., Fusi, P., Ristori, G.G., 2000. Interaction of catalase with montmorillonite homoionic to cations with different hydrophobicity: effect on enzymatic activity and microbial utilization. Soil Biol. Biochem. 32 (6), 815–823.

Cao, J., Li, Y., Tu, N., Lv, Y., Chen, Q., Dong, H., 2016. Novel enzyme/exfoliated bentonite nanohybrids as highly efficient and recyclable biocatalysts in hydrolytic reaction. J. Mol. Catal. B: Enzym. 132, 41–46.

Carrasco, M.S., Rad, J.C., Gonzalez-Carcedo, S., 1995. Immobilization of alkaline-phosphatase by sorption on Na-sepiolite. Bioresour. Technol. 51 (2–3), 175–181.

Causserand, C., Jover, K., Aimar, P., Meireles, M., 1997. Modification of clay cake permeability by adsorption of protein. J. Membr. Sci. 137 (1–2), 31–44.

Cengiz, S., Çavaş, L., Yurdakoç, K., 2012. Bentonite and sepiolite as supporting media: immobilization of catalase. Appl. Clay Sci. 65–66, 114–120.

Cetinus, S.A., Oztop, H.N., 2000. Immobilization of catalase on chitosan film. Enzyme Microb. Technol. 26 (7), 497–501.

Chang, Y.T., Lee, J.F., Liu, K.H., Liao, Y.F., Yang, V., 2016. Immobilization of fungal laccase onto a nonionic surfactant-modified clay material: application to PAH degradation. Environ. Sci. Pollut. Res. Int. 23 (5), 4024–4035.

Chien, S.H., Clayton, W.R., 1980. Application of Elovich equation to the kinetics of phosphate release and sorption in soils. Soil Sci. Soc. Am. J. 44 (2), 265.

Compton, S.J., Jones, C.G., 1985. Mechanism of dye response and interference in the Bradford protein assay. Anal. Biochem. 151 (2), 369–374.

Daly, S.M., Przybycien, T.M., Tilton, R.D., 2003. Coverage-dependent orientation of lysozyme adsorbed on silica. Langmuir 19 (9), 3848–3857.

Demirbaş, Ö., Alkan, M., Demirbaş, A., 2013. Surface properties of catalase and casein on kaolinite and design of experiments. Microporous Mesoporous Mater. 172, 151–160.

Díaz Ramos, M., Giraldo Gómez, G.I., Sanabria González, N., 2014. Immobilization of Candida rugosa lipase on bentonite modified with benzyltriethylammonium chloride. J. Mol. Catal. B: Enzym. 99, 79–84.

Dong, H., Li, Y., Li, J., Sheng, G., Chen, H., 2013a. Comparative study on lipases immobilized onto bentonite and modified bentonites and their catalytic properties. Ind. Eng. Chem. Res. 52 (26), 9030–9037.

Dong, H., Li, Y., Sheng, G., Hu, L., 2013b. The study on effective immobilization of lipase on functionalized bentonites and their properties. J. Mol. Catal. B: Enzym. 95, 9–15.

Duarte-Silva, R., Villa-García, M.A., Rendueles, M., Díaz, M., 2014. Structural, textural and protein adsorption properties of kaolinite and surface modified kaolinite adsorbents. Appl. Clay Sci. 90, 73–80.

Farahnaky, A., Dadfar, S.M.M., Shahbazi, M., 2014. Physical and mechanical properties of gelatin-clay nanocomposite. J. Food Eng. 122, 78–83.

Fernandes, F.M., Ruiz, A.I., Darder, M., Aranda, P., Ruiz-Hitzky, E., 2009. Gelatin-clay bio-nanocomposites: structural and functional properties as advanced materials. J. Nanosci. Nanotechnol. 9 (1), 221–229.

Fu, Q.L., Wang, W.Q., Hu, H.Q., Chen, S.W., 2008. Adsorption of the insecticidal protein of Bacillus thuringiensis subsp. kurstaki by minerals: effects of inorganic salts. Eur. J. Soil Sci. 59, 216–221.

Fu, Q.L., Deng, Y.L., Li, H.S., Liu, J., Hu, H.Q., Chen, S.W., Sa, T.M., 2009. Equilibrium, kinetic and thermodynamic studies on the adsorption of the toxins of Bacillus thuringiensis subsp. kurstaki by clay minerals. Appl. Surf. Sci. 255 (8), 4551–4557.

Fusi, P., Ristori, G.G., Calamai, L., Stotzky, G., 1989. Adsorption and binding of protein on "clean" (homoionic) and "dirty" (coated with Fe oxyhydroxides) montmorillonite, illite and kaolinite. Soil Biol. Biochem. 21 (7), 911–920.

Ganley, W.J., Ryan, P.T., van Duijneveldt, J.S., 2017. Stabilisation of water-in-water emulsions by montmorillonite platelets. J. Colloid Interface Sci. 505, 39–147.

Garland, A., Shen, L., Zhu, X., 2012. Mobile precursor mediated protein adsorption on solid surfaces. Prog. Surf. Sci. 87 (1–4), 1–22.

Ghiaci, M., Aghaei, H., Soleimanian, S., Sedaghat, M.E., 2009a. Enzyme immobilization: part 2. Immobilization of alkaline phosphatase on Na-bentonite and modified bentonite. Appl. Clay Sci. 43 (3–4), 308–316.

Ghiaci, M., Aghaei, H., Soleimanian, S., Sedaghat, M.E., 2009b. Enzyme immobilization. Part 1. Modified bentonite as a new and efficient support for immobilization of Candida rugosa lipase. Appl. Clay Sci. 43 (3–4), 289–295.

Glassford, S.E., Byrne, B., Kazarian, S.G., 2013. Recent applications of ATR FTIR spectroscopy and imaging to proteins. Biochim. Biophys. Acta Protein Proteomics 1834 (12), 2849–2858.

Gopinath, S., Sugunan, S., 2007. Enzymes immobilized on montmorillonite K 10: effect of adsorption and grafting on the surface properties and the enzyme activity. Appl. Clay Sci. 35 (1–2), 67–75.

Gougeon, R.D., Reinholdt, M., Delmotte, L., Miehé-Brendlé, J., Jeandet, P., 2006. Solid-state NMR investigation on the interactions between a synthetic montmorillonite and two homopolypeptides. Solid State Nucl. Magn. Reson. 29, 322–329.

Haynes, C.A., Norde, W., 1994. Globular proteins at solid/liquid interfaces. Colloids Surf. B Biointerfaces 2, 517–566.

Helassa, N., Quiquampoix, H., Noinville, S., Szponarski, W., Staunton, S., 2009. Adsorption and desorption of monomeric Bt (Bacillus thuringiensis) Cry1Aa toxin on montmorillonite and kaolinite. Soil Biol. Biochem. 41, 498–504.

Helassa, N., M'Charek, A., Quiquampoix, H., Noinville, S., Déjardin, P., Frutos, R., Staunton, S., 2011a. Effects of physicochemical interactions and microbial activity on the persistence of Cry1Aa Bt (Bacillus thuringiensis) toxin in soil. Soil Biol. Biochem. 43 (5), 1089–1097.

Helassa, N., Revault, M., Quiquampoix, H., Déjardin, P., Staunton, S., Noinville, S., 2011b. Adsorption on montmorillonite prevents oligomerization of Bt Cry1Aa toxin. J. Colloid Interface Sci. 356 (2), 718–725.

Hristova, S.H., Zhivkov, A.M., 2015. Adsorption of cytochrome c on montmorillonite nanoplates: protein concentration dependence. J. Colloid Interface Sci. 446, 252–262.

Hu, X., Zhao, X., Hwang, H.M., 2007. Comparative study of immobilized Trametes versicolor laccase on nanoparticles and kaolinite. Chemosphere 66 (9), 1618–1626.

Jahn, T.R., Radford, S.E., 2005. The Yin and Yang of protein folding. FEBS J. 272 (23), 5962–5970.

Jaynes, W.F., Zartman, R.E., Green, C.J., San Francisco, M.J., Zak, J.C., 2005. Castor toxin adsorption to clay minerals. Clay Clay Miner. 53 (3), 268–277.

Johnston, C.T., Premachandra, G.S., Szabo, T., Lok, J., Schoonheydt, R.A., 2012. Interaction of biological molecules with clay minerals: a combined spectroscopic and sorption study of lysozyme on saponite. Langmuir 28 (1), 611–619.

Kalburcu, T., Tabak, A., Ozturk, N., Tuzmen, N., Akgol, S., Caglar, B., Denizli, A., 2015. Adsorption of lysozyme from aqueous solutions by a novel bentonite-tryptophane (Bent-Trp) microcomposite affinity sorbent. J. Mol. Struct. 1083, 156–162.

Kasemo, B., 2002. Biological surface science. Surf. Sci. 500, 656–677.

Koo, J., Erlkamp, M., Grobelny, S., Steitz, R., Czeslik, C., 2013. Pressure-induced protein adsorption at aqueous-solid interfaces. Langmuir 29 (25), 8025–8030.

Lepoitevin, M., Jaber, M., Guegan, R., Janot, J.-M., Dejardin, P., Henn, F., Balme, S., 2014. BSA and lysozyme adsorption on homoionic montmorillonite: influence of the interlayer cation. Appl. Clay Sci. 95, 396–402.

Leprince, F., Quiquampoix, H., 1996. Extracellular enzyme activity in soil: effect of pH and ionic strength on the interaction with montmorillonite of two acid phosphatases secreted by the ectomycorrhizal fungus Hebeloma cylindrosporum. Eur. J. Soil Sci. 47 (4), 511–522.

Li, P., Zheng, J.P., Ma, Y.L., De Yao, K., 2003. Gelatin/montmorillonite hybrid nanocomposite. II. Swelling behavior. J. Appl. Polym. Sci. 88 (2), 322–326.

Lowry, O.H., Rosebrough, N.J., Farr, A.L., Randall, R.J., 1951. Protein measurement with the folin-phenol reagent. J. Biol. Chem. 193 (1), 265–275.

Lozzi, I., Calamai, L., Fusi, P., Bosetto, M., Stotzky, G., 2001. Interaction of horseradish peroxidase with montmorillonite homoionic to Na 1 and Ca 2 1: effects on enzymatic activity and microbial degradation. Soil Biol. Biochem. 33 (7–8), 1021–1028.

Lozzi, I., Pucci, A., Pantani, O.L., D'Acqui, L.P., Calamai, L., 2008. Interferences of suspended clay fraction in protein quantitation by several determination methods. Anal. Biochem. 376 (1), 108–114.

Lyklema, J., Norde, W., 1996. Interfacial behaviour of biomacromolecules. Progr. Colloid Polym. Sci. 4, 9–17.

Martucci, J.F., Ruseckaite, R.A., 2017. Antibacterial activity of gelatin/copper (II)-exchanged montmorillonite films. Food Hydrocoll. 64, 70–77.

McGrath, J.L., 2007. Cell spreading: the power to simplify. Curr. Biol. 17 (10), R357–R358.

Mehta, G., Kiel, M.J., Lee, J.W., Kotov, N., Linderman, J.J., Takayama, S., 2007. Polyelectrolyte-clay-protein layer films on microfluidic PDMS bioreactor surfaces for primary murine bone marrow culture. Adv. Funct. Mater. 17, 2701–2709.

Nakanishi, K., Sakiyama, T., Imamura, K., 2001. On the adsorption of proteins on solid surfaces, a common but very complicated phenomenon. J. Biosci. Bioeng. 91 (3), 233–244.

Nishimoto, E., Yamashita, S., Szabo, A.G., Imoto, T., 1998. Internal motion of lysozyme studied by time-resolved fluorescence depolarization of tryptophan residues. Biochemistry 37 (16), 5599–5607.

Nishimoto, E., Yamashita, S., Yamasaki, N., Imoto, T., 1999. Resolution and characterization of tryptophyl fluorescence of hen egg-white lysozyme by quenching- and time-resolved spectroscopy. Biosci. Biotechnol. Biochem. 63 (2), 329–336.

Oberg, K.A., Fink, A.L., 1998. A new attenuated total reflectance Fourier transform infrared spectroscopy method for the study of proteins in solution. Anal. Biochem. 256 (1), 92–106.

Öztürk, N., Tabak, A., Akgöl, S., Denizli, A., 2007. Newly synthesized bentonite-histidine (Bent-His) micro-composite affinity sorbents for IgG adsorption. Colloids Surf. A Physicochem. Eng. Asp. 301 (1–3), 490–497.

Pagel-Wieder, S., Gessler, F., Niemeyer, J., Schroder, D., 2004. Adsorption of the Bacillus thuringiensis toxin (Cry1 Ab) on Na-montmorillonite and on the clay fractions of different soils. J. Plant Nutr. Soil Sci. 167 (2), 184–188.

Pelton, J.T., McLean, L.R., 2000. Spectroscopic methods for analysis of protein secondary structure. Anal. Biochem. 277, 167–176.

Quiquampoix, H., 1987. A stepwise approach to the understanding of extracellular enzyme activity in soil. I. Effect of electrostatic interactions on the conformation of a beta-D-glucosidase adsorbed on different mineral surfaces. Biochimie 69 (6–7), 753–763.

Quiquampoix, H., Ratcliffe, R.G., 1992. A ^{31}P NMR study of the adsorption of bovine serum albumin on montmorillonite using phosphate and the paramagnetic cation Mn^{2+}: modification of conformation with pH. J. Colloid Interface Sci. 148 (2), 343–352.

Rabe, M., Verdes, D., Seeger, S., 2011. Understanding protein adsorption phenomena at solid surfaces. Adv. Colloid Interface Sci. 162, 87–106.

Rangel-Porras, G., Rangel-Rivera, P., Pfeiffer-Perea, H., Gonzalez-Muñoz, P., 2015. Changes in the characteristics of acid-treated clay after the inclusion of proteins. Surf. Interface Anal. 47 (1), 135–141.

Rao, Y.Q., 2007. Gelatin-clay nanocomposites of improved properties. Polymer 48 (18), 5369–5375.

Rao, M.A., Violante, A., Gianfreda, L., 2000. Interaction of acid phosphatase with clays, organic molecules and organo-mineral complexes: kinetics and stability. Soil Biol. Biochem. 32 (7), 1007–1014.

Reshmi, R., Sugunan, S., 2013. Superior activities of lipase immobilized on pure and hydrophobic clay supports: characterization and catalytic activity studies. J. Mol. Catal. B: Enzym. 97, 36–44.

Rigou, P., Rezaei, H., Grosclaude, J., Staunton, S., Quiquampoix, H., 2006. Fate of prions in soil: adsorption and extraction by electroelution of recombinant ovine prion protein from montmorillonite and natural soils. Environ. Sci. Technol. 40 (5), 1497–1503.

Rotenberg, B., Patel, A.J., Chandler, D., 2011. Molecular explanation for why talc surfaces can be both hydrophilic and hydrophobic swelling materials for protein adsorption from white wines. Int. J. Food Sci. Technol. 35 (1), 41–47.

Rytwo, G., Mendelovits, A., Eliyahu, D., Pitcovski, J., Aizenshtein, E., 2010. Adsoption of two vaccine-related proteins to montmorillonite and organo-montmorillonite. Appl. Clay Sci. 50, 569–575.

Salazar, F.N., Achaerandio, I., Labbe, M.A., Guell, C., Lopez, F., 2006. Comparative study of protein stabilization in white wine using zirconia and bentonite: physicochemical and wine sensory analysis. J. Agric. Food Chem. 54, 9955–9958.

Sarmento, M.R., Oliveira, J.C., Boulton, R.B., 2000. Selection of low swelling materials for protein adsorption from white wines. Int. J. Food Sci. Technol. 35, 41–47.
Schmidt, M.P., Martínez, C.E., 2016. Kinetic and conformational insights of protein adsorption onto montmorillonite revealed using in situ ATR-FTIR/2D-COS. Langmuir 32 (31), 7719–7729.
Sedaghat, M.E., Ghiaci, M., Aghaei, H., Soleimanian-Zad, S., 2009a. Enzyme immobilization. Part 3. Immobilization of α-amylase on Na-bentonite and modified bentonite. Appl. Clay Sci. 46 (2), 125–130.
Sedaghat, M.E., Ghiaci, M., Aghaei, H., Soleimanian-Zad, S., 2009b. Enzyme immobilization. Part 4. Immobilization of alkaline phosphatase on Na-sepiolite and modified sepiolite. Appl. Clay Sci. 46 (2), 131–135.
Šekuljica, N.Ž., Prlainović, N.Ž., Jovanović, J.R., Stefanović, A.B., Grbavčić, S.Ž., Mijin, D.Ž., Knežević-Jugović, Z.D., 2016. Immobilization of horseradish peroxidase onto kaolin by glutaraldehyde method and its application in decolorization of anthraquinone dye. Hemijska Industrija 70 (2), 217–224.
Servagent-Noinville, S., Revault, M., Quiquampoix, H., Baron, M., 2000. Conformational changes of bovine serum albumin induced by adsorption on different clay surfaces: FTIR analysis. J. Colloid Interface Sci. 221 (2), 273–283.
Shamsuddin, R.M., Verbeek, C.J.R., Lay, M.C., 2014. Producing protein intercalated bentonite—equilibrium, kinetics and physical properties of gelatin-bentonite system. Appl. Clay Sci. 87, 52–60.
Smith, P.K., Krohn, R.I., Hermanson, G.T., Mallia, A.K., Gartner, F.H., Provenzano, M.D., Fujimoto, E.K., Goeke, N.M., Olson, B.J., Klenk, D.C., 1985. Measurement of protein using bicinchoninic acid. Anal. Biochem. 150 (1), 76–85.
Songurtekin, D., Yalcinkaya, E.E., Ag, D., Seleci, M., Demirkol, D.O., Timur, S., 2013. Histidine modified montmorillonite: laccase immobilization and application to flow injection analysis of phenols. Appl. Clay Sci. 86, 64–69.
Sposito, G., 1980. Derivation of the Freundlich equation for ion exchange reactions in soils. Soil Sci. Soc. Am. J. 44 (3), 652.
Stackhouse, S., Coveney, P.V., Benoit, D.M., 2004. Density functional based theory of the dehydroxylation behavior of the aluminous dioctahedral 2:1 layer type minerals. J. Phys. Chem. B 108 (28), 9685–9694.
Staunton, S., Quiquampoix, H., 1994. Adsorption and conformation of bovine serum-albumin on montmorillonite—modification of the balance between hydrophobic and electrostatic interactions by protein methylation and pH variation. J. Colloid Interface Sci. 166 (1), 89–94.
Strickland, E.H., Billups, C., 1973. Oscillator strengths of the 1La and 1Lb absorption bands of tryptophan and several other indoles. Biopolymers 12 (9), 1989–1995.
Sun, X., Li, C., Wu, Z., Xu, X., Ren, L., Zhao, H., 2007. Adsorption of protein from model wine solution by different bentonites. Chin. J. Chem. Eng. 15 (5), 632–638.
Szabó, T., Szekeres, M., Dékány, I., Jackers, C., Feyter, S.D., Johnston, C.T., Schoonheydt, R.A., 2007. Layer-by-layer construction of ultrathin hybrid films with proteins and clay minerals. J. Phys. Chem. C 111, 12730–12740.
Szabó, T., Mitea, R., Leeman, H., Premachandra, G.S., Johnston, C.T., Szekeres, M., Dékány, I., Schoonheydt, R.A., 2008. Adsorption of protamine and papain proteins on saponite. Clay Clay Miner. 56 (5), 494–504.
Talibudeen, O., 1950. Interlamellar adsorption of protein monolayers on pure montmorillonoid clays. Nature 166 (4214), 236.

Tang, X., Shen, Y.-f., Suib, S.L., Coughlinb, R.W., Vinopal, R., 1993. Immobilization of alkaline phosphatase on protamine—intercalated bentonite. Microporous Mater. 2, 65–71.

Tran, A.T.T., James, B.J., 2012. A study the interaction forces between the bovine serum albumin protein and montmorillonite surface. Colloids Surf. A Physicochem. Eng. Asp. 414, 104–114.

Tran, A.T.T., Patterson, D.A., James, B.J., 2012. Investigating the feasibility of using polysulfone-montmorillonite composite membranes for protein adsorption. J. Food Eng. 112 (1–2), 38–49.

Tzialla, A.A., Pavlidis, I.V., Felicissimo, M.P., Rudolf, P., Gournis, D., Stamatis, H., 2010. Lipase immobilization on smectite nanoclays: characterization and application to the epoxidation of α-pinene. Bioresour. Technol. 101 (6), 1587–1594.

Venyaminov, S.Y., Prendergast, F.G., 1997. Water (H_2O and D_2O) molar absorptivity in the 1000–4000 cm^{-1} range and quantitative infrared spectroscopy of aqueous solutions. Anal. Biochem. 248 (2), 234–245.

Vettori, C., Calamai, L., Yoder, M., Stotzky, G., Gallori, E., 1999. Adsorption and binding of AmpliTaq® DNA polymerase on the clay minerals, montmorillonite and kaolinite. Soil Biol. Biochem. 31 (4), 587–593.

Wertz, C.F., Santore, M.M., 2002. Adsorption and reorientation kinetics of lysozyme on hydrophobic surfaces. Langmuir 18 (4), 1190–1199.

Xavier, J.R., Thakur, T., Desai, P., Jaiswal, M.K., Sears, N., Cosgriff-Hernandez, E., Kaunas, R., Gaharwar, A.K., 2015. Bioactive nanoengineered hydrogels for bone tissue engineering: a growth-factor-free approach. ACS Nano 9 (3), 3109–3118.

Yeşiloğlu, Y., 2005. Utilization of bentonite as a support material for immobilization of Candida rugosa lipase. Process Biochem. 40 (6), 2155–2159.

Yu, W.H., Li, N., Tong, D.S., Zhou, C.H., Lin, C.X., Xu, C.Y., 2013. Adsorption of proteins and nucleic acids on clay minerals and their interactions: a review. Appl. Clay Sci. 80–81, 443–452.

Zheng, J.P., Li, P., Ma, Y.L., De Yao, K., 2002. Gelatin/montmorillonite hybrid nanocomposite. I. Preparation and properties. J. Appl. Polym. Sci. 86 (5), 1189–1194.

Zheng, J.P., Wang, C.Z., Wang, X.X., Wang, H.Y., Zhuang, H., De Yao, K., 2007. Preparation of biomimetic three-dimensional gelatin/montmorillonite-chitosan scaffold for tissue engineering. React. Funct. Polym. 67, 780–788.

Zhou, X.Y., 2011. Adsorption formulation of the insecticidal protein from Bacillus thuringiensis. In: Hu, J. (Ed.), Advances in Biomedical Engineering. Information Engineering Research Inst, Newark, DE.

Zhou, Y., Li, Z., Hu, N., Zeng, Y., Rusling, J.F., 2002. Layer-by-layer assembly of ultrathin films of hemoglobin and clay nanoparticles with electrochemical and catalytic activity. Langmuir 18 (22), 8573–8579.

Zhou, X., Huang, Q., Chen, S., Yu, Z., 2005. Adsorption of the insecticidal protein of Bacillus thuringiensis on montmorillonite, kaolinite, silica, goethite and red soil. Appl. Clay Sci. 30 (2), 87–93.

Zhou, X.Y., Gao, J.B., Chen, S., Wang, B.J., 2008. Adsorption and anti-ultraviolet light characteristics of the protoxin from Bacillus thuringiensis on montmorillonite, kaolinite, zinc oxide and rectorite. Biocontrol Sci. Technol. 18 (10), 1051–1061.

Zhuang, H., Zheng, J.P., Gao, H., Yao, K.D., 2007. In vitro biodegradation and biocompatibility of gelatin/montmorillonite-chitosan intercalated nanocomposite. J. Mater. Sci. Mater. Med. 18, 951–957.

Zor, T., Selinger, Z., 1996. Linearization of the Bradford protein assay increases its sensitivity: theoretical and experimental studies. Anal. Biochem. 236, 302–308.

Clay mineral catalysts

9

Dipak Kumar Dutta
Advanced Materials Group, Materials Science and Technology Division,
CSIR-North East Institute of Science and Technology, Jorhat, Assam, India

9.1 Introduction

Clay minerals are layered aluminosilicates showing a large variety of physicochemical properties, such as expansion of the interlayer space, cation exchange, small particle size, high surface area, and so forth. They are naturally occurring, cheap, eco-friendly, and abundantly available. The most studied clay mineral of the smectite group is montmorillonite (Mt), which is considered to be an excellent catalyst for, for example, the catalysed formation of biopolymers of the first life (Ferris and Ertem, 1993; Ferris et al., 1996; Hill et al., 1998; Liu and Orgel, 1998; Ferris, 2002), and an efficient support of heterogeneous catalysts with several advantages, such as high activity and selectivity (Ballantine et al., 1984; Clark, 1994; Phukan et al., 2003; Ahmed and Dutta, 2005).

Solid acid catalysts can be developed from clay minerals, such as Mt, a hydrated 2:1 layered dioctahedral aluminosilicate of the smectite group having cation exchange capacity (CEC) 70–130 meq per 100 g and high swelling properties for adequate intercalation and intersalation (accommodation of ion-pair salt into the interlayer space). Exchange of interlayer Na^+ with high charge cations such as Al^{3+}, Zn^{2+}, and so forth, leads to acidity as high as 10 mol dm^{-3} (Ballantine et al., 1984; Clark, 1994). Synergistic effects generated from the supporting materials and the reagents can lead to unexpectedly high activity (Adams, 1987; Phukan et al., 2003; Ahmed and Dutta, 2005).

Acid activation of clay minerals by mineral acids, under controlled conditions, leads to partial dissolution of the clay minerals, giving rise to a mixture of protonated clay minerals and nanoporous amorphous hydrous silica phases (Tkác et al., 1994; Breen et al., 1995; Komadel et al., 1996; Madejová et al., 1998; Komadel, 1999) with increased surface area, porosity, and Brønsted acidity (Adams, 1987; Rhodes and Brown, 1992, 1993; Gates et al., 2000). Modified Mt-based products, such as Mt K10 (surface area: 220–270 m^2/g) and Mt KSF (surface area: 20–40 m^2/g), are commercially available for use as catalysts for organic synthesis. Clay mineral-based layered nanocomposites, as well as nonlayered nanoporous (micro- and mesoporous) matrices, find potential applications as catalyst supports and catalysts for industrial organic reactions such as: esterification (Bhorodwaj and Dutta, 2010, 2011), alkylation (Rhodes and Brown, 1993; Vaccari, 1999; Yadav et al., 2004a), acylation (Jasra, 2003), isomerisation (Yadav et al., 2004c), Diels–Alder (Clark, 1994; Nagendrappa, 2002), coupling (Adams, 1987), and polymerisation reactions (Rhodes and Brown, 1994; Komadel et al., 1997; Aslya et al., 2016). Replacement of environmentally unacceptable anhydrous $AlCl_3$, an efficient catalyst for Friedel–Crafts reactions, by metal salt supported

Developments in Clay Science, Vol. 9. https://doi.org/10.1016/B978-0-08-102432-4.00009-3

Mt K10 has been claimed (Clark et al., 1989, 1995; Rhodes et al., 1991). Metal salts and/or metal ions supported on Mt K10, for example, zinc chloride (clayzic), copper chloride (claycue), iron nitrate (clayfen), and so forth, and metal ions such as Zn^{2+}, Cu^{2+}, and so forth, are used as solid acid catalysts in various organic reactions (Laszlo and Mathy, 1987; Clark et al., 1989; Soma and Soma, 1989; Rhodes et al., 1991; Massam and Brown, 1995; Clark and Macquarrie, 1996; Clark et al., 1997; Ramchandani et al., 1997; Phukan et al., 1999). Acid-treated clay minerals can be further modified by pillaring (Mokaya and Jones, 1994; Bovey and Jones, 1996; Bovey et al., 1996), or by intercalation of organic cations (Breen et al., 1997; Breen and Watson, 1998). Attempts are also made to incorporate nanoparticles of metals into such modified layered clay minerals (Ahmed and Dutta, 2003a,b).

Advances in the intercalation and intersalation reactions of metal complexes (Loeppert et al., 1979; Berry et al., 1990; Petridis et al., 1992; Sato et al., 1992; Laired et al., 1995; Borah et al., 2001) or other organic compounds with Mt have paved the way for the development of new classes of supported (heterogeneous) catalysts (Pinnavaia, 1983; Nakai et al., 1986; Laszlo, 1987; Choudary et al., 1989; Ruiz-Hitzky and Aranda, 1990; Aranda and Ruiz-Hitzky, 1992; Lagaly, 1994; Takada et al., 1995; Johnston et al., 2002). Most of the intercalation reactions have been carried out in solid–liquid or solid–gas systems. Solid–solid intercalation reactions between metal complexes and Na^+-Mt have also been successfully carried out (Ogawa et al., 1991, 1993; Ogawa and Kuroda, 1995). Until recently, the layered clay mineral supports are, in general, modified by expanding the clay mineral layers through intercalation or intersalation reactions (Loeppert et al., 1979; Laszlo, 1987; Berry et al., 1990; Ruiz-Hitzky and Aranda, 1990; Petridis et al., 1992; Takada et al., 1995; Schoonheydt et al., 1999; Bergaya and Lagaly, 2001; Borah et al., 2001). By applying these techniques, only limited amounts of materials could be loaded into the clay mineral interlayer spaces for generating micropores (<2 nm). But, in practice, high demand is for mesoporous (>2–50 nm) heterogeneous catalysts, because very large reacting molecules cannot enter into the micropores of zeolites (pore diameters of ZSM-5 and X,Y-Mordenite are ~5.5 and ~7 Å, respectively). However, the discovery of a new family of ordered mesoporous materials such as MCM-41, with a pore diameter in the range 2–10 nm, has stimulated research in the area of catalysis. In view of this, Mt (surface area: ~100 m^2/g; pore size: ~20 Å) can be modified by controlled activation with acids, for generating nanopores, that is, micro- (<2 nm) as well as mesopores (2–8 nm) (Dutta and Pathak, 2009) with high surface area (>300 m^2/g) for use as nanoporous heterogeneous catalysts, and also as supports for preparing metal nanoparticles–clay mineral nanocomposites. Recently, nanoscience and nanotechnology have become an important and exciting area of research (Goodman et al., 2005; Regan et al., 2005), and catalytic processes have become key areas of research because of their direct impact on academic and industrial developments (Somorjai and McCrea, 2001; Grunes et al., 2003; Kalidindi and Jagirdar, 2012). Catalysis by metal nanoparticles has attracted considerable attention, due to their high activity and selectivity. A new generation of heterogeneous catalysts: 'nanocatalysts', based on supported metal nanoparticles, has been developed. The nanocatalysts are superior to the conventional bulk catalysts due to:

(i) high surface-to-volume ratio; (ii) higher activity because of more active surface atoms; (iii) higher selectivity; (iv) they are recyclable, and (v) the recovery of the catalyst from the reaction mixture is easy. The catalytic characteristics are significantly dependent on particle size, shape of the crystal (Aiken and Finke, 1999; Gonsalves et al., 2000; Kralik and Biffis, 2001; Bronstein et al., 2004; Shchukin and Sukhorukov, 2004), and the chemical nature of the environment, that is, the support of the metal nanoparticles (Burda et al., 2005; Noguez, 2007; Astruc, 2008; Klasovsky and Claus, 2008).

Metal complexes and metal salts have long played a major role in metal-based catalysts for a wide range of reactions, but they suffer from poor recyclability and separation. For example, the Heck reaction catalysed by Pd^{2+} salts or complexes is usually carried out at high temperatures in the presence of organic ligands, but the catalytic system cannot be reused (Toebes et al., 2001; Corain et al., 2003; de Vries, 2006; Cui et al., 2007). Thus, there is a need for the development of efficient metal-based catalysts that could remove these weaknesses. The distinct advantages of metal nanoparticles catalysts, having characteristics of high catalytic activity, good recyclability and ease of separation, will make them a promising group of efficient heterogeneous catalysts (Schmid, 1992; Lewis, 1993; Toshima and Yonezawa, 1998; Ye et al., 2008; Dutta et al., 2015). It has been demonstrated (Wang et al., 2009) that Sn-nanoparticles (20–30 nm) promoted allylation of carbonyl compounds in water to the corresponding products with greater efficiency (yield 93% in 1.5 h), compared with only 13% yield shown by the bulk Sn metal. The control over the nanoparticle size and shape is of fundamental importance for catalytic applications (Bennett and Che, 1989; Coq and Figueras, 1998; Schmid et al., 1999; Corma and Garcia, 2008). Highly active catalysts with greater selectivity by shape control of the particles have been developed (Bronstein et al., 2004; Burda et al., 2005; Liz-Marzán, 2006; Mahmoud et al., 2008; Tao et al., 2008). For example, hexagonal Pt(111) surfaces are about 3–7 times more active than cubic Pt(100) surfaces for aromatisation reactions (Lee et al., 2006; Rioux et al., 2006). Metal nanoparticles are generally prepared on supports such as metal oxides, zeolites, carbon, clay minerals, polymers, organic ligands, and so forth (Aiken and Finke, 1999; Gonsalves et al., 2000; Kralik and Biffis, 2001). The synthesis of stabilised metal nanoparticles is generally carried out by any of the following procedures: (i) metal salt reduction; (ii) organic ligand reduction; (iii) thermal decomposition; (iv) metal vapour synthesis, and (v) electrochemical synthesis. Several reviews on metal nanoparticles stabilised on supports such as Mt emphasised the importance of such clay mineral supported nanocatalysts (Davis et al., 1984; Semagina and Kiwi-Minsker, 2009).

Some of the earlier books or reviews (Theng, 1974; Laszlo, 1986; McCabe, 1996; Adams and McCabe, 2006) mostly cover the acid-treated clay minerals or cation-exchanged clay minerals for catalytic applications, in addition to other aspects. In this chapter, the focus is on Mt because of its importance due to its multifaceted catalytic applications. More particularly, this chapter attempts to emphasise the recent developments, starting from the basic structure of Mt, its modifications and metal nanoparticles stabilisation by different procedures and their catalytic applications for fine chemicals synthesis.

9.2 Structural formula of some 2:1 clay minerals

9.2.1 Smectite group of clay minerals

Among the different clay minerals of the 2:1 group, the smectite group, particularly its dioctahedral member, Mt, has attracted a great deal of attention because of its unique properties. It consists of negatively charged layers formed by a 2:1 condensation of sheets of linked $M_x(OH)_6$ octahedra, where 'M' is either a divalent or a trivalent cation, (Mg^{2+}, Al^{3+}, Fe^{2+} and also Li^+, Ti^{4+}, etc.), being sandwiched between two sheets of inward pointing $Si(O,OH)_4$ tetrahedra, giving a layer with the formula of $M_xSi_4O_{10}(OH)_2$ (Grim, 1953; Theng, 1974; Suquet et al., 1975; Brindly and Brown, 1980; Choy and Park, 2004). The tetrahedral and octahedral sheets are combined so that the tips of the tetrahedra of each silica sheet and one of the hydroxyl layers of the octahedral sheet form a common layer. The atoms common to both the tetrahedral and octahedral layers become O instead of OH.

The layered silicates, constituting one of the most versatile groups of catalysts, are considered to have played a crucial role in prebiotic chemistry and chemical evolution (Ferris and Ertem, 1993; Ferris et al., 1996; Ferris, 2002). Their ability to catalyse diverse types of reactions arises from unique structural features: high surface area; optimum layer charge of 0.4–1.2 per Si_8O_{20} unit; high CEC, easily intercalated exchangeable cations; capability to intercalate metallic cations, organic cations, cationic metal complexes or organometallic cations in the interlayer; high swelling capacity; and capability for intercalating different types of polar solvents in the interlayers.

9.2.2 Structure of Mt

The basic building blocks of the clay mineral are tetrahedral and octahedral sheets (Rhodes et al., 1991; Tkác et al., 1994). The tetrahedral sheets consist of silica tetrahedra linked via three corners to form a hexagonal mesh (Fig. 9.1), and the fourth corner of each tetrahedron (normal to the plane of the sheet) is shared with octahedra in adjacent sheets. Octahedral sheets consist of flat sheets of edge-sharing octahedra, each formally containing cations at its centre (usually Mg^{2+} or Al^{3+}) and OH^- or O^{2-} at its apices.

9.3 Properties of Mt

9.3.1 Isomorphous substitution and CEC

Sometimes in clay minerals, Si^{4+} is partly replaced by Al^{3+} in tetrahedral sheets and Al^{3+} or Fe^{3+} by Mg^{2+} or Fe^{2+} in the octahedral sheets. Such substitutions by ions of lesser charge produce negative charges in the layers that are mainly balanced by adsorbed interlayer cations. The interlayer cations are mainly Na^+, Ca^{2+} or Mg^{2+} ions, but a wide range of other cations, including organic ions, can also be introduced by

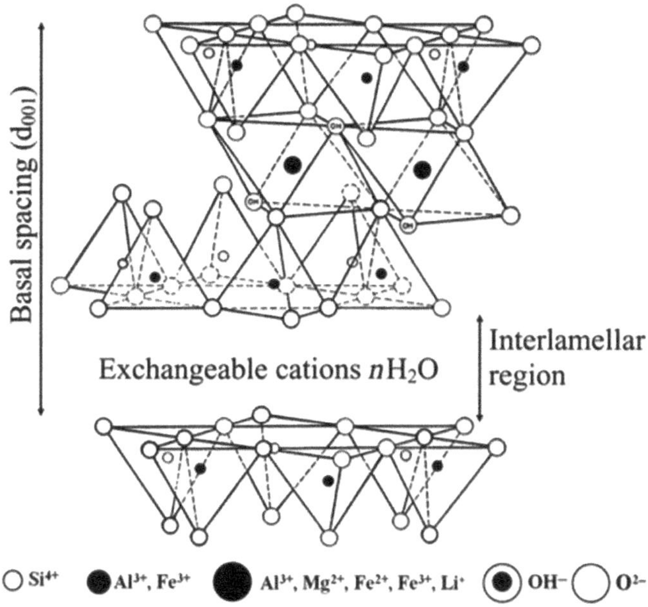

Fig. 9.1 Layer structure of Mt.

cation exchange reactions (MacEwan et al., 1961). In general, the quantity of cations, reversibly adsorbed per unit weight of the mineral is called CEC and can be expressed in milliequivalents (meq) per 100 g of dried clay mineral.

9.3.2 Acidity of the clay mineral

Most of the chemical reactions occurring on smectites, especially on Mt, are based on the acidity of the clay mineral (Adams, 1987). Clay minerals are solid acidic catalysts that can function as both Brønsted and Lewis acids in their natural and modified forms. The interlayer cations contribute to the acidity of clay minerals. The Brønsted acid character of clay minerals arises mainly due to the dissociation of the intercalated water molecules coordinated to the interlayer cations (Clark et al., 1989; McBride, 1994). The released proton immediately attaches to any reacting species in the interlayer space:

$$[M(OH_2)_n]^{m+} \xrightarrow[\text{(M = metal)}]{} [M(OH_2)_{n-1}OH]^{(m-1)+} + H^+$$

Both protons and polarising cations (e.g. Al^{3+}) give rise to strong Brønsted acidity (Clark, 1994). Higher levels of Brønsted acidity are achieved with highly polarising ions such as Al^{3+}, Fe^{3+}, Cr^{3+}, and so forth, and low water content. For example, Al^{3+} reacts with water to form the hydroxyl-Al-complex.

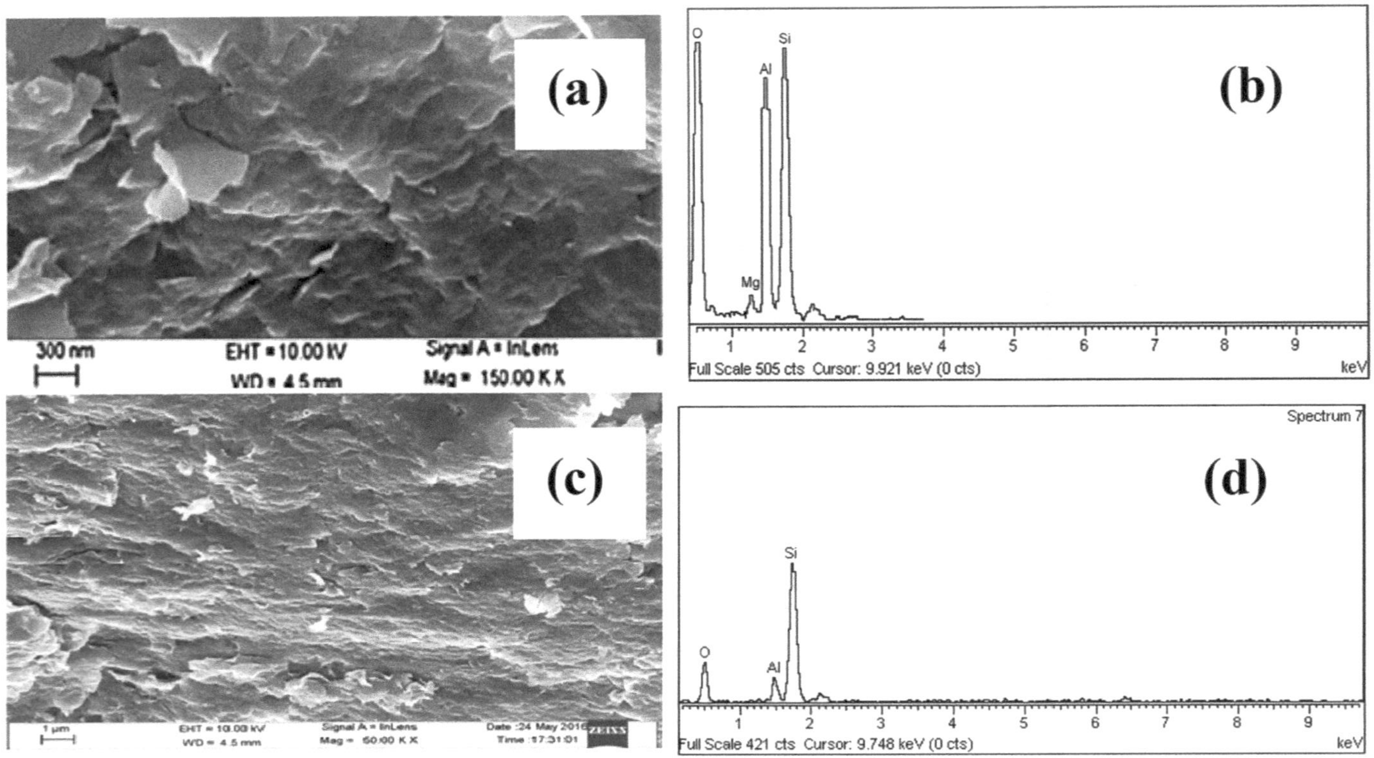

Fig. 9.2 (A) SEM image and (B) EDX analysis of untreated Mt; (C) SEM image and (D) EDX analysis of acid-activated Mt.

$$[Al(H_2O)_6]^{3+} \longrightarrow [Al(H_2O)_5OH]^{2+} + H^+$$

The Brønsted acid strength depends on the polarising power of the exchangeable cations (Mortland and Raman, 1968; Cornelis and Laszlo, 1985; Balogh and Laszlo, 1993) in the order:

$$H^+ > Al^{3+}, Fe^{3+} > Mg^{2+} > Ca^{2+} > Na^+ > K^+$$

Lewis acids are derived from aluminium or iron species located at crystal edges (Theng, 1974). By heating the clay mineral to about 325°C, almost all the adsorbed water is removed. If a reaction is then carried out in an anhydrous nonpolar organic liquid, the Lewis acidity predominates. The surface acidity of natural clay minerals with Na^+ or $NH_4{}^+$ as interstitial cations ranges from +1.5 to −3 (Hammett acidity function). Washing of the clay mineral with mineral acid, such as HCl, brings down the Hammett (Ho) function from −6 to −8, which is between that of concentrated HNO_3 (−5) and H_2SO_4 (−12).

9.3.3 Metal ions and metal complexes exchanged Mt

Metal ions and metal complex cations can be exchanged into the interlayer space of clay minerals, depending upon the CEC (Adams et al., 1979; Ballantine et al., 1985; Adams, 1987; Borah et al., 2001). Metal salts and/or metal ions supported on Mt K10, such as zinc chloride (clayzic), copper chloride (claycue), iron nitrate (clayfen), and so forth, and metal ions, such as Al^{3+}, Zn^{2+}, Fe^{2+}, and so forth, are used as solid acid catalysts in various organic reactions, namely, Friedel–Crafts, Diels–Alder, esterification, and so forth (Traynor et al., 1978; Laszlo and Mathy, 1987; Rhodes et al., 1991; Massam and Brown, 1995; Clark and Macquarrie, 1996; Brown and Rhodes, 1997; Clark et al., 1997). The interlayer space of Mt can be increased by incorporating organic or inorganic cationic compounds, depending upon their size and shape (McBride and Mortland, 1973; Theng, 1974; Barrer, 1985, 1989a,b; Booij et al., 1996). Organic cations such as tetraalkylammonium ions are intercalated into Mt during the preparation of chromatographic materials for separation of mixtures of gases (Thielmann and McAtee, 1975; Pinnavaia, 1983; Favre, 1991; Lao et al., 1991; Lao and Detellier, 1994). Remarkable chiral recognition phenomena have been observed with metal complexes such as $[M(phen)_3]^{n+}$, $[M(bpy)_3]^{n+}$, $M = Fe^{2+}$, Ru^{2+}, Ni^{2+} as guest molecules intercalated in the interlayer space of clay minerals (Yamagisi and Soma, 1981; Villemure, 1990; Yamagishi, 1993). A newer type of metal salt–metal ion-exchanged Mt, MCl_2-M^{n+}-Mt ($M = Ni, Cu, Zn$; $n = 2$) has been evaluated as a solid acid catalyst for Friedel–Crafts reactions, particularly for benzylation of benzene (Phukan et al., 2003).

Being stronger acids, heteropoly acids (HPA) will have significantly higher catalytic activity than conventional catalysts, such as mineral acids, mixed oxides, zeolites, and so forth. In particular, in organic media, the molar catalytic activity of

HPA is often 100–1000 times higher than that of H_2SO_4 (Kaur et al., 2002), but the main disadvantages of such HPA as catalysts lie in their relatively low surface area (1–10 m^2/g), and separation from the reaction mixture (Kozhevnikov, 1998). Therefore, fine and uniform dispersion of HPA on the large surface area of mesoporous acid-activated Mt increases the overall acidity and catalytic activity of HPA for different organic reactions (Bhorodwaj et al., 2009; Bhorodwaj and Dutta, 2010, 2011).

9.3.4 Acid-modified nanoporous Mt

Acid activation leads to modification of Mt with improved properties such as enhanced surface area, pore diameters, number of acid sites, and catalytic activities (Ravichandran and Sivasankar, 1997). The treatment of the naturally occurring and purified Mt with hot mineral acid replaces exchangeable cations with H^+ ions. Gradual leaching of Al^{3+} out of both tetrahedral and octahedral sites occurs, but the silicate groups remain largely intact. Such acid-activated clay minerals are, therefore, partially delaminated and exhibit higher surface area, pore volume, pore diameter (Traynor et al., 1978; Ballantine et al., 1981; Adams, 1987; Rhodes et al., 1991; Breen and Watson, 1998), and higher surface acidity (Hart and Brown, 2004). Acid activation results in several significant changes in Mt: (i) exchange of the hydrated interlayer charge compensating cations for H^+, (ii) delamination of particles and aggregates of Mt into individual clay mineral layers, (iii) partial dissolution of the individual clay mineral layers and release of the constituent cations into solution, and (iv) formation of an amorphous, porous, protonated, and hydrated silica. The acid-activated Mt exhibits micro- (<2 nm) and mesopores (>2 nm) during the initial period of acid activation, but on prolonging the acid activation, the surface area decreases because of the collapse of some pore walls without significant change of the overall specific pore volume. The SEM image of the acid-activated Mt (Fig. 9.2) indicates the formation of a porous surface upon acid treatment, which is not observed on the untreated one. The EDX patterns (Fig. 9.2) show a decrease in Al content upon acid activation, which clearly indicates the leaching of aluminum from the clay mineral structure.

The parent clay mineral exhibits an intense XRD reflection at $2\theta = 7.06$ degrees, corresponding to a basal spacing (d_{001}) of 12.5 Å, which decreases in intensity during acid activation, and no basal spacing (d_{001}) is observed for prolonged activation, indicating disordering of Mt layers in the *c*-direction. A low, intense, broad reflection in the 2θ range of 20–30 degrees suggests the formation of amorphous silica (Wang et al., 2006). The use of clay mineral-based materials as catalysts in organic synthesis has been extensively studied (Benesi and Winquest, 1978; Roberts and Khalaf, 1984; Olah et al., 1991; Tateiwa et al., 1994, 1995; Breen and Watson, 1998; Nikalje et al., 2000; Clark, 2002; Varma, 2002; Kappe and Dallinger, 2006; Loupy, 2006). Clay minerals or clay mineral-based catalysts absorb microwave energy and are excellent catalysts for microwave-assisted organic synthesis. Mt, acid activated for 1 h, showed micro- and mesopores with pore diameters up to about 8 nm, a high specific surface area up to 580 m^2/g, and a large specific pore volume of about 0.7 cm^3/g (Borah et al., 2010). The adsorption-desorption isotherms are of type-IV with an H3 hysteresis loop at P/P_o 0.4–0.8, indicating mesoporous solids (Borah et al., 2010).

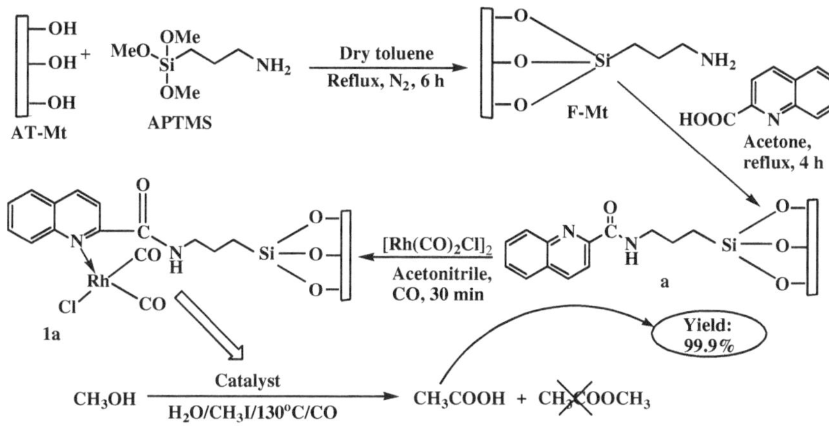

Scheme 9.1 Different steps for synthesising the composite [Rh(CO)$_2$Cl-ligand(L)-APTMS-Mt].

Formic acid-activated Mt is reported (Saikia et al., 2016a) to be functionalised with APTMS = (3-aminopropyl)-trimethoxysilane (F-Mt), followed by immobilisation of different quinoline carboxylic acids and reaction with $[Rh(CO)_2Cl]^-$ to form [Rh(CO)$_2$Cl-ligand(L)-APTMS-Mt] for use as catalysts. The preparation steps of the catalyst are shown in Scheme 9.1. The catalyst is evaluated in carbonylation of methanol to acetic acid and found to exhibit conversion up to 99.9%. The catalysts were recovered and reused for several runs without any significant loss of catalytic efficiency.

9.3.5 Acid-modified Mt as support for metal nanoparticles

For the development of eco-friendly and sustainable catalysts and catalytic processes (Dutta and Sarmah, 2016), Mt has been selected for stabilising metal nanoparticles. The different steps involved, starting from the raw Mt to the metal nanoparticle supported on Mt are: (i) purification of raw Mt; (ii) conversion into Na^+-Mt by treating with NaCl; (iii) activation with mineral or organic acids; (iv) incipient wetness impregnation of metal salts, and (v) reduction with suitable agents to produce metal nanoparticles on Mt (Scheme 9.2). In general, metal nanoparticles stabilised on Mt have been synthesised in situ by the reduction of respective metal salts impregnated clay mineral samples with reducing agents such as $NaBH_4$, sodium citrate, ethylene glycol, H_2, hydrazine, and so forth (Dutta et al., 2011a,b). Formation of Pt nanoparticles on the Mt surface was carried out by hydrogen reduction of tetraamine Pt-complex intercalated into Mt at 140°C (Harrison et al., 1988). Rh-nanoparticles in the interlayer space of Mt were synthesised (Kotkar and Thakkar, 1997) by oxidative degradation of intercalated tris(phenanthroline)-Rh(III), followed by reduction under a hydrogen atmosphere. Pt^o-nanoparticles stabilised on Mt were prepared (Vicente and Lambert, 2001) by reduction of $[Pt(NH_3)_4]^{2+}$ on Al-pillared Mt. $RuCl_3$ loaded

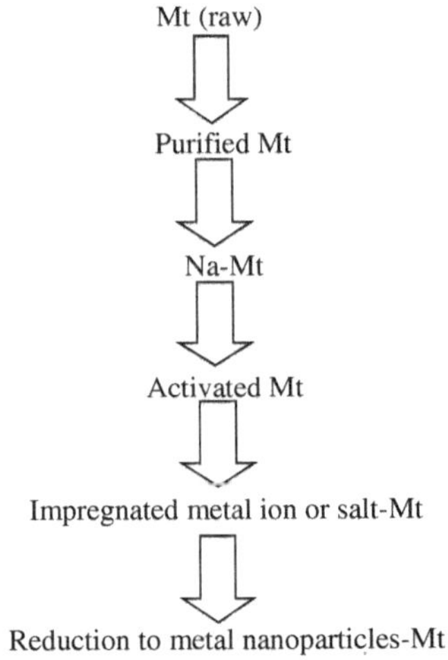

Scheme 9.2 General protocol for the synthesis of metal nanoparticles supported on Mt.

on Na^+-Mt is reduced to Ru^o by hydrogen at 220°C within 2 h, as indicated by the XPS results (Miao et al., 2006). Dutta and Dutta (2013) recently reported on the preparation of Pt^o-nanoparticles of 0–10 nm size by in situ impregnation of $H_2PtCl_6 \cdot 6H_2O$ into the nanopores of modified Mt, followed by reduction with different reducing agents such as hydrogen, ethylene glycol, and so forth.

Sodium borohydride ($NaBH_4$) is a very active and promising reducing agent for converting metal cations into their respective metal nanoparticles supported on a clay mineral matrix (Mastalir et al., 2002; Mitsudome et al., 2007; Borah et al., 2010, 2011; Agarwal and Ganguli, 2013; Varadwaj and Parida, 2013). Dutta et al. (2011a,b) reported the techniques for preparation, characterisation, and for studying reactivities of different metal (Cu, Ni, Au, Ru, Pt, etc.) nanoparticles. For example, the surface area of the acid-activated Mt, loaded with Cu^o-nanoparticles, was reduced from 680 to 307 m^2/g, indicating that the metal nanoparticles occupied the pores of the modified clay mineral matrix. This is also substantiated by a decrease of the pore volume from 0.69 to 0.40 cm^3/g. Agarwal and Ganguli (2013) reported Rh(0)-nanoparticles generation in the nanopores of Mt K10 by a reduction of $RhCl_3 \cdot 3H_2O$ with $NaBH_4$. The Rh^o-nanoparticles supported on the clay mineral are a highly active and selective heterogeneous catalyst for hydrogenation of monoterpenes at ambient temperature and pressure. Borah et al. (2011) also reported in situ generation of Cu^o-nanoparticles in the nanopores of Mt, acid activated with H_2SO_4 under controlled conditions. $Cu(CH_3COO)_2$ was loaded on acid-activated Mt through incipient wetness impregnation, followed by reduction with $NaBH_4$. A TEM study revealed that Cu^o-nanoparticles <10 nm size were evenly distributed on the support and exhibited face-centred cubic (fcc) lattice geometry. Zhou et al. (2006) reported pillaring of Mt with Fe^{3+}polycations, followed by reduction with KBH_4 to afford dispersed zero-valent iron nanoparticles with a diameter in the range of 3–10 nm in the Mt matrix. Mitsudome et al. (2007) reported a stable Pd nanocluster entrapped within the interlayer space of Mt by reduction of a [Pd(dba)$_2$] (dba = *trans,trans*-dibenzylideneacetone) complex, ion exchanged on Mt, with KBH_4. Water-soluble gold nanoparticles were prepared (Paek et al., 2006) by a $NaBH_4$ reduction of an Au precursor in the presence of *N,N,N*-trimethyl(11-mercapto-undecyl)ammonium (HS $(CH_2)_{11}N$-Me_3C), acting as a protecting ligand. Manikandan et al. (2007a,b) reported the preparation of Pt-nanoparticles in Mt utilising the surfactant hexadecyl trimethylammonium bromide, hexachloroplatinic acid as Pt precursor and $NaBH_4$ as reducing agent. They also reported the synthesis of Pt- and Ru-nanoparticles in the interlayer space of Mt by a similar method (Manikandan et al., 2008). Mt ion exchanged with hexamine Rh(III), and reduced with $NaBH_4$ afforded Rh-nanoparticles in the interlayer space of Mt (Sidhpuria et al., 2009). Fan et al. (2009) described the $NaBH_4$ reduction of ferric trichloride hexahydrate to iron nanoparticles in the presence of Mt. Stable Ag-nanoparticles on Mt were prepared in *n*-hexanol using $NaBH_4$ (Miyoshi et al., 2010). Fe-nanoparticles were prepared from hexadecyl trimethyl ammonium bromide surfactant-modified Mt by borohydride reduction (Li et al., 2010; Wu et al., 2012). Gu et al. (2010) reported the preparation of subnanosized zerovalent iron (ZVI) using smectite layers as templates, and $NaBH_4$ reduction of the exchangeable Fe(III) ions present in the interlayer space of Mt. In situ generation of iron nanoparticle on Mt by chemical reduction with $NaBH_4$ was described by Zhang et al. (2011). Kadu et al. (2011) reported bimetallic Fe, Ni-nanoparticle formation and dispersion by in situ generation of the nanoparticles on a clay mineral matrix with the help of chemical reduction of the metal precursors. Organophilic worm-like Ru-nanoparticles were prepared in the interlayer space of surfactant-modified Mt by reduction with $NaBH_4$ (Zhou et al., 2013). Rh^o-nanoparticles were generated in the nanopores of Mt K10 by reduction of $RhCl_3{\cdot}3H_2O$ impregnated Mt K10 with $NaBH_4$ (Agarwal and Ganguli, 2013).

There are several reports for preparing metal nanoparticles-supported Mt by a surfactant-assisted synthesis route, followed by reduction with hydrazine (Kiraly et al., 1999; Mastalir et al., 2000, 2004). The preparation of Pd particles onto the clay mineral layers involves a Pd hydrosol in the micellar assembly of a cationic surfactant, which is reduced with hydrazine (Kiraly et al., 1999; Mastalir et al., 2000). Dhakshinamoorthy and Pitchumani (2008) described the preparation of Ni^o-nanoparticles immobilised on Mt by hydrazine reduction in an aqueous alkaline medium using Ni^{2+}-exchanged Mt K10 as a precursor. Borah and Dutta (2013) recently reported the in situ generation of Pd^o-nanoparticles into the nanopores of modified Mt and their catalytic performance in carbon–carbon bond formation, namely, the Heck and Sonogashira reactions. Fievet et al. (1989), Im et al. (2005), and Xiong et al. (2005) developed a polyol process as a general methodology for controlling the shape of metallic structures, where polyol served both as solvent and as reducing agent. Ravindranathan et al. (1990) reported the preparation of Cu^o-nanoparticles in interlayers of Mt by a polyol reduction route by refluxing a hydroxyl copper acetate intercalated Mt. Later, they reported the same reduction method for preparation of copper metal clusters of 4–5 Å in the interlayer space and on the external surface of Mt (Malla et al., 1991, 1992). A polyol process was reported (Crocker et al., 1993) to generate Pd-nanoparticles of 2–3 nm size in different environments by the reduction of electrophilic Pd(II)-complexes intercalated in Mt at 0°C in MeOH. Ayyappan et al. (1996) reported in situ reduction of Ni^{2+} and Ag^{+} in the interlayers of Mt using boiling ethylene glycol. They had taken nickel acetate hydrate and silver nitrate as the metal precursors of Ni and Ag respectively, and found that a small fraction of metal clusters of 4–5 Å size remained inside the clay mineral. Ahmed and Dutta (2003b) reported preparation of Ni^{o} and Zn^{o}-Mt nanocomposites, wherein intercalation/intersalation reactions of metal acetate hydrates had been performed on Mt, followed by reduction of the intercalated/intersalated metal ions by a polyol process. Ahmed and Dutta (2003a) also reported in situ generation of Ni^{o}-Mt K10 (composite I) with 26 nm Ni° particle size and Co^{o}-Mt K10 (composite II) with 40 nm Co° particle size by polyol reduction of nickel acetate and cobalt acetate impregnated Mt K10. Dutta et al. (2011a,b) also described the in situ preparation of Ni°-nanoparticles of 0–8 nm size having fcc lattice into the nanopores of modified Mt. Sarmah and Dutta (2012) reported the preparation of Ru-nanoparticles by incipient impregnation of $RuCl_3$ into the nanopores of acid-activated Mt, followed by polyol reduction. Typical Mt supported Ru^{o}-nanoparticles with particle size distribution (inset) having a mean size of 4 nm are shown in Fig. 9.3A. The selected area electron diffraction (SAED) of Ru^{o}-nanoparticles showed a hexagonal pattern (Fig. 9.3B), and exhibited interplanar lattice fringe spacing of 0.21 nm of the fcc lattice of Ru^{o} crystals (Fig. 9.3C).

Reducing agents such as dimethyl aminoborane, tetrahydrofuran, and ligands with suitable reducing functional groups are also reported for preparing different metals' (Pd, Rh, Au, Ag, Ni) nanoparticles supported in the interlayer space of layered silicates, such as Mt (Papp and Dekany, 2002; Papp et al., 2004; Chen and Kuo, 2006; Liu et al., 2007; Yamauchi et al., 2012). Highly stable dispersions of nanosized copper particles with an average particle size less than 2 nm were synthesised using

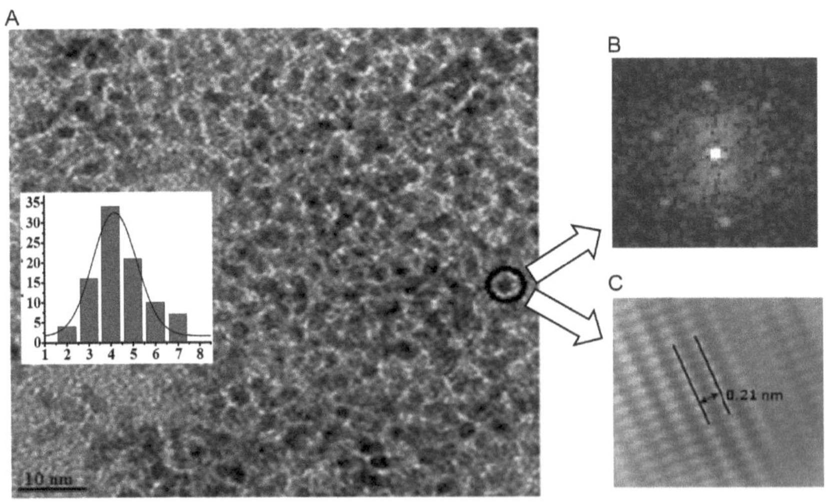

Fig. 9.3 (A) TEM image and particle size histograms with a Gaussian curve fitting (*inset*); (B) selected area electron diffraction (SAED) of Ru^{o}-nanoparticles and (C) interplanar lattice fringe spacing of Ru^{o}-nanoparticles.

nontoxic L-ascorbic acid as both a reducing and capping agent precursor in an aqueous medium (Xiong et al., 2011). Citrate-capped metal nanoparticles of gold nanorods with aspect ratios of 4.6 ± 1.2, 13 ± 2 and 18 ± 2.5 nm (all with 16 ± 3 nm short axis) were prepared (Jana et al., 2001) by a seeding growth approach in the presence of an aqueous micellar template. Glucose can be used as the reductant in the presence of a stabiliser for preparation of nanogold particles supported on Mt from the most common anionic gold precursor $HAuCl_4 \cdot 3H_2O$ by deposition-precipitation (Suraja et al., 2011).

There are physical methods for preparation of metal nanoparticles supported on Mt such as: ultrasonic intercalations (Belova et al., 2008, 2009a,b), supercritical fluids technique (Pan et al., 2009), and so forth. There are very interesting reports (Valaskova et al., 2008; Tokarsky et al., 2010a,b) for the preparation of Ag-nanoparticles on the surface of Mt without the use of reducing agents.

9.4 Modified Mt for solid acid catalysis

9.4.1 *Friedel–Crafts alkylations and acylations*

The Friedel–Crafts reaction is one of the most important reactions in organic chemistry (Clark, 1994). The newer type of metal ion-exchanged Mt, MCl_2-M^{n+}-Mt ($M = Ni, Cu, Zn$; $n = 2$) catalysed the benzylation of benzene (Scheme 9.3) efficiently (conversion 97%). The activity of such composites depends upon the loading of metal salts on Mt, and 1.5 mmol loading per mole of the support showed the highest activity (Phukan et al., 2003). During the initial stages of the benzylation reaction, Cd^{2+}-exchanged composites, in general, show higher activity, but in the long run, Zn^{2+}-exchanged clay minerals exhibit higher catalytic activities. Acid-treated metal ion-exchanged clay mineral composites exhibit higher catalytic activity than nonacid-treated ones because the former show a higher surface area, as well as a higher number of acid sites (Ahmed and Dutta, 2005).

Mt K10 catalysed benzylation of toluene to produce benzyl toluenes through bromination (Dasgupta and Török, 2008). The products have been obtained in good yields (87%) in a continuous process. During the reaction, the formed HBr is recycled and reoxidised to bromine with air (Scheme 9.4).

Mt KSF served as an efficient catalyst (conversion 80%) for the alkylation of indoles with *α,β*-unsaturated ketones and esters, and the reaction was considered an important step in the synthesis of constrained pseudo peptides (Scheme 9.5) (Ecija et al., 2003).

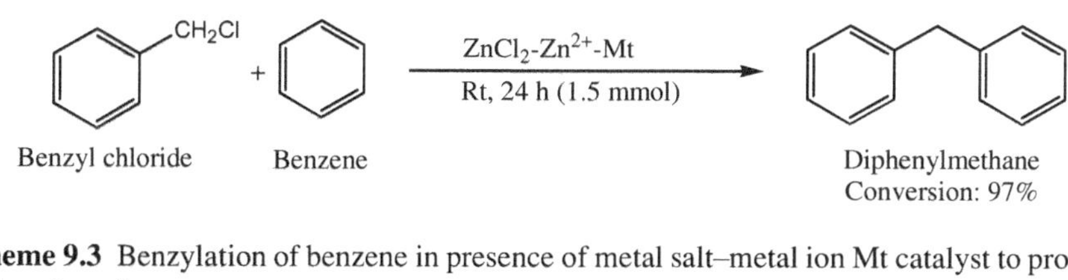

Scheme 9.3 Benzylation of benzene in presence of metal salt–metal ion Mt catalyst to produce diphenyl methane.

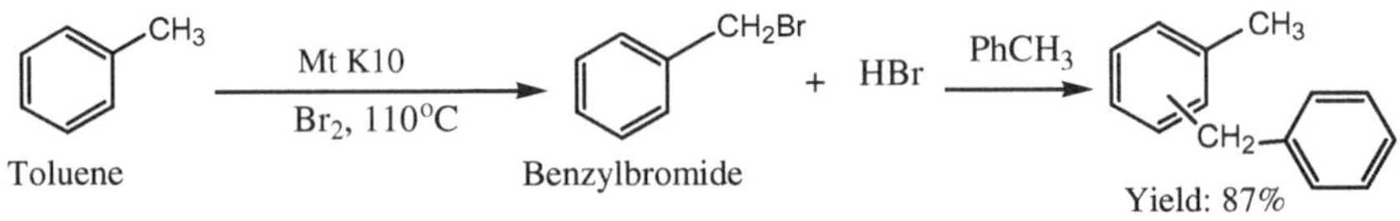

Scheme 9.4 Benzylation of toluene through bromination in presence of a clay mineral catalyst.

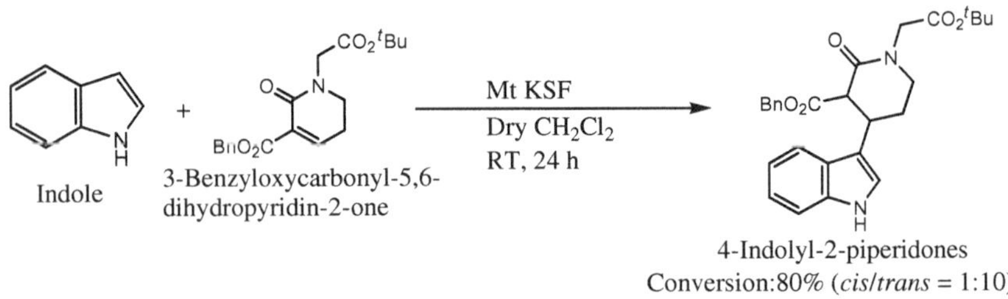

Scheme 9.5 Alkylation of indoles with α,β-unsaturated ketones in presence of a clay mineral catalyst.

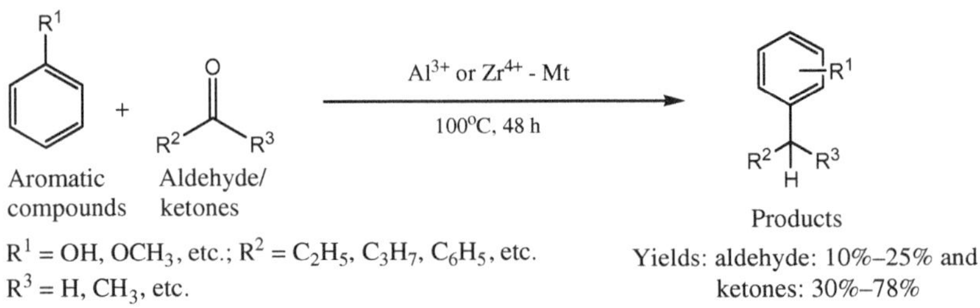

Scheme 9.6 Alkylation of aromatic compounds with aldehydes and ketones.

M^{n+}-Mt (M=Al, Zr) efficiently catalysed the alkylation of activated aromatic compounds (phenol, anisole, etc.) with aldehydes and ketones (Scheme 9.6). Al^{3+}-Mt exhibited significant activity (Tateiwa et al., 1997).

The biologically active 3-diarylvinylindoles were synthesised in good yields (88%) using phosphoryl chloride on Mt K10 under microwave conditions (Scheme 9.7) (Jaisankar and Srinivasan, 2005).

Dodecatungstophosphoric acid, $H_3[PW_{12}O_{40}]$ (20% w/w), supported on Mt K10 was used to synthesise (yield 40%) industrially important 4-*tert*-butylated dihydroxybenzenes from catechol, resorcinol, and so forth, and methyl *tert*-butyl ether (Scheme 9.8) (Yadav et al., 2001).

The cesium salt of dodecatungstophosphoric acid supported on Mt K10 acted as a catalyst for the acylation of anisole with benzoyl chloride. The main advantage of this catalyst is that it can be reused without any chemical reactivation (Scheme 9.9) (Yadav et al., 2003).

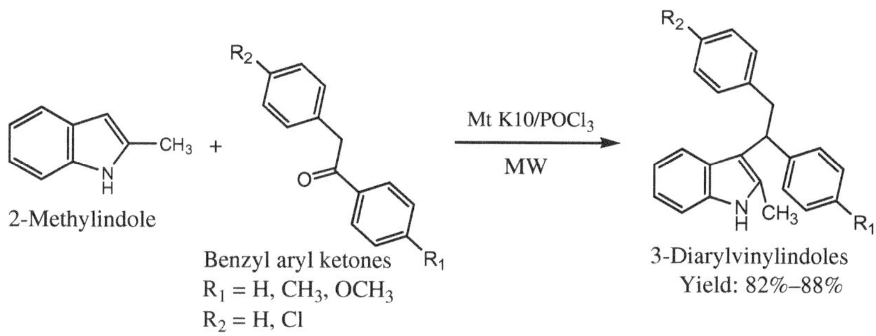

Scheme 9.7 Synthesis of 3-diarylvinylindoles using Mt K10 under microwave.

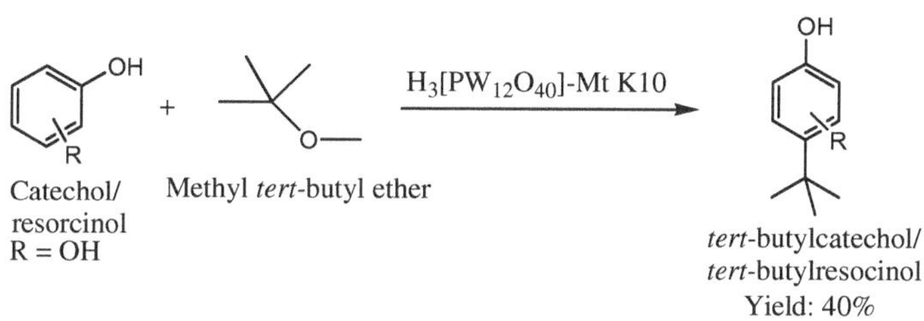

Scheme 9.8 Dodecatungstophosphoric acid, $H_3[PW_{12}O_{40}]$ (20% w/w), supported on Mt K10 for industrial preparations.

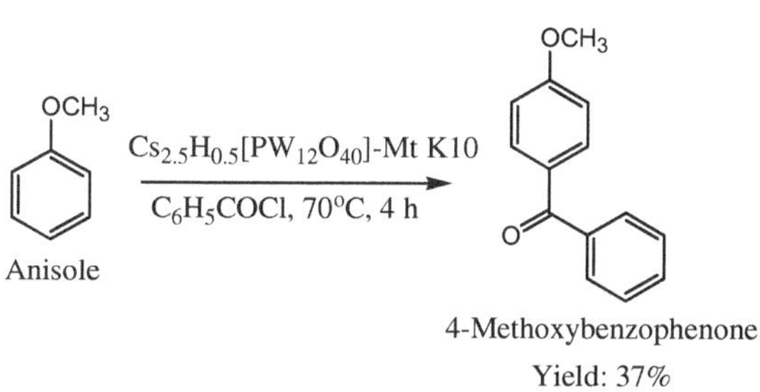

Scheme 9.9 Cesium salt of dodecatungstophosphoric acid supported on Mt K10 acted as a catalyst for the acylation of anisole with benzoyl chloride.

9.4.2 Other substitutions reactions

Al^{3+}-Mt was used efficiently as a catalyst in the presence of SO_2Cl_2 in 2,2,4-trimethylpentane for the selective 4-chlorination of phenols at room temperature (Scheme 9.10). When this methodology was extended to *ortho*-cresol and anisole, it gave 98% and 94% conversions, respectively (Gnaim and Sheldon, 2004).

Bromination of aromatic substrates using zinc bromide Mt K10 catalyst (clayzib) has been described as an environmentally friendly alternative for the synthesis of bromoaromatics (Scheme 9.11). The catalytic system gave excellent yields under very

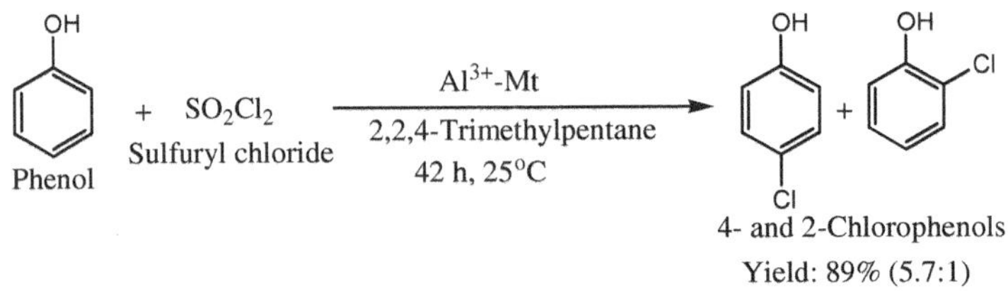

Scheme 9.10 The Al^{3+}-Mt catalysed chlorination of phenols.

Scheme 9.11 Zinc bromide-K10 catalysed bromination of aromatic substrates.

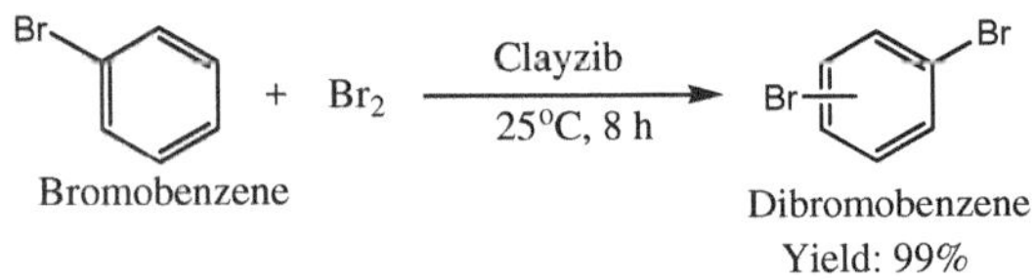

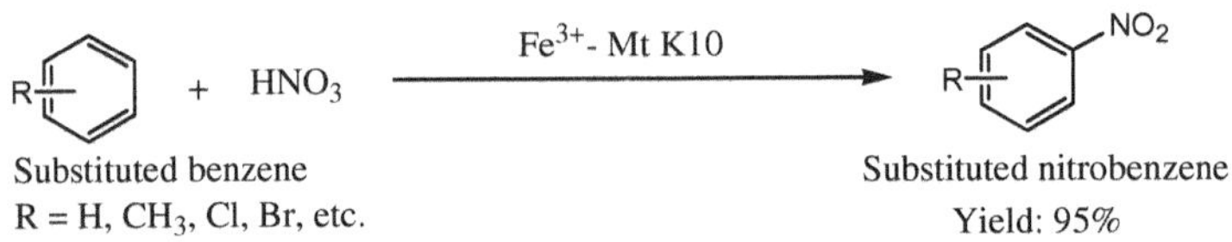

Scheme 9.12 Nitration of aromatic compounds in presence of Fe^{3+} or Al^{3+}-ion-exchanged Mt catalysts.

mild reaction conditions. Though the bromination proceeded well with clayzib, mesoporous silica gel gave the best selectivities, and good reaction rates with other substituted benzenes (Clark et al., 1997).

Several organic intermediates required in large quantities in the fine chemical industry are produced by nitration of aromatic compounds. Selective nitration of aromatic compounds can be carried out with high yields (95%) using nitric acid and modified clay minerals. Among these, a Fe^{3+}-ion-exchanged Mt catalyst was the most effective. In the case of substituted aromatics, the 4-substituted products formed in excess (Scheme 9.12) (Choudary et al., 2000b).

Sulfonylation of aromatic compounds to sulfones was carried out using various metal ion-exchanged Mt K10. The modified Mt K10 served as a mild and efficient catalyst (yield 60%–94%) for these transformations (Scheme 9.13) (Choudary et al., 2000a).

The acylation of a wide range of aromatics with carboxylic acids was performed in the presence of cations H^+, Al^{3+}, Cu^{2+}, Zr^{4+}, and so forth, exchanged on Mt (Scheme 9.14) and the H^+, Al^{3+}-exchanged Mt showed the highest activities (Chiche et al., 1987). The process suffered from limitations such as the requirement of a stoichiometric quantity of the catalyst, and only longer chain carboxylic acids were active in the reactions.

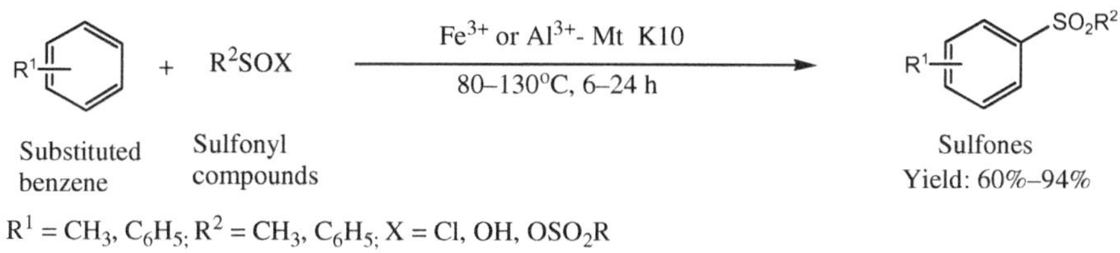

Scheme 9.13 Metal (Fe, Al) ion-exchanged Mt K10 catalysed sulfonylation of aromatic compounds to sulfones.

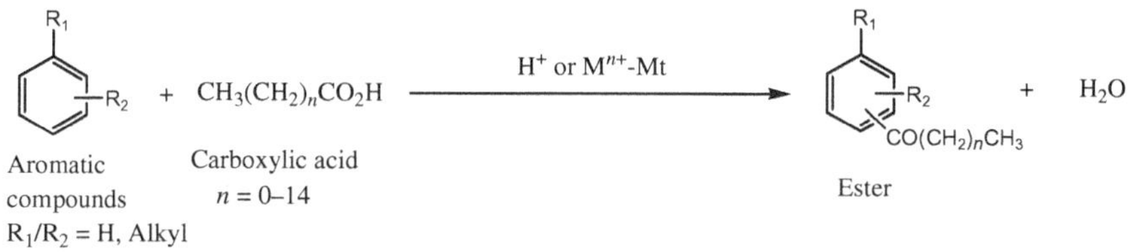

Scheme 9.14 Acylation reaction of aromatics with carboxylic acids in presence of cations exchanged Mt.

9.4.3 Cycloaddition reactions

Mt K10 has been used as a catalyst for the intermolecular hetero Diels–Alder reaction between unactivated alkenes and in situ generated 2-quinomethanes in $LiClO_4$-CH_3NO_2 solution (Scheme 9.15) (Chiba et al., 1999).

The synthesis of various quinoline or naphthyridine derivatives from arylamines and enecarbamates was carried out in the presence of Mt KSF as a catalyst (Scheme 9.16). The products were obtained in excellent yields (79%–91%), and moderate diastereoselectivity (Yadav et al., 2004b).

Bifunctional aromatic and aliphatic compounds also readily undergo cyclisations in the presence of clay mineral catalysts. Mt KSF or Mt K10 catalysts were used for the reaction of catechols and pyrogallols, with aldehydes and ketones to synthesise a series of 2-substituted and 2,2-disubstituted 1,3-benzodioxoles (Scheme 9.17) (Li et al., 1998).

9.4.4 Ring-opening and condensation reactions

The ring-opening polymerisation of δ-valerolactone to polymeric products with controlled molecular weights was catalysed efficiently by Sn^{4+}-exchanged Mt in a solvent-free system (conversion 98%), although a minute amount of ethanol was used to initiate the reaction. The Al^{3+}- and Fe^{3+}-exchanged Mt were also effective for the polymerisation of δ-valerolactone (Scheme 9.18) (Kadokawa et al., 2002).

Chemoselective condensation of indoles and imines were carried out in the presence of Mt K10 or Mt KSF at room temperature to give diindolyl arylmethanes (Scheme 9.19) (Feng et al., 2005).

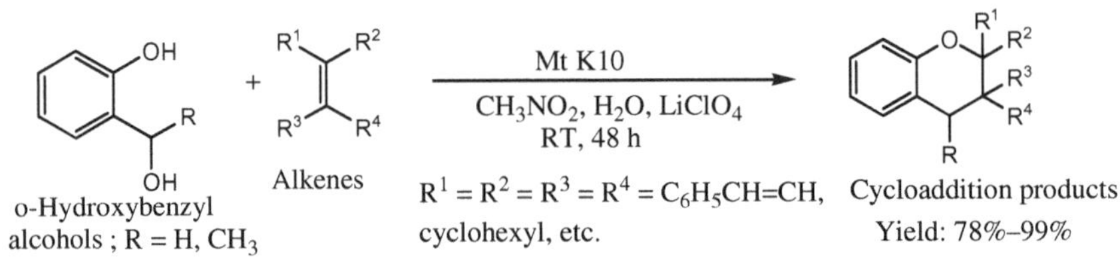

Scheme 9.15 Mt K10 catalysed Diels–Alder reactions.

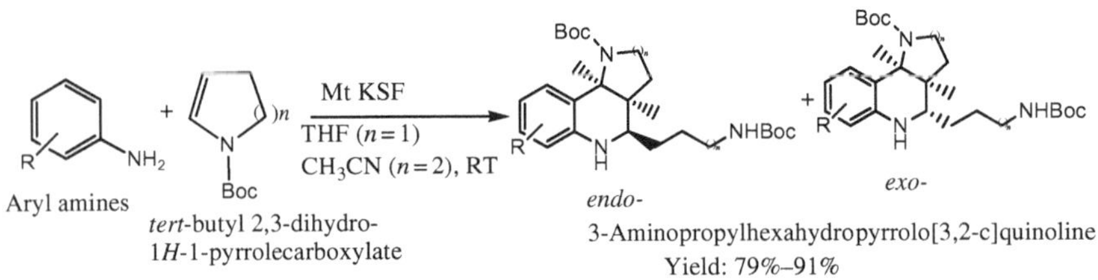

Scheme 9.16 Mt KSF catalysed reaction for quinoline preparation.

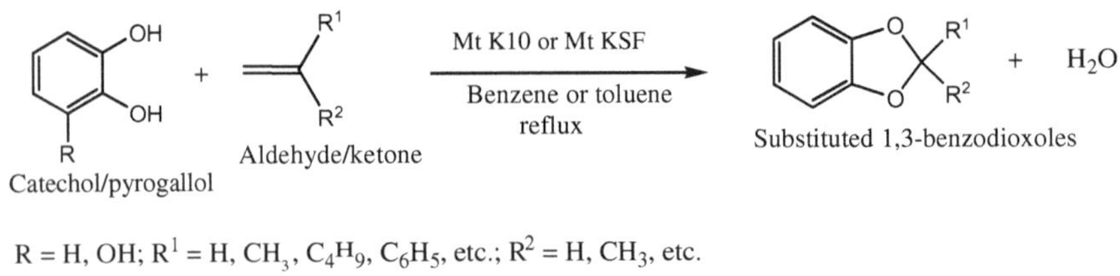

Scheme 9.17 Mt K10 or Mt KSF catalysed cyclization reactions to produce 1,3-benzodioxoles.

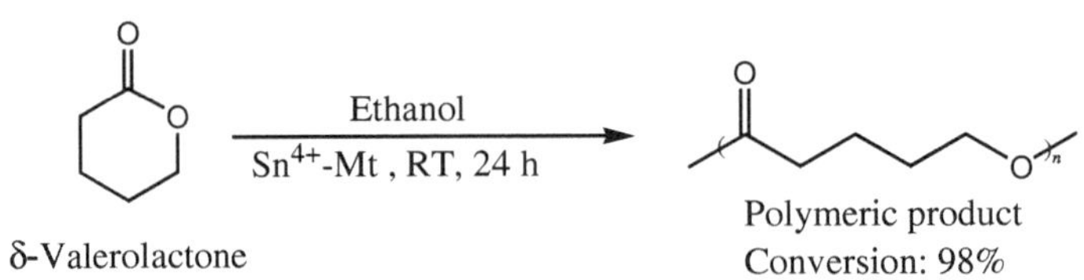

Scheme 9.18 Sn^{4+}-Mt catalysed ring-opening reaction.

9.4.5 Heck and other reactions

The Heck vinylation reaction between aryl halides and acrylates or styrenes to form alkyl (*E*)-cinnamates or (*E*)-stilbenes was efficiently catalysed by palladium and copper ion-exchanged Mt K10 in the presence of potassium carbonate in dimethylformamide under refluxing conditions (Scheme 9.20) (Ramchandani et al., 1997).

Michael reaction of 1,3-dicarbonyl compounds with enones was catalysed by Sc^{3+}-ion-exchanged Mt K10 efficiently (Yields 70%–99%) under solvent-free conditions.

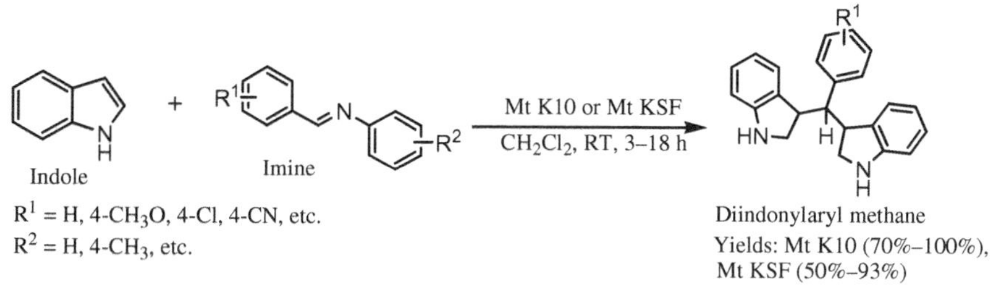

Scheme 9.19 Mt K10 or Mt KSF catalysed condensation reactions to produce diindolyl arylmethane.

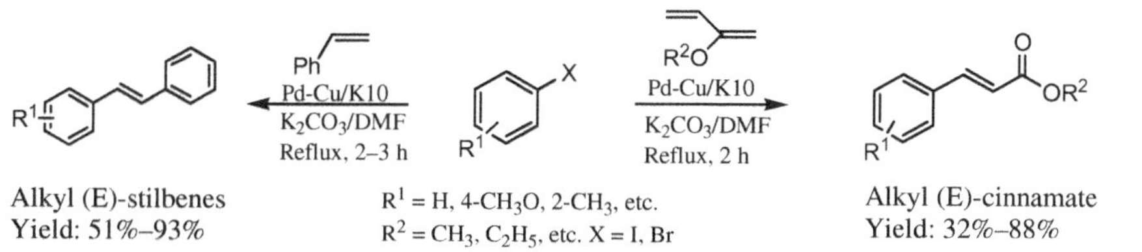

Scheme 9.20 Palladium and copper ion-exchanged Mt K10 catalysed Heck vinylation reaction between aryl halides and acrylates.

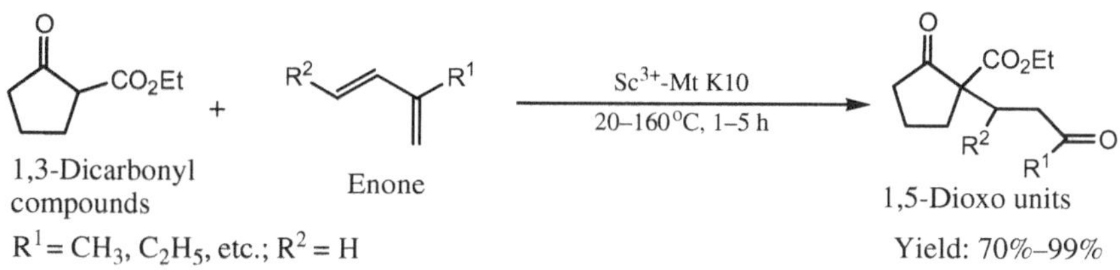

Scheme 9.21 Sc^{3+} ion-exchanged Mt K10 efficiently catalysed the Michael reaction of 1,3-dicarbonyl with enones.

The catalyst was reusable without any appreciable loss of its high activity and selectivity (Scheme 9.21) (Kawabata et al., 2003).

Iridium and rhodium complex–clay mineral catalysts were prepared by immobilising organometallic complexes on clay minerals such as Mt K10, sodium bentonite, and lithium hectorite (Ht). The catalysts were used for the hydrogenation of *N*-benzylideneaniline. The catalytic activity of iridium (I) systems was superior (yields 55%–99%) (Scheme 9.22) to the other catalysts used in this study (Claver et al., 2001).

Chemoselective hydrogenation is an important and useful reaction for the synthesis of fine chemicals used in pharmaceuticals, perfumes, and so forth; 5% platinum on Mt K10 catalysed a highly selective hydrogenation (99%) of cinnamaldehyde to cinnamyl alcohol (Scheme 9.23). The process was described as bifunctional catalysis, where the

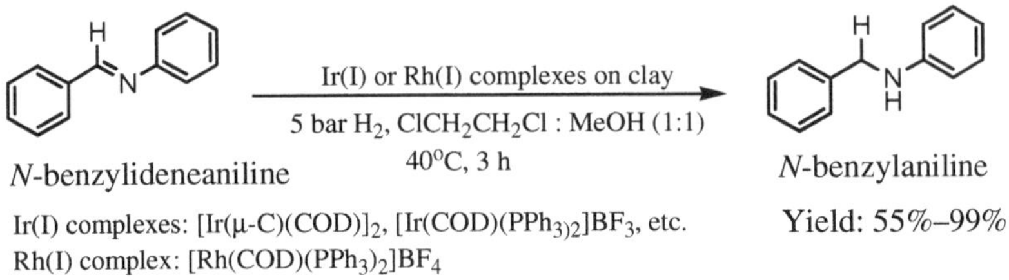

Scheme 9.22 Ir or Rh complexes-Mt K10 catalysed selective hydrogenation reactions.

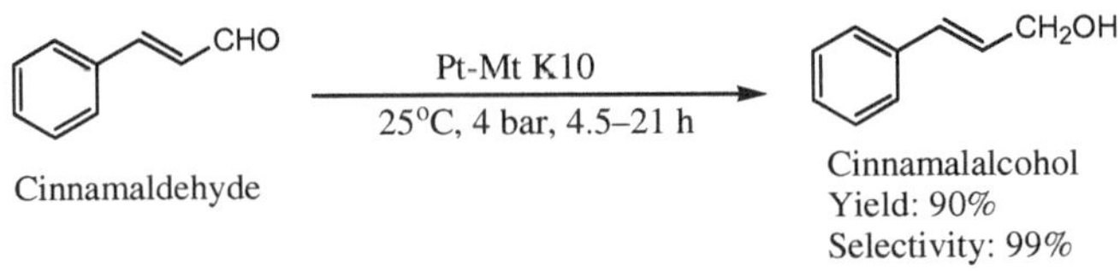

Scheme 9.23 Pt-Mt K10 catalysed highly selective hydrogenation of cinnamaldehyde to cinnamyl alcohol.

Scheme 9.24 Ru^{3+}-Mt catalysed hydrogenation of benzene.

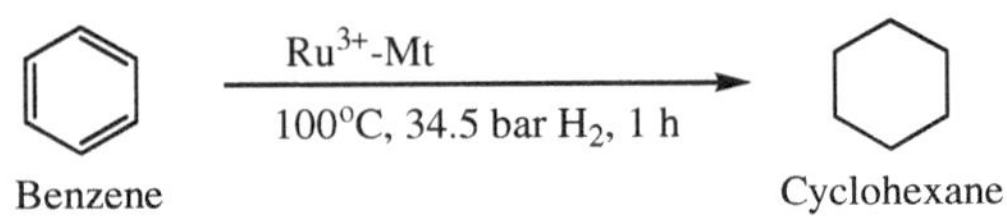

acidic centres of the clay mineral ensured the selective adsorption of the strongly nucleophilic carbonyl oxygen, making it an easier target for the hydrogenation (Dasgupta and Török, 2008).

Benzene is present in petroleum and diesel fuels, and therefore needs to be removed due to its carcinogenic effect, but removal of benzene by solvent extraction methods decreases the octane number of the gasoline. Ru^{3+}-exchanged Mt was used to hydrogenate benzene at 100°C and 34.5 bars of hydrogen pressure (Boricha et al., 2006). Cu^{2+}-salts impregnated on Mt increased the sulfur tolerance of Ru^{3+}-exchanged Mt (Scheme 9.24). The hydrogenation may likely be associated with the formation of Ru metal particles (Yang et al., 2003).

9.4.6 *Esterification reactions*

The esterification of acetic acid with butanol is an electrophilic substitution reaction that is relatively slow, and needs activation either by high temperature or by a catalyst. Esterification of dicarboxylic acids with various alcohols and phenols in the presence of a metal-exchanged Mt catalyst (M^{n+}-Mt; $M^{n+} = Al^{3+}$, Fe^{3+}, Cr^{3+}, Zn^{2+}, Mn^{2+} and Ni^{2+}) was studied (Scheme 9.25). Among the catalysts used, Al^{3+}-Mt was found to be the most effective under mild reaction conditions. The heterogeneous catalysts can be regenerated and reused. All these features indicate a high potential of the reaction in '*Green chemistry*' processes (Reddy et al., 2005).

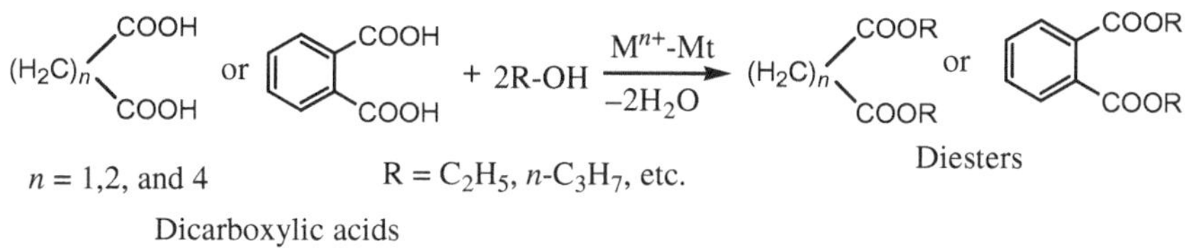

Scheme 9.25 Metal cations exchanged Mt catalysed esterification of dicarboxylic acids with various alcohols.

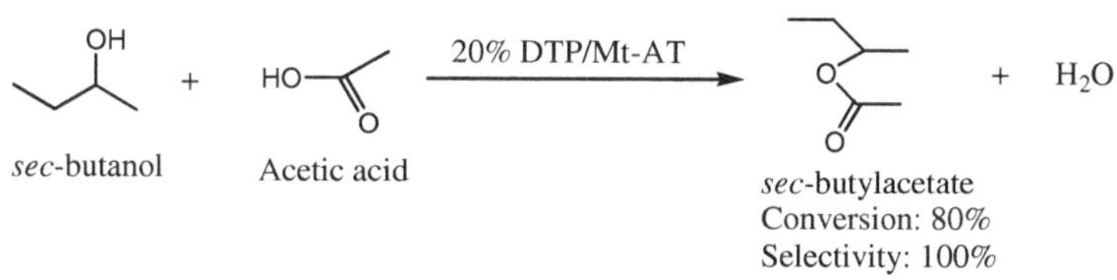

Scheme 9.26 Esterifications of acetic acid with *sec*-butanol catalysed by 20% supported dodecatungstophosphoric acid $H_3PW_{12}O_{40}$ (DTP) on Mt-AT.

Esterification of acetic acid with *n*-butanol catalysed by supported HPA dodecatungstophosphoric acid, $H_3PW_{12}O_{40}$ (DTP), on acid-modified Mt (Mt-AT) matrix was carried out (Haber et al., 2003; Kozhevnikov, 2003; Timofeeva et al., 2004; Yadav, 2005; Sawant et al., 2007a,b). Among different catalysts, 20% DTP supported onto Mt-AT (15 min) showed the highest catalytic activity, with 88% conversion and nearly100% selectivity toward *n*-butyl acetate (Bhorodwaj et al., 2009). Esterifications of acetic acid with *sec*-butanol catalysed by 20% supported dodecatungstophosphoric acid $H_3PW_{12}O_{40}$ (DTP) on Mt-AT (Scheme 9.26) showed catalytic activity with about 80% conversion, with nearly 100% selectivity toward *sec*-butyl acetate.

A requirement for use of porous solid acid catalysts in liquid-phase processes is that the pores should be large enough to allow rapid diffusion of reactants in, and products out of the pores. Thus, the catalysts should be mesoporous (pore diameter 2–50 nm) (Clark et al., 1989). Therefore, despite the actual dominant use of microporous aluminosilicate catalysts, such as zeolites (pore sizes typically <1 nm) in modern petrochemical processes, there is a wide interest in the use of mesoporous aluminosilicates such as acid-modified clay minerals that possess tailorable pore sizes in the mesoporous range (typically 2–50 nm) that may be advantageous for different catalytic reactions (Vaccari, 1998; Ding et al., 2001; Phukan et al., 2017a,b). The suitability of mesoporous aluminosilicate catalysts obtained by acid activation of clay minerals, particularly smectites, to promote Brønsted or Lewis acid catalysed reactions depends on the extent of leaching of the clay mineral, which in turn depends upon the acid activation conditions (Komadel et al., 1997). Smectites can acquire a high degree of Brønsted acidity, either through the presence of hydronium ions at the exchange sites, or through the polarising effect of exchangeable cations (e.g. Al^{3+}), while Lewis acidity in smectites arises mainly from the Al^{3+} or Mg^{2+} exposed at the clay mineral edges, and from the dehydration of clay minerals at high temperature (McBride, 1994; Clark, 1994).

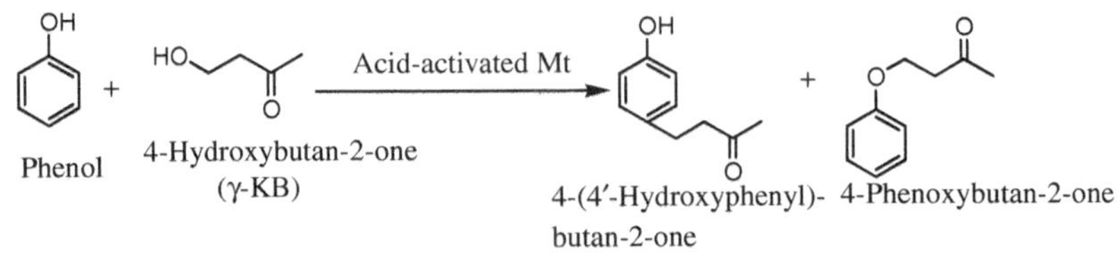

Scheme 9.27 Friedel–Crafts alkylation between phenol and 4-hydroxybutan-2-one in presence of the acid-activated Mt catalysts.

9.4.7 Acid-activated nanoporous Mt and pillared clay mineral catalysts

The Friedel–Crafts alkylation reaction between phenol and 4-hydroxybutan-2-one in the presence of the acid-activated Mt catalysts exhibiting layered clay mineral structures (d_{001} ranging from about 10 to 13.5 Å), high surface area, high porosity including mesopores, produces 4-(4-hydroxyphenyl)butan-2-one (Raspberry ketone) with a conversion of about 35%–55% and a high selectivity in the range of 75%–81% (Scheme 9.27) (Dutta and Pathak, 2009).

The activity of acid-treated or cation-exchanged clay minerals is relatively poor at temperatures above 150°C because of dehydration and collapse of the clay mineral layers. To overcome this problem, the so-called pillared clays were produced. Inorganic polyoxocations are intercalated by ion exchange, and upon calcination, make the polyoxocations, called ‘pillars’ in the interlayer space. The pillars prevent collapse of the clay mineral layers. The material is known as ‘pillared layered clay’ (PILC) (Vaughan et al., 1979; Pinnavaia, 1986; Poncelet and Schutz, 1986; Molina et al., 2011; Vicente et al., 2013). PILC has the advantage that the pillars themselves may be catalytically active (Suib et al., 1986). Pillaring also confers some degree of size/shape selectivity in catalytic reactions (Kikuchi and Matsuda, 1988). PILC possesses several interesting properties, such as large surface area, high pore volume, and tunable pore size (from micropore to mesopore), high thermal stability, strong surface acidity, and catalytically active substrates/metal oxide pillars. These unique characteristics make PILC an attractive material in catalytic reactions (Kloprogge, 1998; Vaccari, 1998). The activity and porosity of the PILC are highly dependent on the method of preparation (Sapag and Mendioroz, 2001; Salerno et al., 2001).

9.5 Nanoporous Mt supported metal nanoparticles catalysts

Nanoporous Mt can be utilised for the development of stable, durable, and sustainable heterogeneous catalysts. The stabilisation of metals nanoparticles into the nanopores of modified clay minerals is widely used for the preparation of stable, environmentally friendly heterogeneous catalysts for organic synthesis. The nanoporous clay mineral support may also act as an acid catalyst by itself.

Catalytic oxidation is a very important process for controlling pollution of water caused by industrial and agricultural wastes. Metal-modified Mt has performed a pivotal role for the oxidation of toxic organic pollutants, such as phenol, substituted phenols, and dyes, in aqueous solutions under mild conditions. Gu et al. (2010) reported that nanostructured metallic iron particles in a Mt matrix could be prepared at ambient temperature by reduction of iron pillared Mt with potassium borohydride. The zero-valent iron nanoparticles are highly dispersed in the clay mineral matrix, have diameters in the range of 3–10 nm, and are catalysts for phenol oxidation with hydrogen peroxide with a conversion of 49.5% and a selectivity of 67.4% to carbon dioxide and tar. The iron species dispersed in the Mt matrix provide the catalytic active sites, and the size of iron species has an effect on selectivity. Gorobinskii et al. (2006) reported Pt dispersed into acid-activated pillared Mt K10, prepared by impregnation with a chloroplatinic acid solution followed by hydrogen reduction. The catalytic activity of the Pt-containing clay minerals in CO oxidation in excess hydrogen was efficient.

The Fe_3O_4 magnetic nanoparticles developed into the nanopores of modified Mt exhibited high catalytic efficiency for the Baeyer–Villiger oxidation of various cyclic and aromatic ketones under solvent-free conditions, and showed conversions of up to 98% (Scheme 9.28) (Saikia et al., 2016b). The catalyst can be magnetically recovered and recycled for several runs without any significant loss of efficiency.

Anilines are a very important class of intermediates for preparation of halo-derivatives. The anilines are obtained from the corresponding nitro precursors by liquid or gas-phase hydrogenation using metals such as Au or bimetallic Au-Pd supported on alumina in the presence of a dechlorination inhibitor (Cardenas-Lizana et al., 2009). Such processes are associated with high pressure (40–150 bar) and temperature (80–250°C). Metal nanoparticles supported on Mt have been extensively investigated toward hydrogenation of various substrates. Hydrogenation of nitrobenzene to aniline catalysed by Pt-nanoparticles immobilised on Mt carried out under a flow of H_2 atmospheric pressure and at 25°C was reported by Pan et al. (2009). The selectivity to aniline of 82%–99%, the 100% conversion to nitrobenzene within 200 min, and the turnover frequency (TOF) of 1073, were attributed to Pt-nanoparticles with sizes in the range of 4–8 nm, highly dispersed over the Mt support (surface area of about 43 m^2/g). In the catalytic hydrogenation of nitrobenzene, it was shown by Zhao et al. (2004) that both the conversion and the selectivity decreased with the increase in the Pt particle size from 7.2 to 19.6 nm. Recently, Dutta and Dutta (2013) have reported that supported Pt^o-nanoparticles (<10 nm) show efficient catalytic activity for the selective reduction of chloronitrobenzenes (CNB) at a H_2 pressure of 5–20 bar, temp. 40 ± 5°C for a period of 5–240 min to the corresponding chloroanilines

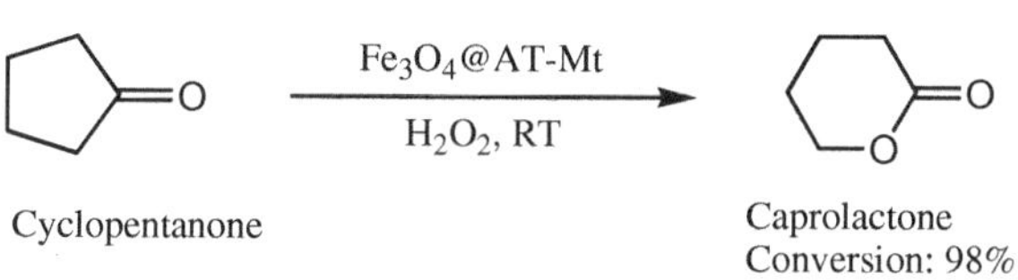

Scheme 9.28 Baeyer–Villiger oxidation of cyclic ketone catalysed by Fe_3O_4@AT-Mt under solvent-free condition at room temperature.

with 43%–100% conversion and selectivity in the range 97%–100% and very negligible C–Cl bond cleavage. *p*-Chloroaniline (*p*-CAN) is an important intermediate for the synthesis of organic fine chemicals, such as dyes, analgesics, herbicides, pesticides, and antipyretic drugs (Figueras and Coq, 2001; Wang et al., 2007). For such hydrogenation reactions, various precious metal catalysts (Pt, Pd, Ru, and Rh) have been studied (Figueras and Coq, 2001; Zuo et al., 2004; Zhang et al., 2005; Wang et al., 2007; Dutta and Dutta, 2013). However, it is observed that the hydrogenation process is accompanied by hydrogenolysis of the carbon–halogen bond over most of the catalysts, and that hydrogen halide (e.g. HCl) produced from this process is corrosive to the reactor. Much effort has been made to overcome this problem, including introduction of special additives (promoters and inhibitors), modification of the active metal with a second metal, and tuning the particle sizes of the metal particles. Hydrogenation of *p*-CNB is usually carried out at 90–150°C, with Au- and Ag-nanoparticles, and at 30–60°C with Pt catalysts, and the selectivity to the undesired aniline was less than 1% (Xiong et al., 2007). From both an economic and environmental point of view, there has recently been increasing interest in hydrogenation of CNB with different chlorine substituent sites (*o*-, *m*-, or *p*-CNB) to the corresponding CAN catalysed by less costly metals such as nickel. Liu et al. (2010) reported that nickel nanoparticles supported on bentonite showed selective hydrogenation of *p*-CNB to *p*-CAN under 15 bar H_2 and 100°C. After 1 h reaction time, the conversion and selectivity were about 92% and 98%, respectively. This high activity of the catalysts is attributed to the small Ni-B particles and greater content of structural water enhancing the hydrophilicity of the catalyst. In view of these developments, another recent report (Dutta and Dutta, 2014) states that the Mt supported Ni^o-nanoparticles serve as efficient and selective heterogeneous catalysts for hydrogenation of *p*-CNB to corresponding *p*-CAN with a conversion of 97%, and selectivity of 99% under mild reaction conditions: H_2 pressure of 10 bar, a temperature of 45°C, and a reaction time of 1 h. The metal nanocatalysts could be recycled and reused several times without significant loss of their catalytic activities. The reason for showing higher catalytic activities may be due to the synergistic effect of high surface area (290 m^2/g) of the modified Mt, and highly exposed and stabilised Ni^o-nanoparticles.

Hydrogenation of nitrobenzene to aniline catalysed by cheaper metals such as ZVI supported (Gu et al., 2010) into the interlayer spaces of Mt has also been reported. The supported ZVI were found to exhibit much higher reactivity in the reduction of nitrobenzene to aniline involving Fe concentration of only 0.33 g/L, and initial nitrobenzene concentration of 2.1 mmol for a conversion of about 80% within 1 min. The layer structure of Mt prevents the agglomeration of ZVI to bigger particles, and hence sustained the high reactivity. The high reduction efficiency of the clay mineral-templated ZVI is plausibly attributable to the discrete distribution of Fe(0) on the surface of Mt, with isolated subnanosized (<0.5 nm) Fe(0) particles. Thus, the clay mineral-templated ZVI showed superior reactivity and efficiency compared with some other previously reported studies (Agrawal and Tratnyek, 1996; Choe et al., 2001). Sarmah and Dutta (2012) reported high activity of Ru^o-nanoparticles immobilised on acid-activated Mt toward the transfer hydrogenation (hydrogenation with a source different from H_2) of substituted nitrobenzenes. The high catalytic activities in the chemoselective transfer hydrogenation reactions of substituted nitrobenzenes to corresponding anilines with high conversion (76%–97%) and selectivity (91%–100%) depend upon the nature of the substituents.

Hydrogenation of aromatic molecules is another class of industrially important reactions. Ir-nanoparticles synthesised into the nanopores of modified Mt has shown good efficiency in such types of reactions. High surface area (327 m^2/g) of the catalyst and nanoparticles of about 4 nm, uniformly distributed on the activated supports, were used for such a reaction (Das et al., 2016). The catalyst showed a maximum conversion up to 100%, and a TOF up to 79 h^{-1}. Dutta et al. (2011a,b) reported transfer hydrogenation of acetophenone catalysed by Ni^{o} supported acid-activated Mt. These supported Ni^{o}-nanoparticles show efficient catalytic activity in transfer hydrogenation of acetophenone to 1-phenylethanol with about 98% conversion, having nearly 100% selectivity. Mt supported Rh^{o}-nanoparticles (about 5 nm) were also reported (Sarmah and Dutta, 2014) to show good catalytic activity in reduction of aromatic carbonyl compounds to corresponding alcohols through the transfer hydrogenation route with high conversion (>56%) and selectivity up to 100%.

High activity and selectivity of the hydrogenation of quinoline was reported by Zhou et al. (2013) using worm-like Ru^{o}-nanoparticles intercalated in Mt. Vicente and Lambert (2001) reported that Pt^{o}-pillared Mt showed efficient hydrogenation of benzene. Structural differences associated with the preparation procedure resulted in different activities. The hydrogenation of benzene by Pt^{o}-nanoparticles supported on Mt was reported by Pan et al. (2009). They proposed that small Ru^{o}-nanoparticles formed in the interlayers of Mt are responsible for the high efficiency toward benzene reduction. Sidhpuria et al. (2009) reported the hydrogenation of benzene and some other aromatic compounds commonly present in petroleum feedstocks using Rh-nanoparticles intercalated Mt. Asymmetric hydrogenation of ethyl pyruvate to (*R*)-ethyl lactate by Pt-nanoparticles dispersed on Na^{+}-Mt and Ht was reported by Mastalir et al. (2000, 2004). They also reported chemoselective hydrogenation of 2-cyclohexene-1-one and asymmetric hydrogenation of ethyl pyruvate using Pt-bentonite. Manikandan et al. (2007a,b) reported that at low temperature and moderate H_2 pressure, efficient hydrogenation of cinnamaldehyde to cinnamyl alcohol occurred in high yield using low loaded Pt-nanoparticles in the interlayer of Mt. Kotkar and Thakkar (1997) reported the hydrogenation of cycloalkane using Rh-nanoparticles supported Mt. Pd-nanoparticles immobilised in an organo-Mt was catalytically effective toward the hydrogenation of various olefins as reported by Kiraly et al. (1999). An eco-friendly and economic catalyst for hydrogenation of olefins was reported by Dhakshinamoorthy and Pitchumani (2008). They reduced a Ni^{2+}-exchanged Mt K10 with hydrazine in aqueous alkaline medium to obtain Ni^{o}-nanoparticles. With this catalyst, 100% yield was obtained within 8 h in the hydrogenation of styrene, while Mt K10 and Ni^{2+}-exchanged Mt K10 showed no catalytic activity.

Recently, Dutta et al. (2011a,b, 2018) and Dutta and Dutta (2013), reported several important coupling reactions catalysed by metal nanoparticles stabilised on modified Mt. They reported Cu^{o}-nanoparticles with high activity in promoting 1,3-dipolar cycloaddition reactions between terminal alkynes and azides to synthesise highly regioselective 1,4-disubstituted mono- and bis-1,2,3-triazoles (Scheme 9.29) in

Scheme 9.29 1,3-Dipolar cycloaddition of azides and alkynes catalysed by Cu°-nanoparticles supported on Mt.

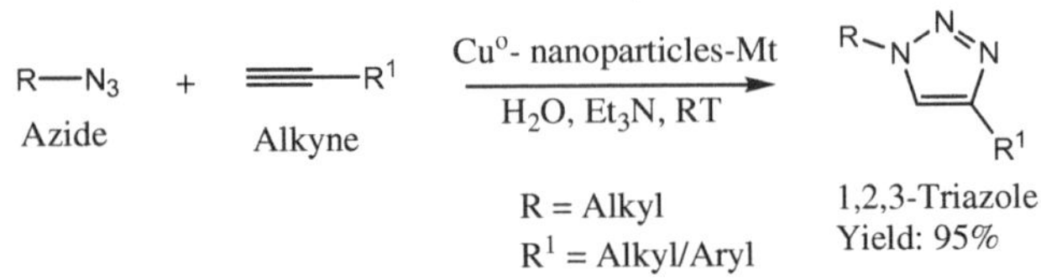

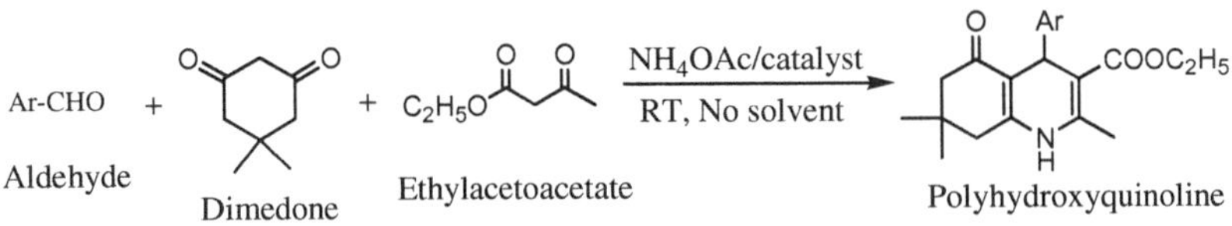

Scheme 9.30 Multicomponent one-pot synthesis of Hantzsch polyhydroquinoline catalysed by Mt supported Ni°-nanoparticles.

excellent yields (up to 95%) at room temperature in water (Borah et al., 2011). The nanocatalysts were reused without significant loss of catalytic activity under the same reaction conditions.

A green and efficient multicomponent one-pot synthesis of Hantzsch polyhydroquinoline was achieved (Saikia et al., 2012) by the condensation of aldehydes, dimedone, ethylacetoacetate, and ammonium acetate at room temperature using Ni°-nanoparticles supported on environmentally benign modified Mt as catalyst. The catalyst has promising features for the reaction, such as easy removal, solvent-free reaction conditions, short reaction time, high product yields (about 95%), and easy workup (Scheme 9.30).

Borah and Dutta (2013) reported the Pd°-nanoparticles supported on Mt as catalysts for efficient Heck and Sonogashira reactions under mild reaction conditions. Vinylation of aryl halides with olefins results in cross-coupling products with maximum 96% yield and 99% *trans* selectivity. In the alkynylation of aryl halides with terminal alkynes, that is, in the Sonogashira coupling reaction, a maximum of 94% yield with 100% selectively of cross-coupling products was observed (Schemes 9.31 and 9.32).

Multicomponent reactions such as A^3 coupling (aldehyde-amine-alkyne) via C–H activation is developed for preparing products such as propargylamines (Dyker, 1999), which are versatile intermediates for the preparation of various biologically important nitrogen-containing compounds (Binda et al., 2004). Earlier, propargylamines were prepared by the addition of metallated alkynes to an imine (Harada et al., 2000). In recent years, various mid- and late-transition metal catalysts supported on different matrixes have been used for A^3 coupling reactions (Peshkov et al., 2012). These catalytic systems often show several disadvantages, such as poisoning of the catalytically active sites. Borah et al. (2014a,b) reported (Scheme 9.33) the efficient and green catalytic system, that is, Cu° and Au°-Mt for the synthesis of propargylamines with excellent yields of about 94%, and selectivity of 100%, under mild reaction conditions, without requiring any additives or an inert atmosphere. The main advantages of the catalytic system over the existing processes are that the support, that is, Mt does not require any surface functionalisation, and is also cheap, environmentally benign, and robust.

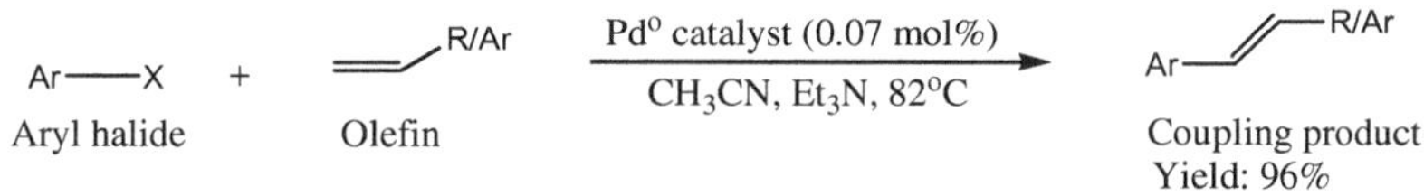

Scheme 9.31 Heck coupling reaction with aryl halides and alkenes catalysed by Mt supported Pd°-nanoparticles.

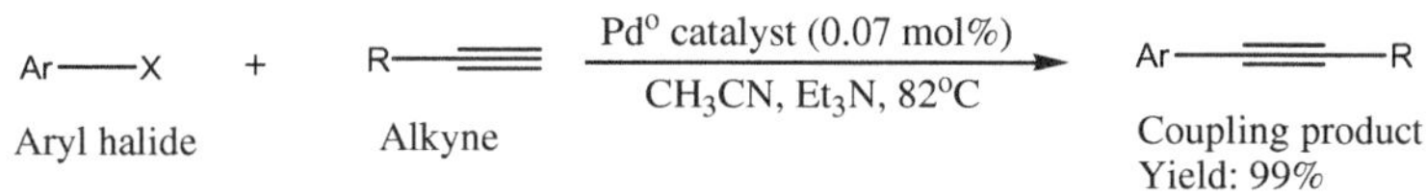

Scheme 9.32 Sonogashira coupling reaction with aryl halides and alkynes catalysed by Mt supported Pd°-nanoparticles.

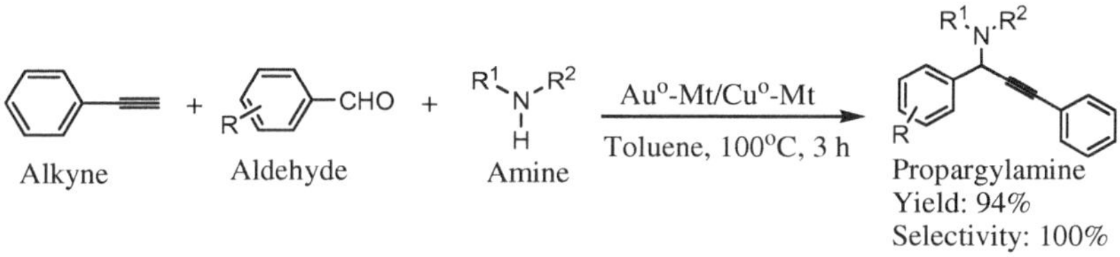

Scheme 9.33 Three-component coupling of aldehyde, alkyne, and amine for the synthesis of propargylamines.

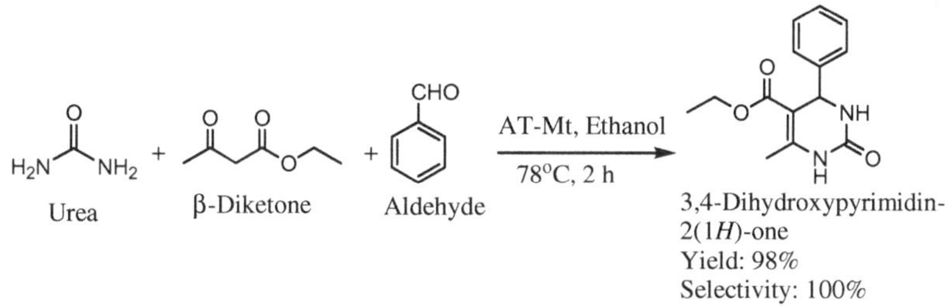

Scheme 9.34 Reaction between aldehyde, β-ketoester and urea for preparation of 3,4-dihydropyrimidin-2(1H)-ones.

Recently, three component coupling reactions of aldehyde, β-ketoester (or β-diketone), and urea (or substituted urea) have been reported (Phukan et al., 2017b). The facile synthesis of 3,4-dihydropyrimidin-2(1H)-ones via the Biginelli reaction under mild reaction conditions with Mt-AT as a catalyst with a specific surface area of 400 m^2/g is shown in Scheme 9.34. Yields of 98% with 100% selectivity were observed.

9.6 Conclusion

Mt, a clay mineral of the smectite group, is a highly attractive raw material for various catalytic applications. Its expandable nature, CEC, small particle size, and high surface area are the most important properties. In addition, it is naturally occurring, cheap, eco-friendly, and abundantly available. The layered Mt, as well as its modified forms, are highly efficient supports for different catalytically active agents. This chapter focuses on the clay mineral catalysed formation of biopolymers of the first life, and recent developments on cation-exchanged and acid-activated Mt, PILC, clay mineral supported metal ions or salts, metal-complexes, heteropolyacids, metal nanoparticles, and their catalytic applications. Tailorable mesoporous Mt may be advantageous over microporous zeolite for different catalytic reactions. Particular emphasis has been placed on the significant progress made for the generation of new heterogeneous 'Nanocatalysts' based on metal nanoparticles supported Mt with innovative characteristics and their expected applications in fine and bulk chemical industries, pharmaceuticals, petroleum refineries, environmental pollution abatement, and many other fields. Therefore, there is a huge opportunity for further development of such Mt-based heterogeneous catalysts and their applications.

References

Adams, J.M., 1987. Synthetic organic chemistry using pillared, cation-exchanged and acid-treated montmorillonite catalysts: a review. Appl. Clay Sci. 2, 309–342.

Adams, J.M., McCabe, R.W., 2006. Clay minerals as catalysts. In: Bergaya, F., Theng, B.K.G., Lagaly, G. (Eds.), In: Handbook of Clay Science. Developments in Clay Science, vol. 1. Elsevier, pp. 541–581 (Chapter 10.2).

Adams, J.M., Ballantine, J.A., Grahan, S.H., Laub, R.J., Purnel, J.H., Reid, P.I., Shanar, W.Y.M., Thomas, J.M., 1979. Selective chemical conversions using sheet silicate intercalates: low temperature addition of water to 1-alkenes. J. Catal. 58, 238–252.

Agarwal, S., Ganguli, J.N., 2013. Selective hydrogenation of monoterpenes on rhodium(0) nanoparticles stabilized in montmorillonite K-10 clay. J. Mol. Catal. A. Chem. 372, 44–50.

Agrawal, A., Tratnyek, P.G., 1996. Reduction of nitro aromatic compounds by zerovalent iron metal. Environ. Sci. Technol. 30, 153–160.

Ahmed, O.S., Dutta, D.K., 2003a. Generation of metal nanoparticles on montmorillonite K-10 and their characterization. Langmuir 19 (13), 5540–5541.

Ahmed, O.S., Dutta, D.K., 2003b. In situ generation of metal clusters in interlamellar spacing of montmorillonite clay and their thermal behaviour. Thermochim. Acta. 395 (1–2), 209–216.

Ahmed, O.S., Dutta, D.K., 2005. Friedel-Crafts benzylation of benzene using Zn and Cd ions exchanged clay composites. J. Mol. Catal. A: Chem. 229, 227–231.

Aiken III, J.D., Finke, R.G., 1999. A review of modern transition-metal nanoclusters: their synthesis, characterization, and applications in catalysis. J. Mol. Catal. A: Chem. 145 (1–2), 1–44.

Aranda, P., Ruiz-Hitzky, E., 1992. Poly(ethylene oxide)-silicate intercalation materials. Chem. Mater. 4 (6), 1395–1403.

Aslya, E., Harrane, A., Belbachir, M., 2016. Polymerization of DL-Lactide induced by protonated montmorillonite clay as a solid catalyst: mechanism study. Mater. Res. 19 (1), 132–137.

Astruc, D., 2008. Astruc, D. (Ed.), Transition-metal nanoparticles in catalysis: from historical background to the state-of-the-art. Nanoparticles and Catalysis Wiley-VCH Verlag GmbH & Co. KGaA, pp. 1–48 (Chapter 1).
Ayyappan, S., Subbanna, G.N., Gopalan, R.S., Rao, C.N.R., 1996. Nanoparticles of nickel and silver produced by the polyol reduction of the metal salts intercalated in montmorillonite. Solid State Ion. 84, 271–281.
Ballantine, J.A., Purnel, J.H., Rayanakorn, M., Thomas, J.M., Williams, K.F.C., 1981. Chemical conversions using sheet silicates: novel intermolecular elimination of ammonia from amines. J. Chem. Soc. Chem. Commun. 1, 9–10.
Ballantine, J.A., Punell, J.H., Thomas, J.M., 1984. Sheet silicates: broad spectrum catalysts for organic synthesis. J. Mol. Catal. A: Chem. 27, 157–167.
Ballantine, J.A., Purnel, J.H., Rayanakorn, M., Thomas, J.M., Williams, K.F., 1985. Organic reactions catalyzed by sheet silicates: intermolecular elimination of ammonia from primary amines. J. Mol. Catal. A: Chem. 30, 373–388.
Balogh, M., Laszlo, P., 1993. Organic Chemistry Using Clays. Springer-Verlag, Berlin, pp. 140–143.
Barrer, R.M., 1985. Porous crystals: a perspective. Pure Appl. Chem. 58, 1317–1322.
Barrer, R.M., 1989a. Clay minerals as selective and shape-selective sorbents. Pure Appl. Chem. 61, 1903–1912.
Barrer, R.M., 1989b. Shape-selective sorbents based on clay minerals: a review. Clay Clay Miner. 37, 385–395.
Belova, V., Mohwald, H., Shchukin, D.G., 2008. Sonochemical intercalation of preformed gold nanoparticles into multilayered clay. Langmuir 24 (17), 9747–9753.
Belova, V., Andreeva, D.V., Mohwald, H., Shchukin, D.G., 2009a. Correlation between ion diffusional motion and ionic conductivity for different electrolytes based on ionic liquid. J. Phys. Chem. C. 113 (16), 5381–5390.
Belova, V., Mohwald, H., Shchukin, D.G., 2009b. Ultrasonic intercalation of gold nanoparticles into a clay matrix in the presence of surface-active materials. Part II: negative sodium dodecylsulfate and positive cetyltrimethylammonium bromide. J. Phys. Chem. C. 113 (16), 6751–6760.
Benesi, H.A., Winquest, B.H.C., 1978. Surface acidity of solid catalysts. Adv. Catal. 27, 82–97.
Bennett, C.O., Che, M., 1989. Some geometric aspects of structure sensitivity. J. Catal. 120 (2), 293–302.
Bergaya, F., Lagaly, G., 2001. Surface modification of clay minerals. Appl. Clay Sci. 19, 1–3.
Berry, F.J., Hayes, M.B., Jones, S.L., 1990. The intercalation of 2,2-bipyridine and 1,10-phenanthroline and their complexes with iron within the interlayer regions of montmorillonite clay. Inorg. Chim. Acta 178 (2), 203–208.
Bhorodwaj, S.K., Dutta, D.K., 2010. Heteropoly acid supported modified montmorillonite clay: an effective catalyst for the esterification of acetic acid with sec-butanol. Appl. Catal. A: Gen. 378, 221–226.
Bhorodwaj, S.K., Dutta, D.K., 2011. Activated clay supported heteropoly acid catalysts for esterification of acetic acid with butanol. Appl. Clay. Sci. 53, 347–352.
Bhorodwaj, S.K., Pathak, M.G., Dutta, D.K., 2009. Esterification of acetic acid with n-butanol using heteropoly acid supported modified clay catalyst. Catal. Lett. 133, 185–191.
Binda, C., Hubalek, F., Li, M., Herzig, Y., Sterling, J., Edmondson, D.E., Mattevi, A., 2004. Crystal structures of monoamine oxidase B in complex with four inhibitors of the N-propargylaminoindan class. J. Med. Chem. 47, 1767–1774.
Booij, E., Kloprogge, J.T., van Veen, J.A.R., 1996. Preparation, structural characteristics and catalytic properties of large-pore rare earth element (Ce, La)/Al-pillared smectites. Clay Clay Miner. 44, 774–782.

Borah, B.J., Dutta, D.K., 2013. In situ stabilization of Pd-nanoparticles into the nanopores of modified montmorillonite: efficient heterogeneous catalysts for Heck and Sonogashira coupling reactions. J. Mol. Catal. A: Chem. 366, 202–209.

Borah, M., Ganguli, J.N., Dutta, D.K., 2001. Intersalation reactions of trisdiimine metal complexes with montmorillonite clay: a new approach. J. Colloid Interf. Sci. 233 (2), 171–179.

Borah, B.J., Dutta, D., Dutta, D.K., 2010. Controlled nanopore formation and stabilization of gold nanocrystals in acid-activated montmorillonite. Appl. Clay Sci. 49, 317–323.

Borah, B.J., Dutta, D., Saikia, P.P., Barua, N.C., Dutta, D.K., 2011. Stabilization of Cu(0)-nanoparticles into the nanopores of modified montmorillonite: an implication on the catalytic approach for "Click" reaction between azides and terminal alkynes. Green Chem. 13, 3453–3460.

Borah, B.J., Borah, S.J., Saikia, L., Dutta, D.K., 2014a. Efficient one-pot three-components coupling reactions catalyzed by Cuo nano-particles stabilized on modified montmorillonite clay. Catal. Sci. Technol. 4, 1047–1054.

Borah, B.J., Borah, S.J., Saikia, L., Dutta, D.K., 2014b. Efficient one-pot synthesis of propargylamines catalysed by gold nanocatalysts stabilized on montmorillonite. Catal. Sci. Technol. 4, 4001–4009.

Boricha, A.B., Mody, H.M., Bajaj, H.C., Jasra, R.V., 2006. Hydrogenation of benzene over ruthenium-exchanged montmorillonite in the presence of thiophene. Appl. Clay Sci. 31, 120–125.

Bovey, J., Jones, W., 1996. Characterisation of Al-pillared acid activated clay catalysts. J. Mater. Chem. 4, 2027–2035.

Bovey, J., Kooli, F., Jones, W., 1996. Preparation and characterisation of Ti-pillared acid-activated clay catalysts. Clay Miner. 31, 501–506.

Breen, C., Watson, R., 1998. Acid-activated organoclays: preparation, characterisation and catalytic activity of polycation-treated bentonites. Appl. Clay Sci. 12, 479–494.

Breen, C., Madejová, J., Komadel, P., 1995. Characterisation of moderately acid treated, size-fractionated montmorillonites using IR and MAS NMR spectroscopy and thermal analysis. J. Mater. Chem. 5, 469–474.

Breen, C., Watson, R., Madejová, J., Komadel, P., Klapyta, Z., 1997. Acid activated organoclays: preparation, characterisation and catalytic activity of acid-treated tetra-alkylammonium exchanged smectites. Langmuir 13, 6473–6479.

Brindly, G.W., Brown, G., 1980. Crystal Structure of Clay Minerals and Their X-ray Identification. Mineralogical Society, London, pp. 305–359.

Bronstein, L.M., Sidorov, S.N., Valetsky, P.M., 2004. Nanostructured polymeric systems as nanoreactors for nanoparticle formation. Russ. Chem. Rev. 73 (5), 501–515.

Brown, D.R., Rhodes, C.N., 1997. Brønsted and Lewis acid catalysis with ion-exchanged clays. Catal. Lett. 45, 35–40.

Burda, C., Chen, X., Narayanan, R., El-Sayed, M.A., 2005. Chemistry and properties of nanocrystals of different shapes. Chem. Rev. 105 (4), 1025–1102.

Cardenas-Lizana, F., Gomez-Quero, S., Hugon, A., Delannoy, L., Louis, C., Keane, M.A., 2009. Pd-promoted selective gas phase hydrogenation of p-chloronitrobenzene over alumina supported Au. J. Catal. 262 (2), 235–243.

Chen, C.C., Kuo, P.L., 2006. Gold nanoparticles prepared using polyethylenimine adsorbed onto montmorillonite. J. Colloid Interf. Sci. 293 (1), 101–107.

Chiba, K., Hirano, T., Kitano, Y., Tada, M., 1999. Montmorillonite-mediated hetero-Diels–Alder reaction of alkenes and o-quinomethanes generated in situ by dehydration of o-hydroxybenzyl alcohols. Chem. Commun. 691–692.

Chiche, B., Finiels, A., Gauthier, C., Geneste, P., Graille, J., Pioch, D., 1987. Acylation over cation-exchanged montmorillonite. J. Mol. Catal. A: Chem. 42, 229–235.
Choe, S., Lee, S.H., Chang, Y.Y., Hwang, K.Y., Khim, J., 2001. Rapid reductive destruction of hazardous organic compounds by nanoscale Fe-0. Chemosphere 42, 367–372.
Choudary, B.M., Prasad, B., Kantam, M.L., 1989. Metal complexes of glufosinate. J. Agr. Food Chem. 37 (5), 1442–1444.
Choudary, B.M., Chowdari, N.S., Kantam, M.L., 2000a. Friedel–Crafts sulfonylation of aromatics catalysed by solid acids: an eco-friendly route for sulfone synthesis. J. Chem. Soc. Perkin Trans. 1, 2689–2694.
Choudary, B.M., Sateesh, M., Kantam, M.L., Rao, K.K., Prasad, K.V.R., Raghavan, K.V., Sarma, J.A.R.P., 2000b. Selective nitration of aromatic compounds by solid acid catalysts. Chem. Commun. 25–26.
Choy, J.-N., Park, M., 2004. Clay Surface: fundamentals and applications. In: Wypych, E., Satyanarayana, K.G. (Eds.), Interface and Technology. Elsevier, UK, pp. 404–407.
Clark, J.H., 1994. Catalysis of Organic Reactions by Supported Inorganic Reagents. VCH Publishers, New York, pp. 40–51.
Clark, J.H., 2002. Solid acids for green chemistry. Acc. Chem. Res. 35, 791–797.
Clark, J.H., Macquarrie, D.J., 1996. Environmentally friendly catalytic methods. Chem. Soc. Rev. 25, 303–310.
Clark, J.H., Kybett, A.P., Macquarrie, D., Barlow, J.S.J., Landon, P., 1989. Montmorillonite supported transition metal salts as Friedel–Crafts alkylation catalysts. J. Chem. Soc. Chem. Commun. 10, 1353–1354.
Clark, J.H., Martin, K., Teasdale, A.J., Barlow, S.J., 1995. Environmentally friendly catalysis using supported reagents: evolution of a highly active form of immobilised aluminium chloride. J. Chem. Soc. Chem. Commun. (19), 2037–2040.
Clark, J.H., Ross, J.C., Macquarrie, D.J., Barlow, S.J., Bastock, T.W., 1997. Environmentally friendly catalysis using supported reagents: the fast and selective bromination of aromatic substrates using supported zinc bromide. Chem. Commun. 1203–1204.
Claver, C., Fernandez, E., Catala, R.M., Medina, F., Salagre, P., Sueiras, J.E., 2001. Studies on the characterization of several iridium and rhodium-clay catalysts and their activity in imine hydrogenation. J. Catal. 201, 70–79.
Coq, B., Figueras, F., 1998. Structure-activity relationships in catalysis by metals: some aspects of particle size, bimetallic and supports effects. Coordin. Chem. Rev. 178–180, 1753–1783.
Corain, B., Centomo, P., Lora, S., Kralik, M., 2003. Functional resins as innovative supports for catalytically active metal nanoclusters. J. Mol. Catal. A: Chem. 204–205, 755–762.
Corma, A., Garcia, H., 2008. Supported gold nanoparticles as catalysts for organic reactions. Chem. Soc. Rev. 37 (9), 2096–2126.
Cornelis, A., Laszlo, P., 1985. Clay-supported copper(II) and iron(III) nitrates: novel multipurpose reagents for organic synthesis. Synthesis, 909–919.
Crocker, M., Buglass, J.G., Herold, R.H.M., 1993. Synthesis and characterization of palladium crystallites intercalated in montmorillonite. Chem. Mater. 5, 105–109.
Cui, X., Li, J., Zhang, Z.P., Fu, Y., Liu, L., Guo, Q.X., 2007. Pd(quinoline-8-carboxylate)$_2$ as a low-priced, phosphine-free catalyst for Heck and Suzuki reactions. J. Org. Chem. 72 (24), 9342–9345.
Das, P., Sarmah, P.P., Borah, B.J., Saikia, L., Dutta, D.K., 2016. Aromatic ring hydrogenation catalysed by nanoporous montmorillonite supported Ir(0)-nanoparticle composites under solvent free conditions. New J. Chem. 40, 2850–2855.

Dasgupta, S., Török, B., 2008. Application of clay catalysts in organic synthesis. A review. Org. Prep. Proced. Int. 40 (1), 1–65.

Davis, S.M., Zaera, F., Somorja, G.A., 1984. Surface structure and temperature dependence of *n*-hexane skeletal rearrangement reactions catalyzed over platinum single crystal surfaces: marked structure sensitivity of aromatization. J. Catal. 85 (1), 206–223.

de Vries, J.G., 2006. A unifying mechanism for all high-temperature Heck reactions. The role of palladium colloids and anionic species. Dalton Trans. 421–429.

Dhakshinamoorthy, A., Pitchumani, K., 2008. Clay entrapped nickel nanoparticles as efficient and recyclable catalysts for hydrogenation of olefins. Tetrahedron Lett. 49, 1818–1823.

Ding, Z., Kloprogge, J.T., Frost, R.L., Lu, G.Q., Zhu, H.Y., 2001. Porous clays and pillared clay-based catalysts. Part 2: a review of the catalytic and molecular sieve applications. J. Porous Mat. 8, 273–293.

Dutta D.K. and Dutta D., A Process for Reducing Chloronitrobenzene Catalysed by Platinum-Nanoparticles Stabilized on Modified Montmorillonite Clay, 2013, (i) Office of the Controller General of Patents, Design & Trade Marks; Government of India, for Indian Patents; (ii) The United States Patent and Trademark Office (USPTO) for USA patents, (iii) World Intellectual Property Organisation (WIPO) for PCT and (iv) Intellectual Property Office, UK, for Great Britain Patents, Indian Patent: Application No. 2183DEL2013, PCT/IN2013/000698, International Publication Number: 0077NF2013/WO; US Patent No.9284259, UK Patent Publication No. GB2537450.

Dutta, D., Dutta, D.K., 2014. Selective and efficient hydrogenation of halonitro-benzene catalyzed by clay supported Ni-nanoparticles. Appl. Catal. A Gen. 487, 158–164.

Dutta, D.K., Pathak, M.G., 2009. Process for the Preparation of 4-(4-Hydroxyphenyl) Butan-2-One Using Solid Acid Clay Catalyst. The United States Patent and Trademark Office (USPTO). PCT/IN2009/000178; US Patent Number: US 8471068.

Dutta, D., Phukan, A., Dutta, D.K., 2018. Nanoporous montmorillonite clay stabilized copper nanoparticles: Efficient and reusable catalyst for oxidation of alcohols. Mol. Catal. 451, 178–185.

Dutta, D.K., Sarmah, P.P., 2016. Nanoporous montmorillonite clay: sustainable catalyst and catalysts support. In: Morton, T. (Ed.), Nanopores and Nanoporous Material. NOVA Science Publisher Inc, USA, pp. 27–54 (Chapter 2).

Dutta, D., Borah, B.J., Saikia, L., Pathak, M.G., Sengupta, P., Dutta, D.K., 2011a. Synthesis and catalytic activity of Ni-acid activated montmorillonite nanoparticles. Appl. Clay Sci. 53 (4), 650–656.

Dutta, D.K., Dutta, D., Sarmah, P.P., Bhorodwaj, S.K., Borah, B.J., 2011b. Metal nanoparticles: in-situ stabilization, characterization and reactivities. J. Biomed. Nanotechnol. 7, 76–77.

Dutta, D.K., Borah, B.J., Sarmah, P.P., 2015. Recent advances in metal nanoparticles stabilization into nanopores of montmorillonite and their catalytic applications for fine chemicals synthesis. Catal. Rev. Sci. Eng. 57 (3), 257–305.

Dyker, G., 1999. Transition metal catalyzed coupling reactions under C – H activation. Angew Chem. Int. Edit 38, 1698–1712.

Ecija, M., Diez, A., Rubiralta, M., Casamitjana, N., Kogan, M.J., Giral, E., 2003. Synthesis of 3-aminolactams as X-Gly constrained pseudodipeptides and conformational study of a Trp-Gly surrogate. J. Org. Chem. 68, 9541–9553.

Fan, M., Yuan, P., Zhu, J., Chen, T., Yuan, A., He, H., Chen, K., Liu, D., 2009. Core-shell structured iron nanoparticles well dispersed on montmorillonite. J. Magn. Magn. Mater. 321, 3515–3519.

Favre, L.G., 1991. Organo-bentonites with quaternary alkylammonium ions. Clay Miner. 26, 19–32.

Feng, X.L., Zhang, Y., Lin, Z.H., Zhao, C.X., 2005. Synthesis of diindolylmethanes-montmorillonite clay K 10 and KSF catalysed condensation of indole with imines. Heterocycl. Commun. 11, 427–432.

Ferris, J.P., 2002. Montmorillonite catalysis of 30-50 mer oligonucleotides: laboratory demonstration of potential steps in the origin of the RNA world. Origins Life Evol. B 32, 311–322.

Ferris, J.P., Ertem, G., 1993. Montmorillonite catalysis of RNA oligomer formation in aqueous fssolution. A model for the prebiotic formation of RNA. J. Am. Chem. Soc. 115, 12270–12275.

Ferris, J.P., Hill Jr., A.R., Liu, R., Orgel, L.E., 1996. Synthesis of long prebiotic oligomers on mineral surfaces. Nature 381, 59–61.

Fievet, F., Lagier, J.P., Blin, B., Beaudoin, B., Figlarz, M., 1989. Homogeneous and heterogeneous nucleation in the polyol process for the preparation of metal nanoparticles. Solid State Ion. 32–33, 198–205.

Figueras, F., Coq, B., 2001. Bimetallic palladium catalysts: influence of the co-metal on the catalyst performance. J. Mol. Catal. A: Chem. 173, 117–134.

Gates, W.P., Anderson, J.S., Raven, M.D., Churchman, G.J., 2000. Mineralogy of a bentonite from Miles, Queensland, Australia and characterisation of its acid activation product. Appl. Clay Sci. 20, 189–197.

Gnaim, J.M., Sheldon, R.A., 2004. Shape-selective para-chlorination of phenol using sulfuryl chloride with the aid of microporous catalysts. Tetrahedron Lett. 45, 9397–9399.

Gonsalves, K.E., Li, H., Perez, R., Santiago, P., Jose-Yacaman, M., 2000. Synthesis of nanostructured metals and metal alloys from organometallic. Coordin. Chem. Rev. 206, 607–630.

Goodman, R.P., Schaap, I.A.T., Tardin, C.F., Erben, C.M., Berry, R.M., Schmidt, C.F., Turberfield, A.J., 2005. Rapid chiral assembly of rigid DNA building blocks for molecular nanofabrication. Science 310, 1661–1665.

Gorobinskii, L.V., Firsova, A.A., Efimova, N.N., Korchak, V.N., 2006. Pt-containing pillared clay catalysts for CO oxidation. Kinet. Catal. 47, 395–399.

Grim, R.E., 1953. Geological Sciences. McGraw Hill Book Company, Inc, pp. 340–342.

Grunes, J., Zhu, J., Somorjai, G.A., 2003. Catalysis and nanoscience. Chem. Commun. 2257–2260.

Gu, C., Jia, H., Li, H., Teppen, B.J., Boyd, S.A., 2010. Synthesis of highly reactive subnanosized zero-valent iron using smectite clay templates. Environ. Sci. Technol. 44, 4258–4263.

Haber, J., Pamin, K., Matachowski, L., Mucha, D., 2003. Catalytic performance of the dodecatungstophosphoric acid on different supports. Appl. Catal. A: Gen. 256, 141–152.

Harada, T., Fujiwara, T., Iwazaki, K., Oku, A., 2000. Tandem cyclization of alkynylmetals bearing remote leaving group via cycloalkylidene carbenes. Org. Lett. 2, 1855–1857.

Harrison, J.B., Berkheiser, V.E., Erdos, G.W., 1988. Hydrogen reduction of $Pt(NH_3)_4^{2+}$ supported on montmorillonite. J. Catal. 112 (1), 126–134.

Hart, M.P., Brown, D.R., 2004. Surface acidities and catalytic activities of acid activated clays. J. Mol. Catal. A: Chem. 212, 315–321.

Hill Jr., A.R., Bohler, C., Orgel, L.E., 1998. Polymerization on the rocks: negatively charged α-amino acids. Origins Life Evol B. 28, 235–243.

Im, S.H., Lee, Y.T., Wiley, B., Xia, Y., 2005. Large-scale synthesis of silver nanocubes: the role of HCl in promoting cube perfection and monodispersity. Angew Chem. Int. Edit. 44, 2154–2157.

Jaisankar, P., Srinivasan, P.C., 2005. Facile synthesis of 3(α,β-diaryl vinyl)indoles under conventional and solvent-free microwave irradiation. Synthetic Commun. 35, 923–928.

Jana, N.R., Gearheart, L., Murphy, C.J., 2001. Wet chemical synthesis of high aspect ratio cylindrical gold nanorods. J. Phys. Chem. B. 105 (19), 4065–4067.

Jasra, R.V., 2003. Solid acid catalysts for acylation of aromatics. Bull. Catal. Soc. Ind. 2 (5), 157–183.

Johnston, C.T., Sheng, G., Teppen, B.J., Boyd, S.A., de Oliveira, M.F., 2002. Spectroscopic study of dinitrophenol herbicide sorption on smectite. Environ Sci. Technol. 36 (23), 5067–5074.

Kadokawa, J., Iwasaki, Y., Tagaya, H., 2002. Ring-opening polymerization of lactones catalyzed by ion-exchanged clay montmorillonite. Green Chem. 4, 14–16.

Kadu, B.S., Sathe, Y.D., Ingle, A.B., Chikate, R.C., Patil, K.R., Rode, C.V., 2011. Efficiency and recycling capability of montmorillonite supported Fe-Ni bimetallic nanocomposites towards hexavalent chromium remediation. Appl. Catal. B. Environ. 104, 407–414.

Kalidindi, S.B., Jagirdar, B.R., 2012. Nanocatalysis and prospects of green chemistry. ChemSusChem 5 (1), 65–75.

Kappe, C.O., Dallinger, D., 2006. The impact of microwave synthesis on drug discovery. Nature Rev. Drug Discov. 5 (1), 51–63.

Kaur, J., Griffin, K., Harrison, B., Kozhevnikov, I.V., 2002. Friedel-Crafts acylation catalysed by heteropoly acids. J. Catal. 208 (2), 448–455.

Kawabata, T., Mizugaki, T., Ebitani, K., Kaneda, K., 2003. A novel montmorillonite-enwrapped scandium as a heterogeneous catalyst for Michael reaction. J. Am. Chem. Soc. 125, 10486–10487.

Kikuchi, E., Matsuda, T., 1988. Shape selective acid catalysis by pillared clays. Catal. Today. 2, 297–307.

Kiraly, Z., Veisz, B., Mastalir, A., Razga, Z., Dekany, I., 1999. Preparation of an organophilic palladium montmorillonite catalyst in a micellar system. Chem. Commun. 1925–1926.

Klasovsky, F., Claus, P., 2008. Metal nanoclusters in catalysis: effects of nanoparticle size, shape, and structure. In: Corain, B., Schmid, G., Toshima, N. (Eds.), Metal Nanoclusters in Catalysis and Materials Science: The Issue of Size Control. Elsevier, New York, pp. 167–181 (Chapter 8).

Kloprogge, J.T., 1998. Synthesis of smectites and porous pillared clay catalysts: a review. J. Porous Mat. 5, 5–41.

Komadel, P., 1999. Structure and chemical characterisation of modified clays. In: Misaelides, P., Macáscek, F., Pinnavaia, T.J., Colella, C. (Eds.), Natural Microporous Materials in Environmental Technology. Kluwer Academic Publisher, Netherlands, pp. 3–18.

Komadel, P., Bujdák, J., Madejová, J., Scucha, V., Elsass, F., 1996. Effect of non-swelling layers on the dissolution of reduced charged montmorillonite in hydrochloric acid. Clay Miner. 31, 333–345.

Komadel, P., Janek, M., Madejová, J., Weekes, A., Breen, C., 1997. Acidity and catalytic activity of mildly acid-treated Mg-rich montmorillonite and hectorite. J. Chem. Soc. Faraday Trans. 93, 4207–4210.

Kotkar, D., Thakkar, N.V., 1997. Hydrogenation of cycloalkenes on Rh/montmorillonite. Proc. Indian Acad. Sci. (Chem. Sci.). 109, 99–104.

Kozhevnikov, I.V., 1998. Catalysis by heteropoly acids and multicomponent polyoxometalates in liquid-phase reactions. Chem. Rev. 98, 171–198.

Kozhevnikov, I.V., 2003. Friedel–Crafts acylation and related reactions catalysed by heteropoly acids. Appl. Catal. A: Gen. 256, 3–18.

Kralik, M., Biffis, A., 2001. Effect of additional metal ions on catalyses of polymer stabilized metal nanoclusters. J. Mol. Catal. A: Chem. 177 (1), 113–138.

Lagaly, G., 1994. Bentonites: adsorbents of toxic substances. Prog. Coll. Pol. Sci. 95, 61–72.

Laired, D.A., Shang, C., Thompson, M.L., 1995. Hysteresis in crystalline swelling of smectites. J. Colloid Interf. Sci. 171 (1), 240–245.

Lao, H., Detellier, C., 1994. Gas chromatographic separation of linear hydrocarbons on microporous organo-smectites. Clay Clay Miner. 42, 477–481.

Lao, H., Latieule, S., Detellier, C., 1991. Molecular recognition in microporous organominerals. Shape-specific interaction of carbon dioxide in functionalized organo montmorillonite micro cavities. Chem. Mater. 3, 1009–1011.

Laszlo, P., 1986. Catalysis of organic reactions by inorganic solids. Acc. Chem. Res. 19, 121–127.

Laszlo, P., 1987. Chemical reactions on clays. Science 235, 1473–1477.

Laszlo, P., Mathy, A., 1987. Catalysis of Friedel-Crafts alkylation by a montmorillonite doped with transition-metal cations. Helv. Chim. Acta. 70, 577–586.

Lee, H., Habas, S.E., Kweskin, S., Butcher, D., Somorjai, G.A., Yang, P., 2006. Morphological control of catalytically active platinum nanocrystals. Angew. Chem. 118 (46), 7988–7992.

Lewis, L.N., 1993. Chemical catalysis by colloids and clusters. Chem. Rev. 93 (8), 2693–2730.

Li, T.-S., Li, L.-J., Lu, B., Yang, F., 1998. Montmorillonite clay catalysis. Part 14.1 A facile synthesis of 2-substituted and 2,2-disubstituted 1,3-benzodioxoles. J. Chem. Soc. Perkin Trans. 1, 3561–3564.

Li, S., Wu, P., Li, H., Zhu, N., Li, P., Wu, J., Wang, X., Dang, Z., 2010. Synthesis and characterization of organo-montmorillonite supported iron nanoparticles. Appl. Clay Sci. 50, 330–336.

Liu, R., Orgel, L.E., 1998. Polymerization on the rocks: β-amino acids and arginine. Origins Life Evol. B 28, 245–257.

Liu, J., Li, X., Zuo, S., Yu, Y., 2007. Preparation and photocatalytic activity of silver and TiO_2 nanoparticles/montmorillonite composites. Appl. Clay Sci. 37 (3–4), 275–280.

Liu, H., Deng, J., Li, W., 2010. Synthesis of nanoparticles supported on boehmite for selective hydrogenation of p-nitrophenol and p-chloronitrobenzene. Catal. Lett. 137, 261–266.

Liz-Marzán, L.M., 2006. Tailoring surface plasmons through the morphology and assembly of metal nanoparticles. Langmuir 2 (1), 32–41.

Loeppert, R.H., Mortland, M.M., Pinnavaia, T.J., 1979. Synthesis and properties of heat-stable expanded smectite and vermiculite. Clay Clay Miner. 27 (3), 201–208.

Loupy, A., 2006. Microwaves in Organic Synthesis, second ed. Wiley-VCH, Weinheim, pp. 134–218.

MacEwan, D.M.C., Ruiz, A.A., Brown, G., 1961. Interstratified clay minerals. In: Brown, G. (Ed.), The X-Ray Identification and Crystal Structures of Clay Minerals. Mineralogical Society, London, pp. 345–393.

Madejová, J., Bujdák, J., Janek, M., Komadel, P., 1998. Comparative FT-IR study of structural modifications during acid treatment of dioctahedral smectites and hectorite. Spectrochim. Acta A 54, 1397–1406.

Mahmoud, M.A., Tabor, C.E., Ding, Y., Wang, Z.L., El-Sayed, M.A., 2008. A new catalytically active colloidal platinum nanocatalyst: the multiarmed nanostar single crystal. J. Am. Chem. Soc. 130 (14), 4590–4591.

Malla, P.B., Ravindranathan, P., Komarneni, S., Roy, R., 1991. Intercalation of copper metal clusters in montmorillonite. Nature 351, 555–557.

Malla, P.B., Ravindranathan, P., Komarneni, S., Breval, E., Roy, R., 1992. Reduction of copper acetate hydroxide hydrate interlayers in montmorillonite by a polyol process. A new approach in the preparation of metal-supported catalysts. J. Mater. Chem. 2, 559–565.

Manikandan, D., Divakar, D., Sivakumar, T., 2007a. Utilization of clay minerals for developing Pt nanoparticles and their catalytic activity in the selective hydrogenation of cinnamaldehyde. Catal. Commun. 8, 1781–1786.

Manikandan, D., Divakar, D., Rupa, A.V., Revathi, S., Preethi, M.E.L., Sivakumar, T., 2007b. Synthesis of platinum nanoparticles in montmorillonite and their catalytic behaviour. Appl. Clay Sci. 37, 193–200.

Manikandan, D., Divakar, D., Sivakumar, T., 2008. Selective hydrogenation of citral over noble metals intercalated montmorillonite catalysts. Catal. Lett. 123, 107–114.

Massam, J., Brown, D.R., 1995. The influence of microwave heating on the morphology and benzene hydrogenation activity of alumina- and silica-supported palladium catalysts. Catal. Lett. 35, 335–351.

Mastalir, A., Kiraly, Z., Szollosia, G., Bartok, M., 2000. Preparation of organophilic Pd-montmorillonite, an efficient catalyst in alkyne semihydrogenation. J. Catal. 194, 146–152.

Mastalir, A., Szollosi, G., Kiraly, Z., Razga, Z., 2002. Preparation and characterization of platinum nanoparticles immobilized in dihydrocinchonidine modified montmorillonite and hectorite. Appl. Clay Sci. 22, 9–16.

Mastalir, A., Kiraly, Z., Berger, F., 2004. Comparative study of size-quantized Pd-montmorillonite catalysts in liquid-phase semihydrogenations of alkynes. Appl. Catal. A. Chem. 269, 161–168.

McBride, M.B., 1994. Environmental Chemistry of Soils. Oxford University Press, New York, pp. 406–420.

McBride, M.B., Mortland, M.M., 1973. Segregation and exchange properties of alkyl-ammonium ion in a smectite and vermiculite. Clay Clay Miner. 21, 323–329.

McCabe, R.W., 1996. Clay chemistry. In: Bruce, D.W., O'Hare, D. (Eds.), Inorganic Materials second ed. Wiley, Chichester, pp. 313–376.

Miao, S., Liu, Z., Han, B., Huang, J., Sun, Z., Zhang, J., Jiang, T., 2006. Ru nanoparticles immobilized on montmorillonite by ionic liquids: a highly efficient heterogeneous catalyst for the hydrogenation of benzene. Angew Chem. Int. Edit. 45 (2), 266–269.

Mitsudome, T., Nose, K., Mori, K., Mizugaki, T., Ebitani, K., Jitsukawa, K., Kaneda, K., 2007. Montmorillonite-entrapped sub-nanoordered Pd clusters as a heterogeneous catalyst for allylic substitution reactions. Angew Chem. Int. Edit. 46, 3288–3290.

Miyoshi, H., Ohno, H., Sakai, K., Okamura, N., Kourai, H., 2010. Characterization and photochemical and antibacterial properties of highly stable silver nanoparticles prepared on montmorillonite clay in n-hexanol. J. Colloid Interf. Sci. 345, 433–441.

Mokaya, R., Jones, W., 1994. Pillared acid-activated clay catalysts. J. Chem. Soc. Chem. Commun. 929–930.

Molina, C.B., Casas, J.A., Pizarro, A.H., Rodriguez, J.J., 2011. Pillared clays as green chemistry catalysis: application to wastewater treatment. In: Humphrey, J.P., Boyd, D.E. (Eds.), Clay Types, properties and Uses. NOVA Science Publishers Inc., USA, pp. 435–474 (Chapter 16).

Mortland, M.M., Raman, K.V., 1968. Surface acidity of smectites in relation to hydration, exchangeable cation, and structure. Clay Clay Miner. 16 (5), 393–398.

Nagendrappa, G., 2002. Organic synthesis using clay catalysts. Resonance 7 (1), 64–67.

Nakai, Y., Yamamoto, K., Kerada, K., Oguchi, T., Izumikawa, S., 1986. Interactions between crystalline medicinals and porous clay. Chem. Pharm. Bull. 34 (11), 4760–4766.

Nikalje, M.D., Phukan, P., Sudalai, A., 2000. Recent advances in clay-catalyzed organic transformations. Org. Prep. Proc. Int. 32, 1–40.

Noguez, C., 2007. Surface plasmons on metal nanoparticles: the influence of shape and physical environment. J. Phys. Chem. C 111 (10), 3806–3819.

Ogawa, M., Kuroda, K., 1995. Photofunctions of intercalation compounds. Chem. Rev. 95 (2), 399–438.

Ogawa, M., Hashizume, T., Kuroda, K., Kato, C., 1991. Intercalation of 2,2′-bipyridine and complex formation in the interlayer space of montmorillonite by solid-solid reactions. Inorg. Chem. 30 (3), 584–585.

Ogawa, M., Aono, T., Kuroda, K., Kato, C., 1993. Photophysical probe study of alkyl-ammonium-montmorillonites. Langmuir 9 (6), 1529–1533.

Olah, G.A., Krisnamurthy, R., Prakash, G.K.S., 1991. Friedel-Crafts alkylation. In: Trost, B.M. (Ed.), first ed. In: Comprehensive Organic Synthesis, vol. 3. Pergamon Press, Oxford, pp. 239–293 (Chapter 1.8).

Paek, S.M., Jang, J.U., Hwang, S.J., Choy, J.H., 2006. Exfoliation-restacking route to Au nanoparticle-clay nanohybrids. J. Phys. Chem. Solids 67, 1020–1023.

Pan, J., Liu, J., Guo, S., Yang, Z., 2009. Preparation of platinum/montmorillonite nanocomposites in supercritical methanol and their application in the hydrogenation of nitrobenzene. Catal. Lett. 131 (1–2), 179–183.

Papp, S., Dekany, I., 2002. Growth of Pd nanoparticles on layer silicates on disaggregated kaolilite lamellae. Prog. Colloid Polym. Sci. 280 (10), 956–962.

Papp, S., Szel, J., Oszko, A., Dekany, I., 2004. Synthesis of polymer-stabilized nano sized rhodium particles in the interlayer space of layered silicates. Chem. Mater. 16 (9), 1674–1685.

Peshkov, V.A., Pereshivko, O.P., Eycken, E.V.V., 2012. A walk around the A^3–coupling. Chem. Soc. Rev. 41, 3790–3797.

Petridis, D., Falaras, P., Pinnavaia, T.J., 1992. Self-assembly of ion-paired electron transfer centers in a clay-modified electrode. Inorg. Chem. 31 (17), 3530–3533.

Phukan, P., Mohan, J.M., Sudalai, A., 1999. Reaction of methyl diazoacetate with aldehydes, amines, thiols, alcohols and acids over transition metal-exchanged clays. J. Chem. Soc. Perkin Trans. 1, 3685–3689.

Phukan, A., Ganguli, J.N., Dutta, D.K., 2003. $ZnCl_2$-Zn^{2+}-montmorillonite composite: efficient solid acid catalyst for benzylation of benzene. J. Mol. Catal. A: Chem. 202, 279–287.

Phukan, A., Bhorodwaj, S.K., Sharma, P.P., Dutta, D.K., 2017a. Mesoporous aluminosilicate: efficient and reusable catalysts for esterification of *sec*-butanol with acetic acid. J. Porous Mat. https://doi.org/10.1007/s10934-017-0426-x.

Phukan, A., Borah, S.J., Bhorodwaj, S.K., Bordoloi, P., Sharma, K., Borah, B.J., Sharma, P.P., Dutta, D.K., 2017b. An efficient and robust heterogeneous mesoporous montmorillonite clay catalyst for the Biginelli type reactions. Adv. Powder Technol. 28, 1585–1592.

Pinnavaia, T.J., 1983. Intercalated clay catalysts. Science 220, 365–371.

Pinnavaia, T.J., 1986. Pillared clays: synthesis and structural features. In: Setton, R. (Ed.), Chemical Reactions in Organic and Inorganic Constrained Systems. Reidel, New York, pp. 151–164.

Poncelet, G., Schutz, A., 1986. Pillared montmorillonite and beidellite. In: Setton, R. (Ed.), Chemical Reactions in Organic and Inorganic Constrained Systems. Reidel, New York, pp. 165–178.

Ramchandani, R.K., Uphade, B.S., Vinod, M.P., Wakharkar, R.D., Choudhary, V.R., Sudalai, A., 1997. Pd-Cu-exchanged montmorillonite K10 clay: an efficient and reusable heterogeneous catalyst for vinylation of aryl halides. Chem. Commun. 2071–2072.

Ravichandran, J., Sivasankar, B., 1997. Properties and catalytic activity of acid modified montmorillonite and vermiculite. Clay Clay Miner. 45 (6), 854–858.

Ravindranathan, P., Malla, P.B., Komarneni, S., Roy, R., 1990. Preparation of metal supported montmorillonite catalyst: A new approach. Catal. Lett. 6, 401–407.

Reddy, C.R., Iyengar, P., Nagendrappa, G., Prakash, B.S., 2005. Esterification of dicarboxylic acids to diesters over M^{n+}-montmorillonite clay catalysts. Catal. Lett. 101, 87–91.
Regan, B.C., Aloni, S., Jensen, K., Ritchie, R.O., Zettl, A., 2005. Nanocrystal-powered nanomotor. Nano Lett. 5 (9), 1730–1733.
Rhodes, C.N., Brown, D.R., 1992. Structural characterization and optimization of acid treated montmorillonite and high porosity silica supports for $ZnCl_2$ alkylation catalysts. J. Chem. Soc. Faraday Trans. 88, 2269–2274.
Rhodes, C.N., Brown, D.R., 1993. Surface properties and porosities of silica and acid treated montmorillonite catalyst supports: influence on activities of supported $ZnCl_2$ alkylation catalysts. J. Chem. Soc. Faraday Trans. 89, 1387–1391.
Rhodes, C.N., Brown, R., 1994. Catalytic activity of acid-treated montmorillonite in polar and non-polar reaction media. Catal. Lett. 24 (3–4), 285–291.
Rhodes, C.N., Franks, M., Parkes, G.M.B., Brown, D.R., 1991. The effect of acid treatment on the activity of clay supports for $ZnCl_2$ alkylation catalysts. J. Chem. Soc. Chem. Commun. 804–807.
Rioux, R.M., Song, H., Grass, M., Habas, S., Niesz, K., Hoefelmeyer, J.D., Yang, P., Somorjai, G.A., 2006. Monodisperse platinum nanoparticles of well-defined shape: synthesis, characterization, catalytic properties and future prospects. Top. Catal. 39, 167–174.
Roberts, R.M., Khalaf, A.A., 1984. Friedel-Crafts Alkylation Chemistry. Marcel Dekker, New York, pp. 404–520.
Ruiz-Hitzky, E., Aranda, P., 1990. Polymer-salt intercalation complexes in layer silicates. Adv. Mater. 2 (11), 545–547.
Saikia, L., Dutta, D., Dutta, D.K., 2012. Efficient clay supported Ni^{o} nanoparticles as heterogeneous catalyst for solvent-free synthesis of Hantzsch polyhydroquinoline. Catal. Commun. 19, 1–4.
Saikia, P.K., Sarmah, P.P., Borah, B.J., Saikia, L., Dutta, D.K., 2016a. Functionalized montmorillonite supported rhodium complexes: efficient catalysts for carbonylation of methanol. J. Mol. Catal. A: Chem. 412, 27–33.
Saikia, P.K., Sarmah, P.P., Borah, B.J., Saikia, K., Saikia, L., Dutta, D.K., 2016b. Stabilized Fe_3O_4 magnetic nanoparticles into nanopores of modified montmorillonite clay: a highly efficient catalyst for the Baeyer-Villiger oxidation under solvent free conditions. Green Chem. 18, 2843–2850.
Salerno, P., Asenjo, M.B., Mendioroz, S., 2001. Influence of preparation method on thermal stability and acidity of Al-PILCs. Thermochim. Acta. 379, 101–109.
Sapag, K., Mendioroz, S., 2001. Synthesis and characterization of micro-mesoporous solids: pillared clays. Coll. Surf. A-Physicochem. Eng. Aspects. 187, 141–149.
Sarmah, P.P., Dutta, D.K., 2012. Chemo-selective reduction of a nitro group through transfer hydrogenation catalysed by Ru^{o}-nanoparticles stabilized on modified montmorillonite clay. Green Chem. 14, 1086–1093.
Sarmah, P.P., Dutta, D.K., 2014. Stabilized Rh^{o}-nanoparticles-montmorillonite clay composite: synthesis and catalytic transfer hydrogenation reaction. Appl. Catal. A Gen. 470, 355–360.
Sato, H., Yamagishi, A., Kato, S.J., 1992. Submicron-range attraction between hydro-phobic surfaces of monolayer-modified mica in water. J. Am. Chem. Soc. 114 (27), 10927–10933.
Sawant, D.P., Vinu, A., Justus, J., Srinivasu, P., Halligudi, S.B., 2007a. Catalytic performances of silicotungstic acid/zirconia supported SBA-15 in an esterification of benzyl alcohol with acetic acid. J. Mol. Catal. A: Chem. 276, 150–157.
Sawant, D.P., Vinu, A., Lefebvre, F., Halligudi, S.B., 2007b. Tungstophosphoric acid supported over zirconia in mesoporous channels of MCM-41 as catalyst in veratrole acetylation. J. Mol. Catal. A: Chem. 262, 98–108.

Schmid, G., 1992. Large clusters and colloids. Metals in the embryonic state. Chem. Rev. 92 (8), 1709–1727.
Schmid, G., Bäumle, M., Geerkens, M., Heim, I., Osemann, C., Sawitowski, T., 1999. Current and future applications of nanoclusters. Chem. Soc. Rev. 28 (3), 179–185.
Schoonheydt, R.A., Pinnavaia, T., Lagaly, G., Gangas, N., 1999. Pillared clays and pillared layered solids. Pure Appl. Chem. 71 (12), 2367–2371.
Semagina, N., Kiwi-Minsker, L., 2009. Recent advances in the liquid-phase synthesis of metal nanostructures with controlled shape and size for catalysis. Catal. Rev. 51 (2), 147–217.
Shchukin, D.G., Sukhorukov, G.B., 2004. Nanoparticle synthesis in engineered organic nanoscale reactors. Adv. Mater. 16 (8), 671–682.
Sidhpuria, K.B., Patel, H.A., Parikh, P.A., Bahadur, P., Bajaj, H.C., Jasra, R.V., 2009. Rhodium nanoparticles intercalated into montmorillonite for hydrogenation of aromatic compounds in the presence of thiophene. Appl. Clay Sci. 42, 386–390.
Soma, Y., Soma, U., 1989. Chemical reactions of organic compounds on clay surfaces. Environ. Health Persp. 83, 205–214.
Somorjai, G.A., McCrea, K., 2001. Road map for catalysis science in the 21st century: a personal view of building the future on past and present accomplishments. Appl. Catal. A: Gen. 222 (1–2), 3–18.
Suib, S.L., Tanguy, J.F., Occelli, M.F., 1986. Comparison of the photochemical and photophysical properties of clays, pillared clays and zeolites. J. Am. Chem. Soc. 108, 6972–6977.
Suquet, H., de la Calle, C., Pezerat, H., 1975. Swelling and structural organization of saponite. Clay Clay Miner. 23, 1–9.
Suraja, P.V., Binitha, N.N., Yaakob, Z., Silija, P.P., 2011. Preparation and characterization of nano gold supported over montmorillonite clays. IOP Conf. Ser. Mater. Sci. Eng. 17, 012019/1–012019/4.
Takada, K., Tatsuma, T., Oyama, N.J., 1995. Electrochemical control of montmorillonite clay swelling. J. Chem. Soc. Faraday 91 (10), 1547–1550.
Tao, A.R., Habas, S., Yang, P., 2008. Shape control of colloidal metal nanocrystals. Small 4 (3), 310–325.
Tateiwa, J., Nishimura, T., Horiuchi, H., Wemura, S., 1994. Rearrangement of alkyl phenyl ethers to alkylphenols in the presence of cation-exchanged montmorillonite (Mn^+-mont). J. Chem. Soc. Parkin Trans. 1 (23), 3367–3371.
Tateiwa, J., Horiuchi, H., Wemura, S., 1995. Ce^{3+}-exchanged montmorillonite (Ce^{3+}-Mont) as a useful substrate-selective acetalization catalyst. J. Org. Chem. 60 (13), 4039–4043.
Tateiwa, J., Hayama, E., Nishimura, T., Uemura, S., 1997. Metal cation-exchanged montmorillonite (M^{n+}-mont)-catalysed aromatic alkylation with aldehydes and ketones. J. Chem. Soc. Perkin Trans. 1, 1923–1928.
Theng, B.K.G., 1974. The Chemistry of Clay Organic Reaction. Adam Hilger, London, pp. 09–11.
Thielmann, V.J., McAtee, J.L., 1975. Gas chromatographic behavior of metal-tri(ethylenediamine) complex cation-exchanged montmorillonites. J. Chromatogr. A 105, 115–123.
Timofeeva, M.N., Matrosova, M.M., Reshetenko, T.V., Avdeeva, L.B., Budneva, A.A., Ayupov, A.B., Paukshtis, E.A., Chuvilin, A.L., Volodin, A.V., Likholobov, V.A., 2004. Filamentous carbons as a support for heteropoly acid. J. Mol. Catal. A. 211, 131–137.
Tkác, I., Komadel, P., Muller, D., 1994. Acid-treated montmorillonites: a study by ^{29}Si and ^{27}Al MAS-NMR. Clay Miner. 29, 11–19.
Toebes, M.L., van Dillen, J.A., de Jong, K.P., 2001. Synthesis of supported palladium catalyst. J. Mol. Catal. A: Chem. 173 (1–2), 75–98.

Tokarsky, J., Capkova, P., Klemm, V., Rafaja, D., Kukutschova, J., 2010a. Adhesion of silver nanoparticles on the montmorillonite surface. J. Phys. Chem. Solids. 71 (4), 634–637.
Tokarsky, J., Capkova, P., Rafaja, D., Klemm, V., Valaskova, M., Kukutschova, J., Tomasek, V., 2010b. Adhesion of silver nanoparticles on the clay substrates; modelling and experiment. Appl. Surf. Sci. 256 (9), 2841–2848.
Toshima, N., Yonezawa, T., 1998. Bimetallic nanoparticles-novel materials for chemical and physical applications. New J. Chem. 22 (11), 1179–1201.
Traynor, M., Mortland, M.M., Pinnavaia, T.J., 1978. Ion exchange and intersalation reactions of hectorite with tris-bipyridyl metal complexes. Clay Clay Miner. 26 (5), 318–326.
Vaccari, A., 1998. Preparation and catalytic properties of cationic and anionic clays. Catal. Today. 41, 53–71.
Vaccari, A., 1999. Clays and catalysis: a promising future. Appl. Clay Sci. 14, 161–198.
Valaskova, M., Martynkova, G.S., Leskova, J., Capkova, P., Klemm, V., Rafaja, D., 2008. Silver nanoparticles/montmorillonite composites prepared using nitrating reagent at water and glycerol. J. Nanosci. Nanotechno. 8 (6), 3050–3058.
Varadwaj, G.B.B., Parida, K.M., 2013. Montmorillonite supported metal nanoparticles: an update on syntheses and applications. RSC. Adv. 3 (33), 13583–13593.
Varma, R.S., 2002. Clay and clay-supported reagents in organic synthesis. Tetrahedron 58, 1235–1255.
Vaughan, D.E.W., Lussier, R.J., Magee, J.S., 1979. Pillared Interlayer Clay Products Useful as Catalysts and Sorbents. The United States Patent and Trademark Office (USPTO). US Patent, 4176090.
Vicente, M.A., Lambert, J.-F., 2001. Synthesis of Pt pillared clay nanocomposite catalysts from $[Pt^{II}(NH_3)_4]Cl_2$ precursor. Phys. Chem. Chem. Phys. 3, 4843–4852.
Vicente, M.A., Gil, A., Bergaya, F., 2013. Pillared clays and clay minerals. Chapter 10.5, In: Handbook of Clay Science. Developments in Clay Science. A Fundamentals, vol. 5, Elsevier, pp. 523–557.
Villemure, G., 1990. Effect of negative surface-charge densities of smectite clays on the adsorption isotherms of racemic and enantiomeric tris(2,2′-bipyridyl)ruthenium(II) chloride. Clay Clay Miner. 38, 623–630.
Wang, L., Lu, A., Wang, C., Zheng, X., Zhao, D., Liu, R., 2006. Nano-fibriform production of silica from natural chrysotile. J. Colloid Interf. Sci. 295, 436–439.
Wang, X., Liang, M., Zhang, J., Wang, Y., 2007. Selective hydrogenation of aromatic cholonitro compounds. Curr. Org. Chem. 11 (3), 299–314.
Wang, S., Wang, Z., Zha, Z., 2009. Metal nanoparticles or metal oxide nanoparticles, an efficient and promising family of novel heterogeneous catalysts in organic synthesis. Dalton Trans. 9363–9373.
Wu, P., Li, S., Ju, L., Zhu, N., Wu, J., Li, P., Dang, Z., 2012. Mechanism of the reduction of hexavalent chromium by organo-montmorillonite supported iron nanoparticles. J. Hazard Mater. 219–220, 283–288.
Xiong, Y., Wiley, B.J., Chen, J., Li, Z.-Y., Xia, Y., 2005. Corrosion-based synthesis of single-crystal Pd nanoboxes and nanocages and their surface plasmon properties. Angew Chem. Int. Edit. 44, 7913–7917.
Xiong, J., Chen, J., Zhang, J., 2007. Liquid-phase hydrogenation of o-chloronitrobenzene over supported nickel catalysts. Catal. Commun. 8, 345–350.
Xiong, J., Ye, Q., Xue, W., Wu, X., 2011. Synthesis of highly stable dispersions of nanosized copper particles using L-ascorbic acid. Green Chem. 13, 900–904.
Yadav, G.D., 2005. Synergism of clay and heteropoly acids as nano-catalysts for the development of green processes with potential industrial applications. Catal. Surv. Asia 9, 117–137.

Yadav, G.D., Goel, P.K., Joshi, A.V., 2001. Alkylation of dihydroxybenzenes and anisole with methyl-tert-butyl ether (MTBE) over solid acid catalysts. Green Chem. 3, 92–99.

Yadav, G.D., Asthana, N.S., Kamble, V.S., 2003. Cesium-substituted dodeca-tungstophosphoric acid on K-10 clay for benzoylation of anisole with benzoyl chloride. J. Catal. 217, 88–89.

Yadav, J.S., Subba Reddy, B.V., Satheesh, S., 2004a. Montmorillonite clay catalyzed alkylation of pyrroles and indoles with cyclic hemi-acetals. Tetrahedron Lett. 45 (18), 3673–3676.

Yadav, J.S., Reddy, B.V.S., Sunitha, V., Reddy, K.S., Ramakrishna, K.V.S., 2004b. Montmorillonite KSF-catalyzed one-pot synthesis of hexahydro-1H-pyrrolo[3,2-c]quinoline derivatives. Tetrahedron Lett. 45, 7947–7950.

Yadav, M.K., Chudasama, C.D., Jasra, R.V., 2004c. Isomerisation of α-pinene using modified montmorillonite clays. J. Mol. Catal. A: Chem. 216, 51–59.

Yamagishi, A., 1993. Chirality recognition by a clay surface modified with an optically active metal chelate. In: Tamaru, K. (Ed.), Dynamic Processes on Solid Surfaces. Plenum Press, New York, pp. 307–347 (Chapter 12).

Yamagisi, A., Soma, M., 1981. Optical resolution of metal chelates by use of adsorption on a colloidal clay. J. Am. Chem. Soc. 103, 4640–4642.

Yamauchi, Y., Itagaki, T., Yokoshima, T., Kuroda, K., 2012. Preparation of Ni nanoparticles between montmorillonite layers utilizing dimethylaminoborane as reducing agent. Dalton Trans. 41 (4), 1210–1215.

Yang, B., Lu, Q., Wang, Y., Zhuang, L., Lu, J., Liu, P., Wang, J., Wang, R., 2003. Simple and low-cost preparation method for highly dispersed PtRu/C catalysts. Chem. Mater. 15, 3552–3557.

Ye, J., Chen, W., Wang, D., 2008. Synthesis, structural characterization, and catalytic behaviour in Heck coupling of palladium(II) complexes containing pyrimidine functionalized N-heterocyclic carbenes. Dalton Trans. 4015–4022.

Zhang, J., Wang, Y., Ji, H., Wei, Y., Wu, N., Zuo, B., Wang, Q., 2005. Magnetic nanocomposite catalysts with high activity and selectivity for selective hydrogenation of ortho-chloronitrobenzene. J. Catal. 229, 114–118.

Zhang, L.-Y., Li, J., Hu, L., Zheng, X., 2011. Enhanced removal of nitrate by a novel composite: nanoscale zero valent iron supported on pillared clay. Chem. Eng. J. 171, 526–531.

Zhao, F., Ikushima, Y., Arai, M., 2004. Hydrogenation of nitrobenzene with supported platinum catalysts in supercritical carbon dioxide: effects of pressure, solvent, and metal particle size. J. Catal. 224, 479–483.

Zhou, C.H., Tong, D.S., Bao, M., Du, Z.X., Ge, Z.H., Li, X.N., 2006. Generation and characterization of catalytic nanocomposite materials of highly isolated iron nanoparticles dispersed in clays. Top Catal. 39 (3–4), 213–219.

Zhou, L., Qi, X., Jiang, X., Zhou, Y., Fu, H., Chen, H., 2013. Organophilic wormlike ruthenium nanoparticles catalysts by the modification of CTAB on montmorillonite supports. J. Colloid Interface Sci. 392, 201–205.

Zuo, B., Wang, Y., Wang, Q., Zhang, J., Wu, N., Peng, L., Gui, L., Wang, X., Wang, R., Yu, D., 2004. An efficient ruthenium catalyst for selective hydrogenation of ortho-chloronitrobenzene prepared via assembling ruthenium and tin oxide nanoparticles. J. Catal. 222 (2), 493–498.

From polymers to clay polymer nanocomposites

10

Fernando Wypych, Faïza Bergaya†, Robert A. Schoonheydt‡*
*Department of Chemistry, Universidade Federal do Paraná, Curitiba, Brazil, †Interfaces, Confinement, Matériaux et Nanostructures (ICMN), Centre National de la Recherche Scientifique (CNRS), Orléans, France, ‡Centre for Surface Chemistry and Catalysis, Catholic University of Leuven, Leuven, Belgium

10.1 Introduction

By the definition provided in the Gold Book of IUPAC (International Union of Pure and Applied Chemistry (Work et al., 2004)), a composite is a 'multicomponent material comprising multiple (nongaseous) phase domains in which at least one type of phase domain is a continuous phase'. By consequence, polymer composites, or polymer-based composites, are composites in which at least one component is a polymer. Since the 1940s, the first generation of polymer-based composites using glass-fibre-reinforced polymers (GFRP) has now become a relatively mature research topic (Sathishkumar and Naveen, 2014). Polymer-based composites were first applied in the marine industry, especially with glass-fibre-reinforced polyesters. With the development of space programmes in the 1960s, the search for new materials began to grow rapidly. One of the first papers on clay mineral polymer composites was on the adsorption of poly(vinyl alcohols) onto montmorillonite (Mt) (Greenland, 1963).

Many years later, especially with the advance of characterization techniques at nanometre level, a series of polymer-based composites (and particularly, clay mineral polymer nanocomposites [CPN]) began to be developed (Gao, 2004). By definition, CPN have one component with at least one dimension of the order of nanometres, which corresponds to a scale from 1 to 100 nm (Schoonheydt and Bergaya, 2011). Clay mineral structures are a repetition of the half unit cell (e.g. $Mg_3Si_4O_{10}(OH)_2$ for a trioctahedral clay mineral) considered as monomer, and could be called 'solid layered inorganic nanopolymers'. Consequently, a CPN could be considered as belonging to the category of (inorganic) nanopolymer (organic) polymer nanocomposites.

The term nanocomposite was first used in the 1980s to define solids containing two (or more) nanometre-size domains which differ in composition and/or structure (Roy et al., 1984). This term is often reported without mentioning the second constituent added to the polymer, which could be a second polymer, a clay mineral, or any other nanofiller.

In case of CPN, any filler morphology (layered, fibres, nanotubes) can be dispersed into the polymeric matrix. The clay minerals with different morphologies are classified as natural or synthetic, two-dimensional (2D) materials, because one of their

Developments in Clay Science, Vol. 9. https://doi.org/10.1016/B978-0-08-102432-4.00010-X

dimensions is always at the nanometric scale. What makes these materials interesting is that their properties can be significantly different from the properties of the polymers, resulting in some cases in dramatic synergetic effects (Bergaya et al., 2013).

10.2 Clay minerals used in clay mineral polymer nanocomposites

Montmorillonite (Mt) was the first clay mineral used to prepare CPN in the 1980s by the researchers from the Toyota Research Center in Japan. Their seminal papers are considered to be pioneer work in the field of CPN (Fukushima and Inagaki, 1987; Kojima et al., 1993; Usuki et al., 1993). Mt is generally considered the most important clay mineral of the smectite group for applications in CPN. Mt is available worldwide and it is relatively inexpensive. The main problem encountered is the unusually wide range of chemical compositions, which presents a challenge in production of CPN with uniform properties (Schoonheydt and Bergaya, 2011). In spite of this lack of homogeneity, the layered structure gives rise to the characteristic layer-like morphology with high aspect ratio (length/thickness) and possible cleavage along the layers, being well-suited for use as functional fillers.

The other clay minerals often used as fillers of CPN are from the kaolin group. Kaolinite (the most abundant member of the kaolin group) is also one of the most abundant clay minerals on earth (Murray, 2006). Kaolinite is a nonswelling clay mineral. The kaolinite particles have submicrometre dimensions along the plane of the particles and nanometre dimensions perpendicular to the plane (Fig. 10.1).

More recently, fibrous clay minerals (sepiolite and palygorskite) have also been used as fillers in CPN (Inagaki et al., 1995; Ruiz-Hitzky et al., 2011, 2013). Finally, anionic clay minerals or layered double hydroxides (LDH) have also been studied in

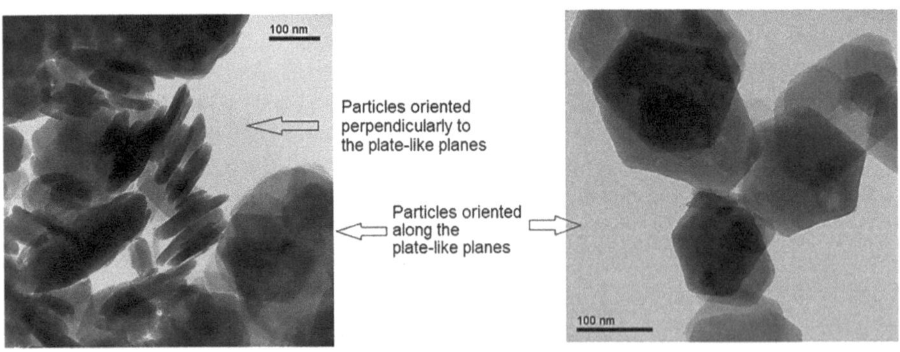

Fig. 10.1 Transmission microscopic images of a sample of kaolinite from the Rio Capim Basin in the state of Pará—Brazil, showing the submicrometric particles oriented in two different ways.

CPN (Forano et al., 2013). The research on LDH-polymer nanocomposites is published in a wide range of journals (Costantino et al., 2013) and is not considered in this chapter.

10.3 Structures and surface properties of clay minerals

In a simplified way, all clay minerals structures are based on the brucite-like ($Mg(OH)_2$) or gibbsite-like ($Al(OH)_3$) layers and layers of Si^{+4}, coordinated tetrahedrically to four oxygen anions (O^{-2}). A 2:1 clay mineral structure with the recommended terminology for the different clay minerals parts can be seen in Fig. 10.2.

In brucite, the octahedral sheets (O) in the clay mineral structure are produced by edge sharing slightly distorted octahedra, in the centre of which, a Mg^{+2} cation is coordinated to six hydroxyl anions. In this case (Fig. 10.3A), all the octahedral sites in the O sheets are occupied, and the structure is called *trioctahedral*. As the charge of Mg^{+2} is shared by six hydroxyl groups, and each hydroxyl is shared by three Mg^{+2} cations, the sheet is electrically neutral.

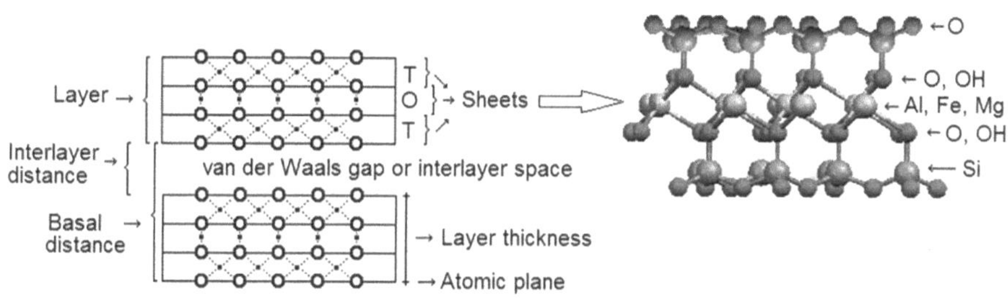

Fig. 10.2 Different parts of the side view of a generic 2:1 clay mineral structure.

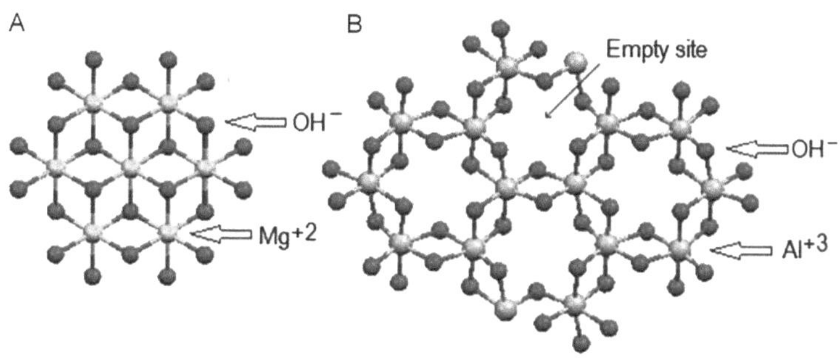

Fig. 10.3 Schematic representation of the brucite (A) and gibbsite layers (B): top view. Hydrogen atoms were removed to facilitate the visualization. Figure built with Mercury—Crystal Structure Visualization—The Cambridge Crystallographic Data Centre—CCDC.

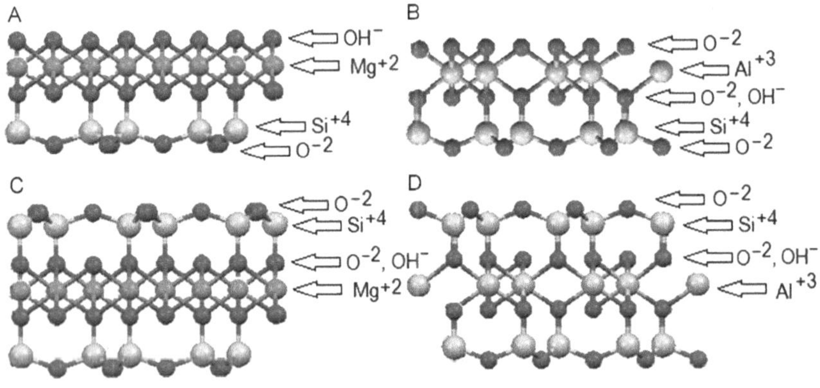

Fig. 10.4 Schematic representation of the chrysotile (A), kaolinite (B), talc (D) and pyrophyllite layers (D). Hydrogen atoms were removed to facilitate the visualization. Figure built with Mercury—Crystal Structure Visualization—The Cambridge Crystallographic Data Centre—CCDC.

In gibbsite, a similar arrangement of the Al^{+3} in the centre of the octahedra having six hydroxyls in the corners and sharing corners results in the gibbsite layer. As the charge of the central metal is +3, only two-thirds of the octahedra are occupied, and the structure is called *dioctahedral* (Fig. 10.3B).

When a brucite-like layer is connected to one tetrahedral Si sheet, the 1:1 neutral layers of the serpentine group are obtained (e.g. chrysotile—$Mg_3Si_2O_5(OH)_4$ in Fig. 10.4A). When the gibbsite-like layer is connected to one tetrahedral Si sheet (denominated as tetrahedral sheets (T) in a clay mineral structure), the kaolin group is obtained (e.g. kaolinite—$Al_2Si_2O_5(OH)_4$ in Fig. 10.4B). When a gibbsite-like layer is packed between two tetrahedral Si sheets, a 2:1 dioctahedral clay mineral structure is obtained (pyrophyllite $Al_2Si_4O_{10}(OH)_2$ in Fig. 10.4C). When one sheet of brucite-like is packed between two tetrahedral Si sheets, a 2:1 trioctahedral clay mineral (talc—$Mg_3Si_4O_{10}(OH)_2$ in Fig. 10.4D) is obtained.

For dioctahedral sheets, not only can Al^{+3} in the octahedral layers be substituted, but also Si^{+4} in the tetrahedral sheets. In the first situation, when part of the Al^{+3} in the octahedral sheet is replaced by Mg^{+2}, an excess of negative charge is created in the layer. To compensate for this charge, more or fewer hydrated cations need to be intercalated between the layers (e.g. montmorillonite—$(M_y^{+}\cdot nH_2O)(Al_{2-y}Mg_y)Si_4O_{10}(OH)_2$). In the second case, when Si^{+4} in the tetrahedral sheets is partially substituted by Al^{+3}, an excess of negative charge is created in the layers, but this charge is more localized at the surface. Thus, a beidellite structure $(M_x^{+}\cdot nH_2O)Al_2(Si_{4-x}Al_x)O_{10}(OH)_2$ is obtained. When Al^{+3} is almost totally substituted by Fe^{+3}, and Si^{+4} is partially substituted by Al^{+3}, the structure of nontronite $(M_x^{+}\cdot nH_2O)Fe_2^{+3}(Si_{4-x}Al_x)O_{10}(OH)_2$ is obtained.

In the case of trioctahedral layers, a similar situation is obtained. When part of the Mg^{+2} in the octahedral sheets is substituted by Al^{+3}, and/or Fe^{+3} and Si^{+4} by Al^{+3}, an excess of negative charge is created in the layers, and hydrated cations need to be intercalated between the layers, for example, saponite-($M_{x-y}{}^{+}\cdot nH_2O$)Mg_{3-y}(Al, Fe)$_y$($Si_{4-x}Al_x$)$O_{10}(OH)_2$. When Mg^{+2} is partially substituted by Li^{+}, the hectorite-($M_y{}^{+}\cdot nH_2O$)($Mg_{3-y}Li_y$)$Si_4O_{10}(OH)_2$ is obtained. In some cases, part of the hydroxyl anions can be replaced by fluor, like in fluorohectorite-($M_y{}^{+}\cdot nH_2O$)($Mg_{3-y}Li_y$) $Si_4O_{10}(OH)_{2-x}F_x$.

Because of the presence of the localized (or delocalized) layer charges and the presence of more or less hydrated cations in the interlayer space, the surfaces of the smectites are hydrophilic, which is not compatible with the majority of hydrophobic polymers. Clay minerals should be chemically modified to match the hydrophobicity of these polymer matrixes.

10.4 CPN obtained by cation exchange of a hydrophilic polymer with long alkyl chain/or a cationic monomer

Direct intercalation of water soluble polymers occurs easily in aqueous clay mineral dispersion, either by cation exchange with the polar heads of the polymers or by exchange of the cationic monomers that are grown into polymers. Hereafter are a few examples:

Example: *Replacement of hydrated sodium cations by vinylbenzyltrimethylammonium, followed by polymerization (Moet and Akelah, 1993).*

Na^{+}-Mt was dispersed in water and mixed with vinylbenzyltrimethylammonium chloride, and the mixture was stirred for 3 h. The white precipitate was filtered, washed with water, methanol, and ether, and dried under vacuum at room temperature. The dried material was dispersed in acetonitrile, mixed with styrene and AIBN (azobisisobutyronitrile) polymerization initiator, and the mixture stirred under nitrogen atmosphere at 80°C for 5 h. After cooling, the material was precipitated with methanol. The result is the polymerization of the vinyl monomer intercalated into Mt to polystyrene with a molecular mass of 22,000 g/mol. A good dispersion was achieved, and a good adhesion was observed between the polymer and hydrophobic Mt.

Example: *Replacement of hydrated sodium cations by two cationic initiators (quaternary ammonium bromides) followed by polymerization (Fan et al., 2003).*

Na^{+}-Mt was dispersed in deionized water and stirred vigorously overnight at room temperature. The clay mineral dispersion was then ultrasonicated while stirring for 24 h. To obtain a homogeneous dispersion, the amounts of polymerization initiators were calculated to fully replace the sodium cations of the clay mineral used. The mixture was stirred and ultrasonicated for another 24 h, and the white precipitate was separated by filtration and then dispersed in deionized water. This filtration/dispersion process was repeated 10 times in order to remove the free initiator. The products were dried in a vacuum oven at room temperature for 3 days to yield a yellowish powder. The initiator-intercalated clay mineral was dispersed in tetrahydrofuran (THF), ultrasonicated and vigorously stirred (6–10 h). To avoid decomposition of the initiator, ice was added to the sonication water bath. The styrene monomer was then added to the dispersion, and the mixture was stirred for 1 h under N_2 flow, the temperature was increased to 60°C, and the polymerization reactions continued for 72 h under N_2 atmosphere.

The surface-initiated polymerization product of the bicationic initiator contained some intercalated Mt layers, while a exfoliated Mt structure was achieved with the monocationic initiator. Size exclusion chromatography showed that the surface-initiated polymerization product of the monocationic initiator had a higher MW for both free and bound polymer.

10.5 CPN obtained by grafting of organophilic polymers with a hydrophilic group

In order to compatibilize the hydrophilicity of the clay mineral basal surfaces with the hydrophobicity of organic polymers, a possibility not very often explored is to graft polymer chains with a hydrophobic group. However, as the clay minerals particles only have hydroxyl groups at the edges, available for reaction, grafting the clay minerals with silanes like aminopropyl trimethoxysilane (APTS) is a necessary first step and the resultant silanized particles can be used to graft the end-functionalized polymer chains (Tsubokawa, 2007; Achilleos and Vamvakaki, 2010).

The clay mineral particles (before or after exfoliation) react with APTS (Fig. 10.5) to generate the reactive group at the edge surfaces (Steps A, G) or at the basal surfaces (Steps B, F).

In the case of silane coupling/modification of clay mineral surfaces, special attention should be given to the distance between reactive sites on the grafted polymer chains and the grafting sites of the clay minerals' layers. Sometimes, due to the steric constraints, not all reactive groups on the polymers chains match the positions of the reactive sites on the clay mineral surfaces, and only a small number of covalent bonds can be established. These coupling reactions are rather complex, and the grafting of molecules to the clay mineral surfaces is not necessarily straightforward. In the

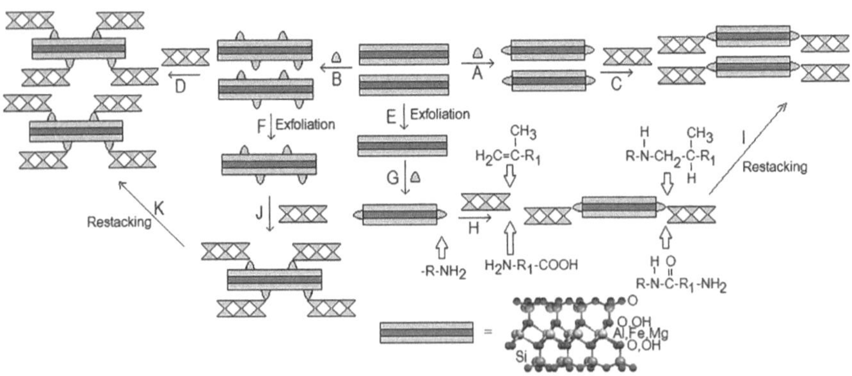

Fig. 10.5 Potential strategies to produce clay minerals/polymers nanocomposites with edge and basal surface grafting. In this figure, a generic 2:1 clay mineral is considered and to facilitate visualization, the intercalated hydrated exchange cations were removed.

sequence of Fig. 10.5, the reactive groups on the edges react with an alkene (edge surface: Steps C for the clay mineral particles, Step H for the individual clay mineral layers; basal surface: Step D for the clay mineral particles, Step J for the individual clay mineral layers). Also, the grafted clay minerals can eventually be restacked as indicated in Steps I and K. Steps E and F refer to the exfoliation.

10.6 CPN obtained by melt intercalation of a pristine clay mineral with hydrophilic or organophilic polymer and surfactant

In some cases, the use of a solvent can be avoided using solid-state synthesis. Interaction of pristine clay minerals with hydrophilic or organophilic polymers in solid-state reaction, mainly with molten rubber polymer, leads to a CPN via a melt intercalation process (Vaia and Giannelis, 1999). The CPN is usually performed at a temperature slightly above the melting point of the polymer. Higher temperatures should be avoided to prevent polymer molecular changes or even degradation.

10.7 Other strategies of CPN synthesis with organophilic polymers

When the hydrophilic 2:1 clay minerals cannot be dispersed into hydrophobic polymers, they should be hydrophobized before addition of the polymer or during the in situ reaction in a one-pot synthesis in which all components are mixed together (clay mineral, organic component, and polymer).

10.7.1 Hydrophobization of clay minerals by exchange with cationic species with long alkyl chain or with monomers followed by in situ polymerization

Transforming the surface of hydrophilic expandable clay minerals to a hydrophobic surface (i.e. hydrophobization) can occur by cation exchange with quaternary ammonium cations (QAC), also known as quaternary cationic surfactants or ammonium salts. This leads to what is known as organoclay (OC). In Fig. 10.6, an example of a generic 2:1 clay mineral exchanged with dodecylammonium is shown. This QAC has a long linear alkyl chain which is ideal for exchange without perturbation of the clay mineral structure.

Depending on the chain length, degree of branching, type of head group, and the amount of cations between the layers, the QAC can be arranged vertically in a single layer (Fig. 10.6A), horizontally in a paraffin-like arrangement (Fig. 10.6B), vertically in an interdigitated way (Fig. 10.6C) or in a double layer (Fig. 10.6D). This reaction is very easy and is classified as soft chemistry. A 2:1 clay mineral, usually Mt, is

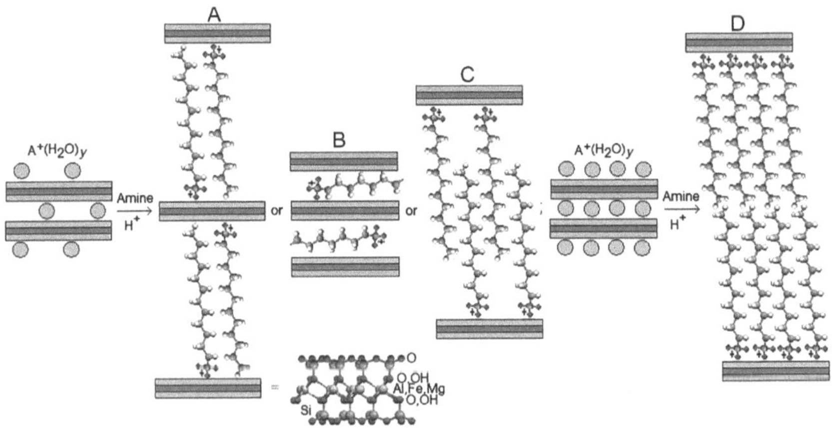

Fig. 10.6 Hydrophobization of a 2:1 clay mineral with a dodecylamine salt. (A) Vertical single layer; (B) horizontal parallel paraffin-like; (C) vertical interdigitated double layer and (D) vertical double layer.

dispersed in water, and the QAC is added in excess, in the presence of an acid, necessary to protonate the amine groups and produce a salt ($-NH_2$ to $-NH_3Cl$, in the case of hydrochloric acid). The formed ammonium cation exchanges the original cation present in the clay mineral. The effect of this process is the maintenance of the 'a' and 'b' cell parameters and the expansion of the basal distance, characterized by the shift of the d_{001} reflexion to lower angles. The longer is the organic chain, the higher the basal distance. The increase of the interlayer distance weakens the electrostatic interaction between two adjacent clay minerals layers. This increases the hydrophobicity of the pristine clay mineral and consequently favours the migration of the polymeric molecules and their location between the OC layers.

Example: *Polymerization of epoxy resin on the single layers of hydrophobic Mt (Wang and Pinnavaia, 1994).*

Na^+ Mt was exchanged by protonated primary diamines $H_2N(CH_2)_nNH_2$ and primary amines $CH_3(CH_2)_nNH_2$ (n=6 and 12) in aqueous dispersion at 60°C for 3 h. The cation-exchanged Mt was separated by centrifugation and washing was repeated until chloride free (using 0.1 mol/L $AgNO_3$). The washed amounts of hydrophobic Mt were added to epoxy resin and the mixture was magnetically stirred in a 250 mL beaker at 75°C for 30 min. The beaker was sealed with aluminium foil, and the temperature was raised. The Mt epoxide nanocomposite was obtained within 1 min. Mt was exfoliated into the polymeric matrix, but due to the powdery texture, the mechanical properties of the nanocomposites could not be measured.

Example: *Direct intercalation of ethylene-vinyl acetate copolymer into organomodified Mt (Alexandre et al., 2001).*

The Mt was exchanged with dioctadecylammonium bromide (DODABr), using the clay mineral dispersion and the aqueous organic salt solution. Three EVA-based nanocomposites filled with the same amount of organomodified Mt (ca. 3.5 wt% of clay mineral) were prepared on a two-roll mill operating at 130°C for 7 min. The CPN were prepared by mixing the organomodified Mt with the molten ethylene-vinyl acetate copolymer. The results indicated that the polymer was intercalated between the organically modified clay mineral layers.

In the case of bulky QAC like cetyl trimethylammonium ($CTMA^+$), or tetrabutylammonium (TBA^+), partial exchange and increase of the basal distance with increased organic cation concentration were observed. For linear amines (*n*-alkyl ammonium), other orientations of the long chain alkyls in the interlayer space can occur, as already reported by Lagaly (2013). The arrangements of the intercalated surfactants in the OC depend on the length of the alkyl chains and on the layer charge. These arrangements from intercalated monolayers, bilayers, pseudo-trimolecular layers to paraffin-type layers arrangements with different tilting angles of the alkyl chains, depend also on the ratio of the amount of added surfactant with respect to the CEC. These tilting angles are not only dependent on the functional groups of the cations that approach the clay mineral layer, but also on the ratio of the amount of added surfactant with respect to the CEC and on the degree of hydration of the polar groups. The alkyl chains are almost perpendicular to the basal surface of the clay mineral for clay minerals with high charge density and for amounts of surfactant slightly exceeding the CEC, about 1.2 CEC (Mandalia and Bergaya, 2006).

10.7.2 Hydrophobization by covalent grafting of a compatibilizer on basal surfaces and/or at the edges of clay minerals

Interfacial bonding is one of the most important parameters to obtain CPN with optimized properties. Hydrophobization of the clay minerals is the only way to permit the compatibilization with the hydrophobic polymers, which is done with coupling agents.

In the case of 2:1 clay minerals having a layered domain morphology (morphology in which phase domains have shapes with two dimensions much larger than the third dimension), the coupling agent is intercalated between the individual clay mineral layers. Note that plate-like phase domains have the appearance of extended planes that are often oriented essentially parallel to one another. The coupling agent, intercalated between the individual clay mineral layers will be present at the surface of the clay mineral layers acting as a bridge between the clay mineral layers and the polymeric matrix.

Hydrophobization by intercalation is one way to achieve this objective, but as the cationic species are mobile and can be easily exchanged, the grafting of clay mineral surfaces is more suitable to permanently change the properties of these surfaces.

Example: *In situ grafting of polymethylmethacrylate into single layers of Mt (Shah et al., 2005).*

Na^+ Mt was exchanged with the initiator molecules (a QAC). Methyl methacrylate monomer and THF were added to the clay mineral dispersion and the mixture stirred for 30 min under N_2 atmosphere. The reaction medium was exposed to UV light for 40 h, at room temperature. The mixture was diluted with THF and precipitated into methanol. On drying, the precipitate was obtained as a white powder. The surface grafting allowed the exfoliation and stabilization of single layers and the study of their structures and dynamics.

In Figs. 10.7 and 10.8, different potential strategies are reported for hydrophobization of clay minerals. They could be considered for CPN production by intercalation/grafting on basal surfaces and by grafting on edge surfaces. Some of the proposed mechanisms indicated in these figures are not yet reported in the literature.

One basic procedure for interaction of an organic molecule with a clay mineral particle is to intercalate monomers carrying a reactive group between the layers, replacing the hydration shell of the cations (Step A in Fig. 10.7) (Wypych and Satyanarayana, 2005; Mévellec et al., 2007; Wypych et al., 2008).

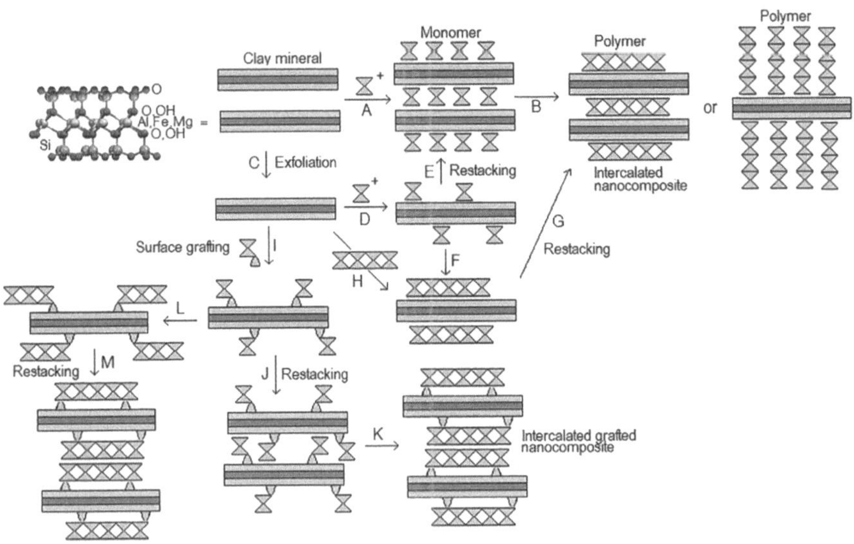

Fig. 10.7 Potential strategies to produce CPN on internal (by in situ growth) and external basal surfaces. In this figure, a generic 2:1 clay mineral is considered and to facilitate the visualization, the intercalated hydrated exchange cations were removed.

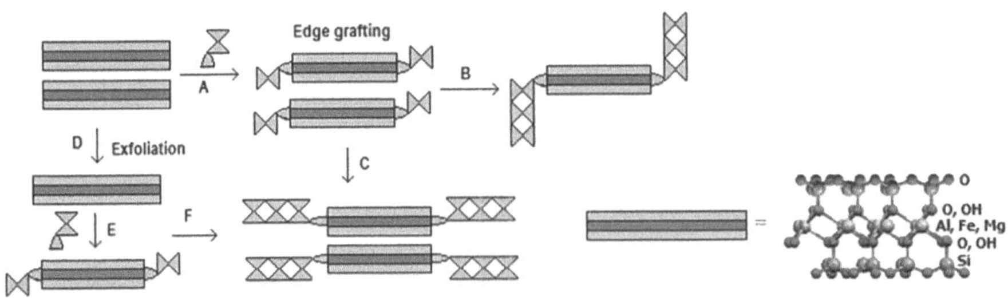

Fig. 10.8 Potential strategies to produce CPN by edge grafting. In this figure, a generic 2:1 clay mineral is considered and to facilitate the visualization, the intercalated hydrated exchange cations are removed.

In the schematic representation of Fig. 10.7 the monomers are physisorbed onto the clay mineral particles surface (Step A) but they can also replace the cationic intercalated species (also in Step A), as frequently reported in the literature (Moet and Akelah, 1993; Fan et al., 2003; Viville et al., 2004; Tsubokawa, 2007; Achilleos and Vamvakaki, 2010). These monomers can be polymerized (polymer-directed synthesis or in situ growth) and polymer chains are grown from the surface as indicated in Step B and an intercalated CPN can be obtained. Polymerization can occur by heating, by UV or gamma irradiation, or through catalysis by an initiator molecule. Step D refers to the cation exchange reaction after exfoliation.

The polymer chains are physisorbed, but more than one layer can be polymerized, and the clay mineral particles can be trapped. In Step C, the clay mineral is exfoliated (if it is delaminated, the same procedure of Step A should be followed), and again, the single layers adsorb the monomers, and the latter are polymerized, as indicated in Step F. Also, the single layers can be restacked before polymerization (as indicated in Step E) or polymerized (as indicated in Step F) and then restacked, as indicated in Step G. The polymer molecules can also be adsorbed to single layers (as indicated in Step H) that can be restacked as indicated in Step G. In the restacked layers, the polymer can be trapped (intercalated) between the clay mineral layers without the possibility of being easily removed. The advantage of these methods is the one-pot synthesis, because the clay mineral particles are present in the reaction medium during the adsorption of the monomer and polymerization. The same is valid for the adsorption of the polymer.

Single layers are desired to maximize the properties of the CPN, and the restacking is not needed in the CPN formulation. If necessary, single layers can be restacked before polymerization as indicated in Step E, or after polymerization, as indicated in Step F, and then restacked as indicated in Step G. The polymer molecules can also be adsorbed on the single layers as indicated in Step H, which can be restacked as indicated in Step G.

As the basal surfaces of clay minerals consist of oxygen atoms, the grafting (normally, alkoxysilanes are used, but other organic species can also be used) is limited. Alkaline leaching removes part of the Si atoms from the outer surfaces of the layers, and the grafting of monomers having initiating groups is possible (Step I). The single layers populated with the monomers can be restacked (Step J) followed by polymerization (Step K) or first polymerized (Step L) and then restacked (Step M). Polymers can also be adsorbed at the surface of single layers when the restacking is unlikely or on the surface of clay mineral aggregates (not indicated). The most effective interaction between the clay minerals particles and the coupling agent will be by covalent grafting of the organic molecule to the basal surface, when during mechanical stress, covalent bonds must be broken as the polymer chain is detached from surface of the clay mineral particles (Wypych, 2004; Wypych and Satyanarayana, 2005).

As shown in Fig. 10.8, the grafting procedure is more likely to happen at the edge surfaces of particles (Step A) and single layers (Step E) where hydroxyl groups are available. In this case, depending on how the polymer is formed, the clay minerals can retain the stacked structure (Step C), can be exfoliated and not restacked (Step B), or can be restacked (Step F).

All the obtained materials can be used as master batches for further dispersion in the same or other miscible polymers. The main advantages of the CPN are the low loading of the fillers used in the nanocomposites formulations (normally less than 5% by mass) in comparison with the traditional polymer composites, the possibility of changing the nanoparticles' surfaces to maximize the interaction, and consequently, the properties of CPN (without mentioning the unique nanoscale size-dependent properties expected for this class of materials).

Example: *Exchange of* Na^+ *by QAC, grafting by silanes and polymerization of polyacrylamide onto organophilic Mt (Mansoori et al., 2010).*

Raw Mt was dispersed into 300 mL of acetone containing $NaHCO_3$*, with vigorous stirring under* N_2 *atmosphere for 15 min. Part of the dispersion was reacted with vinyl trichlorosilane added dropwise over 3 h. The mixture was gently refluxed for 8 h and the product was filtered, successively washed with acetone, acetone/*H_2O *(50/50), filtered, and dried under vacuum at room temperature. The polymerization reaction was performed by mixing the silylated Mt with acrylamide and AIBN in 25 mL of methanol. The flask was purged by dry, oxygen-free nitrogen for 15 min. The mixture was heated while magnetically stirred at 70°C for 4 h. Then, the reaction mixture was transferred into a Soxhlet apparatus and extracted with 50/50 dimethylformamide/water, and the resulting material was dried at 60°C for 24 h in a vacuum oven.*

A similar procedure was used to prepare clay mineral/polyacrylamide grafted nanocomposites (Mansoori et al., 2012).

Example: *Direct grafting of semicrystalline poly(e-caprolactone)-poly(lactic acid) diblock copolymers with defined compositions onto clay mineral single layers (Viville et al., 2004).*

Na^+*Mt was exchanged in water at 85°C for 24 h by 2-hydroxyethyl(hexadecyl) dimethylammonium and hexadecyltrimethylammonium cations and the organomodified Mt were dried in a ventilated oven at 70°C for one night. The dried organomodified clay mineral was then allowed to swell in 100 mL of dry toluene for 1 h at 70°C under a* N_2 *atmosphere.*

A given amount of e-caprolactone (CL) was then added under N_2 *atmosphere, and the reaction was allowed to proceed at 70°C for 1–7 days, depending on the type of organomodified clay mineral. The polymerization reaction was stopped by the addition of diluted HCl aqueous solution, and the resulting clay mineral/polycaprolactone was recovered after solvent and residual monomer removal, performed by drying the mixture at 85°C under reduced pressure for 24 h.*

Poly[caprolactone/lactic acid] diblock copolymers were synthesized via sequential copolymerization of caprolactone and lactic acid monomers, conducted in dry toluene. A given amount of caprolactone was added under N_2 *atmosphere, and the reaction was carried out at 40°C for 1 h. A desired amount of lactic acid, previously dissolved in hot and dry toluene, was then added to the reaction medium, and the polymerization was allowed to proceed at 70°C for 7 days. The polymerization reaction was stopped by the addition of diluted HCl, and the resulting copolymer-grafted nanohybrids were immediately recovered by precipitation in cold methanol, filtration, and drying in vacuum, resulting in clay mineral-surface-grafted (co)polyesters.*

10.7.3 Pillaring and plasma treatments of clay minerals for use in CPN applications

The modifications of the clay minerals reported in the previous sections are most often based on organic treatments of the pristine clay minerals. However, two other treatments of clay minerals can also lead to increased hydrophobicity of the clay minerals and to optimize clay mineral-polymer compatibility. These are pillaring to produce pillared clays (PILC) and the more recent and less studied plasma treatment of clay minerals.

Pillaring is the irreversible intercalation of bulky inorganic polycations by cation exchange of clay minerals, followed by calcination. The most studied polycation is the Keggin ion with the formula $[Al_{13}O_4(OH)_{24}(H_2O)_{12}]^{7+}(Al_{13})$. Exchange of Al_{13}, particularly in concentrated media (Aouad, 2004; Aouad et al., 2005) leads to a PILC with an interlayer distance of about 1 nm. Pillaring by mixed Al-Fe species creates open, thermally stable mesopores, allowing polymers to fill these mesopores. After further organic treatment with hexadecyl trimethylammonium bromide, the obtained solid organo-PILC is mixed for 10 min with a molten polymer. These two-step treatments (pillaring followed by surfactant addition) lead to polymer adsorption in the mesopores present between the OC mineral particles (composed of about four layers). The obtained CPN shows less-degraded mechanical properties and good thermal and firing properties (Bergaya et al., 2005).

Cold plasma acetylene treatment of powdered clay minerals (Mt) in an adapted reactor (Célini, 2004) modifies the hydrophilic surface properties of the clay mineral and improves the dispersion of these plasma-treated Mt in polymer (Célini et al., 2007). The plasma surface treatment of the clay minerals is used to increase its surface adhesion towards the polymer by grafting hydrocarbon moieties on external surfaces of smectites. It is an alternative dry and environmentally friendly technique to obtain an organoclay. The plasma-treated clay minerals used as fillers in polyethylene decrease the flammability of the obtained CPN, although decreasing the mechanical properties. A compromise is found when the mechanical properties of CPN are less degraded, and the fire protective properties improved with a heat release rate that is one-third that of the pristine polymer. The plasma procedure is not yet widespread in clay mineral and CPN research but great potential is expected.

10.8 Challenge: Exfoliation of clay minerals in CPN

Delamination of the clay minerals causes them to aggregate into particles, and exfoliation of these particles into individual layers is the ideal basic requirement for CPN preparation. If this condition is not reached, the mixture of clay minerals aggregates, and polymer molecules will result in a microcomposite with two separate phases, which was (until recently) the traditional way to reinforce polymers (Alexandre and Dubois, 2000; Gao, 2004).

After changing the clay mineral particles from hydrophilic to hydrophobic, the particles acquire the ability to be delaminated or exfoliated in different solvents or melted polymers (Fig. 10.9). These two terms, often confused in the literature, were clearly defined in recent papers (Bergaya and Lagaly, 2011, 2013). The term delamination is used to indicate separation between the planar faces of adjacent layers, without losing the orientation of the layers with respect to each other. The term exfoliation is used

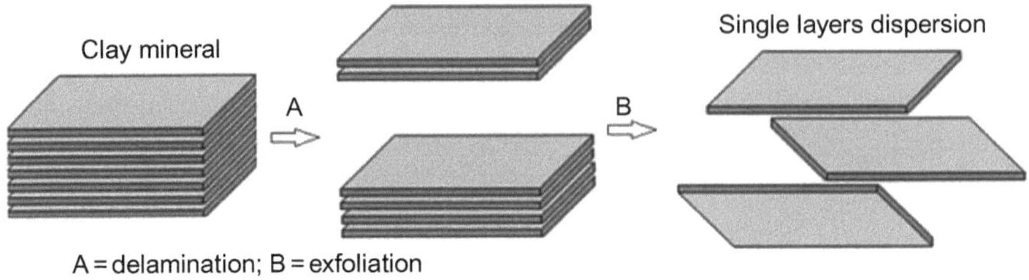

Fig. 10.9 Delamination and exfoliation processes of a clay mineral particle.

when at some point, no further interaction occurs between individual layers, which become independently mobile in a liquid phase.

Interfacial adhesion (adhesion in which interfaces between phases or components are maintained by intermolecular forces, chain entanglements, or both, across the interfaces) and dispersion (finely divided phase domains, often in the colloidal size range, distributed throughout a continuous phase domain) are both important aspects to be considered in CPN formulation. In other words, agglomeration of nanoparticles should be prevented at all costs. In the context of CPN preparation, the individual clay mineral layers are generally stacked in particles on the order of (at most) a few nanometres in thickness, thus, still remaining at the nanometre scale.

Several methods and even the mechanical properties of CPN, themselves, can be used to evidence delamination and exfoliation as significant variations in physical and mechanical properties. X-ray diffraction is also frequently used, but determination of the d_{001} reflexion at very low angles and the often-low intensity of these reflexions are limiting factors on the determination of the distribution of clay mineral layers in the CPN. The only way to directly show the delamination/exfoliation is to use transmission electron microscopy, where direct images of the CPN can be obtained. Even so, several ranges of the sample cut with a microtome machine should be evaluated. The time of exposure of the samples should be kept as low as possible, and refrigerated sample holders should preferentially be used to avoid sample decomposition (Monticelli et al., 2007).

10.9 Properties and potential applications of CPN

CPN are a class of materials formed by the dispersion of layered clay mineral nanoparticles throughout a polymer matrix to reinforce the properties of these polymers. In industry, the main focus is on the use of glass- or carbon-fibre-reinforced polymers, where the largest market is centred in the automotive sector with the main objective of reducing vehicle weight. Recent CPN development efforts sought

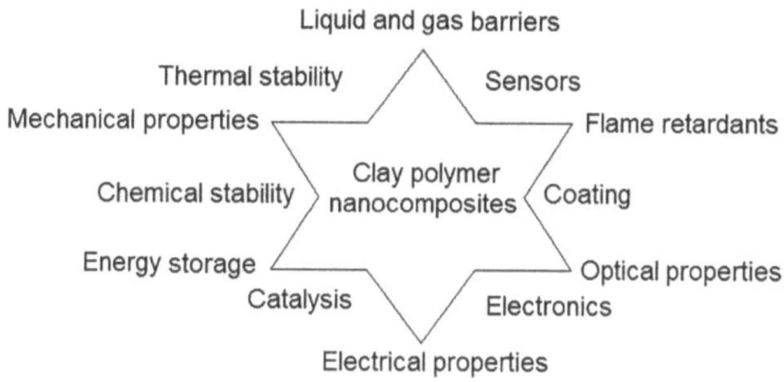

Fig. 10.10 Potential applications of clay polymer nanocomposites (CPN).

to obtain new challenging mechanical and other properties (Fig. 10.10). For polymer composites, the content of filler can reach up to 60% by mass. For CPN, high-level performance at low clay mineral filler content (between 2% and 8%) is needed. Due to CPN' unique mechanical, physical, chemical, and processing properties, their major customers are the automobile, packaging, aeronautic, and sports industries.

The high cost of production is one of the reasons for the weak growth of CPN use in these markets. The professionals involved with CPN development need to centre their efforts on improving the materials' performance and reducing the cost of their production in order to increase global market growth. The key factor for CPN is to achieve exfoliation of the clay minerals into single layers (or a few stacked layers with nanometric thickness), avoiding the tendency of single layers and clay mineral particles to aggregate due to their high surface energy. This will create a large, effective interface between the clay mineral and the polymer, which will not only improve the mechanical properties of the polymer but also its chemical and thermal stabilities, biodegradability or environmental stability, flame retardancy, and optical properties. This large interface also acts as gas and liquid barrier. The resulting CPN have antistatic properties, absorb ultraviolet radiation, and control the electric/dielectric properties. With all these achievements, CPN will conquer the market and assume the role of innovative materials of the future.

The previous sections of this chapter are focused on the preparation of CPN, with emphasis on surface modification of clay minerals so as to obtain maximum dispersion and a match between the surface properties of the clay minerals and those of the polymer matrix. The ultimate goal of the fundamental research on CPN is to obtain innovative materials. In Section 10.10, some examples taken from recent literature illustrate the improved mechanical, thermal, barrier, conductive, and optical properties of CPN obtained at the laboratory scale, as well as drug delivery properties for biomedical applications.

10.10 Improved properties of CPN—Some recent examples

Authors/material	CPN properties
Mechanical properties	
Li et al. (2018)—Anomalous increases in mechanical stiffness and tensile strength were achieved in a poly(*N*-isopropylacrylamide) nanocomposite gel (NC gel) by incorporating a small amount of silica through the sol–gel reaction of tetraethyl orthosilicate in the polymer–clay mineral network	The resulting NC-Si gels, with ternary polymer–clay mineral–silica structures, exhibited very high tensile moduli (~3500 kPa) and strengths (~1700 kPa), as well as well-defined thermoresponsive swelling/deswelling behaviour, through the incorporation of 0.2–2.5 mass% silica (relative to the gel weight). The reinforcing efficiency of in situ formed silica in the NC gel is more than 50 times that of the clay mineral, while the incorporation of preformed silica nanoparticles led to no observable changes in mechanical properties
Asgari et al. (2017)—The present research reveals novel results on the silane functionalization of Na-Mt by comparing the interlayer space and thermal stability of the nanofiller at different initial silane concentrations	The rheological and tensile properties of HDPE/silane functionalized clay mineral suggest that the storage modulus, damping factor, and mechanical behaviour of the composites depend on the clay mineral nanofiller, as well as the nanocomposites preparation procedure (i.e. masterbatch technique). The tensile modulus showed 15% enhancement for 85% HDPE/15% PEMA/2% modified clay mineral over the pure polymer. TEM images revealed well-dispersed clay mineral layers for the HDPE/2 mass% modified clay mineral/PEMA nanocomposites prepared with the masterbatch technique
Shokrieh et al. (2012)—Effects of adding modified nanoclay (Cloisite 30B) on the mechanical properties (tensile strength, compression, flexural and fracture toughness) of epoxy polymer (ML-506) were investigated	The results of mechanical testing showed that the tensile and flexural moduli, compressive strength and fracture toughness were increased by adding 12.5, 13.3, 7.4, and 25.5 mass% of clay mineral, respectively. On the other hand, tensile and flexural strengths as well as strain to failure were decreased
Barrier properties	
Choudalakis et al. (2014)—Incorporation of hydrophilic Li-hectorite nanoparticles in acrylic resin matrix produces nanocomposite membranes with enhanced gas barrier properties	The overall reduction of relative permeability approaches an order of magnitude for volume fractions of less than 4 mass% of clay mineral

Stratigaki et al. (2014)—Waterborne acrylic and polyurethane resins were employed as matrix and hydrophilic or organically modified nanoparticles of montmorillonite as reinforcement	The coatings present improved barrier properties against gases like CO_2. The nanoparticles create tortuous paths in the matrix, hindering the diffusion of the gas molecules through the coating. Various parameters influence the barrier properties of the coating such as the aspect ratio and the volume fraction of the inorganic nanoparticles and the type of the organic modifier present on their surfaces
Dabbaghianamiri et al. (2017)—Development of a coating technique which can be employed to make self-assembled gas barrier films via inkjet printing. The influence of different polymers polyvinylpyrrolidone (PVP) and poly(acrylic acid) (PAA) with Mt dispersions on polyethylene terephthalate (PET) substrate is examined to study their effectiveness as a gas barrier film	The results show the excellent oxygen barrier behaviour of a combination of PVP and Mt nanocomposite with high transparency. These high barrier gas nanocomposite films are good candidates for a variety of food packaging applications
Ionic conductivity	
Dhatarwal et al. (2017)—Poly(ethylene oxide) (PEO) and poly(methyl methacrylate) (PMMA) polymer blend matrix (50/50 mass%) with lithium tetrafluoroborate ($LiBF_4$) as dopant ionic salt (10 mass%), ethylene carbonate (EC) as plasticizer (10 mass%) and Mt as nanofiller (3 mass%)	Exfoliated Mt decreases the dielectric polarization and creates hindrance to the polymer chain segmental dynamics which causes decrease of ionic conductivity in comparison to the ion-dipolar complexes containing intercalated Mt
Prasanth et al. (2013)—Hybrid organic–inorganic polymer gel electrolytes (PGE) based on polymer-clay mineral nanocomposite microporous membrane containing 1 mol/L of $LiPF_6$ in ethylene carbonate/diethyl carbonate (EC/DEC) are fabricated and their electrochemical characteristics in lithium ion batteries are studied	Incorporation of 2 mass% nanoclay in PGE enhances its ionic conductivity from 0.78 to 3.08 $mS\,cm^{-1}$ at room temperature. Li-ion half-cell comprising Li/PVdF-clay minerals (2 mass%)-PGE/$LiMn_2O_4$ delivers high charge–discharge capacity (similar to 120 $mAh\,g^{-1}$) and shows stable cycle performance
Kitajima et al. (2013)—The work was focused on controlling the orientation of clay mineral layers in ion-conductive polymer electrolytes. A strong magnetic field is one of the methods used to control the orientation of clay mineral layers and to improve the ionic conductivity of the clay mineral composites. Two different composite films were obtained using different orientations of the magnetic field: perpendicular ($M_\perp$) and parallel ($M_/$) to the film surface. From two-dimensional wide-angle X-ray diffraction measurements, the Mt layers preferentially oriented along the direction of the magnetic field in the composites	There were substantial correlations between the conductivity and the amount of Mt layers oriented along the direction of the conductivity measurement. The lowest conductivity was observed in the $M_/$ composite, whereas the $M_\perp$ composite showed very good conductivity. The value was higher than that of the original electrolytes ($PMEO_{10}LiClO_4$) at 30°C. These results clearly suggest that the orientation direction of the Mt layers towards the direction parallel to the conductivity measurement causes an improvement in the conductivity of the composites. In particular, the conductivity of the $M_\perp$ composite with 5 mass% LiMt was $1.2 \times 10^{-5}\ S\,cm^{-1}$ at 30°C, which was more than six times higher than that of the original electrolyte

Authors/material	CPN properties
Flame retardancy	
Kim et al. (2018)—A polyethylene vinyl acetic acid (EVA) nanocomposite comprising an intumescent agent that incorporates monobasic ammonium phosphate, mono-pentaerythritol, and melamine with a cationic nanoclay was set up through solution blending and melt blending to assess the fire retardancy utilizing a cone calorimeter	The fire retardancy of the clay mineral–organic intumescent mixture was more successful than that with the same amount of nanoclay and intumescent agent
Wang et al. (2017)—Biobased gelatins were used to improve the compressive properties and flammability of poly(vinyl alcohol)/montmorillonite (PVA/Mt) aerogels, fabricated using a simple and environmentally friendly freeze-drying technique	Combustion testing demonstrated that, with the addition of gelatin, the self-extinguishing time of the aerogel was cut by half and the limiting-oxygen-index values increased to 28.5%. The peak heat release rate, obtained from cone calorimetry, also decreased with the incorporation of gelatin. Thermogravimetric analysis demonstrated that the gelatins slowed a sharp decomposition of the PVA polymer and increased the thermal stability of the aerogels at the major decomposition stage of the composite aerogels
Patra and Swain (2012)—The nanostructured hybrid materials of poly(methylmethacrylate-co-acrylonitrile) copolymer were synthesized with incorporation of two organically modified clay minerals, Cloisite (R) 30B and 93A by in situ intercalative emulsifier-free emulsion polymerization	The results showed significant effect of both clay minerals in enhancing thermal resistance of the materials. Mechanical properties of the composites were also found to be improved at optimum clay mineral loading. Oxygen barrier property of these materials was measured and it was found that the oxygen permeability was reduced almost by half due to incorporation of clay minerals at 3 mass% loading. Further, it was observed that Cloisite (R) 93A was more effective for improvement of properties than Cloisite (R) 30B in the hybrid materials
Thermal properties	
Hatamzadeh et al. (2014)—First poly(styrene-co-4-chloromethyl styrene) [P(St-co-CMSt)] was synthesized via nitroxide-mediated polymerization. Afterwards, the chlorine groups of P(St-co-CMSt) were converted to thiophene groups using the Kumada cross-coupling reaction and thiophene-functionalized PSt multicentre macromonomer (ThPStM) was synthesized. Finally, PSt-g-PTh/Mt nanocomposite was prepared by a solution intercalation method	It was found that the addition of only a small amount of organoclay (5 mass%) was enough to improve the thermal stabilities of the nanocomposite

Karaj-Abad et al. (2016)—Nitroxide-mediated polymerization (NMP) was used for synthesis of graft and block copolymers using cellulose (Cell) as a backbone, and polystyrene (PSt) and poly(methyl methacrylate) (PMMA) as the branches. In addition, Cell-g-(PMMA-b-PSt)/organophilic Mt nanocomposite was prepared through a solution intercalation method	It was demonstrated that the addition of a small percentage of organophilic Mt (3 mass%) was enough to improve the thermal stabilities of the nanocomposite
Hernandez-Guerrero et al. (2017)—The acrylics employed in the present study were based on butyl acrylate (BA), styrene (STY), and methacrylic acid (MAA), and Na-Mt. Na-Mt was blended with the polymer	By increasing the amount of Mt in the polymer matrix concentration the Young's modulus increases by an order of magnitude. Differential scanning calorimetry (DSC) showed an increase in the glass transition temperature of nanocomposites for all weight percentages of Mt. Also, the wetting angle was determined, in order to know how much water is retained on the surface of the nanocomposite; the results showed that increasing the particle size of nanoclay in the polymer matrix increases the hydrophobicity of the nanocomposite
Biosensors	
Emre et al. (2015)—Polymethylmethacrylate (PMMA) layered silicate nanocomposites were prepared by in situ dispersion polymerization by grafting PMMA with Laponite using a suitable grafting agent. Glucose oxidase (Gox, beta-D-glucose: oxygen-1-oxidoreductase, EC 1.1.3.4) was chosen as the model enzyme to prepare a scaffold for glucose sensing. Three different sensing strategies: PCN/Gox, poly(BIPE)/Gox and PCN/poly(BIPE)/Gox were analysed and their biosensor performances were discussed	Through several studies, clay mineral polymer nanocomposites formation was confirmed and change in surface morphology during biosensor construction was revealed. The constructed enzyme electrodes have good K_M^{app} and I_{max} and low LOD values with high sensitivity compared to several literature reports
Electrical properties	
Dhatarwal et al. (2017)—Plasticized nanocomposites solid polymer electrolyte (PNSPE) films, consisting of poly(ethylene oxide) (PEO) and poly(methyl methacrylate) (PMMA) polymer blend matrix (50/50 mass%) with lithium tetrafluoroborate ($LiBF_4$) as dopant ionic salt (10 mass%), ethylene carbonate (EC) as plasticizer (10 mass%) and Mt as nanofiller (3 mass%), have been prepared by classical solution-cast (SC) and ultrasonic-microwave irradiated (US-MW) solution-cast methods	It has been observed that the dielectric and electrical properties of the PNSPE films change notably with the film preparation methods. Results reveal that the exfoliated Mt decreases the dielectric polarization and creates hindrance to the polymer chain segmental dynamics which causes decrease of ionic conductivity in comparison to the ion-dipolar complexes containing intercalated Mt. The room temperature DC ionic conductivity of the PNSPE films are about 10^{-5} S cm^{-1}, which confirms these materials as potential candidates for design and development of flexible all-solid-state lithium-ion conducting electrochromic devices

Authors/material	CPN properties
Fuel cells	
Nicotera et al. (2012)—Synthetic and natural smectites with different structural and physical parameters, were tested as nanofillers for the creation of Nafion nanocomposites. The solution intercalation method has been successfully applied for incorporation of layered materials into the polymer, while the effect of the solvent, temperature and filler loading were examined in order to determine the optimum conditions for the preparation of highly homogeneous composites	A remarkable behaviour at high temperature is observed, where composite membranes maintain stable and unwavering diffusion for many hours and in dry conditions, proving the exceptional water retention property of these materials. Finally, the hybrid membranes are much stiffer and can withstand higher temperatures compared to pure Nafion. Both these characteristics are highly desirable for use in fuel cell applications
Mishra et al. (2012)—Laponite XLS is used as inorganic filler to prepare clay mineral-Nafion nanocomposites membranes using solution mixing	Incorporation of acid-activated Laponite to Nafion improves the mechanical properties, thermal stability, water uptake and proton conductivity of the resulting nanocomposites due to the presence of in situ generated H_3PO_4 on the surface of acid-activated Laponite. The proton conductivity of the nanocomposites membrane containing 3 mass% acid-activated Laponite is found to be 270.2 $mS\,cm^{-1}$ compared to 136.2 $mS\,cm^{-1}$ for the virgin Nafion at 110°C
Coating materials	
Mohamadpour et al. (2011)—Different types of clay mineral polymer nanocomposites were prepared and evaluated for their application as hard and anti-scratch coating materials. Photo-curable nanocomposites, based on 2,2-bis [4-(methacryloxypropoxy)-phenyl] propane (bis-GMA (2,2-bis[4-(methacryloxypropoxy)-phenyl] propane)) and tri(ethylene glycol) dimethacrylate (TEGDMA (tri(ethylene glycol) dimethacrylate)) (50/50 mass %) were prepared using three types of clay minerals at different loading levels	It was concluded that exfoliation or intercalation plays an important role, primarily in surface roughness, which in turn affects adhesion by changing the characteristics of the coating-substrate interface. The highest scratch resistivity with appropriate adhesion quality was obtained for a sample containing 1 mass % of Cloisite (R) 30B in the nanocomposite resin mixture
Building materials	
Scarfato et al. (2012)—Hybrid organic–inorganic systems were prepared by solvent intercalation of an organomodified Mt into two commercial resins: a coating and a pore liner. The obtained nanocomposite systems at 2.4 and 6 mass % of nanoclay were applied on concrete substrates and characterized to evaluate their protection performances in comparison with the plain resins	The results demonstrated that the nanoclay addition can significantly improve the protection effectiveness of both the used plain resins, with chromatic modifications undetectable to the naked eye. However, the extent of the obtained gain strongly depends on the chemical nature of the polymer and therefore on the mechanisms of action of the matrix

Optical properties	
Sapalidis et al. (2012)—This study describes an effective way for the preparation of well-dispersed, bentonite/PVA nanocomposites with improved properties, based on nanoscale interactions. To this end, a series of poly(vinyl alcohol)-bentonite nanocomposites have been prepared with solvent casting	The nanocomposites exhibited increased mechanical, thermal and gas barrier properties, though they retained their optical clarity. Thus, the Young's modulus of the sample containing 20 mass% clay mineral was increased 193 times, while the oxygen permeability was decreased about seven times, with regard to the corresponding values of the neat polymer
Battery electrodes	
Deka et al. (2012)—A novel nanocomposite gel electrolyte based on intercalation of hyperbranched polyurethane (HBPU) into organically modified Mt for application in Li-ion batteries is reported. The nanocomposites have been prepared by solution intercalation with varying clay mineral loading	AC impedance analysis shows that ionic conductivity increases with the increase of clay mineral loading and attains the highest value of $8.3 \times 10^{-3}\ \mathrm{S\,cm^{-1}}$ for 5 mass% clay mineral concentration. Improvement of electrochemical properties, namely, electrochemical potential window and interfacial stability, is also observed in the clay-loaded HBPU samples
Deka and Kumar (2011)—A study is conducted on the electrical and electrochemical properties of nanocomposite polymer electrolytes based on intercalation of poly(vinylidene fluoride) (PVdF) into the interlayer spaces of organically modified Mt. A solution intercalation technique is employed for nanocomposite formation with clay mineral loading varying from 0 to 4 mass%	Examination with AC impedance spectroscopy reveals that the ionic conductivity of the nanocomposite gel increases with increase in clay mineral loading and attains a maximum value of $2.3 \times 10^{-3}\ \mathrm{S\,cm^{-1}}$ for a 4 mass% clay mineral loading at room temperature. The same composition exhibits enhancement in the electrochemical and interfacial properties as compared with those of a clay mineral-free electrolyte system
Biomaterials	
Villaca et al. (2017)—Na-Mt and Na-Mt:chitosan nanocomposite membranes were prepared by casting dispersions onto a glass plate followed by solvent evaporation at 50°C	Membranes with improved mechanical and physical–chemical properties along with excellent biocompatibility could be generated from 33 mass% Na-Mt intercalated with chitosan and containing sodium alendronate, glycerol (20 mass%) and Labrasol (R) (3 mass%) that also supported the induction of Saos-2 cell proliferation and differentiation, exhibiting potential to develop a novel biomaterial device for guided bone tissue regeneration
Bandla et al. (2017)—New process of developing silver nanoparticles (AgNP), using intercalated saponite/polyacrylamide (PAAm) nanocomposite (NC) hydrogels, via the reduction of silver ions with $NaBH_4$ in aqueous solutions	In vitro antibacterial assay showed that pristine saponite could not inhibit the growth of bacteria. However, the saponite/PAAm NC hydrogels had strong antibacterial activity, particularly against Gram-positive bacteria. With increasing amount (and interlayer distance) of the layered silicates in the NC hydrogels, the nanocomposites showed a strong antibacterial effect

Authors/material	CPN properties
Zhai et al. (2017)—Development of a hybrid bioink composed of a hydrogen bonding monomer (*N*-acryloyl glycinamide) (NAGA) and nanoclay. The hybrid ink could be conveniently tailored as a high strength clay mineral-PNAGA composite scaffold under UV light illumination of printed prehydrogel	The hydrogen bonding combined with physical cross-linking of nanoclay contributed to the superior mechanical performances as well as swelling stability of the hydrogels and bioscaffolds. The sustainable release of intrinsic Mg^{2+} and Si^{4+} from the clay mineral-PNAGA scaffold was shown to promote the osteogenic differentiation of primary rat osteoblast (ROB) cells. Importantly, this implantable clay mineral-PNAGA scaffold efficiently facilitated the regeneration of new bone in tibia defects of rats
Environmental applications	
Msaadi et al. (2017)—Ion imprinted Mt-polymer nanocomposite (Mt/IIP) was prepared by the combination of diazonium salt and photopolymerization, and used for the selective removal of Pb(II)	The adsorption capacity was found to be as high as 301 mg Pb/g after about 30 min contact time, and the relative selectivity coefficients of Zn^{2+} and Fe^{3+} were 9.2 and 93, respectively. Compared with nonimprinted clay mineral/polymer nanocomposites, the Mt/IIP exhibited much higher affinity for Pb(II), therefore, stressing the importance of the receptor sites within the nanocomposite
Olad and Aihar (2014)—Starch-Mt/polyaniline (St-Mt/PANI) nanocomposite was synthesized by oxidative polymerization of aniline in the presence of a Mt-starch nanocomposite dispersion. The St-Mt/PANI nanocomposite was used for the adsorption of a reactive dye	The adsorption capacity of reactive dye on St-Mt/PANI nanocomposite was 91.74 mg/g. The experimental data were well fitted to the Langmuir isotherm and pseudo-second-order kinetic model. All these results demonstrated the effectiveness of the hybrid system as an efficient adsorbent for removal of reactive dyes from textile effluents
Urbano et al. (2012)—The synthesis and arsenic-removal properties of nanocomposite ion-exchange resins based on *N*-methyl-D-glucamine groups are presented. First, the *N*-methyl-D-glucamine-based monomer was synthesized by reaction of 4-vinyl benzyl chloride and *N*-methyl-D-glucamine and was subsequently polymerized by in situ radical polymerization in the presence of organically modified Mt (as filler) and *N,N*-methylene-bis-acrylamide as cross-linker reagent	Arsenate adsorption properties were evaluated as a function of pH, initial arsenic concentration, maximum retention capacity, and in the presence of interfering anions, such as phosphate and sulphate. The nanocomposite presented high arsenic adsorption efficiency, reaching final concentrations below the recommendation of the World Health Organization, even in the presence of interfering anions. Equilibrium adsorption results are well described by Langmuir isotherms, and the adsorption capacities were approximately 55 mg As/g resin

Supercapacitors	
Sivaraman et al. (2015)—Nanocomposite solid polymer electrolytes (SPE) have been prepared from polyethylene oxide (PEO), organically modified nanoclay (MNclay), and tetraethylammonium tetrafluoroborate (TEABF(4)) salt. The concentration of the salt has been varied in the respective SPE, wherein MNclay/PEO ratio was kept constant. It has been proposed that three types of complex formation could be operative in the SPEs due to the interactions among PEO, MNclay, and the salt	The specific capacitance obtained at 708°C was found to be 39 F/g. Such supercapacitor could be used for high temperature applications. The ionic conductivity of the SPEs has been correlated with the results obtained from XRD, DSC, and POM analyses. The formation of complex 1 and complex 2 could be responsible for the increase in the ionic conductivity, whereas complex 3 formation might decrease the ionic conductivity. An activated carbon-based supercapacitor has been fabricated using SPEs and characterized by cyclic voltammetry, galvanostatic charge–discharge' behaviour, and impedance spectroscopic analysis
Fatnassi and Es-Souni (2015)—Nanocomposites of Laponite and the conductive polymer polypyrrole are processed from aqueous solutions using dip-coating of a substrate	Supercapacitor electrodes with specific capacitances of up to 360 F/g, high retention rates and long-term durability are obtained. The main advantage of these new electrodes lies in their high materials efficiency because the clay mineral acts as support for the active polymer component that may be downscaled to 10% with little change of properties. Because processing can be achieved in aqueous solutions and temperatures below 100°C are sufficient to cure the films, the electrodes are environmentally friendly and have a low carbon footprint

10.11 Conclusions and perspectives

The topic of clay polymer nanocomposites is very complex, and combinations of different polymers with different unmodified and modified clay minerals led to an explosion in the number of research papers. Several hundred research articles are being published each year in multidisciplinary journals and in journals specializing in clay minerals and/or in polymers, but also in general journals on materials science, chemistry, and physics. Also the number of books, book chapters, and reviews devoted to clay mineral polymer (bio)nanocomposites increased exponentially during this past decade, such as Bergaya and Lagaly (2007), Carrado and Bergaya (2007), Bergaya and Lagaly (2011), Schoonheydt and Bergaya (2011), Galimberti (2011, 2013), Bergaya et al. (2011, 2012), Bergaya et al. (2013), Lambert and Bergaya (2013), Detellier and Letaief (2013), Ruiz-Hitzky et al. (2013), Chakravarthy et al. (2015), Jlassi et al. (2017), Tripathy and Sahoo (2017), to cite a few.

However, the aim of this chapter is to focus on the most relevant structural aspects and physicochemical properties of clay minerals used as fillers of polymers and on the modification of clay mineral surfaces. As polymers and clay minerals have opposite surface properties (hydrophilicity vs hydrophobicity), this chapter describes the existing methods for optimizing the hydrophilic/lipophilic balance (HLB) at the interface of the two components. This corresponds to either modifying the polymer by grafting with reactive groups or modifying the clay mineral through different strategies such as ion exchange/intercalation, in situ polymerization, and grafting of the external or internal basal surfaces. An optimum ratio of surfactant/CEC of the OC is needed to have a sufficient hydrophobicity with the lowest hydrophilicity.

The properties of CPN are not a simple addition of the properties of each separate component, but instead result from a synergetic effect of the interaction between the intercalated polymer (95%) and the dispersed clay mineral particles (5%). To optimize the performance of the CPN, the number of layers per particle should be as low as possible (Annabi-Bergaya, 2008) corresponding to the highest aspect ratio of the nanoparticles. These required geometrical arrangements of the clay minerals allow increase of the adhesion between the polymer chains and the clay mineral surfaces. The obtained properties lead to a large spectrum of applications.

The most expected and promising innovative CPN materials for the future will certainly occur in biomedical (as nanocarriers with tunable properties allowing controlled drug release) and environmental applications (green chemistry) with the developments of biodegradable polymers filled with nanoclay minerals.

References

Achilleos, D.S., Vamvakaki, M., 2010. End-grafted polymer chains onto inorganic nano-objects. Materials (Basel) 3, 1981–2026.

Alexandre, M., Dubois, P., 2000. Polymer-layered silicate nanocomposites: preparation, properties and uses of a new class of materials. Mater. Sci. Eng. R Rep. 28, 1–63.

Alexandre, M., Beyer, G., Henrist, C., Cloots, R., Rulmont, A., Jerome, A., Dubois, P., 2001. "One-pot" preparation of polymer/clay nanocomposites starting from Na^+ montmorillonite. 1. Melt intercalation of ethylene-vinyl acetate copolymer. Chem. Mater. 13, 3830–3832.

Annabi-Bergaya, F., 2008. Layered clay minerals. Basic research and innovative composite applications. Microporous Mesoporous Mater. 107, 141–148.

Aouad, A., 2004. Pontage de bentonites en milieu concentré. Synthèse, Caractérisation physico-chimique et Application. Thesis of the University of Orleans. Spécialité de physico-chimie des matériaux, France.

Aouad, A., Mandalia, T., Bergaya, F., 2005. Nanocomposite based on PE and novel modified raw clays. Colloid Polym. Sci. 283 (7), 773–782.

Asgari, M., Abouelmagd, A., Sundararaj, U., 2017. Silane functionalization of sodium montmorillonite nanoclay and its effect on rheological and mechanical properties of HDPE/clay nanocomposites. Appl. Clay Sci. 146, 439–448.

Bandla, M., Abbavaram, B.R., Kokkarachedu, V., Sadiku, R.E., 2017. Silver nanoparticles incorporated within intercalated clay/polymer nanocomposite hydrogels for antibacterial studies. Polym. Compos. 38, E16–E23.

Bergaya, F., Lagaly, G., 2007. Clay mineral properties responsible for clay-based polymer nanocomposites (CPN). In: Carrado, K.A., Bergaya, F. (Eds.), CMS Workshop Lectures, vol. 15. The Clay Minerals Society, pp. 61–97.

Bergaya, F., Lagaly, G., 2011. Intercalation processes of layered minerals. In: Brigatti, M.F., Mottana, A. (Eds.), Layered Minerals Structures and Their Applications in Advanced Technologies. European Mineralogical Union Notes in Mineralogy, vol. 11. Mineralogical Society, London, pp. 259–284 (Chapter 7).

Bergaya, F., Lagaly, G., 2013. General introduction: clays, clay minerals, and clay science. In: Bergaya, F., Lagaly, G. (Eds.), Handbook of Clay Science. second ed. Developments in Clay Science, vol. 5A. Elsevier, Amsterdam, pp. 1–19 (Chapter 1).

Bergaya, F., Mandalia, T., Amigouët, P., 2005. A brief survey on CLAYPEN and nanocomposites based on unmodified PE and organo-pillared clays. Colloid Polym. Sci. 283, 773–782.

Bergaya, F., Jaber, M., Lambert, J.F., 2011. Organophilic clay minerals. In: Galimberti, M. (Ed.), Rubber-Clay Nanocomposites, Science, Technology and Applications. John Wiley & Sons, pp. 45–86 (Chapter 2).

Bergaya, F., Jaber, M., Lambert, J.F., 2012. Clays and clay minerals as layered nanofillers for (bio)polymers. In: Avérous, L., Pollet, E. (Eds.), Environmental Silicate Nano-biocomposites. Springer, London, pp. 41–75 (Chapter 3).

Bergaya, F., Detellier, C., Lambert, J.-F., Lagaly, G., 2013. Introduction to clay polymer nanocomposites (CPN). In: Bergaya, F., Lagaly, G. (Eds.), Handbook of Clay Science. second ed. Developments in Clay Science, vol. 5A. Elsevier, Amsterdam, pp. 655–677 (Chapter 13.0).

Carrado, K.A., Bergaya, F. (Eds.), 2007. Clay-Based Polymer Nanocomposites (CPN). CMS Workshop Lectures, vol. 15. The Clay Minerals Society.

Célini, N., 2004. Traitement des argiles par plasma froid pour leur utilisation comme charges de nanocomposites argile-polymère. Thesis of University of Maine, France. Spécialité de chimie et physico-chimie despolymères, France.

Célini, N., Bergaya, F., Poncin-Epaillard, F., 2007. Grafting of hydrocarbon moieties on smectites by cold acetylene plasma and characterization of polyethylene plasma-treated clay mineral composites. Polymer 48, 58–67.

Chakravarthy, V.S., Xiang, C., Gupta, N., 2015. Clay/polymer nanocomposites: processing, properties, and applications. In: Kim, C.S., Randow, C., Sano, T. (Eds.), Hybrid and Hierarchical Composite Materials. Springer, Switzerland, pp. 161–200 (Chapter 5).

Choudalakis, G.A., Kalo, H., Breu, J., Gotsis, A.D., 2014. CO_2 gas barrier properties in polymer nanocomposites coatings containing Li-Hectorite clays. J. Appl. Polym. Sci. 131(18) 40805.

Costantino, U., Leroux, F., Nochetti, M., Mousty, C., 2013. LDH in physical, chemical, biochemical, and life sciences. In: Bergaya, F., Lagaly, G. (Eds.), Handbook of Clay Science. second ed. Developments in Clay Science, vol. 5B. Elsevier, Amsterdam, pp. 765–791.

Dabbaghianamiri, M., Das, S., Beall, G.W., 2017. Improvement approach for gas barrier behavior of polymer/clay nanocomposites films. MRS Adv. 2 (57), 3547–3552.

Deka, M., Kumar, A., 2011. Electrical and electrochemical studies of poly(vinylidene fluoride)-clay nanocomposites gel polymer electrolytes for Li-ion batteries. J. Power Sources 196 (3), 1358–1364.

Deka, M., Kumar, A., Deka, H., Karak, N., 2012. Ionic transport studies in hyperbranched polyurethane/clay nanocomposites gel polymer electrolytes. Ionics 18 (1–2), 181–187.

Detellier, C., Letaief, S., 2013. Kaolinite-polymer nanocomposites. In: Bergaya, F., Lagaly, G. (Eds.), Handbook of Clay Science. second ed. Developments in Clay Science, vol. 5A. Elsevier, Amsterdam, pp. 707–719 (Chapter 13.2).

Dhatarwal, P., Sengwa, R.J., Choudhary, S., 2017. Effect of intercalated and exfoliated montmorillonite clay on the structural, dielectric and electrical properties of plasticized nanocomposite solid polymer electrolytes. Compos. Commun. 5, 1–7.

Emre, F.B., Kesik, M., Kanik, F.E., Akpinar, H.Z., Aslan-Gurel, E., Rossi, R.M., Toppare, L., 2015. A benzimidazole-based conducting polymer and a PMMA-clay nanocomposites containing biosensor platform for glucose sensing. Synth. Met. 207, 102–109.

Fan, X., Xia, C., Advincula, R.C., 2003. Grafting of polymers from clay nanoparticles via in situ free radical surface-initiated polymerization: monocationic versus bicationic initiators. Langmuir 19, 4381–4389.

Fatnassi, M., Es-Souni, M., 2015. Nanoscale phase separation in laponite-polypyrrole nanocomposites. Application to electrodes for energy storage. RSC Adv. 5 (28), 21550–21557.

Forano, C., Costantino, U., Prévot, V., Taviot-Gueho, C., 2013. Layered double hydroxides (LDH). In: Bergaya, F., Lagaly, G. (Eds.), Handbook of Clay Science. second ed. Developments in Clay Science, vol. 5B. Elsevier, Amsterdam, pp. 745–782.

Fukushima, Y., Inagaki, S., 1987. Synthesis of an intercalated compound of montmorillonite and 6-polyamide. J. Incl. Phenom. 5 (4), 474–482.

Galimberti, M. (Ed.), 2011. Rubber-Clay Nanocomposites, Science, Technology and Applications. John Wiley & Sons.

Galimberti, M., 2013. Applications of clay polymer nanocomposites. In: Bergaya, F., Lagaly, G. (Eds.), Handbook of Clay Science. second ed. Developments in Clay Science, vol. 5B. Elsevier, Amsterdam, pp. 539–586.

Gao, F., 2004. Clay/polymer composites: the story. Mater. Today 7 (11), 50–55.

Greenland, D.J., 1963. Adsorption of polyvinyl alcohols by montmorillonite. J. Colloid Sci. 18 (7), 647–664.

Hatamzadeh, M., Jaymand, M., Massoumi, B., 2014. Graft copolymerization of thiophene onto polystyrene synthesized via nitroxide-mediated polymerization and its polymer clay nanocomposites. Polym. Int. 63 (3), 402–412.

Hernandez-Guerrero, O., Castillo-Perez, R., Hernandez-Vargas, M.L., Campillo-Illanes, B.F., 2017. Study of thermal and mechanical properties of clay/polymer nanocomposites synthesized via modified solution blending. MRS Adv. 2 (49), 2757–2762.

Inagaki, S., Fukushuma, Y., Miyata, M., 1995. Inclusion polymerization of isoprene in the channels of *sepiolite*. Res. Chem. Intermed. 21, 167–180.

Jlassi, K., Chehimi, M.M., Sabu, T. (Eds.), 2017. Clay Polymer Nanocomposites. Elsevier, Amsterdam.

Karaj-Abad, S.G., Abbasian, M., Jaymand, M., 2016. Grafting of poly[(methyl methacrylate)-block-styrene] onto cellulose via nitroxide-mediated polymerization, and its polymer/clay nanocomposites. Carbohydr. Polym. 152, 297–305.

Kim, H., Park, J.W., Lee, J.H., Jang, S.W., Kim, H.J., Choi, Y., Choy, J.H., Yang, J.H., 2018. Clay-organic intumescent hybrid system for the synergetic flammability of polymer nanocomposites. J. Therm. Anal. Calorim. 132 (3), 2009–2014.

Kitajima, S., Matsuda, M., Yamato, M., Tominaga, Y., 2013. Anisotropic ionic conduction in composite polymer electrolytes filled with clays oriented by a strong magnetic field. Polym. J. 45, 738–743.

Kojima, Y., Usuki, A., Kawasumi, M., Okada, A., Fukushima, Y., Kurauchi, T., Gamigaito, O., 1993. Mechanical properties of nylon 6-clay hybrid. J. Mater. Res. 8 (5), 1185–1189.

Lagaly, G., 2013. Clay mineral organic interactions. In: Bergaya, F., Lagaly, G. (Eds.), Handbook of Clay Science. second ed. Developments in Clay Science, vol. 5A. Elsevier, Amsterdam, pp. 435–505 (Chapter 10.3).

Lambert, J.F., Bergaya, F., 2013. Smectite-polymer nanocomposites. In: Bergaya, F., Lagaly, G. (Eds.), Handbook of Clay Science. second ed. Developments in Clay Science, vol. 5A. Elsevier, Amsterdam, pp. 679–706 (Chapter 13.1).

Li, H., Jiang, H., Haraguchi, K., 2018. Ultrastiff, thermoresponsive nanocomposite hydrogels composed of ternary polymer-clay-silica networks. Macromolecules 51 (2), 529–539.

Mandalia, T., Bergaya, F., 2006. Organo clay mineral–melted polyolefin nanocomposites. Effect of surfactant/CEC ratio. J. Phys. Chem. Solids 67, 836–845.

Mansoori, Y., Atghia, S.V., Zamanloo, M.R., Imanzadeh, G., Sirousazar, M., 2010. Polymer-clay nanocomposites: free-radical grafting of polyacrylamide onto organophilic montmorillonite. Eur. Polym. J. 46 (9), 1844–1853.

Mansoori, Y., Roojaei, K., Zamanloo, M.R., Imanzadeh, G., 2012. Polymer–clay nanocomposites via chemical grafting of polyacrylonitrile onto cloisite 20A. Bull. Mater. Sci. 35 (7), 1063–1070.

Mévellec, V., Roussel, S., Tessier, L., Chancolon, J., Mayne-L'Hermite, M., Deniau, G., Viel, P., Palacin, S., 2007. Grafting polymers on surfaces: a new powerful and versatile diazonium salt-based one-step process in aqueous media. Chem. Mater. 19 (25), 6323–6330.

Mishra, A.K., Kuila, T., Kim, N.H., Lee, J.H., 2012. Effect of peptizer on the properties of Nafion–Laponite clay nanocomposite membranes for polymer electrolyte membrane fuel cells. J. Membr. Sci. 389, 316–323.

Moet, A.S., Akelah, A., 1993. Polymer-clay nanocomposites: polystyrene grafted onto montmorillonite interlayers. Mater. Lett. 18, 97–102.

Mohamadpour, S., Pourabbas, B., Fabbri, P., 2011. Anti-scratch and adhesion properties of photo-curable polymer/clay nanocomposite coatings based on methacrylate monomers. Sci. Iran. 18 (3), 765–771.

Monticelli, O., Musina, Z., Russo, S., Bals, S., 2007. On the use of TEM in the characterization of nanocomposites. Mater. Lett. 61, 3446–3450.

Msaadi, R., Ammar, S., Chehimi, M.M., Yagci, Y., 2017. Diazonium-based ion-imprinted polymer/clay nanocomposite for the selective extraction of lead (II) ions in aqueous media. Eur. Polym. J. 89, 367–380.

Murray, H.H., 2006. Kaolin applications. In: Murray, H.H. (Ed.), Applied Clay Mineralogy. Occurrences, Processing and Application of Kaolins, Bentonites, Palygorskite-Sepiolite, and Common Clays. Developments in Clay Science, vol. 2. Elsevier, Amsterdam, pp. 85–109 (Chapter 5).

Nicotera, I., Enotiadis, A., Angjeli, K., Coppola, L., Gournis, D., 2012. Evaluation of smectite clays as nanofillers for the synthesis of nanocomposites polymer electrolytes for fuel cell applications. Int. J. Hydrogen Energy 37 (7), 6236–6245.

Olad, A., Aihar, F.F., 2014. Eco-friendly biopolymer/clay/conducting polymer nanocomposite: characterization and its application in reactive dye removal. Fibers Polym. 15 (6), 1321–1329.

Patra, S.K., Swain, S.K., 2012. Effect of organoclays on the thermal, mechanical, and oxygen barrier properties of poly(methylmethacrylate-co-acrylonitrile)/clay nanocomposites. Polym. Compos. 33 (5), 796–802.

Prasanth, R., Shubha, N., Hng, H.H., Srinivasan, M., 2013. Effect of nano-clay on ionic conductivity and electrochemical properties of poly(vinylidene fluoride) based nanocomposites porous polymer membranes and their application as polymer electrolyte in lithium ion batteries. Eur. Polym. J. 49 (2), 307–318.

Roy, R., Komarneni, S., Roy, D.M., 1984. Multi-phasic ceramic composites made by sol-gel technique. Mater. Res. Soc. Symp. Proc. 32, 347–359.

Ruiz-Hitzky, E., Aranda, P., Alvarez, A., Santaren, J., Esteban-Cubillo, A., 2011. Advanced materials and new applications on sepiolite. In: Galan, E., Singer, A. (Eds.), Developments in Palygorskite-Sepiolite Research. Developments in Clay Science, vol. 3. Elsevier, oxford, pp. 393–452 (Chapter 17).

Ruiz-Hitzky, E., Aranda, P., Darder, M., Fernandes, F.M., 2013. Fibrous clay mineral-polymer nanocomposites. In: Bergaya, F., Lagaly, G. (Eds.), Handbook of Clay Science. second ed. Developments in Clay Science, vol. 5A. Elsevier, Amsterdam, pp. 721–741 (Chapter 13.3).

Sapalidis, A.A., Katsaros, F.K., Steriotis, T.A., Kanellopoulos, N.K., 2012. Properties of poly(vinyl alcohol)-bentonite clay nanocomposites films in relation to polymer-clay interactions. J. Appl. Polym. Sci. 123 (3), 1812–1821.

Sathishkumar, T.P., Naveen, J., 2014. Glass fiber-reinforced polymer composites—a review. J. Reinf. Plast. Compos. 33 (13), 1258–1275.

Scarfato, P., Di Maio, L., Fariello, M.L., Russo, P., Incarnato, L., 2012. Preparation and evaluation of polymer/clay nanocomposite surface treatments for concrete durability enhancement. Cem. Concr. Compos. 34 (3), 297–305.

Schoonheydt, R.A., Bergaya, F., 2011. Industrial clay minerals as nanomaterials. In: Christidis, G. (Ed.), Advances in the Characterization of Industrial Minerals. EMU Notes in Mineralogy, vol. 9. Miner Soc Publishers, pp. 415–440 (Chapter 10).

Shah, D., Fytas, G., Vlassopoulos, D., Di, J., Sogah, D., Giannelis, E.P., 2005. Structure and dynamics of polymer-grafted clay suspensions. Langmuir 21, 19–25.

Shokrieh, M.M., Kefayati, A.R., Chitsazzadeh, M., 2012. Fabrication and mechanical properties of clay nanocomposites and its polymer concrete. Mater. Des. 40, 443–452.

Sivaraman, P., Shashidhara, K., Thakur, A.P., Samui, A.B., Bhattacharyya, A.R., 2015. Nanocomposite solid polymer electrolytes based on polyethylene oxide, modified nanoclay, and tetraethylammonium tetrafluoroborate for application in solid-state supercapacitor. Polym. Eng. Sci. 55 (7), 1536–1545.

Stratigaki, M., Choudalakis, G., Gotsis, A.D., 2014. Gas transport properties in waterborne polymer nanocomposites coatings containing organomodified clays. J. Coat. Technol. Res. 11 (6), 899–911.

Tripathy, K.D., Sahoo, P.B. (Eds.), 2017. Properties and Applications of Polymer Nanocomposites: Clay and Carbon Based Polymer Nanocomposites. Springer, Berlin, Heidelberg.

Tsubokawa, N., 2007. Surface grafting of polymers onto nanoparticles in a solvent-free dry-system and applications of polymer-grafted nanoparticles as novel functional hybrid materials. Polym. J. 39 (10), 983–1000.

Urbano, B.F., Rivas, B.L., Martinez, F., Alexandratos, S.D., 2012. Water-insoluble polymer-clay nanocomposite ion exchange resin based on N-methyl-D-glucamine ligand groups for arsenic removal. React. Funct. Polym. 72 (9), 642–649.

Usuki, A., Kojima, Y., Kawasumi, M., Okada, A., Fukushima, Y., Kurauchi, T., Gamigaito, O., 1993. Synthesis of nylon 6-clay hybrid. J. Mater. Res. 8 (5), 1179–1184.

Vaia, R.A., Giannelis, E.P., 1999. Polymer melt intercalation in organically modified-layered silicates: model predictions and experiment. Macromolecules 30, 8000–8009.

Villaca, J.C., da Silva, L.C.R.P., de Alexandria, A.K., de Almeida, G.S., Locatelli, F.R., Maia, L.C., Rodrigues, C.R., de Sousa, V.P., Tavares, M.I.B., Cabral, L.M., 2017. Development and characterization of clay-polymer nanocomposite membranes containing sodium alendronate with osteogenic activity. Appl. Clay Sci. 146, 475–486.

Viville, P., Lazzaroni, R., Pollet, E., Alexandre, M., Dubois, P., 2004. Controlled polymer grafting on single clay nanoplatelets. J. Am. Chem. Soc. 126 (29), 9007–9012.

Wang, M.S., Pinnavaia, T.J., 1994. Clay-polymer nanocomposites formed from acidic derivatives of montmorillonite and an epoxy resin. Chem. Mater. 6, 468–474.

Wang, Y.T., Zhao, H.B., Degracia, K., Han, L.X., Sun, H., Sun, M.Z., Wang, Y.Z., Schiraldi, D.A., 2017. Green approach to improving the strength and flame retardancy of poly(vinyl alcohol)/clayaerogels: incorporating biobased gelatin. ACS Appl. Mater. Interfaces 9 (48), 42258–42265.

Work, W.J., Horie, K., Hess, M., Stepto, R.F.T., 2004. Definition of terms related to polymer blends, composites, and multiphase polymeric materials (IUPAC Recommendations 2004). Pure Appl. Chem. 76 (11), 1985–2007.

Wypych, F., 2004. Wypych, F., Satyanarayana, K.G. (Eds.), Clay Surfaces, Fundamentals and Applications. Elsevier, Amsterdam.

Wypych, F., Satyanarayana, K.G., 2005. Functionalization of single layers and nanofibers: a new strategy to produce polymer nanocomposites with optimized properties. J. Colloid Interface Sci. 285 (2), 532–543.

Wypych, F., Arizaga, G.G.C., Satyanarayana, K.G., 2008. Thomas, S., Zaikov, G.E. (Eds.), Polymer Nanocomposite Research Advances. Nova Science Publishers.

Zhai, X., Ma, Y., Hou, C., Gao, F., Zhang, Y., Ruan, C., Pan, H., Weijia Lu, W., Liu, W., 2017. 3D-printed high strength bioactive supramolecular polymer/clay nanocomposite hydrogel scaffold for bone regeneration. ACS Biomater. Sci. Eng. 3 (6), 1109–1118.

From adsorbed dyes to optical materials

11

*Jun Kawamata**, *Yasutaka Suzuki**, *Makoto Tominaga*[†]
*School of Chemistry, Yamaguchi University, Yamaguchi, Japan, [†]Lehrstuhl für Anorganische Chemie I, University of Bayreuth, Bayreuth, Germany

11.1 Introduction

In the modern era, the uses of dyes and pigments, including clay mineral-dye hybrids, have been expanded from painting industries to electronic and optoelectronic devices. One of the oldest clay mineral-dye hybrids utilised as a pigment, Maya blue, is based on an indigo dye, and it exhibits an azure blue colour. Maya blue is known for its remarkable stability against time and harsh weathering conditions. In Maya blue, the host clay mineral, palygorskite, does not contribute to the colour of the pigment, but stabilises the intercalated dye. In the past few decades, hybrid materials consisting of two-dimensional (2D) particles, including clay minerals, and dyes exhibiting characteristic electronic properties, have attracted considerable interest. Owing to the high shape anisotropy of the host 2D particles, hybrids, and thus the confined dyes, exhibit optical responses characteristic of anisotropic media (Hähner et al., 1996; Fischer et al., 1998; Windsor and Tinker, 1999; Yamaoka et al., 2000; Miyamoto et al., 2001; Holzheu and Hoffmann, 2002; Bujdák and Iyi, 2005; Martinez et al., 2006; Arbeloa and Martínez, 2006a,b; Čeklovský et al., 2007; Eguchi et al., 2007; Salleres et al., 2009; Suzuki et al., 2009, 2012; Yan et al., 2011). Characteristic properties of confined dyes can also be attributed to the specific microenvironment provided by the surface and/or interlayer space of 2D materials. By tuning the layer charge density of the host 2D materials and/or the dye loading, the average distance between the confined dyes can be controlled. Furthermore, strong confinement of dyes on 2D material surfaces leads to flattened molecular conformation, and suppresses vibrational and translational motions. These factors impose characteristic optical properties on the confined dyes, different from those in solution and/or in the solid state. Among the 2D materials, clay minerals are attractive host materials. Most clay minerals do not catalyse the photodecomposition of confined dyes differently from the cases of titanate and niobate nanosheets. Furthermore, most clay minerals are transparent in the visible wavelength region, and therefore incident light can fully interact with the confined dyes. In this chapter, clay mineral-dye hybrids exhibiting attractive electronic characteristics arising from the specific microenvironment provided by clay minerals will be outlined.

Properties of clay mineral-dye hybrids dispersed in water are often different from those in solid state. In the former case, the adsorbed dyes are still surrounded by the solvent, and retain some freedom of motion as shown in Fig. 11.1 (left). In the latter case,

Developments in Clay Science, Vol. 9. https://doi.org/10.1016/B978-0-08-102432-4.00011-1

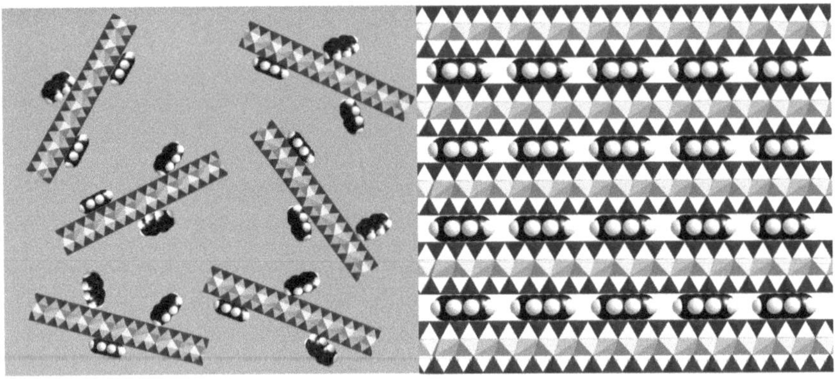

Fig. 11.1 Schematic representation of clay mineral-dye hybrids in water dispersion (*left*) and in solid state (*right*).

the dyes are strongly confined in the interlayer space of the clay minerals (Fig. 11.1, right). However, even in the reviewed articles, these two states have not always been clearly distinguished. The authors recommend that readers pay careful attention to the state of the discussed materials when consulting related articles.

11.2 Controlled adsorption of dyes onto clay minerals

As for dye-based optical materials, in general, an increase of the concentration of the dyes amplifies the optical response. Interlayer spaces of clay minerals can accommodate dyes with high densities. For example, when 4,4′-[4,1-phenylenedi-2, 1-ethynediyl]bis[1-methylpyridinium]diiodide(Ac) is intercalated into the interlayer space of a clay mineral, as illustrated in Fig. 11.1 (right), the calculated density of Ac in the total volume is $1\,mol\,dm^{-3}$. However, dye molecules tend to undergo a phenomenon called self-aggregation (Steiger et al., 2009) under a high concentration, including the crystalline state of the pure dye. Under these aggregated states, the spectroscopic properties of the dyes change drastically due to exciton coupling (Aragó and Troisi, 2015) between transition dipole moments of neighbouring dye molecules. With respect to spatial orientation of dipoles relative to the aggregate axis, the exciton coupling between dye molecules results in H-aggregates with a blue-shifted absorption band with respect to that of the monomer, or in J-aggregates, exhibiting a red-shifted absorption band, as seen in Fig. 11.2.

The H-aggregate is characterised by a sandwich-type dipole geometry. The formation of H-aggregates decreases the excited-state lifetime. Most of the H-aggregates are nonluminescent and nonactive species. The J-aggregate represents the type of molecular assembly with a head-to-tail intermolecular association. In contrast to the H-aggregates, the J-aggregates are interesting supramolecular systems with a broad spectrum of potential applications (Kobayashi, 2012).

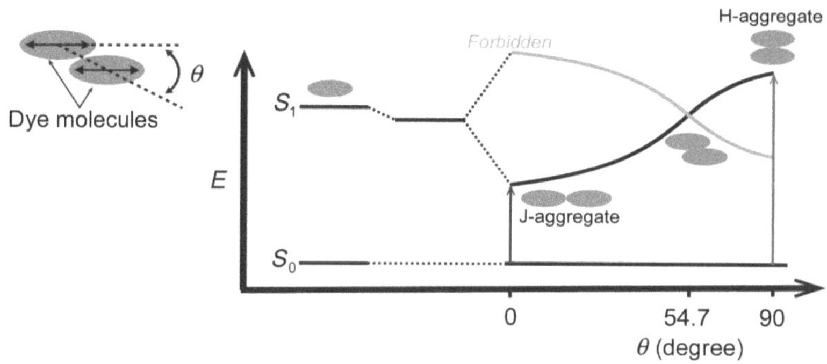

Fig. 11.2 Schematic representation of exciton splitting of the electronic states of dye molecular aggregates depending on θ, the angle between dipole moments of neighbouring molecules. *Black* and *grey lines* represent allowed and forbidden transitions, respectively. 0 and 90 degrees correspond to J- and H-aggregates, respectively. At 54.7 degree, energy separation is the same as in the monomeric state.

Even on clay mineral surfaces, dyes tend to form H-aggregates and/or segregate, when the adsorption density is high. The driving forces are mainly attributed to hydrophobic and van der Waals interactions between the dye molecules (Ghosh and Bard, 1984; Bujdák, 2006). The adsorption of dyes on the clay mineral surfaces are known to strongly depend on the layer charge density (Bujdák and Komadel, 1997; Bujdák et al., 1998; Bujdák, 2006). As seen in Fig. 11.3, by changing the layer charge density, subnanometre to several nanometres order control of intermolecular distance can be achieved, and thus the aggregation state of dye molecules can be controlled (Bujdák and Komadel, 1997; Bujdák, 2006).

Layer charge density of available clay minerals can be controlled by a lithium treatment (Bujdák, 2006). When lithium cations are added to montmorillonite (Mt), the cations compensate for the negative charges of the clay mineral layers, due to isomorphous substitution. The layer charge density of Mt can be reduced by heating (Hofman–Klemen effect), depending on the amount of added lithium cations. Layer charge densities of synthetic clay minerals can easily be controlled in the synthesis process. For example, saponites (Sap) with different layer charges have been synthesised by tuning the ratio Si/Al/Mg (Iiyama and Roy, 1963; Vogels et al., 2005; Egawa et al., 2011).

The aggregation state of dyes on clay minerals can be controlled by tuning the layer charge of host clay minerals. For example, the aggregation state of methylene blue hybridised with reduced charge Mt in aqueous dispersion is reported to depend on the charge density of Mt (Bujdák, 2006). The higher charge density of Mt induced a closer intermolecular distance to adsorbed methylene blue molecules. This effect increased the population of H-aggregates in the hybrid. On the other hand, employment of a lower charge density of Mt increased the intermolecular distance. This caused an increase of the population of J-type aggregates and monomers. By tuning the layer charge densities, control of the aggregation of rhodamins (Bujdák and Iyi, 2005; Bujdák et al., 2007), or triphenylmethane dyes (Bujdák and Iyi, 2002) have also been realised in aqueous dispersions.

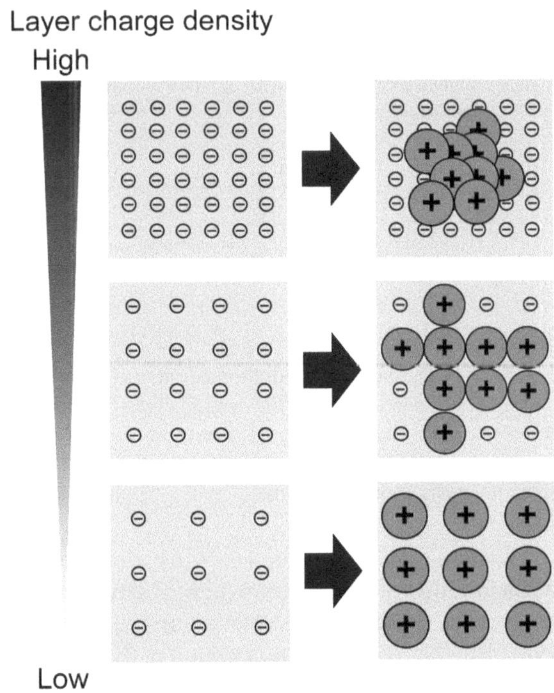

Fig. 11.3 Schematic illustration of differences in dye adsorption on smectites with different layer charge densities.

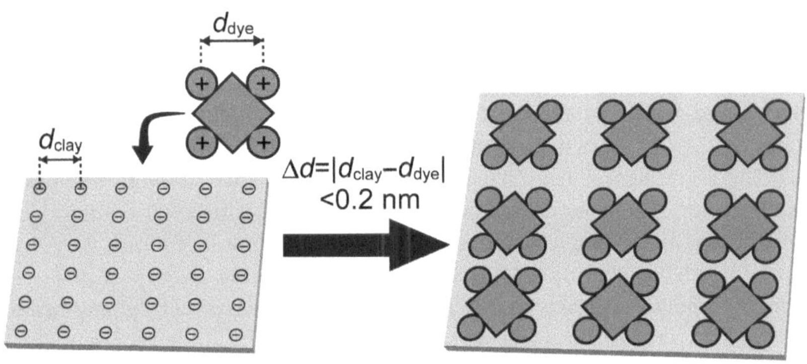

Fig. 11.4 Schematic representation of the *size-matching rule*.

One of the most elegant strategies for realising a high density of dye molecules with monomer adsorption of dyes on clay mineral surfaces is the application of the *size-matching rule* proposed by Takagi et al. (2002, 2013). They found this rule based on systematic experiments with multivalent cationic porphyrin derivatives. When the difference of the distance between cationic sites of the porphyrin derivatives and the averaged distance between anionic sites of a clay mineral is within 0.2 nm, the porphyrin derivatives are adsorbed on clay mineral surfaces without concomitant aggregation in aqueous dispersion. This situation is represented in Fig. 11.4.

In the size-matching condition, adsorbed molecules are placed at anionic sites of the host layer. By increasing the loading of the molecule, the number of occupied anionic sites is gradually increased. Thus, by using this methodology, the average intermolecular distance can be controlled by changing dye loading. This control of distance effectively perturbs the energy and/or electron transfer ability of adsorbed dyes. There are several studies treating energy and/or electron transfer reactions between adsorbed dyes on hybrids in aqueous dispersion (Ishida et al., 2011; Saha et al., 2016; Belušáková et al., 2017; Morimoto et al., 2017). These studies evidence that optimisation of intermolecular distance between adsorbed dyes is an essential factor to maximise photochemical and optical functionalities of the hybrids.

The *size-matching rule* is also effective for clay mineral-dye films. Some of the authors reported that several dyes can be intercalated into the interlayer space of a clay mineral without aggregation up to the space-filling percentage of almost 100% (Suzuki et al., 2011). The dyes exhibited salient nonlinear optical responses attributed to monomeric states.

An alternative means for fabricating clay mineral-dye hybrids containing a relatively high concentration of dyes without aggregation is the use of surfactants, such as tetraalkylammonium cations. Sasai et al. (2009) fabricated a hybrid solid consisting of Rhodamine 6G, tetraalkylammonium, and Laponite. Tetraalkylammonium has been employed to disperse the adsorbed dye molecules, and thus the suppression of aggregation of the dye molecules is realised. The obtained hybrid solid exhibited a very high fluorescence quantum yield of 80%, attributed to monomeric dyes, even at a high concentration of Rhodamine 6G of 10^{-2} mol dm^{-3}.

In solution, J-aggregates are easily formed with dyes that have molecular shapes that are inappropriate for the formation of sandwich-type molecular assemblies. Several clay mineral-dye hybrids containing J-aggregated dyes have also been reported. Compared with the J-aggregated dyes in solution, those in clay mineral films are stable, and the microstructure is well confined. J-aggregates in solid-state materials have been well recognised for exciton transport, and also for electron or hole transport properties. Thus, they can be employed in applications such as organic field effect transistors (Ribierre et al., 2012), or artificial light-harvesting systems (Walker et al., 2010). Therefore, clay mineral films involving J-aggregated dyes are suitable to spectroscopic studies, and will be promising candidates for future practical devices.

Ogawa et al. (1996) and Miyamoto et al. (2000, 2004) studied hybrids consisting of smectites and a pseudoisocyanine dye, 1,1′-diethyl-2,2′-cyanine bromide (PIC). In aqueous dispersions, the cyanine dye cations formed J-aggregates on the surfaces of Mt, while they distributed molecularly on Sap. On the other hand, the dye formed J-aggregates on both Mt and Sap in the cast films. According to the authors, the J-aggregate formation was suppressed when the clay mineral surface was modified by a surfactant, tetramethylammonium. Debnath et al. (2017) fabricated Langmuir–Blodgett (LB) films of *N,N′*-dioctadecylthiacyanine perchlorate (TC18) in the presence and absence of Laponite (Lap). J-aggregates in LB film of TC18 degraded by external stimuli, such as irradiation of light, passage of time, as well as heating. However, in the case of hybrid films, the decrease in the J-aggregate was minimal.

11.3 Flattening of the π-electron system

Confinement of dyes on the surface, or in the interlayer space of clay minerals often changes the conformation of the dyes. Particularly, flattening of dyes significantly affects the electronic properties of dyes. This is because in most cases, the origin of the colour, and thus the electronic properties of dyes are due to delocalisation of π-electrons in their π-conjugated systems. Flattening of the π-conjugated systems enhances the extent of delocalisation of π-electrons as if the π-conjugated system is extended. This flattening results in a spectral red-shift, enhances the absorption coefficient, and therefore enhances optical responses such as nonlinear optical properties of dyes. This functional improvement is especially remarkable for dyes possessing distorted π-conjugated systems. In principle, dyes possessing extended and planar π-conjugated systems exhibit superior electronic properties, such as large oscillator strengths, large hyperpolarisabilities, small electron transfer energies, and so forth, owing to a large delocalisation of π-electrons. However, synthesis and handling of such dyes meets with difficulties due to a strong π–π interaction, and thus very low solubility. On the one hand, synthesis and handling of dyes possessing extended but distorted π-conjugated systems are rather easy. By hybridising such dyes with clay minerals, the distorted π-conjugated systems can be flattened. As a result, the dyes are expected to exhibit optical responses as if the dyes possess extended and planar π-conjugated systems. Thus, the hybridisation of distorted dyes with clay minerals is a promising means to obtain materials possessing superior electronic properties and easy handling.

A typical example is the adsorption of a tetravalent porphyrin derivative, tetramethylpyridylporphyrin (TMPyP, Fig. 11.5A), onto clay mineral surfaces. In a water solution, the methylpyridyl groups are nearly perpendicular to the plane of the porphyrine (Fig. 11.5B). Chernia and Gill (1999) reported a $<30\,\text{nm}$ red-shift in the absorption spectrum of TMPyP adsorbed on outer surfaces of Lap, and a $<60\,\text{nm}$ shift in the TMPyP confined in the interlayer spaces of Lap. Based on systematic studies, the cause of the significant red shifts has been attributed to the flattening of TMPyP, accompanied by the adsorption onto Lap (Fig. 11.5C), or intercalation into the interlayer space of Lap (Fig. 11.5D). This attribution has also been confirmed by other studies (Suzuki et al., 2011; Ishida et al., 2012). The flattening of TMPyP by intercalating between clay mineral layers significantly improves its optical properties. For example, two-photon absorption is one of the third-order nonlinear optical phenomena, efficiency of TMPyP in clay mineral films has been enhanced 13 times compared with that in solution (Suzuki et al., 2011). Similar flattening has been observed, not only for TMPyP, but also for several other dyes, such as viologen and biphenyl derivatives (Villemure et al., 1986; Tominaga et al., 2017).

As described, the dihedral angle of the biphenyl moiety of a dye decreases on the surface or in the interlayer space of clay minerals. A similar decrease of the dihedral angle of biphenyl is also observed when a hydrostatic pressure is applied (Kato et al., 1988). For example, the dihedral angle of biphenyl in solution under ordinary pressure is 45 degrees; whereas at 1.2 GPa, it is 20 degrees. This means that the interlayer space

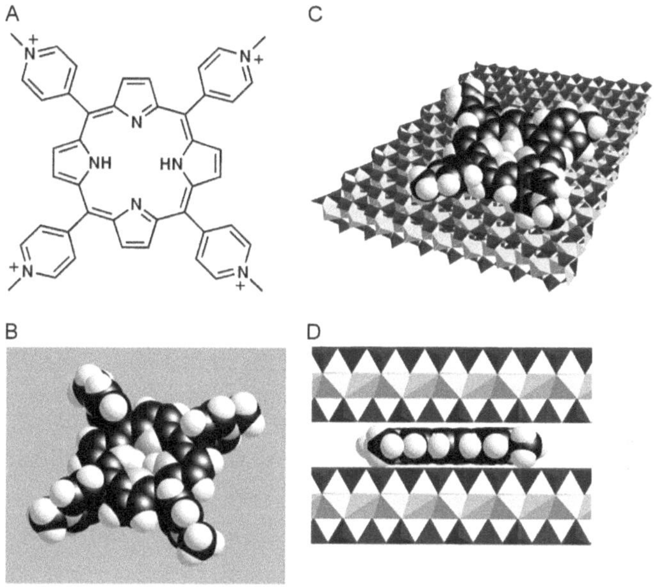

Fig. 11.5 Molecular structure of tetramethylpyridylporphyrin (TMPyP) (A) and schematic illustration of stereostructure of TMPyP in solution (B), adsorbed on a clay mineral (C) and intercalated between clay mineral layers (D).

imposes a flattened conformation similar to that observed in a high-pressure field. The principle force that flattens a dye intercalated in a smectite is the electrostatic force between the intercalated cationic dyes and the anionic sites of the smectite layer (Brigatti et al., 2006). Because the electrostatic force acts in the vertical direction, or perpendicularly to the smectite layer, the force is uniaxial, and the uniaxial force per unit area becomes stronger as the charge density of the smectite increases. When a molecule with a thickness larger than the interlayer space of the smectite is intercalated under specific conditions (depending on the type of intercalated species, loading, humidity, and other factors), the uniaxial force applies to the adsorbed molecules. This uniaxial force can be regarded as a pseudouniaxial pressure. The uniaxial force is applied only for molecules with thicknesses larger than the interlayer space of the pristine smectite. When the thicknesses of the molecules are smaller than the interlayer space of the pristine smectite, the interlayer distance of the smectite should not be expanded by the intercalation, and therefore the uniaxial force is not applied to molecules. The uniaxial force per unit area is determined by the type of smectite, and thus the charge density. The uniaxial force applied to intercalated molecules is shared among all molecules per unit area. Therefore, the uniaxial force applied to one molecule decreases as the number of molecules per unit area increases. Consequently, it is possible for intercalated organic molecules to experience a tunable, pseudo-high-pressure field by varying the charge density or loading.

The authors demonstrated that a biphenyl derivative, 4,4-[(1,1′-biphenyl)-4,4′-diyldi-(1*E*)-2,1-ethenediyl]bis[1-methylpyridinium] diiodide (BP), intercalated in the smectite interlayer space undergoes a conformational change (Tominaga et al., 2017). The molecular thickness of BP in solution at ordinary pressure is calculated to be 0.79 nm. The BP molecule was found to be flattened to 0.45 nm in the interlayer space of a clay mineral. This conformational change depended on the loading and/or layer charge density of the smectite, and was attributed to a pseudouniaxial pressure field that is caused by the electrostatic force that operates between the smectite layer and the interlayer cations. The magnitude of the pseudopressure field was of the order of a few GPa, and was controlled by varying the layer charge density and/or the loading. This methodology provides a means to tune the stereochemical properties of an intercalated molecule, and induces similar changes to those observed in a high-pressure field. The intercalated molecules experience a pseudo-high-pressure field, even under ordinary pressure conditions. Thus, the intercalation of organic molecules into the smectite interlayer space is a novel and simple strategy for obtaining the high-pressure properties of organic molecules.

11.4 Suppression of vibrational motion

In some cases, the confinement of dyes on clay mineral surfaces suppresses their vibrational motion and/or translational mobility. Suppression of vibrational motion is also a factor for enhancing the extent of delocalisation of π-electrons, as well as flattening the π-conjugated system. Thus, this is also a factor for enhancing the oscillator strength and hyperpolarisabilities, and decreasing electron transfer energy. Remarkably improved optical response resulted by suppression of vibrational motion accompanied by confinement between clay mineral layers can be seen in the cases of butadiyne derivatives. The rotational barrier of the diacetylene bridge in 1,4-bis-aromatic butadiyne is estimated to be less than 1 kJ/mol (Kamada et al., 2007), which is considerably smaller than the thermal energy at room temperature ($T = 297\,K$); $N_A kT = 2.5\,kJ/mol$. Therefore, the diacetylene bridge is freely rotating in its solution state. On the one hand, the rotation is suppressed between clay mineral layers, and results in improved optical response, owing to the extended π-conjugated system. In fact, two-photon absorption efficiency of 1,4-bis(2,5-dimethoxy-4-{2-[4-(*N*-methyl)pyridinium] ethenyl}phenyl)butadiyne (MPPBT) (Fig. 11.6) in a clay mineral film was enhanced 5.1 times, compared with that in solution (Suzuki et al., 2011). The suppression of the rotation of the diacetylene bridge was evidenced because the Stokes shift in the hybrid film became smaller than that in solution (Kamada et al., 2007).

Suppression of mobility reduces quenching via collision and other unfavourable relaxation pathways (Ishida et al., 2013). As a result, fluorescence quantum yields of confined dyes are drastically increased in some cases. The phenomenon was named '*surface-fixation induced emission*' (Ishida et al., 2014). For example, the fluorescence quantum yield of a porphyrazine dye intercalated in smectite has been reported to be enhanced 19 times compared with that in water (Ishida et al., 2014).

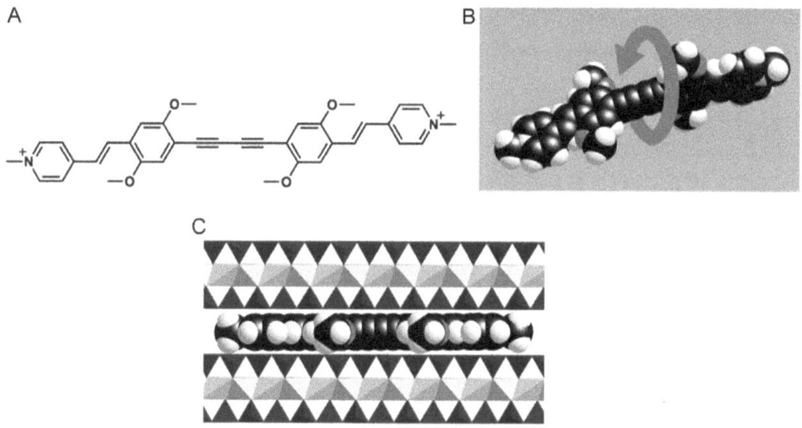

Fig. 11.6 Molecular structure of 1,4-bis(2,5-dimethoxy-4-{2-[4-(*N*-methyl)pyridinium] ethenyl}phenyl)butadiyne (MPPBT) (A) and schematic illustration of stereostructure of MPPBT in solution (B) and in intercalated state (C).

11.5 Switching of the optical properties of the hybrids

Solvatochromism is a phenomenon in which absorption and emission spectra of a dye change depending on the polarity of the solvent. Even when a dye is adsorbed on clay mineral surfaces in aqueous dispersion, most dye molecules are solvated, as shown in Fig. 11.1 (left). Thus, the clay mineral-dye hybrid exhibits solvatochromism to some extent. The polarity of the solvent sometimes affects the adsorption manner of a dye onto the clay mineral surface. Under such a situation, the hybrid exhibits relatively large solvatochromism. Eguchi et al. (2013) reported that due to adsorption of a cationic porphyrin, *cis*-bis(*N*-methylpyridinium-4-yl)diphenylporphin, on a synthetic clay mineral surface results in changes in its orientation as a function of temperature and solvents. For this dye, parallel orientation with respect to the clay mineral surface was stable in protic solvents, whereas a tilted orientation was stable in aprotic solvents. Furthermore, they found that the energy transfer efficiency between dyes can be controlled by the change of relative orientation between dye molecules (Eguchi et al., 2014).

This type of solvatochromism is observed even for clay mineral-dye hybrid films. Swelling is one of the notable characteristics of smectites. Not only pure smectites, but also smectite-dye hybrids can incorporate water or polar organic solvents into their interlayer space. In the swollen state, the surrounding environment of intercalated molecules is different from that in the dried state. Using this property, Takagi et al. (2010) demonstrated the switching of the spectroscopic properties of smectite-dye hybrid films. They reported that the colour of a smectite-dye hybrid film consisting of [5,10,15,20-tetrakis(*N*-methylpyridinium-4-yl)porphyrinato]tin(IV) and a synthetic Sap changed remarkably upon swelling. Furthermore, the colour of the swollen state depended on the type of solvent used for swelling. This colour change was attributed not to conventional solvatochromism caused by the change of dielectric constant, but to the orientational change of organic molecules resulting from the expansion of the interlayer space.

In swollen smectite-dye hybrids, the interlayer space is often expanded by more than 0.8 nm. This interlayer space is large enough for organic molecules to form π–π stackings because it is about twice the thickness of a planar π-conjugated system. Consequently, the organic molecules that exist in monomer form with high density in the dried smectite-dye hybrid can be oriented to form π–π stackings in the swollen state. The swollen interlayer distance can be shrunk to the original interlayer distance by drying. Therefore, simply by drying the swollen smectite-dye hybrid, the intercalated organic molecules can rearrange into their original monomeric form, even when the molecules formed π–π stacking arrangements in the swollen state. Based on this mechanism, several chromic materials have been developed as seen in Fig. 11.7.

Hybrid films consisting of synthetic Sap and fluorescent organic molecules with planar π-conjugated systems were reported to exhibit reversible luminescence colour switching induced by swelling and drying, which cause a changes in interlayer space (Tominaga et al., 2016). The dyes in the films exhibited luminescence attributed to

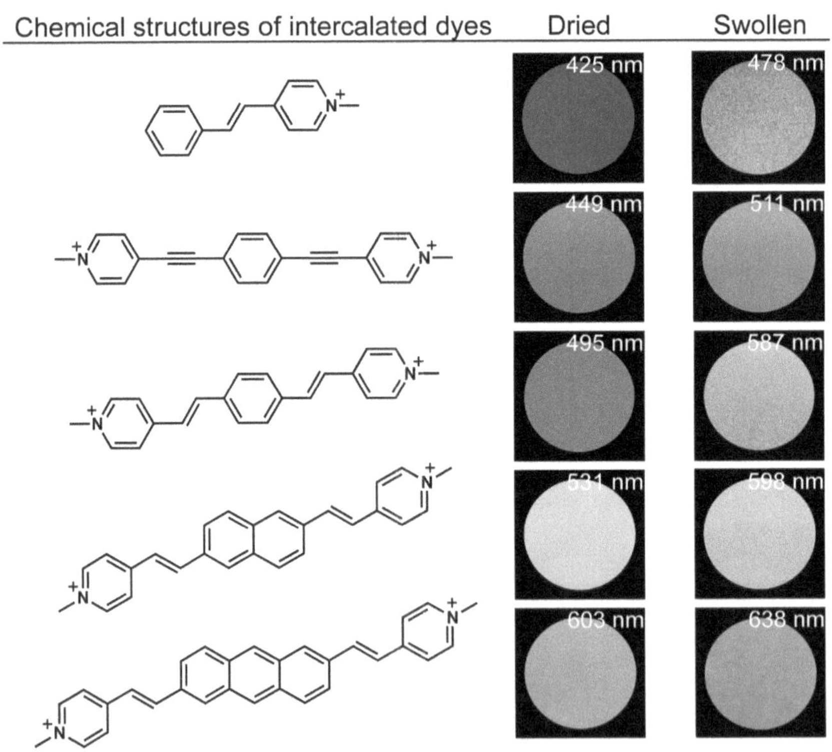

Fig. 11.7 Luminescence colour of dyes in dried and swollen hybrid films. Corresponding luminescence wavelength is also represented.

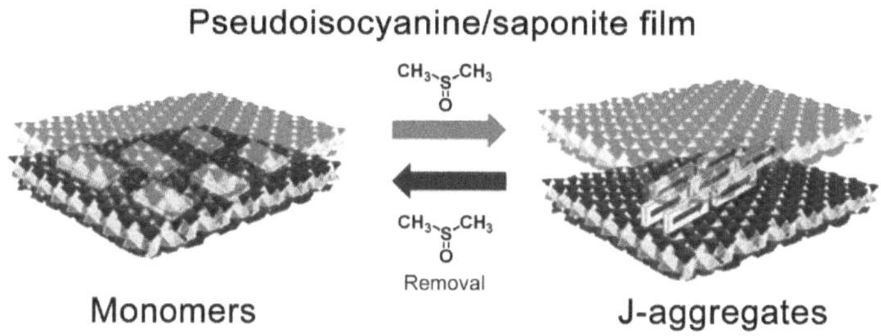

Fig. 11.8 Schematic representation of switching between monomers and J-aggregates of pseudoisocyanine in hybrid film. The switching occurs by addition and removal of a polar solvent, dimethyl sulfoxide.
Reprinted with permission from Matejdes, M., Himeno, D., Suzuki, Y., Kawamata, J., 2017. Controlled formation of pseudoisocyanine J-aggregates in the interlayer space of synthetic saponite. Appl. Clay Sci. 140, 119–123. (Copyright 2017, Elsevier).

monomeric or excimeric states in the dried or swollen films, respectively. As seen in Fig. 11.7, significant changes of fluorescence accompanied by swelling and drying were observed. By repeating the swelling and drying, reversible fluorescence colour changes with almost the same wavelength change were observed for at least 50 cycles.

Switching of aggregation states accompanied by swelling and drying has also been demonstrated. As shown in Fig. 11.8, in a dried clay mineral-pseudoisocyanine hybrid film fabricated at a dye loading level of 70%, most of pseudoisocyanines existed in the monomer form. When the film was swollen using dimethyl sulfoxide, pseudoisocyanines preferred to form J-aggregates (Matejdes et al., 2017). This system is expected to be promising for the development of a practical optical device with controllable properties such as in artificial light harvesting systems.

Colour switching of typical organic dye-based chromic systems accompanies a change in the chemical structure. However, the colour switching of clay mineral-based systems is not due to a change in the chemical structure of the dye. Thus, the repeatability can be higher than in molecular switches, based on chemical structure change. This advantage makes these clay mineral-based systems more attractive for practical applications.

11.6 Future prospective

Numerous hybrids exhibiting attractive optical properties resulting from the characteristics of clay minerals have been developed. This chapter has outlined hybrids utilising specific microenvironments provided by the clay mineral surface and the interlayer space of clay minerals. For further studies, the authors recommend textbooks dealing with a wide variety of 2D materials-dye hybrid composites (Bujdák, 2017; Ishida and Takagi, 2017; Okada and Ogawa, 2017; Schoonheydt and Umemura, 2017).

Typically, clay mineral-dye hybrids have been regarded as light-scattering media. Reduction of light-scattering to a negligible level is an unavoidable requirement for applying this type of material for practical uses. Although several novel means for fabricating low light-scattering hybrids have been obtained by progress in film fabrication, they still do not fully satisfy industrial requirements. The studies on developments of hybrid materials exhibiting novel and/or superior photo functionalities are important. However, improvements of film fabrication protocols should also be a key challenge for obtaining clay mineral-dye hybrids utilisable as optical materials. For clay minerals-based solids, the dominant origin of light scattering is the gaps between clay mineral particles. The size of the gaps must be sufficiently smaller than the wavelength of the light (preferably, $\lambda/10$). The use of clay minerals with a small particle size is indeed effective in this regard. However, in order to minimise the size of the gaps, use of clay minerals with homogeneous size and shape is more important. Therefore, synthetic clay minerals are more promising host materials than natural ones. The authors believe that progress of synthesis will result in synthetic clay minerals with homogeneous size and charge. By utilising such clay minerals, clay mineral-dye hybrids fully satisfying industrial requirements should be obtained in the near future.

References

Aragó, J., Troisi, A., 2015. Dynamics of the excitonic coupling in organic crystals. Phys. Rev. Lett. 114, 026402.

Arbeloa, F.L., Martínez, V.M., 2006a. New fluorescent polarization method to evaluate the orientation of adsorbed molecules in uniaxial 2D layered materials. J. Photochem. Photobiol. A Chem. 181, 44–49.

Arbeloa, F.L., Martínez, V.M., 2006b. Orientation of adsorbed dyes in the interlayer space of clays. 2 Fluorescence polarization of rhodamine 6G in laponite films. Chem. Mater. 18, 1407–1416.

Belušáková, S., Martínez, V.M., Arbeloa, I.L., Bujdák, J., 2017. Resonance energy transfer between dye molecules in colloids of a layered silicate. The effect of dye surface concentration. J. Phys. Chem. C 121, 8300–8309.

Brigatti, M.F., Galan, E., Theng, B.G.K., 2006. Structures and mineralogy of clay minerals. In: Bergaya, F., Theng, B.K.G., Lagaly, G. (Eds.), Handbook of Clay Science. Elsevier, Amsterdam, pp. 19–86.

Bujdák, J., 2006. Effect of the layer charge of clay minerals on optical properties of organic dyes. A review. Appl. Clay Sci. 34, 58–73.

Bujdák, J., 2017. Hybrids with functional dyes. In: Nakato, T., Kawamata, J., Takagi, S. (Eds.), Inorganic Nanosheets and Nanosheet-Based Materials: Fundamentals and Applications of Two-Dimensional Systems. Springer, Japan, pp. 419–466.

Bujdák, J., Iyi, N., 2002. Visible spectroscopy of cationic dyes indispersions with reduced-charge montmorillonites. Clays Clay Miner. 50, 446–454.

Bujdák, J., Iyi, N., 2005. Molecular orientation of rhodamine dyes on surfaces of layered silicates. J. Phys. Chem. B. 109, 4608–4615.

Bujdák, J., Komadel, P., 1997. Interaction of methylene blue with reduced charge montmorillonite. J. Phys. Chem. B. 101, 9065–9068.

Bujdák, J., Janek, M., Madejová, J., Komadel, P., 1998. Influence of the layer charge density of smectites on the interaction with methylene blue. J. Chem. Soc. Faraday Trans. 94, 3487–3492.

Bujdák, J., Martínez, V.M., Arbeloa, F.L., Iyi, N., 2007. Spectral properties of rhodamine 3B adsorbed on the surface of montmorillonites with variable layer charge. Langmuir 23, 1851–1859.

Čeklovský, A., Bujdák, J., Czímerová, A., Iyi, N., 2007. Spectral study on the molecular orientation of a tetracationic porphyrin dye on the surface of layered silicates. Centr. Eur. J. Phys. 5, 236–243.

Chernia, Z., Gill, D., 1999. Flattening of TMPyP adsorbed on laponite. Evidence in observed and calculated UV-vis spectra. Langmuir 15, 1625–1633.

Debnath, P., Chakraborty, S., Deb, S., Nath, J., Dey, B., Bhattacharjee, D., Soda, H., Tominaga, M., Suzuki, Y., Kawamata, J., Hussain, S.A., 2017. Effect of nano clay Laponite on stability of SHG active J-aggregate of a thiacyanine dye onto LB films. Appl. Clay Sci. 147, 105–116.

Egawa, T., Watanabe, H., Fujimura, T., Ishida, Y., Yamato, M., Masui, D., Shimada, T., Tachibana, H., Yoshida, H., Inoue, H., Takagi, S., 2011. Novel methodology to control the adsorption structure of cationic porphyrins on the clay surface using the "size-matching rule" Langmuir 27, 10722–10729.

Eguchi, M., Tachibana, H., Takagi, S., Tryk, D.A., Inoue, H., 2007. Dichroic measurements on dicationic and tetracationic porphyrins on clay surfaces with visible-light-attenuated total reflectance. Bull. Chem. Soc. Jpn. 80, 1350–1356.

Eguchi, M., Shimada, T., Tryk, D.A., Inoue, H., Takagi, S., 2013. Role of hydrophobic interaction in controlling the orientation of dicationic porphyrins on solid surfaces. J. Phys. Chem. C 117, 9245–9251.

Eguchi, M., Watanabe, Y., Ohtani, Y., Shimada, T., Takagi, S., 2014. Switching of energy transfer reaction by the control of orientation factor between porphyrin derivatives on the clay surface. Tetrahedron Lett. 55, 2662–2666.

Fischer, D., Caseri, W.R., Hähner, G., 1998. Orientation and electronic structure of ion exchanged dye molecules on mica: an X-ray absorption study. J. Colloid Interface Sci. 198, 337–346.

Ghosh, P.K., Bard, A.J., 1984. Photochemistry of tris(2,2′-bipyridyl)ruthenium(II) in colloidal clay suspensions. J. Phys. Chem. 88, 5519–5526.

Hähner, G., Marti, A., Spencer, N.D., Caseri, W.R., 1996. Orientation and electronic structure of methylene blue on mica: a near edge X-ray absorption fine structure spectroscopy study. J. Chem. Phys. 104, 7749–7757.

Holzheu, S., Hoffmann, H., 2002. Adsorption study of cationic dyes having a trimethylammonium anchor group on hectorite using electrooptic and spectroscopic methods. J. Colloid Interface Sci. 245, 16–23.

Iiyama, J.T., Roy, R., 1963. Unusually stable saponite in the system Na_2O-MgO-Al_2O_3-SiO_2. Clay Miner. Bull. 5, 161–171.

Ishida, Y., Takagi, S., 2017. Photoenergy conversion. In: Nakato, T., Kawamata, J., Takagi, S. (Eds.), Inorganic Nanosheets and Nanosheet-Based Materials: Fundamentals and Applications of Two-Dimensional Systems. Springer, Japan, pp. 357–372.

Ishida, Y., Shimada, T., Masui, D., Tachibana, H., Inoue, H., Takagi, S., 2011. Efficient excited energy transfer reaction in clay/porphyrin complex toward an artificial light-harvesting system. J. Am. Chem. Soc. 133, 14280–14286.

Ishida, Y., Masui, D., Shimada, T., Tachibana, H., Inoue, H., Takagi, S., 2012. The mechanism of the porphyrin spectral shift on inorganic nanosheets: the molecular flattening induced by the strong host-guest interaction sue to the "size-matching rule" J. Phys. Chem. C 116, 7879–7885.

Ishida, Y., Shimada, T., Tachibana, H., Inoue, H., Takagi, S., 2013. Regulation of the collisional self-quenching of fluorescence in clay/porphyrin complex by strong host-guest interaction. Chem. A Eur. J. 116, 12065–12072.

Ishida, Y., Shimada, T., Takagi, S., 2014. Surface-fixation induced emission of porphyrazine dye by a complexation with inorganic nanosheets. J. Phys. Chem. C 118, 20466–20471.

Kamada, K., Tanamura, Y., Ueno, K., Ohta, K., Misawa, H., 2007. Enhanced two-photon absorption of chromophores confined in two-dimensional nanospace. J. Phys. Chem. C 111, 11193–11198.

Kato, M., Higashi, M., Taniguchi, Y., 1988. Effect of pressure on the internal rotation angle of biphenyl in carbon disulfide. J. Chem. Phys. 89, 5417–5421.

Kobayashi, T., 2012. J-Aggregates. World Scientific, Singapore. Volume 2.

Martinez, V., Salleres, S., Banuelos, J., Arbeloa, F.L., 2006. Application of fluorescence with polarized light to evaluate the orientation of dyes adsorbed in layered materials. J. Fluoresc. 16, 233–240.

Matejdes, M., Himeno, D., Suzuki, Y., Kawamata, J., 2017. Controlled formation of pseudoisocyanine J-aggregates in the interlayer space of synthetic saponite. Appl. Clay Sci. 140, 119–123.

Miyamoto, N., Kawai, R., Kuroda, K., Ogawa, M., 2000. Adsorption and aggregation of a cationic cyanine dye on layered clay minerals. Appl. Clay Sci. 16, 161–170.

Miyamoto, N., Kuroda, K., Ogawa, M., 2001. Uni-directional orientation of cyanine dye aggregates on a $K_4Nb_6O_{17}$ single crystal: toward novel supramolecular assemblies with three-dimensional anisotropy. J. Am. Chem. Soc. 123, 6949–6950.

Miyamoto, N., Kuroda, K., Ogawa, M., 2004. Exfoliation and film preparation of a layered titanate, Na2Ti3O7, and intercalation of pseudoisocyanine dye. J. Mater. Chem. 14, 165–170.

Morimoto, D., Yoshida, H., Sato, K., Saito, K., Yagi, M., Takagi, S., Yui, T., 2017. Light energy accumulation from pyrene derivative to tris(bipyridine)ruthenium on clay surface. Langmuir 33, 3680–3684.

Ogawa, M., Kawai, R., Kuroda, K., 1996. Adsorption and aggregation of a cationic cyanine dye on smectites. J. Phys. Chem. 100, 16218–16221.

Okada, T., Ogawa, M., 2017. Inorganic-organic interactions. In: Nakato, T., Kawamata, J., Takagi, S. (Eds.), Inorganic Nanosheets and Nanosheet-Based Materials: Fundamentals and Applications of Two-Dimensional Systems. Springer, Japan, pp. 163–186.

Ribierre, J.C., Sato, M., Ishizuka, A., Tanaka, T., Watanabe, S., Matsumoto, M., Matsumoto, S., Uchiyama, M., Aoyama, T., 2012. Organic field-effect transistors based on J-aggregate thin films of a bisazomethine dye. Org. Electron. 13, 999–1003.

Saha, J., Dey, D., Roy, A.D., Bhattacharjee, D., Hussain, S.A., 2016. Multi step FRET among three laser dyes pyrene, Acriflavine and Rhodamine B. JOL 172, 168–174.

Salleres, S., Arbeloa, F.L., Martinez, V.M., Arbeloa, T., Arbeloa, I.L., 2009. Improving the fluorescence polarization method to evaluate the orientation of fluorescent systems adsorbed in ordered layered materials. J. Lumin. 129, 1336–1340.

Sasai, R., Itoh, T., Ohmori, W., Itoh, H., Kusunoki, M., 2009. Preparation and characterization of rhodamine 6G/alkyltrimethylammonium/laponite hybrid solid materials with higher emission quantum yield. J. Phys. Chem. C 113, 415–421.

Schoonheydt, R.A., Umemura, Y., 2017. Clay minerals as natural nanosheets. In: Nakato, T., Kawamata, J., Takagi, S. (Eds.), Inorganic Nanosheets and Nanosheet-Based Materials: Fundamentals and Applications of Two-Dimensional Systems. Springer, Japan, pp. 33–54.

Steiger, R., Pugin, R., Heier, J., 2009. J-aggregation of cyanine dyes by self-assembly. Colloids Surf. B Biointerfaces 74, 484–491.

Suzuki, Y., Matsunaga, R., Sato, H., Kogure, T., Yamagishi, A., Kawamata, J., 2009. Non-centrosymmetric behavior of a clay film ion-exchanged with chiral metal complexes. Chem. Commun. 45, 6964–6966.

Suzuki, Y., Tenma, Y., Nishioka, Y., Kamada, K., Ohta, K., Kawamata, J., 2011. Efficient two-photon absorption materials consisting of cationic dyes and clay minerals. J. Phys. Chem. C 115, 20653–20661.

Suzuki, Y., Tenma, Y., Nishioka, Y., Kawamata, J., 2012. Efficient nonlinear optical properties of dyes confined in interlayer nanospaces of clay minerals. Chem. Asian J. 7, 1170–1179.

Takagi, S., Shimada, T., Eguchi, M., Yui, T., Yoshida, H., Tryk, D.A., Inoue, H., 2002. High-density adsorption of cationic porphyrins on clay layer surfaces without aggregation: the size-matching effect. Langmuir 18, 2265–2272.

Takagi, S., Shimada, T., Masui, D., Tachibana, H., Ishida, Y., Tryk, D.A., Inoue, H., 2010. Unique solvatochromism of a membrane composed of a cationic porphyrin-clay complex. Langmuir 26, 4639–4641.

Takagi, S., Shimada, T., Ishida, Y., Fujimura, T., Masui, D., Tachibana, H., Eguchi, M., Inoue, H., 2013. Size-matching effect on inorganic nanosheets: control of distance, alignment, and orientation of molecular adsorption as a bottom-up methodology for nanomaterials. Langmuir 29, 2108–2119.

Tominaga, M., Oniki, Y., Mochida, S., Kasatani, K., Tani, S., Suzuki, Y., Kawamata, J., 2016. Clay-organic hybrid films exhibiting reversible fluorescent color switching induced by swelling and drying of a clay mineral. J. Phys. Chem. C 120, 23813–23822.

Tominaga, M., Nishioka, Y., Tani, S., Suzuki, Y., Kawamata, J., 2017. Tunable high-pressure field operating on a cationic biphenyl derivative intercalated in clay minerals. Sci. Rep. 7, 7651.

Villemure, G., Detellier, C., Szabo, A.G., 1986. Fluorescence of clay-intercalated methylviologen. J. Am. Chem. Soc. 108, 4658–4659.

Vogels, R.J.M.J., Kloprogge, J.T., Geus, J.W., 2005. Synthesis and characterization of saponite clays. Am. Mineral. 90, 931–944.

Walker, B.J., Bulovic, V., Bawendi, M.G., 2010. Quantum dot/J-aggregate blended films for light harvesting and energy transfer. Nano Lett. 10, 3995–3999.

Windsor, S.A., Tinker, M.H., 1999. Electro-fluorescence polarization studies of the interaction of fluorescent dyes with clay minerals in suspensions. Colloids Surf. A Physicochem. Eng. Asp. 148, 61–73.

Yamaoka, K., Sasai, R., Takata, N., 2000. Electric linear dichroism. A powerful method for the ionic chromophore-colloid system as exemplified by dye and montmorillonite suspensions. Colloids Surf. A Physicochem. Eng. Asp. 175, 23–39.

Yan, D., Lu, J., Ma, J., Wei, M., Li, S., Evans, D.G., Duan, X., 2011. Near-infrared absorption and polarized luminescent ultrathin films based on sulfonated cyanines and layered double hydroxide. J. Phys. Chem. C 115, 7939–7946.

Preparation and application of clay mineral films

12

Yasushi Umemura
Department of Applied Chemistry, National Defense Academy, Yokosuka, Kanagawa, Japan

12.1 Introduction

Recently, two-dimensional (2D) materials with layer structures have been extensively studied (Tan et al., 2017). Promising chemical and physical properties of 2D materials such as graphene and MXenes (2D transition metal carbides, carbonitrides, and nitrides) have attracted the attention of many researchers. Potential applications of 2D materials are directed toward energy storage, flexible electronic devices, and sensors, for example, and in such applications, the 2D-spaces between layers play important roles. If 2D materials are exfoliated into discrete layers (sometimes called "nanosheets" or "nanolayers") in solvents, the layers are able to be reassembled in ultrathin monolayer films at a nanometre level by various methods: the self-assembling, layer-by-layer (LbL), Langmuir–Blodgett (LB), and casting methods. One can deposit ultrathin monolayer films on a substrate repeatedly and prepare a multilayer film. It is possible to fabricate a multilayer film using different kinds of nanolayers, and introduce molecules between the layers to add functionalities such as light harvesting, catalytic abilities, and redox properties.

Clay minerals are natural 2D-materials, and they have been used since the birth of ancient civilisations. Clay minerals are generally nontoxic, chemically stable (but chemically modifiable), cation-exchangeable, swelling, and highly adsorptive. Smectites, in particular, are exfoliated easily into discrete clay mineral nanolayers when they are dispersed in water at low concentrations. One nanolayer that is composed of one octahedral sheet and two tetrahedral sheets possesses negative charges, and its thickness is $\sim$1 nm. Smectites are useful and attractive clay minerals for preparation of ultrathin film devices because dispersions of the smectite layers are available without any special techniques. One may say that smectites are model 2D materials in the research field of films.

In this chapter, preparation methods of clay mineral films and their application are described. Particularly, the LbL method and LB (also Langmuir–Schaefer, LS) methods are useful to prepare ultrathin clay mineral nanolayer films; layer orders in the films are able to be designed at the molecular or single nanolayer level.

There have been some excellent review articles on clay mineral films and other 2D materials (Mousty, 2004; Ras et al., 2007; Johnston, 2010; Ma and Sasaki, 2010; Zhou et al., 2011; Miao et al., 2012; Schoonheydt, 2014; Wang and Sasaki, 2014; Guan et al., 2015). These are helpful for further reference.

Developments in Clay Science, Vol. 9. https://doi.org/10.1016/B978-0-08-102432-4.00012-3

12.2 Casting

Casting is very simple, and one can prepare films without any special equipment (Fig. 12.1). A clay mineral is dispersed in a solvent that evaporates slowly, such as water (the solvent may be mixed with other functional materials), and the dispersion is cast in a shallow container with a flat bottom, such as a Petri dish, or is dropped on a flat substrate. As the solvent evaporates, the layers settle on the flat surface in a parallel fashion, and finally, a dry uniform film is obtained. The film can be peeled off from the flat surface. The thickness of the film can be controlled easily by changing the clay mineral concentration in the dispersion, and the amount of the dispersion on the flat surface. However, if the clay mineral concentration is too high, aggregates of the layers settle on the surface, and, as a result, the obtained film is liable to be opaque. To avoid this, a dilute clay mineral dispersion is preferable, because the discrete layers in the dilute dispersion settle uniformly on the surface.

Properties of the films prepared by the casting method are dependent on cations between the layers. Cast films of montmorillonite (Mt) with monovalent and divalent metal cations (Li^+, Na^+, Mg^{2+}, and Ca^{2+}) were smooth and flat, while the film with Al^{3+} cations was shrunken and wavy. The film with Fe^{2+} and Fe^{3+} was torn into curled pieces. The films with the monovalent cations swelled with water sufficiently and showed good gas barrier properties (Nam et al., 2009; Carrier et al., 2013).

Self-standing clay mineral films prepared by the casting method are suitable samples to determine orientation of molecules in interlayer spaces. For example, metal complex cations of $[M(en)_2]^{2+}$ (en = ethylenediamine, M = Co, Ni, and Cu) took square planar structures in the interlayer spaces, and they were parallel to the clay mineral layers (Schoonheydt et al., 1979; Maes et al., 1980; Schoonheydt and Pelgrims, 1983). The dyes of rhodamines and oxazine 4 were incorporated in Laponite (Lap) films, and fluorescence resonance energy transfer (FRET) between the dyes was investigated in relation to the dyes' orientations (Bujdák et al., 2011). Cationic (4+) porphyrin molecules with planar structures were adsorbed on synthetic clay mineral layers (Smecton SA) in the films with parallel orientation to the layers. Interestingly, the amount of adsorbed porphyrin molecules was high, but the molecules did not form aggregates in the interlayer spaces. The distribution of the negative charges on the clay mineral layers matched that of the positive charges in the molecules precisely (size-matching effect) (Takagi et al., 2001, 2002a,b, 2006).

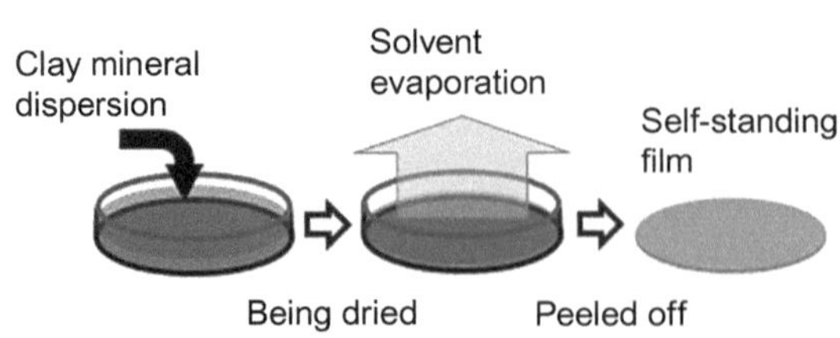

Fig. 12.1 Schematic diagram of a casting method.

Layered double hydroxides (LDH) are anion-exchangeable clay minerals. Cast films of LDH with anionic dye molecules were prepared, and molecular orientations of the dyes were investigated. For example, naphthalene acetate molecules formed inter-digitated bilayers in the LDH interlayer spaces. Due to the orientation in the interlayer spaces, the fluorescence lifetime of the naphthalene acetate increased in the film (Shi et al., 2008).

Clay polymer nanocomposite films can be prepared by casting (Zhou et al., 2011; Chu et al., 2014). Mechanical and thermal stability of organic polymer films were improved by the addition of clay minerals (Alexandre and Dubois, 2000; Kuila and Nandi, 2004; Shikinaka et al., 2012; Wu et al., 2012; Oliveira et al., 2015; Valencia et al., 2016). The film properties such as gas permeability (Yeh et al., 2003; Wang et al., 2004; Bébin et al., 2006; Zhang et al., 2008, 2017; Herrera Alonso et al., 2009; Alix et al., 2012; Wu et al., 2012) and proton conductivity (in clay/Nafion® nanocomposite films) (Rhee et al., 2005; Bébin et al., 2006; Zhang et al., 2008; Herrera Alonso et al., 2009) were affected by the amount of clay minerals. The clay mineral layers that are parallel to the film surface would effectively block the paths of the gas molecules passing through the film. Concerning the proton conductivity, water retention on the clay mineral layer surface would improve proton mobility in the film.

12.3 Spin coating

A clay mineral dispersion is dropped at the centre of a disk (or a substrate), which is rotated at a high rotational speed (Fig. 12.2). The clay mineral dispersion is spread over the disk according to the centrifugal force, and at the same time, the solvent evaporates. The excess amount of the dispersion is removed away from the rim. Finally, a thin clay mineral film (nm–μm thick) is obtained on the substrate. The thickness of the film is dependent on the clay mineral concentration in the dispersion, on the size of the clay mineral layers, on the rotational speed, and on the time of the rotation. For example, Lap films were prepared by dropping the clay mineral dispersions at different concentrations at varying rotational speeds, and the effect of the Lap concentration and the rotational speed on the film thickness was investigated (Fig. 12.3) (Schoonheydt, 2014). The thickness of the film prepared from a 2 mass% dispersion

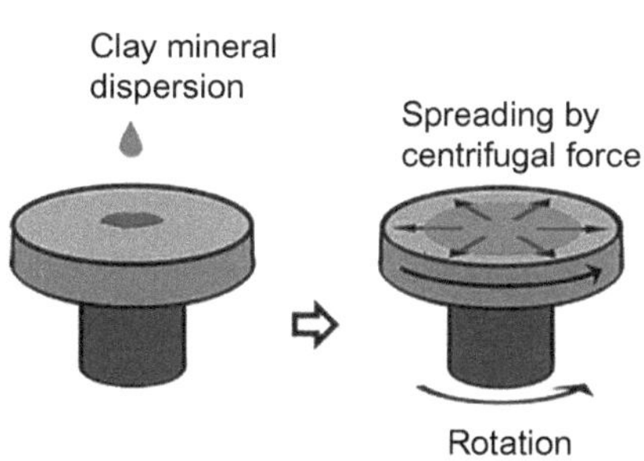

Fig. 12.2 Schematic diagram of the spin coating method.

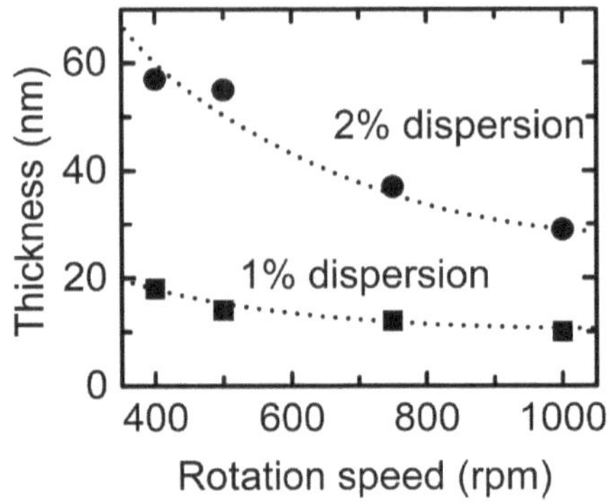

Fig. 12.3 Thickness as a function of rotation speed for spin coating films prepared from 1 and 2 mass% Lap dispersions.
Reproduced with permission from Schoonheydt, R.A., 2014. Functional hybrid clay mineral films. Appl. Clay Sci. 96, 9–21. Copyright 2014, Elsevier.

was two to three times thicker than that prepared from a 1 mass% dispersion, and the films became thinner with the increase in the rotational speed (400–1000 rotations per minute).

Spin coating films of Lap were immersed in a solution of an organic dye (rhodamine 6G, R6G), and molecular orientation of the dye intercalated in the interlayer spaces was determined. When the amount of R6G was very low at 0.1% of the CEC (cation exchange capacity), the dye molecules were present in the interlayer spaces as monomers with an orientation angle of 62 degrees from the normal to the film surface. As the amount increased, the dye molecules formed J-type dimers, H-type dimers, and finally high-order aggregates. The films in which R6G was intercalated were characterised by fluorescence spectroscopy using polarised beams (Martínez Martínez et al., 2005; López Arbeloa and Martínez Martínez, 2006a,b; Martínez et al., 2006). Similarly, spin coated films of clay minerals (fluorohectorite, Mt, and Lap) were immersed in an aqueous solution of a tetracationic porphyrin dye (TMPyP), and spectral analysis of the dye molecules in the films revealed that the pyridinium substituents were parallel to the clay mineral layers (Čeklovský et al., 2009).

Platinum electrodes were modified with clay mineral (Mt or kaolinite)/polyvinyl chloride films by spin coating. The modified electrodes were highly sensitive to Ag^+ with the detection limit of 1.00×10^{-10} mol L^{-1} (Issa et al., 2008). Glassy carbon electrodes modified with smectite films accumulated $[Ru(NH_3)_6]^{3+}$ cations, but not $[Fe(CN)_6]^{3-}$ anions; whereas those modified with films of propylammonium-functionalised smectite gave the opposite results. Clearly, the propylammonium-functionalised smectite carried an excess of propylammonium chloride, and the functionalised smectite was turned into an anion exchanger: Cl^- for $[Fe(CN)_6]^{3-}$. The accumulation on the smectite surfaces lowered the detection limits of the complex cations in the solutions (Tonle et al., 2004). The voltammetric waves for $[M(bpy)_3]^{2+}$($M = Fe$, Ru, and Os) were affected by thickness of clay mineral films on electrodes, and the mechanism was analysed by separating the current waves into two factors, electroactive concentration and effective diffusion in the clay mineral films (Song and Villemure, 2007).

By repeating the film preparation procedure, a multilayered film can be fabricated. Lap and mesoporous oxides (TiO_2 and SiO_2) were deposited alternately to prepare photonic films. The structural colours of the films were tunable by changing the thickness of the Lap and oxide layers. The films were applied to chemo-optical sensing devices for cationic molecules; the thickness change caused by adsorption of the cations on the Lap layers affected the structural colours of the films (Lotsch and Ozin, 2008a,b,c).

Recently, ultrathin films composed of closely packed discrete 2D material layers (graphene and metal oxide nanosheets) have been prepared by the spin coating method (Matsuba et al., 2017). It is expected that ultrathin films of discrete clay mineral layers will be prepared in a similar way.

12.4 LbL deposition

The surfaces of glass, silicon, mica, PET (poly(ethylene terephthalate)), and so forth are negatively charged. When these substrates are immersed in solutions of cationic polymers (whether they are organic- or biopolymers), the polymer molecules are adsorbed on the substrates electrostatically to form thin layers (Fig. 12.4). After being rinsed gently and dried, the substrate surfaces are immersed in dilute clay mineral dispersions. The negatively charged clay mineral layers are adsorbed on the positively charged surfaces of the polymer layers on the substrates. The substrate surfaces covered with the clay mineral layers are negatively charged. By repeating these procedures, multilayers of the alternately deposited layers of the cationic polymer and anionic clay mineral layers are prepared. One can prepare ultrathin films by this method without any special apparatus. In addition to the electrostatic interaction between the adjacent layers, hydrogen bonding (Huang et al., 2009) and cross-linking of a photo-reactive polymer (Vuillaume et al., 2002) were applied to stabilise the layer structures of the films.

The thicknesses of the films increase linearly with the number of deposition cycles (Van Duffel et al., 1999; Priolo et al., 2011). The thickness of each deposition layer and the surface roughness of the film were dependent on the polymer concentration in

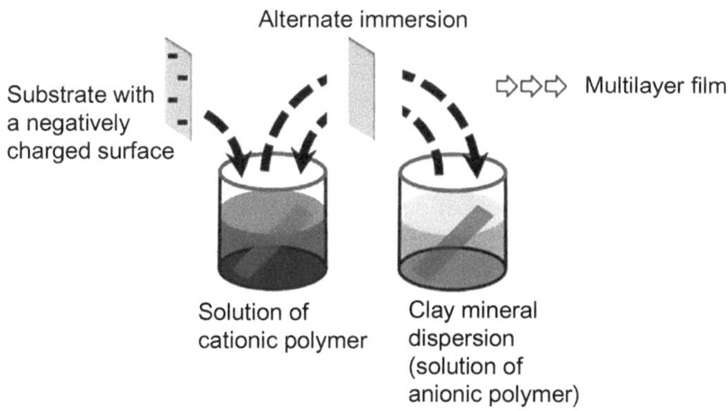

Fig. 12.4 Schematic diagram of an LbL deposition method.

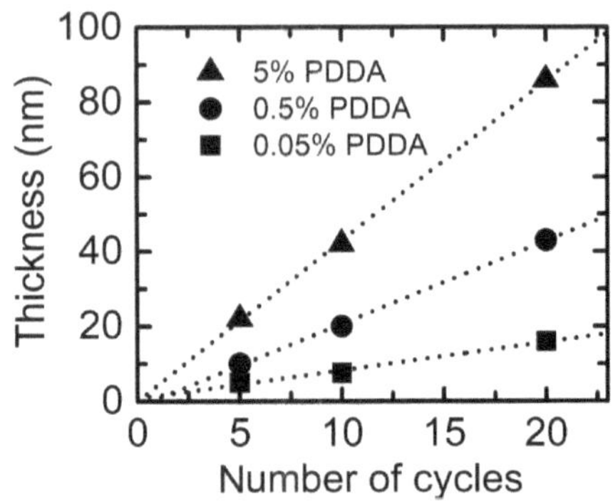

Fig. 12.5 Thickness against number of deposition cycles for Lap and PDDA (poly (diallyldimethylammonium chloride)) films prepared by the LbL deposition method. The Lap concentration in the dispersion was 0.05 mass% and the PDDA concentrations were 0.05, 0.5, and 5 mass%.
Reproduced with permission from Van Duffel, B., Schoonheydt, R.A., Grim, C.P.M., De Schryver, F.C., 1999. Multilayered clay films: atomic force microscopy study and modeling. Langmuir 15, 7520–7529. Copyright 1999, American Chemical Society.

the deposition solution (Fig. 12.5), on the pH and ionic strength of the solution, and on the size of the clay mineral layers (Van Duffel et al., 1999; Li et al., 2009, 2010; Priolo et al., 2010; Yang et al., 2010; Findenig et al., 2013). By introduction of clay mineral layers into films, some properties and functionalities were improved in comparison with polymer films: for example, hardness of the films (Kaushik et al., 2009; Li et al., 2009; Priolo et al., 2010; Humood et al., 2016), gas barrier properties (Jang et al., 2008; Priolo et al., 2010, 2011; Choi et al., 2012; Findenig et al., 2012; Holder et al., 2012; Laufer et al., 2012; Svagan et al., 2012; Hagen et al., 2014; Song et al., 2015), and flame-retardant properties (Laufer et al., 2012). For the purpose of optical device application, luminescent carbon nanodots (Dimos et al., 2017) and a tetra-cationic metalloporphyrin (Banik et al., 2015) were incorporated in Lap films. PET fabrics coated by this method with Mt and oligomeric silsesquioxane showed improvement in flame retardancy properties (Carosio et al., 2018).

Not only negatively charged clay mineral layers, but also positively charged LDH layers (or sheets) can be used as components of the films prepared by the LbL deposition. Reversible photoresponsive films were fabricated with LDH layers and an anionic polymer with an azobenzene moiety. The photoisomerisation of the azobenzene moiety induced by alternate UV and visible light irradiation caused changes in colour and morphologies of the films (Han et al., 2010). Films composed of an anionic polymer and Co–Al–LDH layers showed a magneto-optical effect (Liu et al., 2006), and those of Mg–Al–LDH layers indicated blue fluorescence with high photostability, which was attributed to their multiple quantum-well structures (Yan et al., 2009). Interestingly, in the LbL films, LDH layers were combined with functional nanomaterials such as quantum dots (CD) (Kameyama et al., 2009) and oxide nanosheets ($Ti_{0.91}O_2$ and $Ca_2Nb_3O_{10}$) (Li et al., 2007).

12.5 LB and LS methods

When a solution of water-insoluble amphiphilic molecules in a volatile organic solvent (e.g., chloroform and benzene) is dropped and spread at an air–water interface, the solvent evaporates and the amphiphilic molecules form a floating monomolecular film, which is called a Langmuir film (Gaines, 1966; Petty, 1996; Dynarowicz-Łątka et al., 2001). If, instead of water, an aqueous clay mineral dispersion is used as a subphase and a solution of an amphiphilic cation is spread at an air–clay mineral dispersion interface, the negatively charged clay mineral layers in the dispersion are adsorbed on a floating film of the amphiphilic cations electrostatically (Fig. 12.6) (Shimada et al., 2012). The resultant monolayer of the amphiphilic cations hybridised with the clay mineral layers is compressed by barriers, and then the film is transferred on a substrate repeatedly by vertical dipping (LB method) or by horizontal dipping (LS method); the procedure is exactly the same as that of the conventional LB and LS methods (Fig. 12.7A and B).

Different from the conventional LB and LS films, the hybrid LB and LS films of amphiphilic cations and clay mineral layers show some interesting properties. First, layered structures of the hybrid films are much more stable than those of conventional ones (Umemura et al., 2002, 2003; Toma et al., 2010). In the conventional films, attractive forces between the molecular layers are weak interactions, such as van der Waals interaction and hydrophilic and hydrophobic interactions. In the hybrid films, in addition to these interactions, electrostatic interaction between the amphiphilic cations and the clay mineral layers stabilises the layered structures. Furthermore, the size of the clay mineral layer, which is much larger than the size of the cation, helps to keep the layered structures.

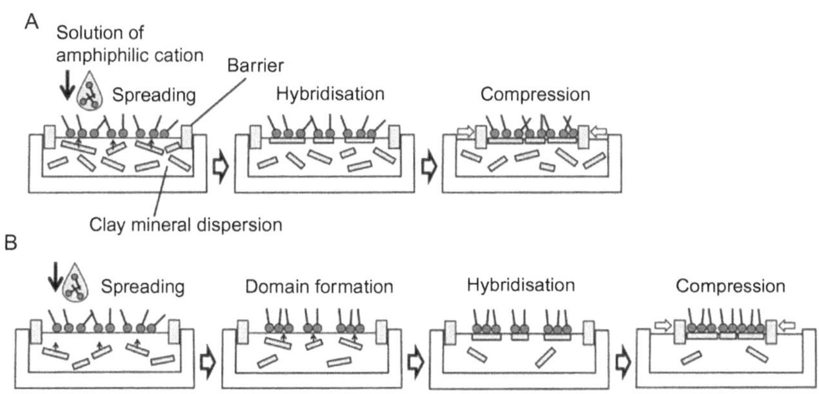

Fig. 12.6 Adsorption of clay mineral layers on a floating monolayer of amphiphilic cations. (A) When a clay mineral concentration is high, an adsorption rate of the clay mineral layers is high, and the density of the amphiphilic cation in the film is low. (B) When a clay mineral concentration is low, an adsorption rate is low, and the clay mineral layers are adsorbed after domain formation of the amphiphilic cations. As a result, the density of the amphiphilic cation in the film is high.

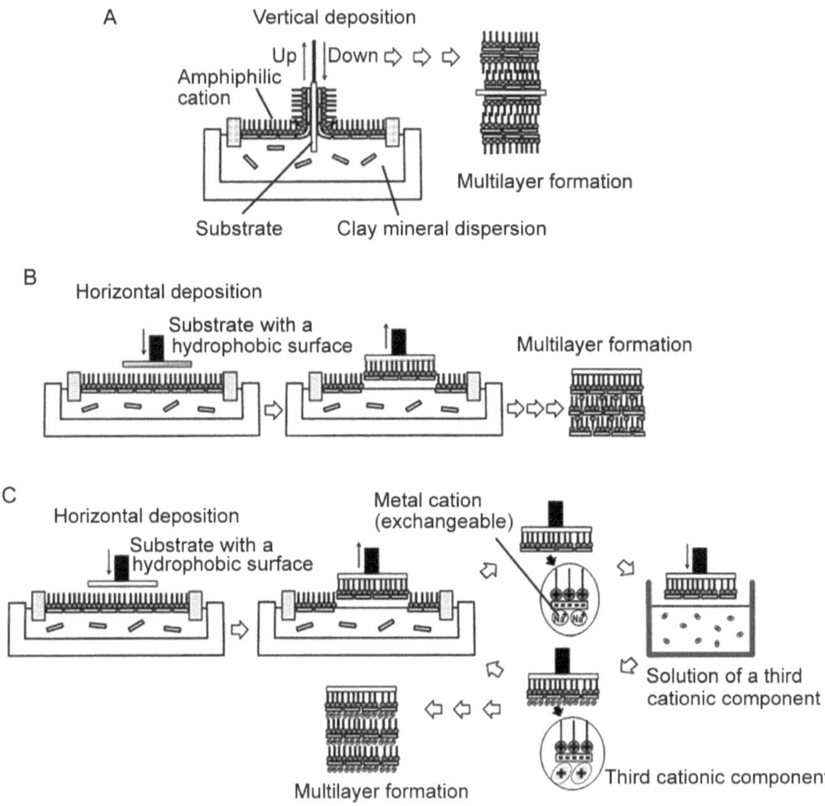

Fig. 12.7 Schematic diagrams of deposition methods and layer structures of multilayers. (A) Vertical deposition method. (B) Horizontal deposition method. (C) Horizontal deposition method for a three-component film.

Second, a density of amphiphilic cations on a clay mineral layer (or a cation's density in a hybrid multilayer film) is controllable by changing the concentration of a clay mineral in a dispersion (Fig. 12.6) (Umemura et al., 2001b; Umemura and Shinohara, 2005; Shimada et al., 2012). The higher the clay mineral concentration in the dispersion is, the lower the density of the amphiphilic cations in the hybrid film is (Fig. 12.8A and B). In a dilute clay mineral dispersion, the adsorption rate of clay mineral layer is low (Fig. 12.6B). When a solution of an amphiphile is dropped at an air–dispersion interface, it spreads over the interface, and after the solvent evaporation, the amphiphilic molecules gather and form their domains. Clay mineral layers in a dilute dispersion are adsorbed on a floating monolayer of amphiphilic cations after the cations form their domains at the air-dispersion interface. In the resultant hybrid film, the density of the amphiphilic cation is high (Fig. 12.6B). In contrast, the adsorption rate of the clay mineral layer is high in a concentrated dispersion (in which the

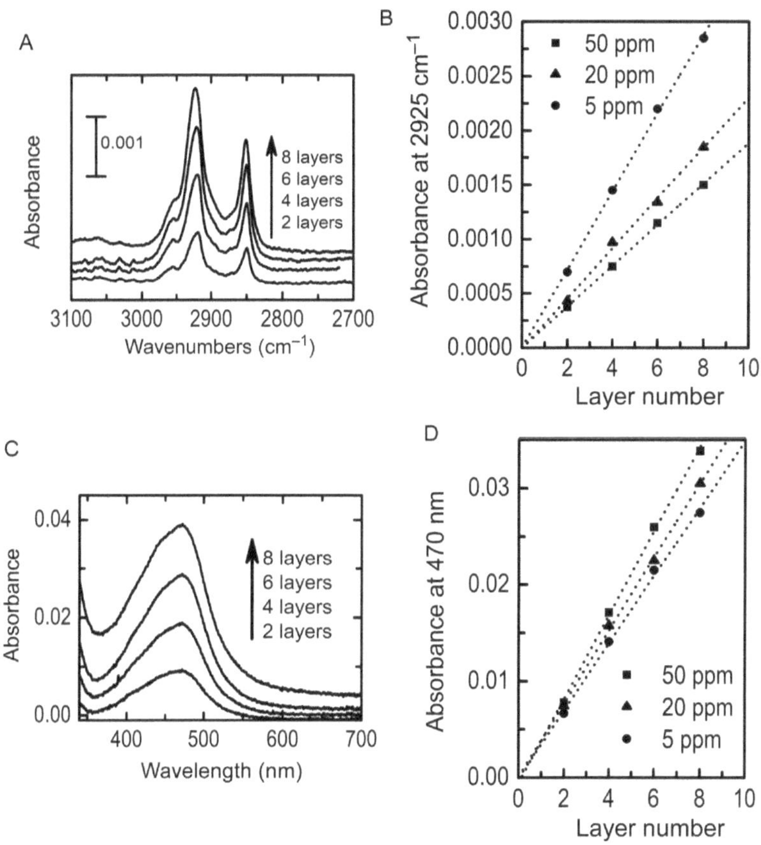

Fig. 12.8 Formation of three-component films prepared by spreading $ODAH^+$ (octadecylammonium cation) on aqueous Mt dispersions at 5, 20, and 50 ppm (ppm = mg dm^{-3}) and then immersing the transferred films in an aqueous $[Ru(dpp)_3]Cl_2$ (dpp = 4,7-diphenyl-1,10-phenanthroline) solution. (A) The IR spectra for the three-component films prepared from the 5-ppm dispersion. The absorption peaks are attributed to the C—H stretching modes of $ODAH^+$. (B) Absorbance at 2925 cm^{-1} as a function of layer number for the three-component films prepared from the 5-, 20-, and 50-ppm dispersions. (C) The visible spectra for the three-component films prepared from the 5-ppm dispersion. The absorptions are attributed to the MLCT (metal to ligand charge transfer) bands of $[Ru(dpp)_3]^{2+}$. (D) Absorbance at 470 nm as a function of layer number for the three-component films prepared from the 5-, 20-, and 50-ppm dispersions.

clay mineral layers are not stacked yet), so that the clay mineral layers are adsorbed on the floating monolayer of the amphiphilic cations before the cations' domains are formed (Fig. 12.6A). The density of the amphiphilic cation in the hybrid film is, therefore, low.

Third, unlike conventional LB and LS films, clay mineral hybrid films are prepared using not only water-insoluble amphiphilic cations, but also water-soluble ones as spreading reagents (Umemura et al., 2001a,b, 2003; Umemura and Shinohara, 2005). When a solution of water-soluble amphiphilic cations is spread at an air-aqueous clay mineral dispersion interface, the cations dissolve from the interface into the dispersion slowly. Some of the cations are adsorbed by the clay mineral layers at the interface before dissolution, and the clay mineral layers hybridised with a number of the amphiphilic cations float at the interface stably.

Fourth, one can add another cationic component in the film; the three-component film is prepared by immersing a hybrid film (composed of clay mineral layers and amphiphilic cations) in a solution of a third cationic component (Figs 12.7C and 12.8) (Umemura, 2002; Umemura et al., 2002). The negative charge density of the clay mineral layers surpasses the positive charge density of the amphiphilic cations in the floating monolayer. When the floating monolayer is transferred on a substrate, some metal cations, such as Na^+, are adsorbed on the outer surface (the other side of the amphiphilic cations) of the transferred monolayer to compensate for the excess negative charges. Because these Na^+ cations are exchangeable, the third cationic component is adsorbed on the film surface by cation exchange when the film surface is immersed in the solution of the third component. The density of the third cationic component in the film is dependent on the clay mineral concentration in the dispersion, because, as described earlier, the density of the amphiphilic cations in the film can be controlled by changing the clay mineral concentration (Fig. 12.8) (Umemura, 2002).

Initially, the preparation of clay mineral films by the LB method was performed by spreading a clay mineral dispersion in an organic solvent at an air–water interface (Inukai et al., 1994; Kotov et al., 1994). The clay minerals (hectorite and saponite) were cation-exchanged with alkylammonium cations, and the surfaces of the clay mineral layers became so hydrophobic that the clay mineral layers floated at the air-water interface.

Orientation of clay mineral layers in the films that were prepared by spreading amphiphilic cations at air–dispersion interfaces were examined by polarised attenuated total reflection Fourier transform infrared (ATR–FTIR) spectroscopy (Ras et al., 2003), and by in-plane X-ray diffraction (Takahashi et al., 2002). Both of the investigations revealed that the clay mineral layers were highly parallel to the substrates.

Conformation and orientation of amphiphilic dye cations in the films were dependent on the clay mineral concentration in the dispersion, and the density of dye cations in the film (Ras et al., 2003, 2004a,b; Chakraborty et al., 2014, 2015; Debnath et al., 2017; Saha et al., 2017). For example, octadecyl rhodamine B (RhB18) was spread at an air–saponite dispersion interface, and the RhB18 molecules were present in the resultant film as monomers and dimers. The ATR–FTIR spectra of the films indicated that both the phenyl group and xanthene group in RhB18 were oriented parallel to the clay mineral surface (and to the substrate). Such conformation was induced by the presence of the clay mineral layers (Ras et al., 2004a).

Amphiphilic metal complex cations with long hydrocarbon chains were also used as spreading reagents at interfaces of the air and clay mineral dispersions (Tamura et al., 1999; Umemura et al., 2002; Yoshida et al., 2006; Chang et al., 2009). The hybrid films were transferred on electrodes to modify their surfaces. Electroanalytical characterisation of the modified electrodes indicated that the clay mineral layers in the films acted as barriers against electron transfer (Yoshida et al., 2006). When amphiphilic Ru(II) complex cations were incorporated in clay mineral films, the electrodes modified with the films worked as photoelectrocatalysts for oxidation of a mononucleotide (guanosine 5′-monophosphate) (Chang et al., 2009), and as oxidation catalysts for nitrite anion (NO_2^-) (Yoshida et al., 2006).

Other functional materials such as donor-(π-electron system)–acceptor zwitterionic molecules (Ogata et al., 2003), enzymes (lysozyme and polyphenol oxidase) (Mbouguen et al., 2007; Miao et al., 2010), and ionic liquids (Ye et al., 2013) were combined with clay mineral layers by the LB method. These film systems were studied with the aim of preparing nonlinear optical devices, electrode modifiers, and sensors.

Layer structures of the three-component films that are composed of the layers of amphiphilic cations, clay mineral layers, and third cationic molecules are intrinsically noncentrosymmetric, even if molecular orientation in each layer is random (Fig. 12.7C). For the appearance of second harmonic generation (one of the nonlinear optical effects) and pyroelectric current generation, noncentrosymmetry is necessary. The three-component clay mineral films showed these properties (Umemura, 2002; Umemura et al., 2002; Hirahara and Umemura, 2015). Bulky molecules and clusters such as octanuclear nickel(II) complex cation and Buckminster fullerene (C_{60}) were successfully incorporated in three-component films as third components (Toma et al., 2011; Gengler et al., 2012a).

Ultrathin layers of Prussian Blue and its analogues were formed in three-component films; amphiphilic alkylammonium cations or amphiphilic azobenzene cations were hybridised with clay mineral layers at an air–dispersion interface, and the hybrid film transferred on a substrate was immersed successively in an aqueous $FeCl_2$ (or $CoCl_2$) solution and in an aqueous $K_3[Fe(CN)_6]$ solution to form the Prussian Blue layer. The films showed photoinduced electron transfer between the metal cations through the bridging cyanide, magnetic anisotropy, and photoinduced magnetisation effect (Yamamoto et al., 2004, 2005; Yamamoto and Einaga, 2007; Gengler et al., 2012b).

Sepiolite, a fibrous clay mineral, possesses rectangular channels along the fibre direction, and has cation exchangeability. Using a sepiolite dispersion, hybrid films were prepared in a similar way and, interestingly, the fibres in the films were aligned perpendicular to the compression direction. The hybrid film was transferred on a substrate and methylene blue cations were adsorbed on the film. The methylene blue cations were present in the fibre channels as monomers and, as a result, the cations were oriented along the fibre direction spontaneously (Umemura et al., 2009).

Instead of negatively charged clay mineral layers, positively charged LDH layers were applied to form hybrid films by the LB method. Amphiphilic Ru(II) complex anions (He et al., 2001; Xiang He et al., 2001), dendrimers (Costa and Imae, 2004), and dyes (Wang et al., 2008) were incorporated in the LDH films. A solution of stearic (octadecanoic) acid was spread on a surface of the aqueous solution containing $Mg(NO_3)_2$ and $Al(NO_3)_3$, and thin crystals of Mg–Al–LDH were formed on the floating stearate monolayer (He et al., 2002).

12.6 Other methods

12.6.1 Electrophoretic deposition

When an electric potential is applied in a clay mineral dispersion, the charged clay mineral layers move to the electrode. Using this phenomenon (electrophoresis), one can deposit clay mineral layers on an electrode to form a uniform and continuous film with several nanometres of thickness. The thickness is controlled easily by deposition time, and the film condition is dependent on the applied potential and on the clay mineral concentration and pH in the dispersion (Shirtcliffe, 1999).

Metal complex cations were adsorbed on films of Mt and kaolinite, and the voltammetric behaviour of the cations was affected by the film thickness (Song and Villemure, 1999, 2007). Films of LDH were prepared in its dispersion by this method (Kim et al., 2008). It is interesting that Zn–Al LDH films were formed on electrodes in aqueous solutions containing $Zn(NO_3)_2$ and $Al(NO_3)_3$; the Zn–Al–LDH layers crystallised near the electrodes, and deposited on them (Yarger et al., 2008).

12.6.2 Chemical deposition (silane coupling)

Silane coupling agents are often used to modify surfaces. This reaction can be used for chemical immobilisation of clay mineral layers on substrates. One example is immobilisation of Mt layers pillared with Al_{13} Keggin ions on SnO_2 and Pt electrodes. Their electrocatalytic behaviour was investigated (Rong et al., 1990; Rong and Mallouk, 1993).

12.6.3 Suction filtration

When a clay mineral dispersion is poured in a suction funnel with a membrane filter and filtered by suction, the clay mineral layers are left on the filter. After being dried, a clay mineral film is obtained by being peeled off from the filter. The films prepared by this method (micrometres thick) are transparent, uniform, and self-standing, so that they are suitable for optical device application. Actually, dispersions of synthetic saponite mixed with some dyes were filtered, and the resultant films showed optical functionalities such as second harmonic generation and two-photon absorption (Takagi et al., 2002a; Suzuki et al., 2008, 2011; Kawamata et al., 2010).

12.7 Concluding remarks

In this chapter, several methods to prepare clay mineral films have been described. Each method has its peculiar features. According to these features, one can choose one of the methods.

The LbL deposition method and the LB (LS) method, in particular, enable the design and preparation of clay mineral films at the molecular and clay mineral layer levels.

Functional molecules can be incorporated in clay mineral films and, in some cases, the functionalities are improved by controlling the orientation and distribution of the functional molecules in each layer.

In the future, these methods will be developed, and other methods will be added. Hopefully, it will be possible to design alignment at two levels in the film: two-dimensional alignment of different molecules at the nanometre level between the clay mineral layers, and two-dimensional alignment of the clay mineral layers themselves.

References

Alexandre, M., Dubois, P., 2000. Polymer-layered silicate nanocomposites: preparation, properties and uses of a new class of materials. Mater. Sci. Eng. R Rep. 28, 1–63.

Alix, S., Follain, N., Tenn, N., Alexandre, B., Bourbigot, S., Soulestin, J., Marais, S., 2012. Effect of highly exfoliated and oriented organoclays on the barrier properties of polyamide 6 based nanocomposites. J. Phys. Chem. C 116, 4937–4947.

Banik, S., Bhattacharjee, J., Hussain, S.A., Bhattacharjee, D., 2015. Clay induced aggregation of a tetra-cationic metalloporphyrin in layer by layer self assembled film. J. Phys. Chem. Solid 87, 128–135.

Bébin, P., Caravanier, M., Galiano, H., 2006. Nafion®/clay-SO_3H membrane for proton exchange membrane fuel cell application. J. Membr. Sci. 278, 35–42.

Bujdák, J., Czímerová, A., Arbeloa, F.L., 2011. Two-step resonance energy transfer between dyes in layered silicate films. J. Colloid Interface Sci. 364, 497–504.

Carosio, F., Di Pierro, A., Alongi, J., Fina, A., Saracco, G., 2018. Controlling the melt dripping of polyester fabrics by tuning the ionic strength of polyhedral oligomeric silsesquioxane and sodium montmorillonite coatings assembled through layer by layer. J. Colloid Interface Sci. 510, 142–151.

Carrier, B., Wang, L., Vandamme, M., Pellenq, R.J.M., Bornert, M., Tanguy, A., Van Damme, H., 2013. ESEM study of the humidity-induced swelling of clay film. Langmuir 29, 12823–12833.

Čeklovský, A., Czímerová, A., Lang, K., Bujdák, J., 2009. Layered silicate films with photochemically active porphyrin cations. Pure Appl. Chem. 81, 1385–1396.

Chakraborty, S., Bhattacharjee, D., Hussain, S.A., 2014. Formation and control of excimer of a coumarin derivative in Langmuir–Blodgett films. J. Lumin. 145, 824–831.

Chakraborty, S., Bhattacharjee, D., Soda, H., Tominaga, M., Suzuki, Y., Kawamata, J., Hussain, S.A., 2015. Temperature and concentration dependence of J-aggregate of a cyanine dye in a Laponite film fabricated by Langmuir–Blodgett technique. Appl. Clay Sci. 104, 245–251.

Chang, X., Wang, S., Lin, D., Guan, W., Zhou, H., Huang, S., 2009. Photoelectrocatalytic oxidation of GMP on an ITO electrode modified with clay/$[Ru(phen)_2(dC18bpy)]^{2+}$ hybrid film. Sci. Chin. B Chem. 52, 318–324.

Choi, J.H., Park, Y.W., Park, T.H., Song, E.H., Lee, H.J., Kim, H., Shin, S.J., Lau Chun Fai, V., Ju, B.-K., 2012. Fuzzy nanoassembly of polyelectrolyte and layered clay multicomposite toward a reliable gas barrier. Langmuir 28, 6826–6831.

Chu, C.-Y., Chen, M.-H., Wu, M.-L., Chen, H.-L., Chiu, Y.-T., Chen, S.-M., Huang, C.-H., 2014. Hierarchical structure and crystal orientation in poly(ethylene oxide)/clay nanocomposite films. Langmuir 30, 2886–2895.

Costa, A.S., Imae, T., 2004. Morphological investigation of hybrid Langmuir–Blodgett films of arachidic acid with a hydrotalcite/dendrimer nanocomposite. Langmuir 20, 8865–8869.

Debnath, P., Chakraborty, S., Deb, S., Nath, J., Dey, B., Bhattacharjee, D., Soda, H., Tominaga, M., Suzuki, Y., Kawamata, J., Hussain, S.A., 2017. Effect of nano clay Laponite on stability of SHG active J-aggregate of a thiacyanine dye onto LB films. Appl. Clay Sci. 147, 105–116.

Dimos, K., Arcudi, F., Kouloumpis, A., Koutselas, I.B., Rudolf, P., Gournis, D., Prato, M., 2017. Top-down and bottom-up approaches to transparent, flexible and luminescent nitrogen-doped carbon nanodot-clay hybrid films. Nanoscale 9, 10256–10262.

Dynarowicz-Łątka, P., Dhanabalan, A., Oliveira, O.N., 2001. Modern physicochemical research on Langmuir monolayers. Adv. Colloid Interface Sci. 91, 221–293.

Findenig, G., Leimgruber, S., Kargl, R., Spirk, S., Stana-Kleinschek, K., Ribitsch, V., 2012. Creating water vapor barrier coatings from hydrophilic components. ACS Appl. Mater. Interfaces 4, 3199–3206.

Findenig, G., Kargl, R., Stana-Kleinschek, K., Ribitsch, V., 2013. Interaction and structure in polyelectrolyte/clay multilayers: a QCM-D study. Langmuir 29, 8544–8553.

Gaines, G.L., 1966. Insoluble Monolayers at Liquid-Gas Interfaces. Interscience Publishers, pp. 136–207.

Gengler, R.Y.N., Gournis, D., Aimon, A.H., Toma, L.M., Rudolf, P., 2012a. The molecularly controlled synthesis of ordered bi-dimensional C 60 arrays. Chemistry 18, 7594–7600.

Gengler, R.Y.N., Toma, L.M., Pardo, E., Lloret, F., Ke, X., Van Tendeloo, G., Gournis, D., Rudolf, P., 2012b. Prussian blue analogues of reduced dimensionality. Small 8, 2532–2540.

Guan, W., Zhou, W., Lu, J., Lu, C., 2015. Luminescent films for chemo- and biosensing. Chem. Soc. Rev. 44, 6981–7009.

Hagen, D.A., Saucier, L., Grunlan, J.C., 2014. Controlling effective aspect ratio and packing of clay with pH for improved gas barrier in nanobrick wall thin films. ACS Appl. Mater. Interfaces 6, 22914–22919.

Han, J., Yan, D., Shi, W., Ma, J., Yan, H., Wei, M., Evans, D.G., Duan, X., 2010. Layer-by-layer ultrathin films of azobenzene-containing polymer/layered double hydroxides with reversible photoresponsive behavior. J. Phys. Chem. B 114, 5678–5685.

He, J.X., Kobayashi, K., Takahashi, M., Villemure, G., Yamagishi, A., 2001. Preparation of hybrid films of an anionic Ru(II) cyanide polypyridyl complex with layered double hydroxides by the Langmuir–Blodgett method and their use as electrode modifiers. Thin Solid Films 397, 255–265.

He, J.X., Yamashita, S., Jones, W., Yamagishi, A., 2002. Templating effects of stearate monolayer on formation of mg – Al-hydrotalcite. Langmuir 18, 1580–1586.

Herrera Alonso, R., Estevez, L., Lian, H., Kelarakis, A., Giannelis, E.P., 2009. Nafion–clay nanocomposite membranes: morphology and properties. Polymer 50, 2402–2410.

Hirahara, M., Umemura, Y., 2015. Fabrication of three-layer-component organoclay hybrid films with reverse deposition orders by a modified Langmuir-Schaefer technique and their pyroelectric currents measured by a noncontact method. Langmuir 31, 8346–8353.

Holder, K.M., Priolo, M.A., Secrist, K.E., Greenlee, S.M., Nolte, A.J., Grunlan, J.C., 2012. Humidity-responsive gas barrier of hydrogen-bonded polymer–clay multilayer thin films. J. Phys. Chem. C 116, 19851–19856.

Huang, S., Cen, X., Peng, H., Guo, S., Wang, W., Liu, T., 2009. Heterogeneous ultrathin films of poly(vinyl alcohol)/layered double hydroxide and montmorillonite nanosheets via layer-by-layer assembly. J. Phys. Chem. B 113, 15225–15230.

Humood, M., Chowdhury, S., Song, Y., Tzeng, P., Grunlan, J.C., Polycarpou, A.A., 2016. Nanomechanical behavior of high gas barrier multilayer thin films. ACS Appl. Mater. Interfaces 8, 11128–11138.

Inukai, K., Hotta, Y., Taniguchi, M., Tomura, S., Yamagishi, A., 1994. Formation of a clay monolayer at an air-water interface. J. Chem. Soc. Chem. Commun. 959.

Issa, A.A., Al-Degs, Y.S., Al-Rabady, N.A., 2008. Deposition of two natural clays on a Pt surface using potentiostatic and spin-coating techniques: a comparative study. Clay Miner. 43, 501–510.

Jang, W.-S., Rawson, I., Grunlan, J.C., 2008. Layer-by-layer assembly of thin film oxygen barrier. Thin Solid Films 516, 4819–4825.

Johnston, C.T., 2010. Probing the nanoscale architecture of clay minerals. Clay Miner. 45, 245–279.

Kameyama, T., Okazaki, K.-i., Takagi, K., Torimoto, T., 2009. Stacked-structure-dependent photoelectrochemical properties of CdS nanoparticle/layered double hydroxide (LDH) nanosheet multilayer films prepared by layer-by-layer accumulation. Phys. Chem. Chem. Phys. 11, 5369–5376.

Kaushik, A.K., Podsiadlo, P., Qin, M., Shaw, C.M., Waas, A.M., Kotov, N.A., Arruda, E.M., 2009. The role of nanoparticle layer separation in the finite deformation response of layered polyurethane-clay nanocomposites. Macromolecules 42, 6588–6595.

Kawamata, J., Suzuki, Y., Tenma, Y., 2010. Fabrication of clay mineral–dye composites as nonlinear optical materials. Philos. Mag. 90, 2519–2527.

Kim, T.W., Sahimi, M., Tsotsis, T.T., 2008. Preparation of hydrotalcite thin films using an electrophoretic technique. Ind. Eng. Chem. Res. 47, 9127–9132.

Kotov, N.A., Meldrum, F.C., Fendler, J.H., Tombacz, E., Dekany, I., 1994. Spreading of clay organocomplexes on aqueous solutions: construction of Langmuir-Blodgett clay organocomplex multilayer films. Langmuir 10, 3797–3804.

Kuila, B.K., Nandi, A.K., 2004. Physical, mechanical, and conductivity properties of poly(3-hexylthiophene)–montmorillonite clay nanocomposites produced by the solvent casting method. Macromolecules 37, 8577–8584.

Laufer, G., Kirkland, C., Cain, A.A., Grunlan, J.C., 2012. Clay–chitosan nanobrick walls: completely renewable gas barrier and flame-retardant nanocoatings. ACS Appl. Mater. Interfaces 4, 1643–1649.

Li, L., Ma, R., Ebina, Y., Fukuda, K., Takada, K., Sasaki, T., 2007. Layer-by-layer assembly and spontaneous flocculation of oppositely charged oxide and hydroxide nanosheets into inorganic sandwich layered materials. J. Am. Chem. Soc. 129, 8000–8007.

Li, Y.-C., Schulz, J., Grunlan, J.C., 2009. Polyelectrolyte/nanosilicate thin-film assemblies: influence of pH on growth, mechanical behavior, and flammability. ACS Appl. Mater. Interfaces 1, 2338–2347.

Li, Y.-C., Schulz, J., Mannen, S., Delhom, C., Condon, B., Chang, S., Zammarano, M., Grunlan, J.C., 2010. Flame retardant behavior of polyelectrolyte – clay thin film assemblies on cotton fabric. ACS Nano 4, 3325–3337.

Liu, Z., Ma, R., Osada, M., Iyi, N., Ebina, Y., Takada, K., Sasaki, T., 2006. Synthesis, anion exchange, and delamination of Co – Al layered double hydroxide: assembly of the exfoliated nanosheet/polyanion composite films and magneto-optical studies. J. Am. Chem. Soc. 128, 4872–4880.

López Arbeloa, F., Martínez Martínez, V., 2006a. New fluorescent polarization method to evaluate the orientation of adsorbed molecules in uniaxial 2D layered materials. J. Photochem. Photobiol. A Chem. 181, 44–49.

López Arbeloa, F., Martínez Martínez, V., 2006b. Orientation of adsorbed dyes in the interlayer space of clays. 2 fluorescence polarization of rhodamine 6G in laponite films. Chem. Mater. 18, 1407–1416.

Lotsch, B.V., Ozin, G.A., 2008a. All-clay photonic crystals. J. Am. Chem. Soc. 130, 15252–15253.

Lotsch, B.V., Ozin, G.A., 2008b. Photonic clays: a new family of functional 1D photonic crystals. ACS Nano 2, 2065–2074.

Lotsch, B.V., Ozin, G.A., 2008c. Clay Bragg stack optical sensors. Adv. Mater. 20, 4079–4084.

Ma, R., Sasaki, T., 2010. Nanosheets of oxides and hydroxides: ultimate 2D charge-bearing functional crystallites. Adv. Mater. 22, 5082–5104.

Maes, A., Schoonheydt, R.A., Cremers, A., Uytterhoeven, J.B., 1980. Spectroscopy of bis(ethylenediamine)copper(2+) on clay surfaces. Surface and charge-density effects. J. Phys. Chem. 84, 2795–2799.

Martínez, V., Salleres, S., Bañuelos, J., Arbeloa, F.L., 2006. Application of fluorescence with polarized light to evaluate the orientation of dyes adsorbed in layered materials. J. Fluoresc. 16, 233–240.

Martínez Martínez, V., López Arbeloa, F., Bañuelos Prieto, J., López Arbeloa, I., 2005. Orientation of adsorbed dyes in the interlayer space of clays. 1. Anisotropy of rhodamine 6G in Laponite films by vis-absorption with polarized light. Chem. Mater. 17, 4134–4141.

Matsuba, K., Wang, C., Saruwatari, K., Uesusuki, Y., Akatsuka, K., Osada, M., Ebina, Y., Ma, R., Sasaki, T., 2017. Neat monolayer tiling of molecularly thin two-dimensional materials in 1 min. Sci. Adv. 3, e1700414.

Mbouguen, J.K., Ngameni, E., Walcarius, A., 2007. Quaternary ammonium functionalized clay film electrodes modified with polyphenol oxidase for the sensitive detection of catechol. Biosens. Bioelectron. 23, 269–275.

Miao, S., Leeman, H., De Feyter, S., Schoonheydt, R.A., 2010. Three-component Langmuir-Blodgett films consisting of surfactant, clay mineral, and lysozyme: construction and characterization. Chemistry 16, 2461–2469.

Miao, S., Qiu, Z., Cui, P., 2012. Clay-protein ultrathin films: design and bio-catalytic performance study. Sci. China Chem. 55, 1842–1855.

Mousty, C., 2004. Sensors and biosensors based on clay-modified electrodes—new trends. Appl. Clay Sci. 27, 159–177.

Nam, H.-J., Ebina, T., Mizukami, F., 2009. Formability and properties of self-standing clay film by montmorillonite with different interlayer cations. Colloids Surf. A Physicochem. Eng. Asp. 346, 158–163.

Ogata, Y., Kawamata, J., Chong, C.-H., Yamagishi, A., Saito, G., 2003. Structural features of a clay film hybridized with a zwiterionic molecule as analyzed by second harmonic generation behavior. Clays Clay Miner. 51, 181–185.

Oliveira, R.L.D., Barud, H.D.S., De Salvi, D.T.B., Perotti, G.F., Ribeiro, S.J.L., Constantino, V.R.L., 2015. Transparent organic–inorganic nanocomposites membranes based on carboxymethylcellulose and synthetic clay. Ind. Crop Prod. 69, 415–423.

Petty, M.C., 1996. Langmuir-Blodgett Films: An Introduction. Cambridge University Press, pp. 12–38.

Priolo, M.A., Gamboa, D., Grunlan, J.C., 2010. Transparent clay – polymer nano brick wall assemblies with tailorable oxygen barrier. ACS Appl. Mater. Interfaces 2, 312–320.

Priolo, M.A., Holder, K.M., Gamboa, D., Grunlan, J.C., 2011. Influence of clay concentration on the gas barrier of clay–polymer nanobrick wall thin film assemblies. Langmuir 27, 12106–12114.

Ras, R.H.A., Johnston, C.T., Franses, E.I., Ramaekers, R., Maes, G., Foubert, P., De Schryver, F.C., Schoonheydt, R.A., 2003. Polarized infrared study of hybrid Langmuir–Blodgett monolayers containing clay mineral nanoparticles. Langmuir 19, 4295–4302.

Ras, R.H.A., Nemeth, J., Johnston, C.T., Dekany, I., Schoonheydt, R.A., 2004a. Orientation and conformation of octadecyl rhodamine B in hybrid Langmuir–Blodgett monolayers containing clay minerals. Phys. Chem. Chem. Phys. 6, 5347–5352.

Ras, R.H.A., Nemeth, J., Johnston, C.T., DiMasi, E., Dekany, I., Schoonheydt, R.A., 2004b. Hybrid Langmuir–Blodgett monolayers containing clay minerals: effect of clay concentration and surface charge density on the film formation. Phys. Chem. Chem. Phys. 6, 4174–4184.

Ras, R.H.A., Umemura, Y., Johnston, C.T., Yamagishi, A., Schoonheydt, R.A., 2007. Ultrathin hybrid films of clay minerals. Phys. Chem. Chem. Phys. 9, 918–932.

Rhee, C.H., Kim, H.K., Chang, H., Lee, J.S., 2005. Nafion/sulfonated montmorillonite composite: a new concept electrolyte membrane for direct methanol fuel cells. Chem. Mater. 17, 1691–1697.

Rong, D., Mallouk, T.E., 1993. Chemical gating of a molecular bilayer rectifier at clay-modified electrodes. Inorg. Chem. 32, 1454–1459.

Rong, D., Kim II, Y., Mallouk, T.E., 1990. Electrochemistry and photoelectrochemistry of pillared-clay-modified electrodes. Inorg. Chem. 29, 1531–1535.

Saha, M., Shil, A., Hussain, S.A., Bhattacharjee, D., 2017. Clay induced changes in the aggregation pattern of Safranine-O in hybrid Langmuir-Blogdgett (LB) films. J. Photochem. Photobiol. A: Chem. 348, 199–208.

Schoonheydt, R.A., 2014. Functional hybrid clay mineral films. Appl. Clay Sci. 96, 9–21.

Schoonheydt, R.A., Pelgrims, J., 1983. Preparation, spectroscopy and reaction with oxygen of $[Co(en)_2]^{2+}$ on the surface of hectorite. J. Chem. Soc. Faraday Trans. 2 79, 1169–1180.

Schoonheydt, R.A., Velghe, F., Uytterhoeven, J.B., 1979. Characterization of $[Ni(en)_x]^{2+}$ ($x = 1, 2, 3$; en = ethylenediamine) on the surface of montmorillonites. Inorg. Chem. 18, 1842–1847.

Shi, W., Wei, M., Lu, J., Li, F., He, J., Evans, D.G., Duan, X., 2008. Molecular orientation and fluorescence studies on naphthalene acetate intercalated Zn_2Al layered double hydroxide. J. Phys. Chem. C 112, 19886–19895.

Shikinaka, K., Aizawa, K., Murakami, Y., Osada, Y., Tokita, M., Watanabe, J., Shigehara, K., 2012. Structural and mechanical properties of Laponite–PEG hybrid films. J. Colloid Interface Sci. 369, 470–476.

Shimada, T., Yamada, H., Umemura, Y., 2012. Surface potential studies on adsorption processes of clay nanosheets onto a floating molecular film of an amphiphilic alkylammonium cation. J. Phys. Chem. B 116, 4484–4491.

Shirtcliffe, N., 1999. Deposition of clays onto a rotating, electrochemical, quartz crystal microbalance. Colloids Surf. A Physicochem. Eng. Asp. 155, 277–285.

Song, C., Villemure, G., 1999. Preparation of clay-modified electrodes by electrophoretic deposition of clay films. J. Electroanal. Chem. 462, 143–149.

Song, C., Villemure, G., 2007. Effect of decreasing film thickness on the electrochemical responses of cations adsorbed in clay-modified electrodes. Electrochim. Acta 52, 6509–6516.

Song, E.-H., Kang, B.-H., Kim, T.-Y., Lee, H.-J., Park, Y.-W., Kim, Y.-C., Ju, B.-K., 2015. Highly oriented gold/nanoclay–polymer nanocomposites for flexible gas barrier films. ACS Appl. Mater. Interfaces 7, 4778–4783.

Suzuki, Y., Kirakawa, S., Sakamoto, Y., Kawamata, J., Kamada, K., Ohta, K., 2008. Hybrid films consisting of a clay and a diacetylenic, two-photon absorptive dye. Clays Clay Miner. 56, 487–493.

Suzuki, Y., Tenma, Y., Nishioka, Y., Kamada, K., Ohta, K., Kawamata, J., 2011. Efficient two-photon absorption materials consisting of cationic dyes and clay minerals. J. Phys. Chem. C 115, 20653–20661.

Svagan, A.J., Åkesson, A., Cárdenas, M., Bulut, S., Knudsen, J.C., Risbo, J., Plackett, D., 2012. Transparent films based on PLA and montmorillonite with tunable oxygen barrier properties. Biomacromolecules 13, 397–405.

Takagi, S., Shimada, T., Yui, T., Inoue, H., 2001. High density adsorption of porphyrins onto clay layer without aggregation: characterization of smectite-cationic porphyrin complex. Chem. Lett. 30, 128–129.

Takagi, S., Shimada, T., Eguchi, M., Yui, T., Yoshida, H., Tryk, D.A., Inoue, H., 2002a. High-density adsorption of cationic porphyrins on clay layer surfaces without aggregation: the size-matching effect. Langmuir 18, 2265–2272.

Takagi, S., Tryk, D.A., Inoue, H., 2002b. Photochemical energy transfer of cationic porphyrin complexes on clay surface. J. Phys. Chem. B 106, 5455–5460.

Takagi, S., Eguchi, M., Tryk, D.A., Inoue, H., 2006. Porphyrin photochemistry in inorganic/organic hybrid materials: clays, layered semiconductors, nanotubes, and mesoporous materials. J. Photochem. Photobiol., C 7, 104–126.

Takahashi, S., Taniguchi, M., Omote, K., Wakabayashi, N., Tanaka, R., Yamagishi, A., 2002. First observation of in-plane X-ray diffraction arising from a single layered inorganic compound film by a grazing incidence X-ray diffraction system with a conventional laboratory X-ray source. Chem. Phys. Lett. 352, 213–219.

Tamura, K., Setsuda, H., Taniguchi, M., Yamagishi, A., 1999. Application of the Langmuir-Blodgett technique to prepare a clay-metal complex hybrid film. Langmuir 15, 6915–6920.

Tan, C., Cao, X., Wu, X.-J., He, Q., Yang, J., Zhang, X., Chen, J., Zhao, W., Han, S., Nam, G.-H., Sindoro, M., Zhang, H., 2017. Recent advances in ultrathin two-dimensional nanomaterials. Chem. Rev. 117, 6225–6331.

Toma, L.M., Gengler, R.Y.N., Prinsen, E.B., Gournis, D., Rudolf, P., 2010. A Langmuir-Schaefer approach for the synthesis of highly ordered organoclay thin films. Phys. Chem. Chem. Phys. 12, 12188–12197.

Toma, L.M., Gengler, R.Y.N., Cangussu, D., Pardo, E., Lloret, F., Rudolf, P., 2011. New magnetic thin film hybrid materials built by the incorporation of octanickel(II)-oxamato clusters between clay mineral platelets. J. Phys. Chem. Lett. 2, 2004–2008.

Tonle, I.K., Ngameni, E., Walcarius, A., 2004. From clay- to organoclay-film modified electrodes: tuning charge selectivity in ion exchange voltammetry. Electrochim. Acta 49, 3435–3443.

Umemura, Y., 2002. Hybrid films of a clay mineral and an iron(II) complex cation prepared by a combined method of the Langmuir-Blodgett and self-assembly techniques. J. Phys. Chem. B 106, 11168–11171.

Umemura, Y., Shinohara, E., 2005. Formation of Langmuir-Blodgett films of a clay and a water-soluble alkylammonium cation. Langmuir 21, 4520–4525.

Umemura, Y., Yamagishi, A., Schoonheydt, R., Persoons, A., De Schryver, F., 2001a. Formation of hybrid monolayers of alkylammonium cations and a clay mineral at an air-water interface: clay as an inorganic stabilizer for water-soluble amphiphiles. Thin Solid Films 388, 5–8.

Umemura, Y., Yamagishi, A., Schoonheydt, R., Persoons, A., De Schryver, F., 2001b. Fabrication of hybrid films of alkylammonium cations ($C_nH_{2n+1}NH^{3+}$; n = 4-18) and a smectite clay by the Langmuir-Blodgett method. Langmuir 17, 449–455.
Umemura, Y., Yamagishi, A., Schoonheydt, R., Persoons, A., De Schryver, F., 2002. Langmuir-Blodgett films of a clay mineral and ruthenium(II) complexes with a noncentrosymmetric structure. J. Am. Chem. Soc. 124, 992–997.
Umemura, Y., Onodera, Y., Yamagishi, A., 2003. Layered structure of hybrid films of an alkylammonium cation and a clay mineral as prepared by the Langmuir-Blodgett method. Thin Solid Films 426, 216–220.
Umemura, Y., Shinohara, E., Schoonheydt, R.A., 2009. Preparation of Langmuir-Blodgett films of aligned sepiolite fibers and orientation of methylene blue molecules adsorbed on the film. Phys. Chem. Chem. Phys. 11, 9804–9810.
Valencia, G.A., Lourenço, R.V., Bittante, A.M.Q.B., do Amaral Sobral, P.J., 2016. Physical and morphological properties of nanocomposite films based on gelatin and Laponite. Appl. Clay Sci. 124–125, 260–266.
Van Duffel, B., Schoonheydt, R.A., Grim, C.P.M., De Schryver, F.C., 1999. Multilayered clay films: atomic force microscopy study and modeling. Langmuir 15, 7520–7529.
Vuillaume, P.Y., Jonas, A.M., Laschewsky, A., 2002. Ordered polyelectrolyte "multilayers". 5. Photo-cross-linking of hybrid films containing an unsaturated and hydrophobized poly(diallylammonium) salt and exfoliated clay. Macromolecules 35, 5004–5012.
Wang, L., Sasaki, T., 2014. Titanium oxide nanosheets: graphene analogues with versatile functionalities. Chem. Rev. 114, 9455–9486.
Wang, Y.-C., Fan, S.-C., Lee, K.-R., Li, C.-L., Huang, S.-H., Tsai, H.-A., Lai, J.-Y., 2004. Polyamide/SDS–clay hybrid nanocomposite membrane application to water–ethanol mixture pervaporation separation. J. Membr. Sci. 239, 219–226.
Wang, J., Ren, X., Feng, X., Liu, S., Sun, D., 2008. Study of assembly of arachidic acid/LDHs hybrid films containing photoactive dyes. J. Colloid Interface Sci. 318, 337–347.
Wu, C.-N., Saito, T., Fujisawa, S., Fukuzumi, H., Isogai, A., 2012. Ultrastrong and high gas-barrier nanocellulose/clay-layered composites. Biomacromolecules 13, 1927–1932.
Xiang He, J., Kobayashi, K., Mei Chen, Y., GillesVillemure, Yamagishi, A., 2001. Electrocatalytic response of GMP on an ITO electrode modified with a hybrid film of Ni(II)–Al(III) layered double hydroxide and amphiphilic Ru(II) cyanide complex. Electr. Commun. 3, 473–477.
Yamamoto, T., Einaga, Y., 2007. Photomagnetic hybrid ultrathin films. J. Solid State Electrochem. 11, 781–790.
Yamamoto, T., Umemura, Y., Sato, O., Einaga, Y., 2004. Photoswitchable magnetic films: Prussian blue intercalated in Langmuir-Blodgett films consisting of an amphiphilic azobenzene and a clay mineral. Chem. Mater. 16, 1195–1201.
Yamamoto, T., Umemura, Y., Sato, O., Einaga, Y., 2005. Observation of the anisotropic photoinduced magnetization effect in Co-Fe Prussian blue thin films fabricated by using clay Langmuir-Blodgett films as a template. J. Am. Chem. Soc. 127, 16065–16073.
Yan, D., Lu, J., Wei, M., Han, J., Ma, J., Li, F., Evans, D.G., Duan, X., 2009. Ordered poly(p-phenylene)/layered double hydroxide ultrathin films with blue luminescence by layer-by-layer assembly. Angew. Chem. Int. Ed. 48, 3073–3076.
Yang, Y.-H., Malek, F.A., Grunlan, J.C., 2010. Influence of deposition time on layer-by-layer growth of clay-based thin films. Ind. Eng. Chem. Res. 49, 8501–8509.
Yarger, M.S., Steinmiller, E.M.P., Choi, K.-S., 2008. Electrochemical synthesis of Zn – Al layered double hydroxide (LDH) films. Inorg. Chem. 47, 5859–5865.

Ye, W., He, S., Ding, L., Yao, Y., Wan, L., Miao, S., Xu, J., 2013. Supported ionic-liquid "semi-heterogeneous catalyst": an interfacial chemical study. J. Phys. Chem. C 117, 7026–7038.
Yeh, J.-M., Yu, M.-Y., Liou, S.-J., 2003. Dehydration of water–alcohol mixtures by vapor permeation through PVA/clay nanocomposite membrane. J. Appl. Polym. Sci. 89, 3632–3638.
Yoshida, J., Saruwatari, K., Kameda, J., Sato, H., Yamagishi, A., Sun, L., Corriea, M., Villemure, G., 2006. Electron transfer through clay monolayer films fabricated by the Langmuir-Blodgett technique. Langmuir 22, 9591–9597.
Zhang, W., Li, M.K.S., Yue, P.-L., Gao, P., 2008. Exfoliated Pt-clay/Nafion nanocomposite membrane for self-humidifying polymer electrolyte fuel cells. Langmuir 24, 2663–2670.
Zhang, Y., Wang, H., Zhang, Y., Ding, X., Liu, J., 2017. Thin film composite membranes functionalized with montmorillonite and hydrotalcite nanosheets for CO_2/N_2 separation. Sep. Purif. Technol. 189, 128–137.
Zhou, C.-H., Shen, Z.-F., Liu, L.-H., Liu, S.-M., 2011. Preparation and functionality of clay-containing films. J. Mater. Chem. 21, 15132–15153.

Index

Note: Page numbers followed by "*f*" indicate figures, "*t*" indicate tables, and "*s*" indicate schemes.

D

E

N

Q

R

T

Kolibri 360 GmbH
Ferdinand-Jühlke-Straße 7
99095 Erfurt, Deutschland
produktsicherheit@kolibri360.de